Bioethics

An Introduction to the History, Methods, and Practice

Third Edition

NANCY S. JECKER, PhD

University of Washington
School of Medicine
Department of Medical History and Ethics

ALBERT R. JONSEN, PhD

University of Washington
School of Medicine
Department of Medical History and Ethics

ROBERT A. PEARLMAN, MD, MPH

University of Washington
School of Medicine
Department of Medicine
and Seattle V.A. Medical Center

JONES & BARTLETT
LEARNING

World Headquarters

Jones & Bartlett Learning
40 Tall Pine Drive
Sudbury, MA 01776
978-443-5000
info@www.jblearning.com
www.jblearning.com

Jones & Bartlett Learning
Canada
6339 Ormindale Way
Mississauga, Ontario L5V 1J2
Canada

Jones & Bartlett Learning
International
Barb House, Barb Mews
London W6 7PA
United Kingdom

Jones & Bartlett Learning books and products are available through most bookstores and online booksellers. To contact Jones & Bartlett Learning directly, call 800-832-0034, fax 978-443-8000, or visit our website, www.jblearning.com.

Substantial discounts on bulk quantities of Jones & Bartlett Learning publications are available to corporations, professional associations, and other qualified organizations. For details and specific discount information, contact the special sales department at Jones & Bartlett Learning via the above contact information or send an email to specialsales@jblearning.com.

Production Credits

Chief Executive Officer: Ty Field
President: James Homer
SVP, Chief Operating Officer:
     Don Jones, Jr.
SVP, Chief Technology Officer:
     Dean Fossella
SVP, Chief Marketing Officer:
     Alison M. Pendergast
SVP, Chief Financial Officer: Ruth Siporin
SVP, Editor-in-Chief: Michael Johnson
Publisher, Higher Education:
     Cathleen Sether

Acquisitions Editor: Molly Steinbach
Senior Associate Editor:
     Megan R. Turner
Editorial Assistant: Rachel Isaacs
Production Manager: Louis C. Bruno, Jr.
Associate Production Editor:
     Leah Corrigan
Senior Marketing Manager:
     Andrea DeFronzo
V.P., Manufacturing and Inventory
     Control: Therese Connell
Composition: Circle Graphics
Cover Design: Kate Ternullo

ISBN-978-0-7637-8552-9

6048

Table of Contents

Part II   The Methods of Ethical Analysis   125

SECTION 1:   The Methods of Philosophy, Casuistry, and Narrative   140

Preface to the Third Edition

Students are often skeptical about ethics. They wonder why the study of ethics appears so different from other fields of study, where textbooks are filled with data and facts. Ethics, it seems, is little more than a collection of opinions and theories. This attitude is especially prevalent among students planning a career in health care, whose primary studies are focused on the physical and life sciences. Skepticism about ethics is perhaps more poignant today than it was in the past, due in part to greater cultural diversity among students pursuing higher education. In the classroom, where a wider range of cultural and religious perspectives are represented, students sometimes are led to doubt their own beliefs, seeing them as just one view among many.

Within the specific area of bioethics, students display not only skepticism, but an absence of any frame of reference, i.e., any memory of the events of the 1960s and 1970s leading to the development of bioethics. Students on campuses today do not recall, for example, the events that took place in Seattle during the 1960s, when treatment for kidney failure first became available and there were not enough kidney machines or trained personnel to provide this life-saving treatment to everyone who needed it. Nor do today's students remember the public outrage in 1972 over the Tuskegee, Alabama, study in which African-American men with syphilis were left untreated while researchers followed the natural course of their disease (and while the men's partners and offspring were exposed to the disease). Nor were most current students alive in 1976, when the New Jersey Supreme Court issued its ruling in the case of Karen Ann Quinlan, a woman in a persistent vegetative state whose parents requested removal of a ventilator.

Our textbook responds to these concerns in a way that no other book currently available does. First, we address the recent history of bioethics, explaining its emergence as discipline and discourse (Part I). Students can read the Quinlan case, learn about the philosophical debate over patient selection for dialysis, and consider the controversies surrounding ethical standards in human subject research that arose during the 1960s. In short, they can see how bioethics came into existence as a formal field of scholarly inquiry. After an initial chapter that relates the history of bioethics from its origins, a selection of articles written by the earliest bioethicists shows how they began to grapple with the initial problems, not as technical or scientific issues, but as ethical concerns.

Second, our book looks carefully at ethical reasoning and the justification of moral beliefs (Part II). We start by asking whether the truth or falsity of ethical claims is determined solely by the culture of the person making the claim or by an outside standard. We then explore various techniques of appealing to reason to justify moral claims: principlism, casuistry, and narrative ethics. We assess the limits of these techniques, which apply primarily to individual cases in the clinical setting of medicine.

This background prepares students to turn a critical eye to the topics of bioethics (Part III), including ethical issues that arise both at the beginning and at the end of life. The book takes up ethical concerns related to reproductive decisions, such as abortion, assisted reproductive technologies, and prenatal genetic testing. It also examines ethical issues at the end of life, including withholding and withdrawing life-sustaining treatment, advance care planning, care of the dying, and the legal and ethical aspects of physician-assisted death.

While the third edition of *Bioethics* retains the three-pronged approach of (1) history, (2) methods, and (3) practice that we used in the first and second editions, it differs from previous editions in key respects. One important difference is that the third edition has dispensed with sections on "Cultural Assumptions" (that were previously included in each of the book's three parts) and replaced these with sections that address not only cultural assumptions, but also broader assumptions related to the historical, philosophical, social, economic, and political contexts in which bioethics exists. The new sections are "Questioning the History of Bioethics," "Questioning the Methods of Bioethics," and "Questioning the Practice of Bioethics."

Other specific changes include a total of 10 new articles distributed throughout the book's three parts. These new articles address a variety of timely topics, including political influences on bioethics, the perspective of disability rights groups, changes in state laws governing end of life decisions, and recent empirical contributions to bioethics. *Part I: The History of Bioethics* offers two new papers addressing the political climate of bioethics. *Part II: The Methods of Ethical Analysis* includes a provocative new piece challenging the professional role and status of ethics consultants. *Part III: The Practice of Bioethics* has a new section on "Care of the Dying," as well as an expanded section on physician-assisted death. We believe our three-pronged introduction to bioethics continues to offer the reader a more sophisticated introduction to the field than any other book available. It is particularly well suited for advanced undergraduate and graduate students who plan to pursue careers in ethics and health care, as well as for practicing professionals.

Having explained what we hope this new edition of *Bioethics* will accomplish, let us tell you something about ourselves. The three of us met as colleagues at the University of Washington School of Medicine, Seattle. Nancy S. Jecker is a philosopher and professor of ethics at the University of Washington School of Medicine, Department of Bioethics and Humanities, where she also holds appointments in the Department of Philosophy and School of Law. She is the author (with Lawrence Schneiderman) of *Wrong Medicine: Doctors, Patients and Futile Treatment* (originally published by Johns Hopkins University Press in 1997, with an updated edition forthcoming in 2011). She is the editor of *Aging and Ethics: Philosophical Problems in Gerontology* (Human Press, 1991). Dr. Jecker has authored over 100 articles and chapters on ethics and health care, which have appeared in *The Journal of the American Medical Association*, the *Hastings*

Center Report, Annals of Internal Medicine, The Journal of Medicine and Philosophy, and other publications. Dr. Jecker has conducted research as a visiting scholar at the Stanford University Center for Biomedical Ethics, the Stanford University Center for Research on Women (now the Michelle R. Clayman Institute for Gender Research), the Georgetown University Kennedy Institute of Ethics, the Hastings Center, and the Princeton University DeCamp Program in Ethics and the Life Sciences, and was twice awarded Rockefeller Resident Fellowships, by the University of Texas Medical Branch Institute for Medical Humanities and by the University of Maryland Center for Philosophy and Public Policy. Her scholarship focuses on theories of justice, with application to the allocation of scarce medical resources; medical futility and related decisions to withhold or withdraw life-sustaining treatment; and ethical issues within families, including intergenerational justice and family caregiving.

Albert R. Jonsen is one of the pioneers in the field of bioethics. He is currently Emeritus Professor of Ethics in Medicine at the School of Medicine, University of Washington, where he was chair of the Department of Medical History and Ethics (now the Department of Bioethics and Humanities) from 1987 to 1999. He is co-director and Senior Ethics Scholar at the Program in Human Values, California Pacific Medical Center, San Francisco. He was formerly chief of the Division of Medical Ethics, University of California, San Francisco. Dr. Jonsen was a commissioner on the National Commission for the Protection of Human Subjects of Biomedical and Behavioral Research (1974–1978) and on The President's Commission for the Study of Ethical Problems in Medicine (1979–1982). Dr. Jonsen is the author, with M. Siegler and W. Winslade, of *Clinical Ethics: A Practical Approach to Ethical Decisions in Clinical Medicine, Sixth Edition* (McGraw-Hill, 2006). Among his other books are *Bioethics Beyond the Headlines: Who Lives? Who Dies? Who Decides?* (Rowman and Littlefield, 2005); *A Short History of Medical Ethics* (Oxford University Press, 1999); *The Birth of Bioethics* (Oxford University Press, 1998); *The New Medicine and the Old Ethics* (Harvard University Press, 1990); and *The Abuse of Casuistry* (with co-author S. Toulmin, University of California Press, 1988).

Robert Pearlman is an adult medicine physician specializing in geriatrics. He is a professor in the Department of Medicine at the University of Washington, School of Medicine, where he also holds appointments in the School of Public Health and Department of Bioethics and Humanities. He is the Ethical Leadership Coordinator (shared with the facility director), Veterans Affairs Puget Sound Health Care System, and chief of the Ethics Evaluation Service, National Center for Ethics in Health Care (Veterans Health Administration). He was a fellow at the Harvard University Program in Ethics and the Professions, and currently is a fellow at the Hastings Center, the American Geriatric Society, and the Gerontological Society of America. Dr. Pearlman has conducted empirical research in bioethics for over 25 years through grant support from the U.S. Department of Health and Human Services, the Department of Veterans Affairs, and private foundations. He has authored over 100 articles and chapters on ethics and health care. He is the author (with H. Starks, K. Cain, W. Cole,

D. Rosengren, and D. Patrick) of a patient-centered workbook on advance care planning entitled *Your Life, Your Choices—Planning for Future Medical Decisions: How to Prepare a Personalized Living Will* [U.S. Department of Commerce (NTIS), 1998]. His scholarly publications, which have appeared in *The Journal of the American Medical Association, Annals of Internal Medicine,* the *Journal of Medicine and Philosophy,* the *Hastings Center Report,* and other publications, explore ethical issues primarily at the end of life, including the role of patient quality of life in medical decision making, advance care planning, and physician-assisted suicide.

PART I

The History of Bioethics

A History of Bioethics
as Discipline and Discourse

Albert R. Jonsen

The word "bioethics" was fashioned in 1970 by a biological scientist, Van Rensselaer Potter, to name his vision of a new conjunction of scientific knowledge and moral appreciation of the converging evolutionary understanding of humans in nature.[1] Potter was writing at a time when many thoughtful scientists, starting with the physicists who created the atomic bomb and progressing to the biologists who were exploring the inner world of the human genome, were expressing concern about the social implications of their science.[2] Hardly had Potter uttered the word "bioethics" to describe this broad question of scientific conscience than it was appropriated to identify a related but much narrower vision: the ethical analysis of a range of moral questions posed to medical practice by the advances in the biomedical sciences and technologies. In 1971, the Kennedy Institute of Ethics was founded at Georgetown University to pioneer the development of a new field of research into medical ethics that its founder, André Hellegers, chose to call "bioethics." *The Encyclopedia of Bioethics*, which the Kennedy Institute began to plan in 1972, defined bioethics as "the study of the ethical dimensions of medicine and the biological sciences."[3] In 1974 Dan Callahan, who had founded the other major research institute in this new field, The Hastings Center, wrote an influential article entitled "Bioethics as a Discipline," in which he suggested that this new field could develop into a unique discipline, using both the traditional methods of philosophical analysis as well as sensitivity to human emotion and to the social and political influences with which medicine was practiced[4] (see the reprinting of Callahan's article in Part I, Section 1). Callahan said this emerging discipline should be designed to serve those faced with the crucial decisions that arise within medicine. Thus, within a few years, a neologism with indeterminate meaning became the focused name for an emerging field of study. That word, and the activity it designated, marked a boundary between the long tradition of medical ethics and a quite distinct approach to moral questions in medicine and science.

For many centuries, in all the great literate cultures—Graeco-Roman, Christian, Buddhist, Hindu, Confucian, and Islamic—the practice of medicine has been associated with certain moral behaviors, qualities, and values. The dominant moral and religious notions of these cultures cast their light over the work of healing and, while these notions are diverse, the duties and decorum of physicians toward their patients are remarkably similar around the world. Moral maxims guided physicians, who tried to carry on their work in a moral manner: the moral physician was competent, dedicated, kept confidences, cared for the sick poor without charge. As healers gradually formed themselves into professions, questions of relationships between practitioners and of economic rivalry also appeared on the agenda of medical ethics. It is not unreasonable to speak of a long tradition of medical ethics prevailing in many places over many centuries.[5,6]

At mid-20th century, this traditional medical morality encountered unprecedented problems. The conditions in which medicine was practiced had changed dramatically during the preceding century. Science had developed much more effective treatments than had ever existed. Many more people had access to trained physicians, and physicians became not only more educated and more competent, but earned more money and social prestige than the profession had previously enjoyed. Old ideas about medical morality, particularly about the relationship with patients, were challenged by these new conditions.

Thomas Percival's *Medical Ethics* (1803), the first book to actually bear that title, came to be seen in England and the United States as a compendium of medical morality. The American Medical Association drew on Percival's work extensively when, in 1842, the association produced its first Code of Ethics for American doctors. However, the book's inadequacy for the sorts of problems confronting modern medicine was noted as early as 1927. In his edition of Percival, Dr. Chauncey Leake commented on the need for a deeper understanding of medical morality, based on explicit theories of moral philosophy. Only with a more philosophically grounded ethics, Leake said, could physicians confront the conditions that even then were disrupting the personal relationship of trust between doctors and their patients.[7] A quarter of a century later, Joseph Fletcher, professor of moral theology at the Episcopal Theological School, Cambridge, Mass., authored a book, *Morals and Medicine* (1954), that utilized a vague form of an explicit philosophical theory of ethics—utilitarianism—to criticize and reformulate medical morality on several traditional topics, namely, euthanasia, telling the truth to patients, abortion, and contraception.[8] Fletcher discussed these issues frankly, providing ethical arguments to support his often controversial contentions. In addition, Fletcher was not himself a physician, whereas almost all of the voluminous writing on medical ethics had come from physicians themselves. *Morals and Medicine* was noted widely as a new direction in medical ethics.

Advances in the biological sciences and in medical technology pushed these questions even harder. The development of artificial ventilators in the 1950s saved many patients from cardiorespiratory death but left them without consciousness. In 1957, an international group of anesthesiologists took the unprecedented step of presenting the ethical problems they encountered in using these technologies to a "moral authority," Pope Pius XII. The pope responded with a lecture that applied an old Catholic teaching—that no one had a moral obligation to sustain life by use of "extraordinary means"—to the new circumstances of ventilatory support. The pope's modernization of that venerable doctrine revealed that the long tradition of Catholic moral theology had accumulated an articulate body of principles and casuistry about medical ethics that could be relevant to quite modern problems, and also stimulated Catholic moral theologians to reflect on the emerging moral issues in medicine.[9]

The artificial kidney, a technology to support loss of kidney function, was another lifesaving intervention that raised unique problems. In 1961, Dr. Belding Scribner, professor of medicine at the University of Washington, Seattle, invented the arteriovenous shunt, a simple device that made possible continued hemodialysis for patients with fatal kidney disease. It was immediately evident that many more patients needed this technology than could be accommodated. A committee of laypersons was formed to choose which patients would be treated and thus live, and those who would be rejected, only to die. Handing over to a "God Committee" of nonphysicians the power to choose who would live and who would die, was unprecedented in medicine. A *Life* magazine article that appeared on November 21, 1962, vividly told the story of the God Committee.[10] In the ensuing public debate, some philosophers and theologians became interested in the problem of selection for lifesaving procedures, and a small literature began to appear on this topic. One of philosophy's major ethical theories, utilitarianism, was again brought to the debate. A philosopher advocated that the selection of patients should be made on the basis of social utility, assessed in light of the social worth of individuals as contributors to society.[11] Several theologians responded that the inherent dignity of individuals required that judgments of social worth be repudiated and that selection be made by random methods, such as lotteries[12,13] (see the reprinting of Childress's article in Part I, Section 1). Scholarly legal analysis sided with the theologians.[14] This was the first major debate of the new bioethics.

The story of the invention of chronic dialysis was a dramatic social event that revealed how medicine's technology could pose previously unaskable questions about life and death. It was also the first example of involving laypersons as essential players in medical decision making. It revealed that what appeared to be medical decision making was actually the making of social policy. It was the first event to catch the attention of academic ethicists and to intrigue them about life and death decision making. The public discussion of this event also initiated a public policy process that concluded with the 1972 passage of federal legislation providing financial support for the treatment of end-stage renal disease by dialysis and transplantation. Thus, the case of chronic hemodialysis stimulated both the discipline and the discourse that was bioethics. The discipline profited from the articulate analysis of the problem by philosophers, theologians, and lawyers. The discourse was advanced from media exposure to formulation of public policy.

Life-sustaining technologies such as hemodialysis and mechanical ventilation, which were developed at about the same time, have a dual effect. They substitute for a damaged organ system and thus keep alive a person who would otherwise die. Usually these technologies are used on a temporary basis, but they can be used as a permanent support when vital organs have failed irreversibly. This leads sometimes to a second effect, the support of biological life when mental life is gone or badly deteriorated, or the prolonging of inevitable death. This effect raises the question of under what conditions life support should be forgone, either not initiated or discontinued. In 1973, a team of physicians who cared for premature newborns exposed their own practice of allowing such babies to die when their prognosis for healthy life was dim.[15] The physicians' article aroused surprise and anger among the public, although they reported a practice familiar to those who treated severely ill babies. Two years later, the story of 21-year-old Karen Ann Quinlan captured the attention of the nation. Brought to a hospital in a coma, she lingered unconscious for months supported by a respirator. When her parents asked that the respirator be stopped, the hospital refused, and the first case to deal with allowing a patient to die went to the Supreme Court of New Jersey. The court determined that her parents had the authority to withdraw life support, offering a thoughtful analysis of this unprecedented problem. Many similar cases have since arisen. One of them, the case of Nancy Cruzan, was adjudicated by the U.S. Supreme Court (1990), which confirmed the series of cases in which courts had permitted withdrawal of life support from patients whose recovery to health and consciousness was highly unlikely, provided there was some evidence that the patient would have chosen such a course. The Supreme Court addressed the question again in 1997. In *Washington v. Glucksberg*, petitioners sought to affirm a "right to die" on the part of terminally ill, competent patients, against state law prohibiting "physician-assisted suicide." The Court found no such right in the Constitution but did allow that states might legislate for or against physician-assisted dying. Bioethicists followed these cases closely and often contributed to the formulation of moral arguments, usually as justification for allowing such patients to die. A substantial body of literature about the appropriate care of the dying was created[16] (see the separate chapters by McCormick and by Rachels in Part I, Section 1).

Medical advancement requires experimentation; experimentation often requires human beings as its objects. When experimental medicine was emerging in the first half of the 19th century, one of its pioneers, the French physiologist Claude Bernard, declared that "it is our duty and right to experiment on man, whenever it can save his life, cure him or gain him some personal benefit. The principle of medical and surgical morality, therefore, consists in never performing on man an experiment which might be harmful to him to any extent, even though the result might be highly advantageous to science, that is, to the health of others."[17] However, this noble rule was unrealistic, for all experimentation is a journey into the unknown and poses some risk. Thus, as the century of experimental medicine moved on, the paradoxes of medical research with humans became increasingly apparent. After World War II, the horrors performed by Nazi physicians in concentration camps revealed how far uncontrolled experimentation could go. The Nuremberg Code (1947), drawn up in the course of the trial of those physicians, was widely acknowledged as the epitome of the ethics of experimentation with humans: medical research must be done only after free and informed consent of the subject and in light of a reasonable relation between risks and benefits.[18] Although Americans could hardly believe that such things would happen here, they learned that, indeed, similar things did happen. In 1968, a distinguished medical researcher, Dr. Henry Beecher, published in the country's most prestigious medical journal an analytic exposé of abuse of human subjects of research.[19] In particular, a study initiated in 1931 by the U.S. Public Health Service that left 400 rural black men without treatment for syphilis, shocked the nation.[20] The federal government, which funded so much biomedical research, had to assure that the research was not abusive to its subjects.

Congress established the National Commission for the Protection of Human Subjects of Biomedical and Behavioral Research (1974–1978) to develop ethical principles to guide research and recommend rules and procedures to protect the rights and welfare of human subjects. During its 4-year tenure, the Commission issued not only specific regulations to govern the conduct of research but also a statement of principle, *The Belmont Report*, which proposed that three principles should guide researchers—respect for persons, beneficence, and justice—and that these principles entailed respectively, informed consent, the assessment of risk in relation to benefit, and the equitable selection of subjects for research.[21] The Commission called upon philosophers and theologians to help clarify the issues as it debated the ethics of research with children, with the mentally infirm, and with incarcerated persons. The thinking of two scholars was particularly influential: that of Hans Jonas, a philosopher, and Jay Katz, a physician and professor of law, both of whom delved deeply into the complex relations among respect for individuals, consent, and the duty to contribute to scientific knowledge[22,23] (see the reprinting of Jonas's article in Part I, Section 1). Again, bioethics manifested itself as discourse and discipline. The widespread public discourse about abuses of experimental subjects was gradually shaped into legislation and regulation. The discipline was improved by scholarly analysis of concepts such as "free and uncoerced consent" and "research versus practice," and of the logic involved in arguing for the rights of research subjects in relation to the common good of society.

One of the most dramatic breakthroughs in the biological sciences was the announcement in 1956 that James Watson and Francis Crick had discerned the "secret" of life, the double helical structure of the DNA molecule. The possibility of learning the most basic lessons about how biological organisms developed, how defects entered that development and how scientists might consciously modify that development, engendered great excitement. This advancement engendered great concern as well, arousing memories of the ill-conceived eugenics that had entranced the nation for the first half of the century, promising improvement of the "stock" by controlled breeding and sterilization. A debate over these issues among scholars erupted in the mid-1960s. Philosophers and theologians endorsed the potential of these sciences to eradicate and alleviate disease, but they questioned the wisdom of modifying human characteristics for which few criteria of perfection exist. In particular, the prospect of cloning humans aroused vigorous debate.[24] However, the scientific prospects of modifying the human genome still seemed far in the future, and these concerns abated somewhat in the 1970s.

Their place was taken by another advance in modern genetics, the ability to test and screen for the presence of genetic diseases and conditions. Again, this was a great medical advantage, but hidden within its techniques lay the potential for discrimination against persons who were learned to have certain genetic conditions. Because few therapies were available to treat the conditions for which testing was available, the prospect of stigmatization and the risk of loss of health insurance and of employment, were troubling. Intense study of the ethical implications of this important medical advance led to guidelines and legislation that defined the permissible scope of screening programs. Just as the ethics of biomedical experimentation stressed the consent of the subject, so the ethics of genetic screening emphasized that programs of premarital and prenatal carrier screening be done with the knowledgeable consent of those screened.[25]

Another dramatic moment in the progress of medicine came in 1968, when Dr. Christiaan Barnard took the beating heart from one dying person and transplanted it into the chest of another, who lived for several months. The era of organ transplantation had begun in the early 1950s when kidney transplantation between genetically identical twins was successful; soon powerful drugs allowed transplantation between unrelated persons by controlling the process of immunological rejection. Physicians, exhilarated by the ability to save lives doomed by kidney failure, also realized that, in order to do so, they had to invade the body of a healthy person, seemingly violating the most ancient ethical maxim of medicine, "do no harm." Also, the obligation of persons to donate an organ to one in need was much debated. Scholars struggled to define the moral nature of the act whereby a part of one human would be used to save another. Finally, the ethical ideal of "donation" was extolled as the justification for placing one person at risk to save another.[26] It quickly became obvious that lifesaving organs were in short supply and that the problem of allocation that had troubled the Seattle dialysis program would be played out on a national stage. Heart transplantation, however, was the drama that propelled the ethical issues associated with that medical advance into public attention. The symbolic value of the heart as the center of human emotion led some to question the morality of transferring one person's heart into another person's body. The still imperfect understanding of the human immune system and the inadequacy of drugs to counter immunological rejection led to tragic disappointments.

Heart transplant, unlike renal transplant, required a new way of thinking about death because the organ, which itself must be physiologically living, must be taken from a person who is already legally dead. From time immemorial, physicians had designated death as the moment when breathing and circulation of blood ceased. New formulations of this definition were attempted; most famous of these was the "Definition of Irreversible Coma: Report of the Ad Hoc Committee at Harvard Medical School to Examine the Definition of Brain Death" (1968).[27] This reformulation stressed the irreversible loss of the neurological activities of the brain. Many states enacted legislation to change their legal criteria for determining death. The Harvard criteria, however, were ambiguous, and 13 years later, the newly appointed President's Commission for the Study of

Ethical Problems in Medicine (1979–1982) issued a more careful definition, defining death as the cessation of all brain activity, including the brain stem.[28] During the long debate over the redefinition of death, bioethicists argued both that the new definitions moved into dangerous territory, opening the possibilities of considering as "dead" persons who retained biological life, and also that the definitions did not go far enough, considering as "alive" human beings who lacked all possibilities for human communication. Although the more conservative formulation of death became law throughout the United States, a valuable literature about the nature of death, of personhood, and of the obligation between persons enriched bioethics.[29]

In subsequent years, debates arose over how to treat patients who were not dead according to brain criteria, but who were without consciousness, in a vegetative state from which they were unlikely to recover. Medical technology was increasingly able to support such persons for many years. The case of Karen Ann Quinlan, mentioned above, was the first case of this sort to reach the courts. Subsequently, medical science worked to clarify the neurological basis for this diagnosis and to strengthen the clinical ability to diagnose its progress. In 2005, the case of Terri Schiavo burst into public attention. Mrs. Schiavo had been in a vegetative state for 15 years when her husband requested that artificial feeding be discontinued. Her parents disagreed, and a complicated battle followed through many courts and even to the U.S. Congress. The jurisprudence developed in the many prior cases was sustained, and the bioethical reasoning of previous cases was generally persuasive. Mr. Schiavo's petition to allow his wife to die was granted.[30] Despite the furious public disputes over this case, the principles and arguments developed in bioethics over the previous 3 decades greatly clarified the central issues.

Human reproduction, long considered a process set immutably by nature, became the object of scientific manipulation. The first successful artificial insemination by donor was performed in 1884 and not reported until 25 years later because those who performed it feared a negative public response. That response did follow and has continued over each subsequent development in reproductive technology. The religious traditions of the Western world have definitive doctrinal positions on matters of sexuality and reproduction, and these aspects of human life are surrounded by strong beliefs concerning public morality. The development of chemical contraceptives in the 1950s made possible a rational means of preventing birth, which contradicted some religious teachings, particularly those preached by the Roman Catholic Church. Ironically, the science that contributed to this means of preventing birth also contributed to the new understandings of the physiology of fertility, making possible the conception of children for previously infertile couples. In 1979, two English physicians reported that they had fertilized a human ovum in a Petri dish and introduced it into the womb of the woman who bore the baby, who was named Louise Brown. This procedure, in vitro fertilization, made possible many other manipulations of the human embryo, all of which have been vigorously debated. The new forms of assisted reproduction have led to new forms of family, because a woman can conceive a child without intercourse or a child can be conceived and carried by a "committee" of progenitors. The identity of "parents" was often contested. The use and disposal of embryos created by this procedure has been a constant topic of bioethical debate and legal regulation. As assisted reproduction has become a widely accepted practice, other ethical questions have been debated, such as the problem posed by implantation of several embryos to increase chances of conception. Often the several embryos develop into multiple fetuses, increasing the risks of premature birth. This ever-expanding technology continually prompts novel ethical problems.[31]

Human reproduction and human genetics are intimately related. The ability to manipulate reproduction was matched scientifically by the ability to manipulate the human genome, the genetic code that designates the biochemical and physical makeup of individuals. The 1953 revelation by Watson and Crick of the structure of that code as a deoxyribonucleic (DNA) molecule initiated a dramatic surge of knowledge as well as the potential for therapeutic and eugenic modification of genes. In 1989, the U.S. Congress authorized an intensive effort to map the human genome. One element of the scientific project was called the Ethical, Legal and Social Implications Program (ELSI). Grants were awarded to bioethicists and social scientists to study these implications. Many questions, raised in the earliest days of bioethics about genetics, now were approaching clinical reality. These questions included dimensions not easily confined to the principles of bioethics first enunciated for issues such as human experimentation and life support. Genetics inevitably includes not only a particular patient but a kinship— all those genetically linked. Each individual may now, in principle, obtain a copy of his or her personal genome, with all the attendant information about his or her health and health risks. Problems of privacy and discrimination attach to almost all aspects of genetics.

In 1997, the world was astonished by the announcement that a newborn lamb named Dolly was the product of cloning, that is, generated not by the joining of male sperm and female egg, but by splicing a skin cell into an egg and implanting the resulting embryo into a ewe. This event resurrected a debate about cloning that had taken place in the earliest years of bioethics, but in the setting not of an abstract scientific idea, but of a successful experiment. Soon after the cloning event, scientists announced that stem cells had been isolated from human embryos: immature but plenipotent cells that could, theoretically, be grown into an infinite variety of human cells and used for research, for therapies, or for creating cloned babies. A rush of scientific excitement was met by deeply held ethical objections, largely from religious communities. In 2001, President George W. Bush announced that public funds could not be used to sponsor stem cell research except in highly restricted circumstances. He appointed a Bioethics Advisory Council, headed by a respected but avowedly conservative bioethicist, Dr. Leon Kass, to advise on cloning and stem cell research. A new area of debate opened for bioethical scholars, in which deeply problematic questions about the control of life's origins and the place of religious values in public policy had to be faced.[32] Although the Obama administration again opened federal support of stem cell research, with less stringent restrictions, discussions about the ethics of this research will continue.

Selection of patients for life and death therapies, experimentation with human subjects, the eugenic and discriminatory potential of molecular biology and the interpersonal complexities of organ transplantation, all of which occurred during the 1960s, raised concerns among professionals and the public and engaged the attention of philosophers, theologians, and legal scholars. These topics were on the first agenda for the nascent bioethics. From 1970 to 1980, a large volume of literature on these topics was published, much of it written by philosophers and theologians. That decade began with a powerfully expressed, rigorously argued, and unabashedly contentious book, *The Patient as Person*, by Princeton theologian Paul Ramsey. This book took up the topics of that era—experimentation, genetic engineering, organ transplantation, and definition of death and care of the dying—and pursued a close ethical analysis, based on the fundamental ethical notion of fidelity between patient and physician. Ramsey's words in that book and on the many other bioethical topics he addressed subsequently echo through the emerging field of bioethics.[33] Also during the 1970s and 1980s, bioethics was enshrined in the academic world. Several research centers, including the Hastings Center (1969) and the Kennedy Institute at Georgetown University (1971), had been established. Professorships to study and teach these subjects had been opened at medical schools. Professors created curricula, wrote textbooks, and explored theoretical approaches to draw together the classical disciplines that deal with morality, philosophy, and theology with the practical questions about the pursuit of science and the practice of medicine. Clearly, by the end of the 1980s, the neologism "bioethics" described more than a vision: it designated a new field of study, with a distinctive literature and the panoply of courses and conferences that academic disciplines engender.

Also from the 1970s onward, public commissions and committees put hard work into the concepts and arguments that surround these issues. In so doing, they not only clarified the issues, but improved their own understanding of how to analyze and argue bioethical questions. Many scholars who had previously known of these issues only as observers, or hardly at all, were attracted to the study of these problems. They became the original bioethicists. In addition to the scholars, many other persons, professional and lay, became peripherally involved in bioethics. Government regulations about research with human subjects required that all research institutions have an Institutional Review Board (IRB) charged with review and approval of all scientific proposals to use human subjects. Hundreds of IRBs came into being, and thousands of physicians, scientists, and others became familiar at first hand with the principles for ethical research, the regulations to protect subjects' welfare and rights, and their application to particular cases. Also, as awareness grew about the ethical complexity of many clinical situations, particularly those involving appropriate care for terminally ill patients, it was suggested that decisions should no longer be left to the discretion of the treating physician alone but informed by the advice of persons familiar with the ethical and legal dimensions of such cases. Thus, during the 1970s, many hospitals established ethics committees to assist physicians with clinical perplexities. These committees, which were advisory to physicians and to the institution, undertook to clarify policies about resuscitation and life support. Again, many persons who served on these committees became familiar with the language, concepts, and literature of bioethics. Around the country, conferences and courses were held and books and journals appeared for this new population of part-time "bioethicists." A distinct field of bioethics, called "clinical ethics" formed around these activities[34] (see articles in Part III).

In the context of this growing application of bioethics to practical problems in the laboratory and the clinic, a debate arose over the use of the classical theories and reasoning of philosophical ethics. Bioethicists had, as mentioned above, worked out a general approach to ethical reasoning by affirming the relevance of a set of principles, namely, respect for autonomy, beneficence, maleficence, and justice. However, it was obvious that general principles require interpretation in specific situations, particularly those arising within the clinical setting. Thus, during the 1980s, a debate arose among bioethicists about whether "principlism" or "casuistry" provided the intellectual tools for analysis of actual ethical dilemmas. In conjunction with these problems of practical reasoning, broader theoretical concerns appeared. In particular, as more women scholars entered the bioethics profession, the historical bias of ethical argumentation in terms of male values was noted and criticized. A distinct "feminist" approach was articulated. Similarly, as bioethics encountered more problems rooted in particular cultural values, and as bioethical concerns spread around the world, authors began to call for deeper appreciation of cultural diversity and, concomitantly, stimulated reflection on the traditional ethical question of relativism of ethical values. During the 1980s and 1990s, a significant body of literature was generated by these debates[35] (see Part II).

In the 21st century a new debate is forming about the political role of bioethics. The debates over cloning and stem cell research were carried out in a heated political climate; the Council on Bioethics appointed by President George W. Bush set what many bioethicists perceived as a "conservative" tone about these issues. Consequently, bioethics scholars began to reflect on the political implications of "doing" public bioethics. For some authors, almost every bioethical issue, from reproduction to care of the dying, took on the tinges of liberalism and conservatism drawn from the world of political debate. While the debates engaged the more philosophically minded bioethicists, their resonance was heard in the courses developed for the part-time bioethicists, and in the consultations of ethics committees[36] (see the articles by Macklin and by Cohen in Part I, Section 2).

Whereas bioethics has evolved into something like an academic discipline, it has not been unified under any dominant theory or methodology. It reflects the circumstances of its evolution: questions about the ethical dimensions of science and medicine were raised and debated by a variety of commentators—scientists and physicians, lawyers, politicians and public policy experts, social scientists, philosophers, and theologians. These highly intradisciplinary discussions brought considerations from every side into the debate. The two disciplines most famil-iar with the logic and rhetoric of ethics—moral philosophy and moral theology—provided their skills to the debates, often shaping the desultory discussion, so common to ethical discourse, into formats with distinct definitions and logical arguments. They did not, however, submerge the intradisciplinary features in any single ethical theory. Even though some bioethicists have attempted to for-mulate overarching theories, the discipline welcomes a wide range of arguments, views, and methods. Moral questions are argued in terms of some rather consis-tent moral rules, such as "do no harm," and those rules are justified by reference to a general conception of personal and social welfare.

One rule in particular, "respect the autonomy of persons," holds a particularly important place in bioethical argumentation. Its dominance can be attributed, perhaps, to the sorts of problems posed to the original bioethics, particularly the use of humans in research and the eugenic threats of the new genetics; perhaps it can be traced to the deep individualism that marks American culture. Whatever the source of the dominance of the principle of personal autonomy, it has exer-cised both a benign and a malign influence on bioethics. Its benign effect has been to overthrow the unjustifiable paternalism of physicians and the arrogance of scientists: decisions about what is best for a person ought to be made by the one whose life will be affected; decisions about the directions of scientific progress ought to be in the hands of the people whose future will be changed. The malign aspect of the dominance of autonomy lies in the neglect of social and communal dimensions of ethical problems. Bioethics has been less success-ful in dealing with issues of public welfare than in resolving the interpersonal problems of clinical medicine.

Bioethics, however, is not simply an academic discipline in which debates over method and theory absorb scholars' attention. It has, from the beginning, aimed toward the guidance of practices and policies. It is not speculative but practical moral philosophy. Thus, in addition to the disciplinary dimensions of method and theory, bioethics is a form of discourse, promoting public debate over substantial questions and encouraging the formation of agreement and consensus about the ways to resolve those questions. The many commissions and committees that have been established over the last 3 decades have carried on that discourse, and it has spilled into the places where regulations are made and legislation is written, as well as places such as hospitals and professional organizations that seek to guide their practitioners through complex problems. The formal disciplinary arguments found in the literature do contribute to those debates, as does the advice of scholarly bioethicists, but the wider discourse draws on the ingenuity and insight of the many individuals whose moral con-science focuses on these questions. In the state of Oregon, for example, during the 1980s a program called Oregon Health Decisions engaged thousands of cit-izens in such discourse over health care priorities; their discussions informed the state legislature about the people's bioethical values (see Part III, Section 2).

The attention to actual clinical situations highlighted the social, cultural, and economic settings that surround them. The need to integrate empirical data about these settings and about the attitudes and opinions of the persons engaged in them became obvious. Bioethics could be neither a simply speculative enterprise, nor a practical version of moral philosophy. The field needed ways to identify and integrate the findings of sociology, anthropology, economics, and even polling into ethical discourse (see the articles in Part II, Section 3).

In the almost half-century that bioethical questions have been raised and debated, this field of study has established itself as an integral branch of practical or applied philosophy and as a valuable adjunct to health policy and medical practice. It has developed some methodologies for dealing with the questions to which it is devoted. Those questions often touch the human condition so deeply that no final answers will ever be given, but continued examination remains necessary. New questions arise as versions of the old questions, as new technical possibilities emerge from science, and as new social arrangements appear. Bioethics today is much better prepared intellectually to study these questions than it was in its infancy. Whether it has attained the disciplinary status originally proposed by one of its founders, Dan Callahan, remains an open question. Still, many intellectual resources have been marshaled to fulfill the tasks that all of its founders envisioned: promoting clarity in the debates, criticizing simple solutions to complex issues, and aiming toward resolution of problems within perspectives on human life that reflect dignity, compassion, and justice.

References

1. Potter, Van Rensselaer. "Bioethics, the Science of Survival," *Perspectives in Biology and Medicine* 14 (1970): 127–153.
2. Jonsen, Albert. *The Birth of Bioethics*. New York: Oxford University Press, 1998. Chapter 1.
3. Reich, Warren. "Introduction," in Reich (ed.), *The Encyclopedia of Bioethics*. New York: The Free Press, 1978, Vol. 1, pp. xix–xxx. Stephen Post (ed.) Encyclopedia of Bioethics. New York: Macmillan Reference USA, 3rd edition, 2004.
4. Callahan, Dan. "Bioethics as a Discipline." *Hastings Center Studies* 1 (1973): 66–73.
5. Jonsen, Albert. *A Short History of Medical Ethics*. New York: Oxford University Press, 2000.
6. Baker, Robert and McCullough, Lawrence. *A History of Medical Ethics*. Cambridge: Cambridge University Press, 2009.
7. Leake, Chauncey (ed.). *Percival's Medical Ethics*. Baltimore: Williams and Wilkins, 1927.
8. Fletcher, Joseph. *Morals and Medicine. The Moral Problems of the Patient's Right to Know the Truth, Contraception, Artificial Insemination, Sterilization, and Euthanasia*. Princeton: Princeton University Press, 1954.
9. Kelly, David. *The Emergence of Roman Catholic Medical Ethics in North America: A Historical-Methodological-Bibliographical Study*. New York and Toronto: The Edwin Mellen Press, 1979.

10. Alexander, Shana. "They Decide Who Lives, Who Dies." *Life* 53 (1962): 102–125.
11. Rescher, Nicholas, "The Allocation of Exotic Medical Lifesaving Therapy." *Ethics* 79 (1969): 173–186.
12. Childress, James. "Who Shall Live When Not All Can Live?" *Soundings* 53 (1970): 339–355.
13. Ramsey, Paul. *The Patient as Person.* New Haven: Yale University Press, 1970, chapter 7.
14. Sanders, David and Dukeminier, Jesse. "Medical advances and legal lag: hemodialysis and kidney transplantation," *UCLA Law Review* 15 (1968): 366–380.
15. Duff, RS, and Campbell, AGM. "Moral and Ethical Dilemmas in the Special-Care Nursery," *New England Journal of Medicine*; 289(1973): 980–984.
16. Veatch, Robert. *Death, Dying and the Biological Revolution: Our Last Quest for Responsibility.* New Haven: Yale University Press, 1976.
17. Bernard, Claude. *Introduction to the Study of Experimental Medicine.* H. Greene (Trans.). New York: Dover, 1957, pp. 191–102.
18. Annas, George and Grodin, Michael (eds.). *The Nazi Doctors and the Nuremberg Code: Human Rights in Human Experimentation.* New York: Oxford University Press, 1992.
19. Beecher, Henry K. "Ethics and Clinical Research," *New England Journal of Medicine* 1966; 274: 1354–1360.
20. Jones, James. *Bad Blood.* New York: The Free Press, 1981.
21. National Commission for the Protection of Human Subjects of Biomedical and Behavioral Research. *The Belmont Report* (45Code of Federal Regulations 46) 1979, 45.102. In Jonsen, Veatch, Walters, *Source Book in Bioethics.* Washington D.C.: Georgetown University Press, 1998. Part I. See Childress, James, Meslin, Eric, Shapiro, Howard. (eds.) *Belmont Revisited. Ethical Principles for Research with Human Subjects.* Washington, D.C.: Georgetown University Press, 2005.
22. Jonas, Hans. "Philosophical reflections on human experimentation," *Daedalus* 98 (1969): 219–247.
23. Katz, Jay. *Experimentation with Human Beings.* New York: Russell Sage Foundation, 1972.
24. Jonsen, *The Birth of Bioethics*, chapter 6.
25. Jonsen, Veatch, Walters, *Source Book in Bioethics*, Part III.
26. Fox, Renée and Swazey, Judith. *The Courage to Fail. A Social View of Organ Transplants and Dialysis.* Chicago: University of Chicago Press, 1974.
27. Report of the Ad Hoc Committee at Harvard Medical School to Examine the Definition of Brain Death. *Journal of the American Medical Association* 205(1968): 337–340.
28. President's Commission for the Study of Ethical Problems in Medicine and in Biomedical and Behavioral Research. *Defining Death: A Report on the Medical, Legal, and Ethical Issues in the Definition of Death.* Washington, D.C.: U.S. Government Printing Office, 1981. In Jonsen, Veatch, Walters, *Source Book in Bioethics*, Part II.

29. See Battin, Margaret, *The Least Worst Death. Essays in Bioethics on the End of Life*. New York: Oxford University Press, 1995.

30. Goodman, Kenneth (ed.). *The Case of Terri Schiavo*. New York: Oxford University Press, 2010.

31. See Jonsen, Veatch, Walters, *Source Book in Bioethics*, part IV. See also Harris, John and S. Holm (eds.) *The Future of Human Reproduction*. New York: Oxford University Press, 1998; and McGee, Glen. *The Perfect Baby: Parenthood in the New World of Cloning and Genetics*. Lanham, MD: Rowman and Littlefield, 2000.

32. President's Council on Bioethics. *Human Cloning and Human Dignity*. Washington, D.C.: Government Printing Office, 2002; *Monitoring Stem Cell Research*, 2004. See Green, Ronald. *Human Embryo Research Debates. Bioethics in the Vortex of Controversy*. New York: Oxford University Press, 2001.

33. Ramsey, Paul. *The Patient as Person*. New Haven: Yale University Press, 1970/2002.

34. Jonsen, Albert, Siegler, Mark, Winslade. *Clinical Ethics. A Practical Approach to Ethical Decisions in Clinical Medicine*. New York, Macmillan, 1982; 6th edition, McGraw-Hill, 2006.

35. See Sugarman, J., and Sulmasy, D. (eds.) *Methods in Medical Ethics*. Washington D.C.: Georgetown University Press, 2001. Khushf, G. (ed.). *Handbook of Bioethics: Taking Stock of the Field From a Philosophical Perspective*. Dordrecht/Boston/London: Kluwer Academic Publishers, 2002.

36. Macklin, R. "The new conservatives in bioethics: who are they and what do they seek?" Cohen, E. "Conservative bioethics and the search for wisdom," *Hastings Center Report* 36(2006):1; 34–43, 44–56.

THE EMERGENCE OF BIOETHICS AS DISCIPLINE AND DISCOURSE

ॐ

Bioethics as a Discipline

Daniel Callahan

After Dan Callahan earned a Ph.D. in philosophy from Harvard University and worked briefly as an editor at the Catholic intellectual weekly, *Commonwealth*, he turned his interests to the questions of the "new biology." In 1969, together with New York physician Willard Gaylin, he founded the Institute for Society, Ethics, and the Life Sciences, now known as the Hastings Center, with the purpose of encouraging scholarly study and public discussion of the ethical issues raised by scientific innovation.

"Bioethics as a Discipline" appeared in the Institute's publication, *Hastings Center Studies* (now known as *Hastings Center Report*). In this article, Callahan foresees the emergence of a discipline that he calls bioethics and reflects on what the role of the ethicist might be in the world of medicine and biology. He asserts that philosophers must learn about that world and adapt their standards of intellectual rigor to the nature of the problems arising in it. To serve the physicians and biologists responsible for making practical decisions, bioethics must define issues, methodological strategies, and decision procedures that are sensitive to specific cases in all the their complexity.

J ust what is the role of the ethicist in trying to make a contribution to the ethical problems of medicine, biology, or population? I resisted, with utter panic, the idea of participating with the physicians in their actual decision. Who me? I much preferred the safety of the profound questions I pushed on them. But I also realized when faced with an actual case—and this is my excuse—that there was nothing whatever in my philosophical training which had prepared me

Abridged from Daniel Callahan, "Bioethics as a Discipline," *Hastings Center Studies*, Volume 1, 1973, pp. 66–73. Reprinted by permission.

to make a flat, clear-cut ethical decision at a given hour on a given afternoon. I had been duly trained in that splendid tradition of good scholarship and careful thinking which allows at least a couple thousand years to work through any problem. . . .

When we ask what the place of bioethics might be, we of course need to know just what the problems are in medicine and biology which raise ethical questions and need ethical answers. I will not retail the whole catalogue of issues here; suffice it to say that they begin with "A" (abortion and amniocentesis) and run all the way to "Z" (the moral significance of zygotes). One evident and first task for the ethicist is simply that of trying to point out and define which problems raise moral issues. A second and no less evident task is providing some systematic means of thinking about, and thinking through, the moral issues which have been discerned. A third, and by far the most difficult, task is that of helping scientists and physicians to make the right decisions; and that requires a willingness to accept the realities of most medical and much scientific life, that is, that at some discrete point in time all the talk has to end and a choice must be made, a choice which had best be right rather than wrong. . . .

I used above the phrase "the realities of life." Another one of these realities is that the ethical issues of medicine and biology rarely present themselves in a way nicely designed to fit the kinds of categories and processes of thought which philosophers and theologians traditionally feel secure about. They almost always start off on the wrong foot by coming encumbered with the technical jargon of some other discipline. And only in text books is one likely to encounter cases which present a clear occasion, say, for deciding on the validity of a deontological or utilitarian ethical solution. The issues come, that is, in a mossy, jumbled form, cutting through many disciplines, gumming up all our clean theoretical engines, festooned with odd streamers and complicated knots.

The fact that this is the case immediately invites the temptation of what can be called "disciplinary reductionism." By that I mean a penchant for distilling out of an essentially complex ethical problem one transcendent issue which is promptly labeled *the* issue. Not coincidentally, this issue usually turns out to be a classic, familiar argument in philosophy or theology. By means of this kind of reductionism, the philosopher or theologian is thus enabled to do what he has been trained to do, deal with those classic disputes in a language and a way he is comfortable with—in a way which allows him to feel he is being a good "professional." The results of this tendency are doleful. It is one reason why most biologists and physicians find the contributions of the professional ethicist of only slight value. Their problems, very real to them in their language and their frame of reference, are promptly made unreal by being transmuted into someone else's language and reference system, in the process usually stripping the original case of all the complex facticity with which it actually presented itself. The whole business becomes positively pitiable when the philosopher or theologian, rebuffed or ignored because of his reductionism, can only respond by charging that his critics are obviously "not serious" about ethics, not interested in "real" ethical thinking.

I stress the problem of "disciplinary reductionism" out of a conviction that if a discipline of bioethics is to be created, it must be created in a way which does not allow this form of evading responsibility, of blaming the students for the faults of the teacher, of changing the nature of the problems to suit the methodologies of professional ethicists.

Toward this end, no subject would seem to me more worthy of investigation than what I will call the "ordinary language of moral thinking and discourse." Most people do not talk about their ethical problems in the language of philosophers. And I have yet to meet one professional ethicist who, when dealing with his own personal moral dilemmas, talks the language of his professional writings; he talks like everyone else, and presumably he is thinking through his own problems in banal everyday language like everyone else. Now of course it might be said that this misses the whole point of a serious professional discipline. Is it not like claiming that there must be nothing to theoretical physics simply because the physicist does not talk about the furniture in his house in terms of molecules and electrons? But the analogy does not work, for it is of the essence of moral decision-making to be couched in ordinary language and dealt with by ordinary, non-professional modes of thinking. The reason for this is apparent. An ethical decision will not be satisfactory to the person whose decision it is unless it is compatible with the way in which the person ordinarily thinks about himself and what he takes his life to be. . . .

In trying to create the discipline of bioethics, the underlying question raised by the foregoing remarks bears on what it should mean to be "rigorous" and "serious" about bioethics. . . . it is common enough for ethicists to gather among themselves after some frustrating interdisciplinary session to mutter about the denseness and inanity of their scientific and medical colleagues.

There are two options open here. One is to continue the muttering, being quite certain that the muttering is being reciprocated back in the scientific lab. That is, one can stick to traditional notions of philosophical and theological rigor, in which case one will rarely if ever encounter it in the interdisciplinary work of bioethics. Or, more wisely, the thought may occur that it is definitions of "rigor" which need adaptation. Not the adaptation of expediency or passivity in the face of careless thinking, but rather a perception that the kind of rigor required for bioethics may be of a different sort than that normally required for the traditional philosophical or scientific disciplines.

This is to say no more than that the methodological rigor should be appropriate to the subject matter. I spoke above of three tasks for the bioethicist: definition of issues, methodological strategies, and procedures for decision-making. Each of these tasks requires a different kind of rigor. The first requires what I will paradoxically call the rigor of an unfettered imagination: an ability to see in, through, and under the surface appearance of things; to envision alternatives; to get under the skin of people's ethical agonies or ethical insensitivities; to look at things from many perspectives simultaneously.

A different kind of rigor is needed for the development of methodological strategies. Here the traditional methodologies of philosophy and theology are indispensable; there are standards of rigor which can and should come into play, bearing on logic, consistency, careful analysis of terms, and the like. Yet at the same time they have to be adapted to the subject matter at hand, and that subject matter is not normally, in concrete ethical cases of medicine and biology, one which can be stuffed into a too-rigidly structured methodological mold.

I am not about to attempt here a full discourse on what should be the proper and specific methodology of bioethics. Some sketchy, general comments will have to do, mainly in the way of assertions. Traditionally, the methodology of ethics has concerned itself with ethical thinking, how to think straight about ethical problems. However, I believe that the province of the bioethicist can legitimately encompass a concern with three areas of ethical activity: thinking, feeling (attitudes), and behavior. The case for including feelings and behavior along with thinking rests on the assumptions (1) that in life both feelings and behavior shape thinking, often helping to explain why defective arguments are nonetheless, for all that, persuasive and pervasive; and (2) that it is legitimate for an ethicist to worry about what people do and not just what they think and say; a passion for the good is not inappropriate for ethicists.

If ethics was nothing other than seeing to it that no logical fallacies were committed in the process of ethical argumentation, it would hardly be worthy of anyone's attention. It is the premises of ethical arguments, the visions behind ethical systems, the feelings which fuel ethical (or non-ethical) behavior, which make the real difference for human life. Verbal formulations and arguments are only the tip of the iceberg. An ethicist can restrict himself to that tip; he will be on safe enough professional grounds if he does so. But I see no reason why he can't dare more than that, out of a recognition that the source and importance of his field lie not in the academy but in private and public human life, where what people think, feel, and do make all the difference there is. . . .

I will only offer one negative and one positive criterion for ethical methodology. The wrong methodology will be used if it is not a methodology which has been specifically developed for ethical problems of medicine and biology. This does not mean it cannot or should not bear many of the traits of general philosophical or theological methodology. But if it bears only those traits one can be assured that it will not deal adequately with specific issues which arise in the life sciences. My positive criterion for a good methodology is this: it must display the fact that bioethics is an interdisciplinary field in which the purely "ethical" dimensions neither can nor should be factored out without remainder from the legal, political, psychological and social dimensions. The critical question, for example, of who should make the ethical decisions in medicine and biology is falsified at the outset if too sharp a distinction is drawn between what, ethically, needs to be decided and who, politically, should be allowed to decide. It is surely important to ethical theory to make this kind of distinction; unfortunately, if pressed too doggedly it may well falsify the reality of the way decisions are and will continue to be made.

The problem of decision-making, which I include as the third task of the bioethicist, cannot be divorced from the methodological question. Actually it makes me realize that I have a second positive criterion to offer as a test of a good bioethical methodology. The methodology ought to be such that it enables those who employ it to reach reasonably specific, clear decisions in those instances which require them—in the case of what is to be done about Mrs. Jones by four o'clock tomorrow afternoon, after which she will either live or die depending upon the decision made. I have already suggested that philosophers are not very good at that sort of thing, and that their weakness in this respect is likely to be altogether vexing to the physician who neither has the right atmosphere nor the time to think through everything the philosopher usually argues *needs* to be thought through.

In proposing that a good methodology should make it possible to reach specific conclusions at specific times, I am proposing a utopian goal. The only kinds of ethical systems I know of which make that possible are those of an essentially deductive kind, with well-established primary and secondary principles and a long history of highly refined casuistical thinking. The Roman Catholic scholastic tradition and the Jewish *responsa* tradition are cases in point. Unfortunately, systems of that kind presuppose a whole variety of cultural conditions and shared world-views which simply do not exist in society at large. In their absence, it has become absolutely urgent that the search for a philosophically viable normative ethic, which can presuppose some commonly shared principles, go forward with all haste. Short of finding that, I do not see how ethical methodologies can be developed which will include methods for reaching quick and viable solutions in specific cases. Instead, we are likely to get only what we now have, a lot of very broad and general thinking, full of vagrant insights, but on the whole of limited use to the practicing physician and scientist.

Much of what I have been saying presupposes that a distinction can be drawn between "ethics" understood broadly and ethics understood narrowly. In its narrow sense, to do "ethics" is to be good at doing what well-trained philosophers and theologians do: analyze concepts, clarify principles, see logical entailments, spot underlying assumptions, and build theoretical systems. There are better and worse ways of doing this kind of thing and that is why philosophers and theologians can spend much of their time arguing with each other. But even the better ways will, I think, not be good enough for the demands of bioethics. That requires understanding "ethics" in a very broad, well-nigh unmanageable sense of the term.

My contention is that the discipline of bioethics should be so designed, and its practitioners so trained, that it will directly—at whatever cost to disciplinary elegance—serve those physicians and biologists whose position demands that they make the practical decisions. This requires, ideally, a number of ingredients as part of the training—which can only be life-long—of the bioethicist: sociological understanding of the medical and biological communities; psychological understanding of the kinds of needs felt by researchers and clinicians, patients and physicians, and the varieties of pressures to which they are subject; historical understanding of the sources of regnant value theories and common practices; requisite scientific training; awareness of and facility with the usual methods of ethical analysis as understood in the philosophical and theological communities—and no less a full awareness of the limitations of those methods when applied to actual cases; and, finally, personal exposure to the kinds of ethical problems which arise in medicine and biology. . . .

One important test of the acceptance of bioethics as a discipline will be the extent to which it is called upon by scientists and physicians. This means that it should be developed inductively, working at least initially from the kinds of problems scientists and physicians believe they face and need assistance on. As often as not, they will be wrong about the real nature of the issues with which they have to wrestle. But no less often the person trained in philosophy and theology will be equally wrong in his understanding of the real issues. Only a continuing, probably tension-ridden dialectic will suffice to bridge the gap, a dialectic which can only be kept alive by a continued exposure to specific cases in all their human dimensions. Many of them will be very unpleasant cases, the kind which make one long for the security of writing elegant articles for professional journals on such manageable issues as recent distinctions between "rules" and "maxims."

The New Biology:
What Price Relieving Man's Estate?

Leon R. Kass

Leon R. Kass is a physician with a strong philosophical bent. Beginning his career as a research scientist at the National Institutes of Health, he is now professor of humanities at the University of Chicago. He was chairperson of President George W. Bush's Council on Bioethics from 2001 to 2005. His book, *The New Biology*, collects many of his trenchant essays on bioethics and the philosophy of medicine.

This essay represents one of the first sweeping reviews of the ethical implications of the "new biology" that was published in a leading scientific journal. The excerpts are but the opening paragraphs and one other section of a long article that presages many of the major questions bioethicists will ponder over the next several decades

Recent advances in biology and medicine suggest that we may be rapidly acquiring the power to modify and control the capacities and activities of men by direct intervention and manipulation of their bodies and minds. Certain means are already in use or at hand, others await the solution of relatively minor technical problems, while yet others, those offering perhaps the most precise kind of control, depend upon further basic research. Biologists who have considered these matters disagree on the question of how much how soon, but all agree that the power for "human engineering," to borrow from the jargon, is coming and that it will probably have profound social consequences.

These developments have been viewed both with enthusiasm and with alarm; they are only just beginning to receive serious attention. Several biologists have undertaken to inform the public about the technical possibilities, present and future. Practitioners of social science "futurology" are attempting to predict and describe the likely social consequences of and public responses to the new technologies. Lawyers and legislators are exploring institutional innovations for assessing new technologies. All of these activities are based upon the hope that we can harness the new technology of man for the betterment of mankind.

Yet this commendable aspiration points to another set of questions, which are, in my view, sorely neglected—questions that inquire into the meaning of phrases such as the "betterment of mankind." A *full* understanding of the new technology of man requires an exploration of ends, values, standards. What ends will or should the new techniques serve? What values should guide society's adjustments? By what standards should agencies assess? Behind these questions lie others: what is a good man, what is a good life for man, what is a good community? This article is an attempt to provoke discussion of these neglected and important questions.

While these questions about ends and ultimate ends are never unimportant or irrelevant, they have rarely been more important or more relevant. That this is so can be seen once we recognize that we are dealing here with a group of technologies that are in a decisive respect unique: the object upon which they operate is man himself. The technologies of energy or food production, of communication, of manufacture, and of motion greatly alter the implements available to man and the conditions in which he uses them. In contrast, the biomedical technology works to change the user himself. To be sure, the printing press, the automobile, the television, and the jet airplane have greatly altered the conditions under which and the way in which men live; but men as biological beings have remained largely unchanged. They have been, and remain, able to accept or reject, to use and abuse these technologies; they choose, whether wisely or foolishly, the ends to which these technologies are means. Biomedical technology may make it possible to change the inherent capacity for choice itself. Indeed, both those who welcome and those who fear the advent of "human engineering" ground their hopes and fears in the same prospect: *that man can for the first time recreate himself.* . . .

After this cursory review of the powers now and soon to be at our disposal, I turn to the questions concerning the use of these powers. First, we must recognize that questions of use of science and technology are always moral and political questions, never simply technical ones. All private or public decisions to develop or to use biomedical technology—and decisions *not* to do so—inevitably contain judgments about value. This is true even if the values guiding those decisions are not articulated or made clear, as indeed they often are not. Secondly, the value judgments cannot be derived from biomedical science. This is true even if scientists themselves make the decisions.

These important points are often overlooked for at least three reasons.

1) They are obscured by those who like to speak of "the control of nature by science." It is men who control, not that abstraction "science." Science may provide the means, but men choose the ends; the choice of ends comes from beyond science.

2) Introduction of new technologies often appears to be the result of no decision whatsoever, or of the culmination of decisions too small or unconscious to be recognized as such. What can be done is done. However, someone is deciding on the basis of some notions of desirability, no matter how self-serving or altruistic.

3) Desires to gain or keep money and power no doubt influence much of what happens, but these desires can also be formulated as reasons and then discussed and debated.

Insofar as our society has tried to deliberate about questions of use, how has it done so? Pragmatists that we are, we prefer a utilitarian calculus: we weigh "benefits" against "risks," and we weigh them for both the individual and "society." We often ignore the fact that the very definitions of "a benefit" and "a risk" are themselves based upon judgments about value. In the biomedical areas just reviewed, the benefits are considered to be self-evident: prolongation of life, control of fertility and of population size, treatment and prevention of genetic disease, the reduction of anxiety and aggressiveness, and the enhancement of memory, intelligence, and pleasure. The assessment of risk is, in general, simply pragmatic—will the technique work effectively and reliably, how much will it cost, will it do detectable bodily harm, and who will complain if we proceed with development? As these questions are familiar and congenial, there is no need to belabor them.

The very pragmatism that makes us sensitive to considerations of economic cost often blinds us to the larger social costs exacted by biomedical advances. For one thing, we seem to be unaware that we may not be able to maximize all the benefits, that several of the goals we are promoting conflict with each other. On the one hand, we seek to control population growth by lowering fertility; on the other hand, we develop techniques to enable every infertile woman to bear a child. On the one hand, we try to extend the lives of individuals with genetic disease; on the other, we wish to eliminate deleterious genes from the human population. I am not urging that we resolve these conflicts in favor of one side or the other, but simply that we recognize that such conflicts exist. Once we do, we are more likely to appreciate that most "progress" is heavily paid for in terms not generally included in the simple utilitarian calculus.

How Medicine Saved the Life of Ethics

Stephen Toulmin

Stephen Toulmin was a well-established philosopher whose work, *Reason in Ethics* (Cambridge University Press, 1950), had long been appreciated when he took leave from the University of Chicago to work as philosophy consultant to the National Commission for the Protection of Human Subjects of Biomedical and Behavioral Research (1974–1978). He is currently Professor of Multiethnic and Transnational Studies at the University of Southern California.

"How Medicine Saved the Life of Ethics" was inspired by Toulmin's service with the National Commission. The article, although published somewhat later than the others in this section, had a broad influence on the emerging field of bioethics. Its main ideas are more fully developed in Toulmin's later book, coauthored with Albert Jonsen, *The Abuse of Casuistry: A History of Moral Reasoning* (Berkeley: University of California, 1988).

D uring the first 60 years or so of the twentieth century, two things characterized the discussion of ethical issues in the United States, and to some extent other English-speaking countries also. On the one hand, the theoretical analyses of moral philosophers concentrated on questions of so-called metaethics. Most professional philosophers assumed that their proper business was not to take sides on substantive ethical questions but rather to consider in a more formal way what *kinds* of issues and judgments are properly classified as moral in the first place. On the other hand, in less academic circles, ethical debates repeatedly ran into stalemate. A hard-line group of dogmatists, who appealed either to a code of universal rules or to the authority of a religious system or teacher, confronted a rival group of relativists and subjectivists, who found in the anthropological and psychological diversity of human attitudes evidence to justify a corresponding diversity in moral convictions and feelings.

For those who sought some "rational" way of settling ethical disagreements, there developed a period of frustration and perplexity. Faced with the spectacle of rival camps taking up sharply opposed ethical positions (e.g., toward premarital sex or anti-Semitism), they turned in vain to the philosophers for guidance. Hoping for intelligent and perceptive comments on the actual substance of such issues, they were offered only analytical classifications, which sought to locate the realm of moral issues, not to decide them. . . .

Abridged from Stephen Toulmin, "How Medicine Saved the Life of Ethics," *Perspectives in Biology and Medicine*, Volume 25, 1982, pp. 736–750. Copyright © 1982 by University of Chicago Press.

How did the fresh attention that philosophers began paying to the ethics of medicine, beginning around 1960, move the ethical debate beyond this stand-off? It did so in four different ways. In place of the earlier concern with attitudes, feelings, and wishes, it substituted a new preoccupation with situations, needs, and interests; it required writers on applied ethics to go beyond the discussion of general principles and rules to a more scrupulous analysis of the particular kinds of "cases" in which they find their application; it redirected that analysis to the professional enterprises within which so many human tasks and duties typically arise; and, finally, it pointed philosophers back to the ideas of "equity," "reasonableness," and "human relationships," which played central roles in the *Ethics* of Aristotle but subsequently dropped out of sight [1 esp. 5.10.1136b30–1137b32]. Here, these four points may be considered in turn. . . .

The new attention to applied ethics (particularly medical ethics) has done much to dispel the miasma of subjectivity that was cast around ethics as a result of its association with anthropology and psychology. At least within broad limits, an ethics of "needs" and "interests" is objective and generalizable in a way that an ethics of "wishes" and "attitudes" cannot be. Stated crudely, the question of whether one person's actions put another person's health at risk is normally a question of ascertainable fact, to which there is a straightforward "yes" or "no" answer, not a question of fashion, custom, or taste, about which (as the saying goes) "there is no arguing." This being so, the objections to that person's actions can be presented and discussed in "objective" terms. So, proper attention to the example of medicine has helped to pave the way for a reintroduction of "objective" standards of good and harm and for a return to methods of practical reasoning about moral issues that are not available to either the dogmatists or the relativists.

The Importance of Cases

One writer who was already contributing to the renewed discussion of applied ethics as early as the 1950s was Joseph Fletcher of the University of Virginia, who has recently been the object of harsh criticism from more dogmatic thinkers for introducing the phrase "situation ethics."[1] . . .

. . . In retrospect, Joseph Fletcher's introduction of the phrase "situation ethics" can be viewed as one further chapter in a history of "the ethics of *cases*," as contrasted with "the ethics of *rules and principles*"; this is another area in which the ethics of medicine has recently given philosophers some useful pointers for the analysis of moral issues.

Let me here mention one of these, which comes out of my own personal experience. From 1975 to 1978 I worked as a consultant and staff member with the

---

[1]Just how much of a pioneer Joseph Fletcher was in opening up the modern discussion of the ethics of medicine is clear from the early publication date (1954) of his first publications on this subject [2–4].

National Commission for the Protection of Human Subjects of Biomedical and Behavioral Research, based in Washington, D.C.; I was struck by the extent to which the commissioners were able to reach agreement in making recommendations about ethical issues of great complexity and delicacy.[2] If the earlier theorists had been right, and ethical considerations really depended on variable cultural attitudes or labile personal feelings, one would have expected 11 people of such different backgrounds as the members of the commission to be far more divided over such moral questions than they ever proved to be in actual fact. Even on such thorny subjects as research involving prisoners, mental patients, and human fetuses, it did not take the commissioners long to identify the crucial issues that they needed to address, and, after patient analysis of these issues, any residual differences of opinion were rarely more than marginal, with different commissioners inclined to be somewhat more conservative, or somewhat more liberal, in their recommendations. Never, as I recall, did their deliberations end in deadlock, with supporters of rival principles locking horns and refusing to budge. The problems that had to be argued through at length arose, not on the level of the principles themselves, but at the point of applying them: when difficult moral balances had to be struck between, for example, the general claims of medical discovery and its future beneficiaries and the present welfare or autonomy of individual research subjects.

How was the Commission's consensus possible? It rested precisely on this last feature of their agenda: namely, its close concentration on specific types of problematic cases. Faced with "hard cases," they inquired what particular conflicts of claim or interest were exemplified in them, and they usually ended by balancing off those claims in very similar ways. Only when the individual members of the Commission went on to explain their own particular "reasons" for supporting the general consensus did they begin to go seriously different ways. For, then, commissioners from different backgrounds and faiths "justified" their votes by appealing to general views and abstract principles which differed far more deeply than their opinions about particular substantive questions. Instead of "deducing" their opinions about particular cases from general principles that could lend strength and conviction to those specific opinions, they showed a far greater certitude about particular cases than they ever achieved about general matters.

This outcome of the Commission's work should not come as any great surprise to physicians who have reflected deeply about the nature of clinical judgment in medicine. In traditional case morality, as in medical practice, the first

---

[2] The work of the national commission generated a whole series of government publications—mainly reports and recommendations on the ethical aspects of research involving research subjects from specially "vulnerable" groups having diminished autonomy, such as young children and prisoners. . . . As a member of the commission, A. R. Jonsen was also struck by the casuistical character of its work, and this led to the research project of which this paper is one product.

indispensable step is to assemble a rich enough "case history." Until that has been done, the wise physician will suspend judgment. If he is too quick to let theoretical considerations influence his clinical analysis, they may prejudice the collection of a full and accurate case record and so distract him from what later turn out to have been crucial clues. Nor would this outcome have been any surprise to Aristotle, either. Ethics and clinical medicine are both prime examples of the concrete fields of thought and reasoning in which (as he insisted) the theoretical rigor of geometrical argument is unattainable: fields in which we should above all strive to be *reasonable* rather than insisting on a kind of *exactness* that "the nature of the case" does not allow [1, 1.3.1094b12–27]. . . .

By taking one step further, indeed, we may view the problems of clinical medicine and the problems of applied ethics as two varieties of a common species. Defined in purely general terms, such ethical categories as "cruelty" and "kindness," "laziness" and "conscientiousness" have a certain abstract, truistical quality: before they can acquire any specific relevance, we have to identify some *actual* person, or piece of conduct, as "kind" or "cruel," "conscientious" or "lazy," and there is often disagreement even about that preliminary step. Similarly, in medicine: if described in general terms alone, diseases too are "abstract entities," and they acquire a practical relevance only for those who have learned the diagnostic art of identifying real-life cases as being cases of one disease rather than another.

In its form (if not entirely in its point) the *art* of practical judgment in ethics thus resembles the art of clinical diagnosis and prescription. In both fields, theoretical generalities are helpful to us only up to a point, and their actual application to particular cases demands, also, a human capacity to recognize the slight but significant features that mark off, say, a "case" of minor muscular strain from a life-threatening disease or a "case" of decent reticence from one of cowardly silence. Once brought to the bedside, so to say, applied ethics and clinical medicine use just the same Aristotelean kinds of "practical reasoning," and a correct choice of therapeutic procedure in medicine is the *right* treatment to pursue, not just as a matter of medical technique but for ethical reasons also.

In the last decades of the nineteenth century, F. H. Bradley of Oxford University expounded an ethical position that placed "duties" in the center of the philosophical picture, and the recent concern of moral philosophers with applied ethics (most specifically, medical ethics) has given them a new insight into his arguments also. It was a mistake (Bradley argued) to discuss moral obligations purely in universalistic terms, as though nobody was subject to moral claims unless they applied to everybody—unless we could, according to the Kantian formula, "will them to become universal laws." On the contrary, different people are subject to different moral claims, depending on where they "stand" toward the other people with whom they have to deal, for example, their families, colleagues, and fellow citizens [5]. . . .

As the modern discussion of medical ethics has taught us, professional affiliations and concerns play a significant part in shaping a physician's obligations and commitments, and this insight has stimulated detailed discussions both about professionalism in general and, more specifically, about the relevance of "the physician/relationship" to the medical practitioner's duties and obligations. . . .[3]

In recent years, as a result, moral philosophers have begun to look specifically and in greater detail at the situations within which ethical problems typically arise and to pay closer attention to the human relationships that are embodied in those situations. In ethics, as elsewhere, the tradition of radical individualism for too long encouraged people to overlook the "mediating structures" and "intermediate institutions" (family, profession, voluntary associations, etc.) which stand between the individual agent and the larger scale context of his actions. So, in political theory, the obligation of the individual toward the state was seen as the only problem worth focusing on; meanwhile, in moral theory, the differences of status (or station) which in practice expose us to different sets of obligations (or duties) were ignored in favor of a theory of justice (or rights) that deliberately concealed these differences behind a "veil of ignorance."[4]

On this alternative view, the only just—even, properly speaking, the only moral—obligations are those that apply to us all equally, regardless of our standing. By undertaking the tasks of a profession, an agent will no doubt accept certain special duties, but so it will be for us all. The obligation to perform those duties is "just" or "moral" only because it exemplifies more general and universalizable obligations of trust, which require us to do what we have undertaken to do. So, any exclusive emphasis on the universal aspects of morality can end by distracting attention from just those things which the student of applied ethics finds most absorbing—namely, the specific tasks and obligations that any profession lays on its practitioners.

Most recently, Alasdair MacIntyre has pursued these considerations further in his new book, *After Virtue.*[9] MacIntyre argues that the public discussion of ethical issues has fallen into a kind of Babel, which largely springs from our losing any sense of the ways in which *community* creates obligations for us. One thing that can help restore that lost sense of community is the recognition that, at the present time, our professional commitments have taken on many of the roles that our communal commitments used to play. Even people who find moral philosophy generally unintelligible usually acknowledge and respect the specific ethical demands associated with their own professions or jobs, and this offers us some kind of a foundation on which to begin reconstructing our view of ethics. For it reminds us that we are in no position to fashion individual lives for ourselves, purely *as individuals*. Rather, we find ourselves born into communities in which the available ways of acting are largely laid out in advance: in which human activity takes on different *Lebensformen*, or "forms of life" (of which the professions are one special case), and our obligations are shaped by the requirements of those forms.

In this respect, the lives and obligations of professionals are no different from those of their lay brethren. Professional obligations arise out of the enterprises of the professions in just the same kinds of way that other general moral obligations arise out of our shared forms of life; if we are at odds about the *theory* of ethics, that is because we have misunderstood the basis which ethics has in our actual *practice*. Once again, in other words, it was medicine—as the first profession to which philosophers paid close attention during the new phase of "applied ethics" that opened during the 1960s—that set the example which was required in order to revive some important, and neglected, lines of argument within moral philosophy itself.

Equity and Intimacy

Two final themes have also attracted special attention as a result of the new interaction between medicine and philosophy. Both themes were presented in clear enough terms by Aristotle in the *Nicomachean Ethics*. But, as so often happens, the full force of Aristotle's concepts and arguments was overlooked by subsequent generations of philosophers, who came to ethics with very different preoccupations. Aristotle's own Greek terms for these notions are *epieikeia* and *philia*, which are commonly translated as "reasonableness" and "friendship," but I shall argue here that they correspond more closely to the modern terms, "equity" and "personal relationship" [1].

Modern readers sometimes have difficulty with the style of Aristotle's *Ethics* and lose patience with the book, because they suspect the author of evading philosophical questions that they have their own reasons for regarding as central. Suppose, for instance, that we go to Aristotle's text in the hope of finding some account of the things that mark off "right" from "wrong": if we attempt to press this question, Aristotle will always slip out of our grasp. What makes one course of action better than another? We can answer that question, he replies, only if we first consider what kind of a person the agent is and what relationships he stands in toward the other people who are involved in his actions; he sets about explaining why the kinds of relationship, and the kinds of conduct, that are possible as between "large-spirited human beings" who share the same social standing are simply not possible as between, say, master and servant, or parent and child [1].

The bond of *philia* between free and equal friends is of one kind, that between father and son of another kind, that between master and slave of a third, and there is no common scale in which we can measure the corresponding kinds of conduct. By emphasizing this point, Aristotle draws attention to an important point about the manner in which "actions" are classified, even before we say anything ethical about them. Within two different relationships the very same deeds, or the very same words, may—from the ethical point of view—represent quite different *acts* or *actions*. Words that would be a perfectly proper command from an officer to an enlisted man, or a straightforward order from a master to a servant, might be a humiliation if uttered by a father to a son, or an insult if exchanged between friends. A judge may likewise have a positive duty to say, from the bench, things that he would never dream of saying in a situation where he was no longer acting *ex officio*, while a physician may have occasion, and even be obliged, to do things to a patient in the course of a medical consultation that he would never be permitted to do in any other context.

. . . For surely, the very deed or utterance by Dr. A toward Mrs. B which would be a routine inquiry or examination within a strictly professional "physician-patient relationship"—for example, during a gynecological consultation—might be grounds for a claim of assault if performed outside that protected context. The *philia* (or relationship) between them will be quite different in the two situations, and, on this account, the "circumstances" do indeed "alter cases" in ways that are directly reflected in the demands of professional ethics.

With this as background, we can turn to Aristotle's ideas about *epieikeia* ("reasonableness" or "equity"). As to this notion, Aristotle pioneered the general doctrine that principles never settle ethical issues by themselves: that is, that we can grasp the moral force of principles only by studying the ways in which they are applied to, and within, particular situations. The need for such a practical approach is most obvious, in judicial practice, in the exercise of "equitable jurisdiction," where the courts are required to decide cases by appeal, not to specific, well-defined laws or statutes, but to general considerations of fairness, of "maxims of equity." In these situations, the courts do not have the benefit of carefully drawn rules, which have been formulated with the specific aim that they should be precise and self-explanatory: rather, they are guided by rough proverbial mottoes—phrases about "clean hands" and the like. The questions at issue in such cases are, in other words, very broad questions—for example, about what would be *just* or *reasonable* as between two or more individuals when all the available facts about their respective situations have been taken into account [10–12].

In ethics and law alike, the two ideas of *philia* ("friendship" or "relationship") and *epieikeia* (or "equity") are closely connected. The expectations that we place on people's lines of conduct will differ markedly depending on who is affected and what relationships the parties stand in toward one another. Far from regarding it as "fair" or "just" to deal with everybody in a precisely *equal* fashion, as the "veil of ignorance" might suggest, we consider it perfectly *equitable*, or *reasonable*, to show some degree of partiality, or favor, in dealing with close friends and relatives whose special needs and concerns we understand. What father, for instance, does not have an eye to his children's individual personalities and tastes? And, apart from downright "favoritism," who would regard such differences of treatment as unjust? Nor, surely, can it be morally offensive to discriminate, within reason, between close friends and distant acquaintances, colleagues and business rivals, neighbors and strangers? We are who we are: we stand in the human relationships we do, and our specific moral duties and obligations can be discussed in practice only at the point at which these questions of personal standing and relationship have been recognized and taken into the account.

Conclusion

From the mid-nineteenth century on, then, British and American moral philosophers treated ethics as a field for general theoretical inquiries and paid little attention to issues of application or particular types of cases. The philosopher who did most to inaugurate this new phase was Henry Sidgwick, and, from an autobiographical note, we know that he was reacting against the work of his contemporary, William Whewell [13, 14]. Whewell had written a textbook for use by undergraduates at Cambridge University that resembled in many respects a traditional manual of casuistics, containing separate sections on the ethics of promises or contracts, family and community, benevolence, and so on [15]. For his part, Sidgwick found Whewell's discussion too messy: there must be some way of introducing into the subject the kinds of rigor, order, and certainty associated with, for example, mathematical reasoning. So, ignoring all of Aristotle's cautions about the differences between the practical modes of reasoning appropriate to ethics and the formal modes appropriate to mathematics, he set out to expound the theoretical principles (or "methods") of ethics in a systematic form.

By the early twentieth century, the new program for moral philosophy had been narrowed down still further, so initiating the era of "metaethics." The philosopher's task was no longer to organize our moral beliefs into comprehensive systems: that would have meant *taking sides* over substantive issues. Rather, it was his duty to stand back from the fray and hold the ring while partisans of different views argued out their differences in accordance with the general rules for the conduct of "rational debate," or the expression of "moral attitudes," as defined in *metaethical* terms. And this was still the general state of affairs in Anglo-American moral philosophy in the late 1950s and the early 1960s, when public attention began to turn to questions of medical ethics. By this time, the central concerns of the philosophers had become so abstract and general—above all, so definitional or analytical—that they had, in effect, lost all touch with the concrete and particular issues that arise in actual practice, whether in medicine or elsewhere.

Once this demand for intelligent discussion of the ethical problems of medical practice and research obliged them to pay fresh attention to applied ethics, however, philosophers found their subject "coming alive again" under their hands. But, now it was no longer a field for academic, theoretical, even mandarin investigation alone. Instead, it had to be debated in practical, concrete, even political terms, and before long moral philosophers (or, as they barbarously began to be called, "ethicists"[5]) found that they were as liable as the economists to be called on to write "op ed" pieces for the *New York Times*, or to testify before congressional committees.

Have philosophers wholly risen to this new occasion? Have they done enough to modify their previous methods of analysis to meet these new practical needs? About those questions there can still be several opinions. Certainly, it would be foolhardy to claim that the discussion of "bioethics" has reached a definitive form, or to rule out the possibility that novel methods will earn a place

---

[5]Once again, the *Oxford English Dictionary* has a point to make. It includes the word "ethicist" but leaves it without the dignity of a definition, beyond the bare ethnology, "ethics + ist."

in the field in the years ahead. At this very moment, indeed, the style of current discussion appears to be shifting away from attempts to relate problematic cases to general theories—whether those of Kant, Rawls, or the utilitarians—to a more direct analysis of the practical cases themselves, using methods more like those of traditional "case morality." ...

Whatever the future may bring, however, these 20 years of interaction with medicine, law, and the other professions have had spectacular and irreversible effects on the methods and content of philosophical ethics. By reintroducing into ethical debate the vexed topics raised by *particular cases*, they have obliged philosophers to address once again the Aristotelean problems of *practical reasoning*, which had been on the sidelines for too long. In this sense, we may indeed say that, during the last 20 years, medicine has "saved the life of ethics," and that it has given back to ethics a seriousness and human relevance which it had seemed—at least, in the writings of the interwar years—to have lost for good.

References

1. Aristotle. *Nicomachean Ethics*.
2. Fletcher, J. *Morals and Medicine*. Princeton, N.J.: Princeton Univ. Press, 1954.
3. Fletcher, J. *Situation Ethics*. Philadelphia: Westminster, 1966.
4. Fletcher, J. *Humanhood*. Buffalo, N.Y.: Prometheus, 1979.
5. Bradley, F. *Ethical Studies*. London, 1876.
6. Bledstein, B. *The Culture of Professionalism*. New York: Norton, 1976.
7. Rawls, J. *A Theory of Justice*. Cambridge, Mass.: Harvard Univ. Press, 1971.
8. Toulmin, S. The tyranny of principles. *Hastings Cent. Rep.* 11:6, 1981.
9. MacIntyre, A. *After Virtue*. South Bend, Ind.: Notre Dame Univ. Press, 1981.
10. Davis, K. *Discretionary Justice*. Urbana: Univ. of Illinois Press, 1969.
11. Newman, R. *Equity and Law*. Dobbs Ferry, N.Y.: Oceana, 1961.
12. Hamburger, M. *Morals and Law: The Growth of Aristotle's Legal Theory*. New Haven, Conn.: Yale Univ. Press, 1951.
13. Sidgwick, H. *The Methods of Ethics*, Introduction to 6th ed. London and New York: Macmillan, 1901.
14. Schneewind, J. *Sidgwick's Ethics and Victorian Moral Philosophy*. Oxford and New York: Oxford Univ. Press, 1977.
15. Whewell, W. *The Elements of Morality*, 4th ed. Cambridge: Bell, 1864.

Who Shall Live When Not All Can Live?

James F. Childress

James F. Childress began his teaching career in the Department of Religion, University of Virginia; he later taught at Georgetown University before returning to the University of Virginia as the Kornfeld Professor of Biomedical Ethics. He is the coauthor, with Tom Beauchamp, of *Principles of Biomedical Ethics*, first published in 1979. This volume, in its fourth edition, is the leading exposition of the philosophical underpinnings and the practical application of biomedical ethics.

"Who Shall Live" comments on the problem of selecting patients for a scarce life-saving resource, as exemplified by the Seattle Selection Committee's policy of using "social worth" criteria to choose patients for dialysis. That policy was strongly criticized by two law professors, David Sanders and Jesse Dukeminier, in their article, "Medical Advance and Legal Lag: Hemodialysis and Kidney Transplantation" (*UCLA Law Review* 1968; 15: 366–380) and implicitly defended by philosopher Nicholas Rescher in one of the earliest philosophical contributions to the field of bioethics, "The Allocation of Exotic Medical Lifesaving Therapy" (*Ethics* 1969; 79: 173–186). Childress agrees with the broad criticism made by Sanders and Dukeminier and attempts to craft an ethical position contrary to the utilitarianism of Rescher.

W ho shall live when not all can live? Although this question has been urgently forced upon us by the dramatic use of artificial internal organs and organ transplantations, it is hardly new.

A significant example of the distribution of scarce medical resources is seen in the use of penicillin shortly after its discovery. Military officers had to determine which soldiers would be treated—those with venereal disease or those wounded in combat.[1] In many respects such decisions have become routine in medical circles. Day after day physicians and others make judgments and decisions "about allocations of medical care to various segments of our population, to various types of hospitalized patients, and to specific individuals,"[2] for example, whether mental illness or cancer will receive the higher proportion of available funds. Nevertheless, the dramatic forms of "Scarce Life-Saving Medical Resources" (hereafter abbreviated as SLMR) such as hemodialysis and kidney and heart transplants have compelled us to examine the moral questions that have been concealed in many routine decisions. . . .

Just as current SLMR decisions are not totally discontinuous with other medical decisions, so we must ask whether some other cases might, at least by

---

Abridged from James F. Childress, "Who Shall Live When Not All Can Live?" *Soundings*, Volume 53, 1970, pp. 339–355. Reprinted by permission.

analogy, help us develop the needed criteria and procedures. Some have looked at the principles at work in our responses to abortion, euthanasia, and artificial insemination.[3] Usually they have concluded that these cases do not cast light on the selection of patients for artificial and transplanted organs. The reason is evident: in abortion, euthanasia, and artificial insemination, there is no conflict of life with life for limited but indispensable resources (with the possible exception of therapeutic abortion). In current SLMR decisions, such a conflict is inescapable, and it makes them so morally perplexing and fascinating. If analogous cases are to be found, I think that we shall locate them in moral conflict situations.

Analogous Conflict Situations

An especially interesting and pertinent one is *U.S. v. Holmes*.[4] In 1841 an American ship, the *William Brown*, which was near Newfoundland on a trip from Liverpool to Philadelphia, struck an iceberg. The crew and half the passengers were able to escape in the two available vessels. One of these, a longboat, carrying too many passengers and leaking seriously, began to founder in the turbulent sea after about twenty-four hours. In a desperate attempt to keep it from sinking, the crew threw overboard fourteen men. Two sisters of one of the men either jumped overboard to join their brother in death or instructed the crew to throw them over. The criteria for determining who should live were "not to part man and wife, and not to throw over any women." Several hours later the others were rescued. Returning to Philadelphia, most of the crew disappeared, but one, Holmes, who had acted upon orders from the mate, was indicted, tried, and convicted on the charge of "unlawful homicide."

We are interested in this case from a moral rather than a legal standpoint, and there are several possible responses to and judgments about it. Without attempting to be exhaustive I shall sketch a few of these. The judge contended that lots should have been cast, for in such conflict situations, there is no other procedure "so consonant both to humanity and to justice." Counsel for Holmes, on the other hand, maintained that the "sailors adopted the only principle of selection which was possible in an emergency like theirs—a principle more humane than lots."

Another version of selection might extend and systematize the maxims of the sailors in the direction of "utility"; those are saved who will contribute to the greatest good for the greatest number. . . .

There are several significant differences between the *Holmes* and SLMR cases, a major one being that the former involves *direct* killing of another person, while the latter involve only *permitting* a person to die when it is not possible to save all. Furthermore, in extreme situations such as *Holmes*, the restraints of civilization have been stripped away, and something approximating a state of nature prevails, in which life is "solitary, poor, nasty, brutish and short." The state of nature does not mean that moral standards are irrelevant and that might should prevail, but it does suggest that much of the matrix which normally supports morality has been removed. Also, the necessary but unfortunate decisions about who shall live and die are made by men who are existentially and personally involved in the outcome. Their survival too is at stake. Even though the institutional role of sailors seems to require greater sacrificial actions, there is obviously no assurance that they will adequately assess the number of sailors required to man the vessel or that they will impartially and objectively weigh the common good at stake. As the judge insisted in his defense of casting lots in the *Holmes* case: "In no other way than this [casting lots] or some like way are those having equal rights put upon an equal footing, and in no other way is it possible to guard against partiality and oppression, violence, and conflict." This difference should not be exaggerated since self-interest, professional pride, and the like obviously affect the outcome of many medical decisions. Nor do the remaining differences cancel *Holmes'* instructiveness.

Criteria of Selection for SLMR

Which set of arrangements should be adopted for SLMR? Two questions are involved: Which standards and criteria should be used? and, Who should make the decision? The first question is basic, since the debate about implementation, e.g., whether by a lay committee or physician, makes little progress until the criteria are determined.

We need two sets of criteria which will be applied at two different stages in the selection of recipients of SLMR. First, medical criteria should be used to exclude those who are not "medically acceptable." Second, from this group of "medically acceptable" applicants, the final selection can be made. Occasionally in current American medical practice, the first stage is omitted, but such an omission is unwarranted. Ethical and social responsibility would seem to require distributing these SLMR only to those who have some reasonable prospect of responding to the treatment. Furthermore, in transplants such medical tests as tissue and blood typing are necessary, although they are hardly fully developed. . . .

The most significant moral questions emerge when we turn to the final selection. Once the pool of medically acceptable applicants has been defined and still the number is larger than the resources, what other criteria should be used? How should the final selection be made? First, I shall examine some of the difficulties that stem from efforts to make the final selection in terms of social value; these difficulties raise serious doubts about the feasibility and justifiability of the utilitarian approach. Then I shall consider the possible justification for random selection or chance.

Occasionally criteria of social worth focus on past contributions but most often they are primarily future-oriented. The patient's potential and probable contribution to the society is stressed, although this obviously cannot be abstracted from his present web of relationships (e.g., dependents) and occupational activities (e.g., nuclear physicist). Indeed, the magnitude of his contribution to society (as an abstraction) is measured in terms of these social roles, relations, and functions. Enough has already been said to suggest the tremendous range of factors that affect social value or worth.[5] Here we encounter the first major difficulty of this approach: How do we determine the relevant criteria of social value?

The difficulties of quantifying various social needs are only too obvious. How does one quantify and compare the needs of the spirit (e.g., education, art, religion), political life, economic activity, technological development? ... I am not convinced that we can ever quantify values, or that we should attempt to do so. But even if the various social and human needs, in principle, could be quantified, how do we determine how much weight we will give to each one? Which will have priority in case of conflict? Or even more basically, in the light of which values and principles do we recognize social "needs"?

One possible way of determining the values which should be emphasized in selection has been proposed by Leo Shatin.[6] He insists that our medical decisions about allocating resources are already based on an unconscious scale of values (usually dominated by material worth). Since there is really no way of escaping this, we should be self-conscious and critical about it. How should we proceed? He recommends that we discover the values that most people in our society hold and then use them as criteria for distributing SLMR. These values can be discovered by attitude or opinion surveys. Presumably if fifty-one percent in this testing period put a greater premium on military needs than technological development, military men would have a greater claim on our SLMR than experimental researchers. But valuations of what is significant change, and the student revolutionary who was denied SLMR in 1970 might be celebrated in 1990 as the greatest American hero since George Washington.

Shatin presumably is seeking criteria that could be applied nationally, but at the present, regional and local as well as individual prejudices tincture the criteria of social value that are used in selection. Nowhere is this more evident than in the deliberations and decisions of the anonymous selection committee of the Seattle Artificial Kidney Center where such factors as church membership and Scout leadership have been deemed significant for determining who shall live.[7] As two critics conclude after examining these criteria and procedures, they rule out "creative nonconformists, who rub the bourgeoisie the wrong way but who historically have contributed so much to the making of America. The Pacific Northwest is no place for a Henry David Thoreau with bad kidneys."[8]

Closely connected to this first problem of determining social values is a second one. Not only is it difficult if not impossible to reach agreement on social values, but it is also rarely easy to predict what our needs will be in a few years and what the consequences of present actions will be. Furthermore it is difficult to predict which persons will fulfill their potential function in society. Admissions committees in colleges and universities experience the frustrations of predicting realization of potential. For these reasons, as someone has indicated, God might be a utilitarian, but we cannot be. We simply lack the capacity to predict very accurately the consequences which we then must evaluate. Our incapacity is never more evident than when we think in societal terms.

Other difficulties make us even less confident that such an approach to SLMR is advisable. Many critics raise the spectre of abuse, but this should not be overemphasized. The fundamental difficulty appears on another level: the utilitarian approach would in effect reduce the person to his social role, relations, and functions. Ultimately it dulls and perhaps even eliminates the sense of the person's transcendence, his dignity as a person which cannot be reduced to his past or future contribution to society. It is not at all clear that we are willing to live with these implications of utilitarian selection. Wilhelm Kolff, who invented the artificial kidney, has asked: "Do we really subscribe to the principle that social standing should determine selection? Do we allow patients to be treated with dialysis only when they are married, go to church, have children, have a job, a good income and give to the Community Chest?"[9] . . .

The Values of Random Selection

My proposal is that we use some form of randomness or chance (either natural, such as "first come, first served," or artificial, such as a lottery) to determine who shall be saved. Many reject randomness as a surrender to non-rationality when responsible and rational judgments can and must be made. Edmond Cahn criticizes "Holmes' judge" who recommended the casting of lots because, as Cahn puts it, "the crisis involves stakes too high for gambling and responsibilities too deep for destiny."[10] Similarly, other critics see randomness as a surrender to "non-human" forces which necessarily vitiates human values. Sometimes these values are identified with the process of decision-making (e.g., it is important to have persons rather than impersonal forces determining who shall live). Sometimes they are identified with the outcome of the process (e.g., the features such as creativity and fullness of being which make human life what it is are to be considered and respected in the decision). Regarding the former, it must be admitted that the use of chance seems cold and impersonal. But presumably the defenders of utilitarian criteria in SLMR want to make their application as objective and impersonal as possible so that subjective bias does not determine who shall live.

Such criticisms, however, ignore the moral and nonmoral values which might be supported by selection by randomness or chance. A more important criticism is that the procedure that I develop draws the relevant moral context too narrowly. That context, so the argument might run, includes the society and its future and not merely the individual with his illness and claim upon SLMR. But my contention is that the values and principles at work in the narrower context may well take precedence over those operative in the broader context both because of their weight and significance and because of the weaknesses of selection in terms of social worth. As Paul Freund rightly insists, "The more nearly total is the estimate to be made of an individual, and the more nearly the consequence determines life and death, the more unfit the judgment becomes for human reckoning. . . . Randomness as a moral principle deserves serious study."[11] Serious study would, I think, point toward its implementation in certain conflict situations, primarily because it preserves a significant degree of *personal dignity* by providing *equality* of opportunity. Thus it cannot be dismissed as a "non-rational" and "non-human" procedure without an inquiry into the reasons, including human values, which might justify it. Paul Ramsey stresses this point about the *Holmes* case:

> Instead of fixing our attention upon "gambling" as the solution—with all the frivolous and often corrupt associations the word raises in our minds—we should think rather of equality of opportunity as the ethical substance of the relations of those individuals to one another that might have been guarded and expressed by casting lots.[12]

The individual's personal and transcendent dignity, which on the utilitarian approach would be submerged in his social role and function, can be protected and witnessed to by a recognition of his equal right to be saved. Such a right is best preserved by procedures which establish equality of opportunity. Thus selection by chance more closely approximates the requirements established by human dignity than does utilitarian calculation. It is not infallibly just, but it is preferable to the alternatives of letting all die or saving only those who have the greatest social responsibilities and potential contribution.

This argument can be extended by examining values other than individual dignity and equality of opportunity. Another basic value in the medical sphere is the relationship of trust between physician and patient. Which selection criteria are most in accord with this relationship of trust? Which will maintain, extend, and deepen it? My contention is that selection by randomness or chance is preferable from this standpoint too.

Trust, which is inextricably bound to respect for human dignity, is an attitude of expectation about another. It is not simply the expectation that another will perform a particular act, but more specifically that another will act toward him in certain ways—which will respect him as a person. . . . This trust cannot be preserved in life-and-death situations when a person expects decisions about him to be made in terms of his social worth, for such decisions violate his status as a person. An applicant rejected on grounds of inadequacy in social value or virtue would have reason for feeling that his "trust" had been betrayed. Indeed, the sense that one is being viewed not as an end in himself but as a means in medical progress or the achievement of a greater social good is incompatible with attitudes and relationships of trust. We recognize this in the billboard which was erected after the first heart transplants: "Drive Carefully. Christiaan Barnard Is Watching You." The relationship of trust between the physician and patient is not only an instrumental value in the sense of being an important factor in the patient's treatment. It is also to be endorsed because of its intrinsic worth as a relationship.

Thus the related values of individual dignity and trust are best maintained in selection by chance. But other factors also buttress the argument for this approach. Which criteria and procedures would men agree upon? We have to suppose a hypothetical situation in which several men are going to determine for themselves and their families the criteria and procedures by which they would want to be admitted to and excluded from SLMR if the need arose.[13] We need to assume two restrictions and then ask which set of criteria and procedures would be chosen as the most rational and, indeed, the fairest. The restrictions are these: (1) The men are *self-interested*. They are interested in their own welfare (and that of members of their families), and this, of course, includes survival. Basically, they are not motivated by altruism. (2) Furthermore, they are *ignorant* of their own talents, abilities, potential, and probable contribution to the social good. They do not know how they would fare in a competitive situation, e.g., the competition for SLMR in terms of social contribution. Under these conditions, which institution would be chosen—letting all die, utilitarian selection, or the use of chance? Which would seem the most rational? the fairest? By which set of criteria would they want to be included in or excluded from the list of those who will be saved? The rational choice in this setting (assuming self-interest and ignorance of one's competitive success) would be random selection or chance since this alone provides equality of opportunity. A possible response is that one would prefer to take a "risk" and therefore choose the utilitarian approach. But I think not, especially since I added that the participants in this hypothetical situation are choosing for their children as well as for themselves; random selection or chance could be more easily justified to the children. It would make more sense for men who are self-interested but uncertain about their relative contribution to society to elect a set of criteria which would build in equality of opportunity. They would consider selection by chance as relatively just and fair.[14]

An important psychological point supplements earlier arguments for using chance or random selection. The psychological stress and strain among those who are rejected would be greater if the rejection is based on insufficient social worth than if it is based on chance. Obviously stress and strain cannot be eliminated in these borderline situations, but they would almost certainly be increased by the opprobrium of being judged relatively "unfit" by society's agents using society's values. . . .

In the framework that I have delineated, are the decrees of chance to be taken without exception? If we recognize exceptions, would we not open Pandora's box again just after we had succeeded in getting it closed? The direction of my argument has been against any exceptions, and I would defend this as the proper way to go. But let me indicate one possible way of admitting exceptions while at the same time circumscribing them so narrowly that they would be very rare indeed.

An obvious advantage of the utilitarian approach is that occasionally circumstances arise which make it necessary to say that one man is practically indispensable for a society in view of a particular set of problems it faces (e.g., the President when the nation is waging a war for survival). Certainly the argument to this point has stressed that the burden of proof would fall on those who think that the social danger in this instance is so great that they simply cannot abide by the outcome of a lottery or a first-come, first-served policy. Also, the reason must be negative rather than positive; that is, we depart from chance in this instance not because we want to take advantage of this person's potential contribution to the improvement of our society, but because his immediate loss would possibly (even probably) be disastrous (again, the President in a grave national emergency). Finally, social value (in the negative sense) should be used as a standard of exception in dialysis, for example, only if it would provide a reason strong enough to warrant removing another person from a kidney machine if all machines were taken. Assuming this strong reluctance to remove anyone once the commitment has been made for him, we would be willing to put this patient ahead of another applicant for a vacant machine only if we would be willing (in circumstances in which all machines are being used) to vacate a machine by removing someone from it. These restrictions would make an exception almost impossible.

While I do not recommend this procedure of recognizing exceptions, I think that one can defend it while accepting my general thesis about selection by randomness or chance. If it is used, a lay committee (perhaps advisory, perhaps even stronger) would be called upon to deal with the alleged exceptions since the doctors or others would in effect be appealing the outcome of chance (either natural or artificial). This lay committee would determine whether this patient was so indispensable at this time and place that he had to be saved even by sacrificing the values preserved by random selection. It would make it quite clear that exception is warranted, if at all, only as the "lesser of two evils." Such a defense would be recognized only rarely, if ever, primarily because chance and randomness preserve so many important moral and nonmoral values in SLMR cases.[15]

Notes

1. Henry K. Beecher, "Scarce Resources and Medical Advancement," *Daedalus* (Spring 1969), pp. 279–280.
2. Leo Shatin, "Medical Care and the Social Worth of a Man," *American Journal of Orthopsychiatry*, 36 (1967), 97.
3. Harry S. Abram and Walter Wadlington, "Selection of Patients for Artificial and Transplanted Organs," *Annals of Internal Medicine*, 69 (September 1968), 615–620.
4. *United States v. Holmes* 26 Fed. Cas. 360 (C.C.E.D. Pa. 1842). All references are to the text of the trial as reprinted in Philip E. Davis, ed., *Moral Duty and Legal Responsibility: A Philosophical-Legal Casebook* (New York, 1966), pp. 102–118.
5. I am excluding from consideration the question of the ability to pay because most of the people involved have to secure funds from other sources, public or private, anyway.
6. Leo Shatin, op. cit., pp. 96–101.
7. For a discussion of the Seattle selection committee, see Shana Alexander, "They Decide Who Lives, Who Dies," *Life*, 53 (Nov. 9, 1962), 102. For an

examination of general selection practices in dialysis see "Scarce Medical Resources," *Columbia Law Review*, 69(1969), 620; and Harry S. Abram and Walter Wadlington, op cit.

8. David Sanders and Jesse Dukeminier, Jr., "Medical Advance and Legal Lag: Hemodialysis and Kidney Transplantation." *UCLA Law Review*, 15: 367–378.

9. "Letters and Comments," *Annals of Internal Medicine*, 61 (Aug. 1964), 360. Dr. G. E. Schreiner contends that "if you really believe in the right of society to make decisions on medical availability on these criteria you should be logical and say that when a man stops going to church or is divorced or loses his job, he ought to be removed from the programme and somebody else who fulfills these criteria substituted. Obviously no one faces up to this logical consequence" (G.E.W. Wolstenholme and Maeve O'Connor, eds. *Ethics in Medical Progress: With Special Reference to Transplantation*. A Ciba Foundation Symposium [Boston, 1966], p. 127).

10. Cahn, op. cit., p. 71.

11. Paul Freund, "Introduction," *Daedalus* (Spring 1969), xiii.

12. Paul Ramsey, *Nine Modern Moralists* (Englewood Cliffs, N.J., 1962), p. 245.

13. My argument is greatly dependent on John Rawls's version of justice as fairness, which is a reinterpretation of social contract theory. Rawls, however, would probably not apply his ideas to "borderline situations." See "Distributive Justice: Some Addenda," *Natural Law Forum*, 13 (1968), 53. For Rawls's general theory, see "Justice as Fairness," *Philosophy, Politics and Society* (Second Series), ed. by Peter Laslett and W. G. Runciman (Oxford, 1962), pp. 132–157 and Rawls's other essays on aspects of this topic.

14. Occasionally someone contends that random selection may reward vice. Leo Shatin (op. cit., p. 100) insists that random selection "would reward socially disvalued qualities by giving their bearers the same special medical care opportunities as those received by the bearers of socially valued qualities. Personally I do not favor such a method." Obviously society must engender certain qualities in its members, but not all of its institutions must be devoted to that purpose. Furthermore, there are strong reasons, I have contended, for exempting SLMR from that sort of function.

15. I read a draft of this paper in a seminar on "Social Implications of Advances in Biomedical Science and Technology: Artificial and Transplanted Internal Organs," sponsored by the Center for the Study of Science, Technology, and Public Policy of the University of Virginia, Spring 1970. I am indebted to the participants in that seminar, and especially to its leaders, Mason Willrich, Professor of Law, and Dr. Harry Abram, Associate Professor of Psychiatry, for criticisms which helped me to sharpen these ideas. Good discussions of the legal questions raised by selection (e.g., equal protection of the law and due process) which I have not considered can be found in "Scarce Medical Resources," *Columbia Law Review*, 69(1969) "Patient Selection for Artificial and Transplanted Organs," *Harvard Law Review*, 82(1969),1322; and Sanders and Dukeminier, op. cit.

Philosophical Reflections on Experimenting with Human Subjects

Hans Jonas

Hans Jonas was Professor of Philosophy at the New School for Social Research in New York City until his death in 1993. As a philosopher of science he became interested in the ethical implications of the new biology and was an early participant in the discussions at the Hastings Center. He was asked to contribute "Philosophical Reflections on Experimenting with Human Subjects" to a conference on ethical aspects of experimenting with human subjects, one of the first on this topic, that was held in Boston on November 3–4, 1967.

This excerpt is from a long, profound essay in which Jonas reflects on the nature of experimenting with human beings, in which the subject is not an inanimate thing but a responsible human relationship. He proposes that a utilitarian ethics is inadequate to deal with the ethics of human experimentation and points out the limitations of a "social contract" ethics. Research progress, desirable as it is, is a "melioristic goal," subordinate to moral obligations of respect for personal freedom and authenticity. Jonas suggests "identification" and "descending order" as rules for selection of subjects. This essay was a seminal exposition of the ethics of medical research.

Experimenting with human subjects is going on in many fields of scientific and technological progress. It is designed to replace the over-all instruction by natural, occasional experience with the selective information from artificial, systematic experiment which physical science has found so effective in dealing with inanimate nature. Of the new experimentation with man, medical is surely the most legitimate; psychological, the most dubious; biological (still to come), the most dangerous. I have chosen here to deal with the first only, where the case *for* it is strongest and the task of adjudicating conflicting claims hardest. . . .

Before going any further, we should give some more articulate voice to the resistance we feel against a merely utilitarian view of the matter. It has to do with a peculiarity of human experimentation quite independent of the question of possible injury to the subject. What is wrong with making a person an experimental subject is not so much that we make him thereby a means (which happens in social contexts of all kinds), as that we make him a thing—a passive thing merely to be acted on, and passive not even for real action, but for token action

---

Abridged from Hans Jonas, "Philosophical Reflections on Experimenting with Human Subjects," *Dædalus*, Journal of the American Academy of Arts and Sciences, from the issue entitled, "Ethical Aspects of Experimentation with Human Subjects," Spring 1969, Volume 98, No. 2.

whose token object he is. His being is reduced to that of a mere token or "sample." This is different from even the most exploitative situations of social life: there the business is real, not fictitious. The subject, however much abused, remains an agent and thus a "subject" in the other sense of the word. The soldier's case is instructive: Subject to most unilateral discipline, forced to risk mutilation and death, conscripted without, perhaps against, his will—he is still conscripted with his capacities to act, to hold his own or fail in situations, to meet real challenges for real stakes. Though a mere "number" to the High Command, he is not a token and not a thing. (Imagine what he would say if it turned out that the war was a game staged to sample observations on his endurance, courage, or cowardice.)

These compensations of personhood are denied to the subject of experimentation, who is acted upon for an extraneous end without being engaged in a real relation where he would be the counterpoint to the other or to circumstance. Mere "consent" (mostly amounting to no more than permission) does not right this reification. Only genuine authenticity of volunteering can possibly redeem the condition of "thinghood" to which the subject submits. . . .

"Individual Versus Society" as the Conceptual Framework

The setting for the conflict most consistently invoked in the literature is the polarity of individual versus society—the possible tension between the individual good and the common good, between private and public welfare. Thus, W. Wolfensberger speaks of "the tension between the long-range interests of society, science, and progress, on one hand, and the rights of the individual on the other.[1] Walsh McDermott says: "In essence, this is a problem of the rights of the individual versus the rights of society."[2] . . . We concede, as a matter of course, to the common good some pragmatically determined measure of precedence over the individual good. In terms of rights, we let some of the basic rights of the individual be overruled by the acknowledged rights of society—as a matter of right and moral justness and not of mere force or dire necessity (much as such necessity may be adduced in defense of that right). But in making that concession, we require a careful clarification of what the needs, interests, and rights of society are, for society—as distinct from any plurality of individuals—is an abstract and, as such, is subject to our definition, while the individual is the primary concrete, prior to all definition, and his basic good is more or less known. Thus the unknown in our problem is the so-called common or public good and its potentially superior claims, to which the individual good must or might sometimes be sacrificed, in circumstances that in turn must also be counted among the unknowns of our question. . . .

Health as a Public Good

The cause invoked is health and, in its more critical aspect, life itself—clearly superlative goods that the physician serves directly by curing and the researcher indirectly by the knowledge gained through his experiments. There is no question about the good served nor about the evil fought—disease and premature death. But a good to whom and an evil to whom? Here the issue tends to become somewhat clouded. In the attempt to give experimentation the proper dignity (on the problematic view that a value becomes greater by being "social" instead of merely individual), the health in question or the disease in question is somehow predicated on the social whole, as if it were society that, in the persons of its members, enjoyed the one and suffered the other. For the purposes of our problem, public interest can then be pitted against private interest, the common good against the individual good. Indeed, I have found health called a national resource, which of course it is, but surely not in the first place.

In trying to resolve some of the complexities and ambiguities lurking in these conceptualizations, I have pondered a particular statement, made in the form of a question, which I found in the *Proceedings* of the earlier *Dædalus* conference: "Can society afford to discard the tissues and organs of the hopelessly unconscious patient when they could be used to restore the otherwise hopelessly ill, but still salvageable individual?" And somewhat later: "A strong case can be made that society can ill afford to discard the tissues and organs of the hopelessly unconscious patient; they are greatly needed for study and experimental trial to help those who can be salvaged."[3] I hasten to add that any suspicion of callousness that the "commodity" language of these statements may suggest is immediately dispelled by the name of the speaker, Dr. Henry K. Beecher, for whose humanity and moral sensibility there can be nothing but admiration. But the use, in all innocence, of this language gives food for thought. Let me, for a moment, take the question literally. "Discarding" implies proprietary rights—nobody can discard what does not belong to him in the first place. Does society then own my body? "Salvaging" implies the same and, moreover, a use-value to the owner. Is the life-extension of certain individuals then a public interest? "Affording" implies a critically vital level of such an interest—that is, of the loss or gain involved. And "society" itself—what is it? When does a need, an aim, an obligation become social? Let us reflect on some of these terms.

What Society Can Afford

"Can society afford . . . ?" Afford what? To let people die intact, thereby withholding something from other people who desperately need it, who in consequence will have to die too? These other, unfortunate people indeed cannot afford not to have a kidney, heart, or other organ of the dying patient, on which they depend for an extension of their lease on life; but does that give them a right to it? And does it oblige society to procure it for them? What is it that *society* can or cannot afford—leaving aside for the moment the question of what it has a *right* to ? It surely can afford to lose members through death; more than that, it is built on the balance of death and birth decreed by the order of life. This is too general, of course, for our question, but perhaps it is well to remember. The specific question seems to be whether society can afford to let some people die whose death might be deferred by particular means if these were authorized by society. Again, if it is merely a question of what society can or cannot afford, rather than of what it ought or ought not to do, the answer must be: Of course, it can. If cancer, heart disease, and other organic, noncontagious ills, especially those tending to strike the old more than the young, continue to exact their toll at the normal rate of incidence (including the toll of private anguish and misery), society can go on flourishing in every way.

Here, by contrast, are some examples of what, in sober truth, society cannot afford. It cannot afford to let an epidemic rage unchecked; a persistent excess of deaths over births, but neither—we must add—too great an excess of births over deaths; too low an average life expectancy even if demographically balanced by fertility, but neither too great a longevity with the necessitated correlative dearth of youth in the social body; a debilitating state of general health; and things of this kind. These are plain cases where the whole condition of society is critically affected, and the public interest can make its imperative claims. The Black Death of the Middle Ages was a *public* calamity of the acute kind; the life-sapping ravages of endemic malaria or sleeping sickness in certain areas are a public calamity of the chronic kind. Such situations a society as a whole can truly not "afford," and they may call for extraordinary remedies, including, perhaps, the invasion of private sacrosanctities.

This is not entirely a matter of numbers and numerical ratios. Society, in a subtler sense, cannot "afford" a single miscarriage of justice, a single inequity in the dispensation of its laws, the violation of the rights of even the tiniest minority, because these undermine the moral basis on which society's existence rests. Nor can it, for a similar reason, afford the absence or atrophy in its midst of compassion and of the effort to alleviate suffering—be it widespread or rare—one form of which is the effort to conquer disease of any kind, whether "socially" significant (by reason of number) or not. And in short, society cannot afford the absence among its members of *virtue* with its readiness for sacrifice beyond defined duty. Since its presence—that is to say, that of personal idealism—is a matter of grace and not of decree, we have the paradox that society depends for its existence on intangibles of nothing less than a religious order, for which it can hope, but which it cannot enforce. All the more must it protect this most precious capital from abuse.

For what objectives connected with the medico-biological sphere should this reserve be drawn upon—for example, in the form of accepting, soliciting, perhaps even imposing the submission of human subjects to experimentation? We postulate that this must be not just a worthy cause, as any promotion of the health of anybody doubtlessly is, but a cause qualifying for transcendent social sanction. Here one thinks first of those cases critically affecting the whole condition, present and future, of the community we have illustrated. Something equivalent to what in the political sphere is called "clear and present danger" may be invoked and a state of emergency proclaimed, thereby suspending certain otherwise inviolable prohibitions and taboos. We may observe that averting a disaster always carries greater weight than promoting a good. Extraordinary danger excuses extraordinary means. This covers human experimentation, which we would like to count, as far as possible, among the extraordinary rather than the ordinary means of serving the common good under public auspices. Naturally, since foresight and responsibility for the future are of the essence of institutional society, averting disaster extends into long-term prevention, although the lesser urgency will warrant less sweeping licenses.

Society and the Cause of Progress

Much weaker is the case where it is a matter not of saving but of improving society. Much of medical research falls into this category. As stated before, a permanent death rate from heart failure or cancer does not threaten society. So long as certain statistical ratios are maintained, the incidence of disease and of disease-induced mortality is not (in the strict sense) a "social" misfortune. I hasten to add that it is not therefore less of a human misfortune, and the call for relief issuing with silent eloquence from each victim and all potential victims is of no lesser dignity. But it is misleading to equate the fundamentally human response to it with what is owed to society: it is owed by man to man—and it is thereby owed by society to the individuals as soon as the adequate ministering to these concerns outgrows (as it progressively does) the scope of private spontaneity and is made a public mandate. It is thus that society assumes responsibility for medical care, research, old age, and innumerable other things not originally of the public realm (in the original "social contract"), and they become duties toward "society" (rather than directly toward one's fellow man) by the fact that they are socially operated.

Indeed, we expect from organized society no longer mere protection against harm and the securing of the conditions of our preservation, but active and constant improvement in all the domains of life: the waging of the battle against nature, the enhancement of the human estate—in short, the promotion of progress. This is an expansive goal, one far surpassing the disaster norm of our previous reflections. It lacks the urgency of the latter, but has the nobility of the free, forward thrust. It surely is worth sacrifices. It is not at all a question of what society can afford, but of what it is committed to, beyond all necessity, by our mandate. Its trusteeship has become an established, ongoing, institutionalized business of the body politic. As eager beneficiaries of its gains, we now owe to "society," as its chief agent, our individual contributions toward its *continued pursuit*. I emphasize "continued pursuit." Maintaining the existing level requires no more than the orthodox means of taxation and enforcement of professional standards that raise no problems. The more optional goal of pushing forward is also more exacting. We have this syndrome: Progress is by our choosing an acknowledged interest of society, in which we have a stake in various degrees; science is a necessary instrument of progress; research is a necessary instrument of science; and in medical science experimentation on human subjects is a necessary instrument of research. Therefore, human experimentation has come to be a societal interest.

The destination of research is essentially melioristic. It does not serve the preservation of the existing good from which I profit myself and to which I am obligated. Unless the present state is intolerable, the melioristic goal is in a sense gratuitous, and this not only from the vantage point of the present. Our descendants have a right to be left an unplundered planet; they do not have a right to new miracle cures. We have sinned against them, if by our doing we have destroyed their inheritance—which we are doing at full blast; we have not sinned against them, if by the time they come around arthritis has not yet been conquered (unless by sheer neglect). And generally, in the matter of progress, as humanity had no claim on a Newton, a Michelangelo, or a St. Francis to appear, and no right to the blessings of their unscheduled deeds, so progress, with all our methodical labor for it, cannot be budgeted in advance and its fruits received as a due. Its coming-about at all and its turning out for good (of which we can never be sure) must rather be regarded as something akin to grace.

The Melioristic Goal, Medical Research, and Individual Duty

Nowhere is the melioristic goal more inherent than in medicine. To the physician, it is not gratuitous. He is committed to curing and thus to improving the power to cure. Gratuitous we called it (outside disaster conditions) as a *social* goal, but noble at the same time. Both the nobility and the gratuitousness must influence the manner in which self-sacrifice for it is elicited, and even its free offer accepted. Freedom is certainly the first condition to be observed here. The surrender of one's body to medical experimentation is entirely outside the enforceable "social contract."

Or can it be construed to fall within its terms—namely, as repayment for benefits from past experimentation that I have enjoyed myself? But I am indebted for these benefits not to society, but to the past "martyrs," to whom society is indebted itself, and society has no right to call in my personal debt by way of adding new to its own. Moreover, gratitude is not an enforceable social obligation; it anyway does not mean that I must emulate the deed. Most of all, if it was wrong to exact such sacrifice in the first place, it does not become right to exact it again with the plea of the profit it has brought me. If, however, it was not exacted, but entirely free, as it ought to have been, then it should remain so, and its precedence must not be used as a social pressure on others for doing the same under the sign of duty.

Indeed, we must look outside the sphere of the social contract, outside the whole realm of public rights and duties, for the motivations and norms by which we can expect ever again the upwelling of a will to give what nobody—neither society, nor fellow man, nor posterity—is entitled to. There are such dimensions in man with trans-social wellsprings of conduct, and I have already pointed to the paradox, or mystery, that society cannot prosper without them, that it must draw on them, but cannot command them.

What about the moral law as such a transcendent motivation of conduct? It goes considerably beyond the public law of the social contract. The latter, we saw, is founded on the rule of enlightened self-interest: *Do ut des*—I give so that I be given to. The law of individual conscience asks more. Under the Golden Rule, for example, I am required to give as I wish to be given to under like circumstances, but not in order that I be given to and not in expectation of return. Reciprocity, essential to the social law, is not a condition of the moral law. One subtle "expectation" and "self-interest," but of the moral order itself, may even then be in my mind: I prefer the environment of a moral society and can expect to contribute to the general morality by my own example. But even if I should always be the dupe, the Golden Rule holds. (If the social law breaks faith with me, I am released from its claim.) . . .

"Identification" as the Principle of Recruitment in General

If the properties we adduced as the particular qualifications of the members of the scientific fraternity itself are taken as general criteria of selection, then one should look for additional subjects where a maximum of identification, understanding, and spontaneity can be expected—that is, among the most highly motivated, the most highly educated, and the least "captive" members of the community. From this naturally scarce resource, a descending order of permissibility leads to greater abundance and ease of supply, whose use should become proportionately more hesitant as the exculpating criteria are relaxed. An inversion of normal "market" behavior is demanded here—namely, to accept the lowest quotation last (and excused only by the greatest pressure of need); to pay the highest price first.

The ruling principle in our considerations is that the "wrong" of reification can only be made "right" by such authentic identification with the cause that it is the subject's as well as the researcher's cause—whereby his role in its service is not just permitted by him, but *willed*. That sovereign will of his which embraces the end as his own restores his personhood to the otherwise depersonalizing context. To be valid it must be autonomous and informed. The latter condition can, outside the research community, only be fulfilled by degrees; but the higher the degree of the understanding regarding the purpose and the technique, the more valid becomes the endorsement of the will. A margin of mere trust inevitably remains. Ultimately, the appeal for volunteers should seek this free and generous endorsement, the appropriation of the research purpose into the person's own scheme of ends. Thus, the appeal is in truth addressed to the one, mysterious, and sacred source of any such generosity of the will—"devotion," whose forms and objects of commitment are various and may invest different motivations in different individuals. The following, for instance, may be responsive to the "call" we are discussing: compassion with human suffering, zeal for humanity, reverence for the Golden

Rule, enthusiasm for progress, homage to the cause of knowledge, even long-ing for sacrificial justification (do not call that "masochism," please). On all these, I say, it is defensible and right to draw when the research objective is worthy enough; and it is a prime duty of the research community (especially in view of what we called the "margin of trust") to see that this sacred source is never abused for frivolous ends. For a less than adequate cause, not even the freest, unsolicited offer should be accepted.

The Rule of the "Descending Order" and Its Counter-Utility Sense

We have laid down what must seem to be a forbidding rule to the number-hungry research industry. Having faith in the transcendent potential of man, I do not fear that the "source" will ever fail a society that does not destroy it—and only such a one is worthy of the blessings of progress. But "elitistic" the rule is (as is the enterprise of progress itself), and elites are by nature small. The com-bined attribute of motivation and information, plus the absence of external pres-sures, tends to be socially so circumscribed that strict adherence to the rule might numerically starve the research process. This is why I spoke of a descend-ing order of permissibility, which is itself permissive, but where the realization that it is a *descending* order is not without pragmatic import. Departing from the august norm, the appeal must needs shift from idealism to docility, from high-mindedness to compliance, from judgment to trust. Consent spreads over the whole spectrum. I will not go into the casuistics of this penumbral area. I merely indicate the principle of the order of preference: The poorer in knowledge, mo-tivation, and freedom of decision (and that, alas, means the more readily avail-able in terms of numbers and possible manipulation), the more sparingly and indeed reluctantly should the reservoir be used, and the more compelling must therefore become the countervailing justification.

Let us note that this is the opposite of a social utility standard, the reverse of the order by "availability and expendability": The most valuable and scarcest, the least expendable elements of the social organism, are to be the first candi-dates for risk and sacrifice. It is the standard of *noblesse oblige*; and with all its counter-utility and seeming "wastefulness," we feel a rightness about it and per-haps even a higher "utility," for the soul of the community lives by this spirit.[4] It is also the opposite of what the day-to-day interests of research clamor for, and for the scientific community to honor it will mean that it will have to fight a strong temptation to go by routine to the readiest sources of supply—the sug-gestible, the ignorant, the dependent, the "captive" in various senses.[5] I do not believe that heightened resistance here must cripple research, which cannot be permitted; but it may indeed slow it down by the smaller numbers fed into ex-perimentation in consequence. This price—a possibly slower rate of progress—may have to be paid for the preservation of the most precious capital of higher communal life. . . .

No Experiments on Patients Unrelated to Their Own Disease

Although my ponderings have, on the whole, yielded points of view rather than definite prescriptions, premises rather than conclusions, they have led me to a few unequivocal yeses and noes. The first is the emphatic rule that patients should be experimented upon, if at all, *only* with reference to *their disease*. Never should there be added to the gratuitousness of the experiment as such the gratuitousness of service to an unrelated cause. This follows simply from what we have found to be the *only* excuse for infracting the special exemption of the sick at all—namely, that the scientific war on disease cannot accomplish its goal without drawing the sufferers from disease into the investigative process. If under this excuse they become subjects of experiment, they do so *because*, and only because, of *their* disease. . . .

. . . Let us not forget that progress is an optional goal, not an unconditional commitment, and that its tempo in particular, compulsive as it may become, has nothing sacred about it. Let us also remember that a slower progress in the conquest of disease would not threaten society, grievous as it is to those who have to deplore that their particular disease be not yet conquered, but that society would indeed be threatened by the erosion of those moral values whose loss, possibly caused by too ruthless a pursuit of scientific progress, would make its most dazzling triumphs not worth having.

References

1. Wolfensberger, "Ethical Issues in Research with Human Subjects," *Proceedings of the Conference on the Ethical Aspects of Experimentation on Human Subjects*, November 3–4, 1967 (Boston, Massachusetts; hereafter called *Proceedings*), p. 48.
2. *Proceedings*, p. 29.
3. *Proceedings*, pp. 50–51.
4. Socially, everyone is expendable relatively—that is, in different degrees; religiously, no one is expendable absolutely: The "image of God" is in all. If it can be enhanced, then not by anyone being expended, but by someone expending himself.
5. This refers to captives of circumstance, not of justice. Prison inmates are, with respect to our problem, in a special class. If we hold to some idea of guilt, and to the supposition that our judicial system is not entirely at fault, they may be held to stand in a special debt to society, and their offer to serve—from whatever motive—may be accepted with a minimum of qualms as a means of reparation.

To Save or Let Die:
The Dilemma of Modern Medicine

Richard A. McCormick

Richard A. McCormick was a Jesuit theologian working in the Roman Catholic tradition of moral theology, which has a long history of interest in medical ethics. Father McCormick was Joseph P. Kennedy Professor of Medical Ethics at the Kennedy Institute for Ethics, Georgetown University, one of the earliest study centers in bioethics. He concluded his career as Professor of Moral Theology at Notre Dame University.

"To Save or Let Die" was among the many responses to the ethical dilemmas posed by Duff and Campbell's revelations and by the paradigm case of an infant with Down's syndrome who was allowed to die at Johns Hopkins Hospital in 1971, a case widely publicized in a short film made by the Joseph P. Kennedy Foundation. In this article, Father McCormick develops a concept of "meaningful life" that, although rooted in theological ideas, can be translated into broadly acceptable secular terms and used as the basis for decisions to sustain newborn life or allow death to come.

. . . In a recent issue of the *New England Journal of Medicine*, Drs. Raymond S. Duff and A. G. M. Campbell[1] reported on 299 deaths in the special-care nursery of the Yale–New Haven Hospital between 1970 and 1972. Of these, 43 (14%) were associated with discontinuance of treatment for children with multiple anomalies, trisomy, cardiopulmonary crippling, meningomyelocele, and other central nervous system defects. After careful consideration of each of these 43 infants, parents and physicians in a group decision concluded that the prognosis for "meaningful life" was extremely poor or hopeless, and therefore rejected further treatment. . . .

In commenting on this study in the *Washington Post* (Oct. 28, 1973), Dr. Lawrence K. Pickett, chief-of-staff at the Yale–New Haven Hospital, admitted that allowing hopelessly ill patients to die "is accepted medical practice." He continued: "This is nothing new. It's just being talked about now."

It has been talked about, it is safe to say, at least since the publicity associated with the famous "Johns Hopkins Case"[2] some three years ago. In this instance, an infant was born with Down's syndrome and duodenal atresia. The blockage is reparable by relatively easy surgery. However, after consultation with

Abridged from Richard A. McCormick, "To Save or Let Die: The Dilemma of Modern Medicine," *The Journal of the American Medical Association*, Volume 229, 1974, pp. 172–176. Copyright © 1974, American Medical Association.

spiritual advisors, the parents refused permission for this corrective surgery, and the child died by starvation in the hospital after 15 days. For to feed him by mouth in this condition would have killed him. Nearly everyone who has commented on this case has disagreed with the decision.

It must be obvious that these instances—and they are frequent—raise the most agonizing and delicate moral problems. The problem is best seen in the ambiguity of the term "hopelessly ill." This used to and still may refer to lives that cannot be saved, that are irretrievably in the dying process. It may also refer to lives that can be saved and sustained, but in a wretched, painful, or deformed condition. With regard to infants, the problem is, which infants, if any, should be allowed to die? On what grounds or according to what criteria as determined by whom? Or again, is there a point at which a life that can be saved is not "meaningful life," as the medical community so often phrases the question. . . .

Thus far, the ethical discussion of these truly terrifying decisions has been less than fully satisfactory. Perhaps this is to be expected since the problems have only recently come to public attention. In a companion article to the Duff–Campbell report,[1] Dr. Anthony Shaw[3] of the Pediatric Division of the Department of Surgery, University of Virginia Medical Center, Charlottesville, speaks of solutions "based on the circumstances of each case rather than by means of a dogmatic formula approach." Are these really the only options available to us? Shaw's statement makes it appear that the ethical alternatives are narrowed to dogmatism (which imposes a formula that prescinds from circumstances) and pure concretism (which denies the possibility or usefulness of any guidelines). . . .

What has brought us to this position of awesome responsibility? Very simply, the sophistication of modern medicine. Contemporary resuscitation and life-sustaining devices have brought a remarkable change in the state of the question. Our duties toward the care and preservation of life have been traditionally stated in terms of the use of ordinary and extraordinary means. For the moment and for purposes of brevity, we may say that, morally speaking, ordinary means are those whose use does not entail grave hardships to the patient. Those that would involve such hardship are extraordinary. Granted the relativity of these terms and the frequent difficulty of their application, still the distinction has had an honored place in medical ethics and medical practice. Indeed, the distinction was recently reiterated by the House of Delegates of the American Medical Association (AMA) in a policy statement. After disowning intentional killing (mercy killing), the AMA statement continues: "The cessation of the employment of extraordinary means to prolong the life of the body when there is irrefutable evidence that biological death is imminent is the decision of the patient and/or his immediate family. The advice and judgment of the physician should be freely available to the patient and/or his immediate family" (*JAMA* 227:728, 1974).

This distinction can take us just so far—and thus the change in the state of the question. The contemporary problem is precisely that the question no longer concerns only those for whom "biological death is imminent" in the sense of the AMA statement. Many infants who would have died a decade ago, whose "biological death was imminent," can be saved. Yesterday's failures are today's successes. Contemporary medicine with its team approaches, staged surgical techniques, monitoring capabilities, ventilatory support systems, and other methods can keep almost anyone alive. This has tended gradually to shift the problem from the means to reverse the dying process to the quality of the life sustained and preserved. The questions, "Is this means too hazardous or difficult to use" and "Does this measure only prolong the patient's dying," while still useful and valid, now often become "Granted that we can easily save the life, what kind of life are we saving?" This is a quality-of-life judgment. And we fear it. And certainly we should. But with increased power goes increased responsibility. Since we have the power, we must face the responsibility.

A Relative Good

In the past, the Judeo-Christian tradition has attempted to walk a balanced middle path between medical vitalism (that preserves life at any cost) and medical pessimism (that kills when life seems frustrating, burdensome, "useless"). Both of these extremes root in an identical idolatry of life—an attitude that, at least by inference, views death as an unmitigated, absolute evil, and life as the absolute good. The middle course that has structured Judeo-Christian attitudes is that life is indeed a basic and precious good, but a good to be preserved precisely as the condition of other values. It is these other values and possibilities that found the duty to preserve physical life and also dictate the limits of this duty. In other words, life is a relative good, and the duty to preserve it a limited one. These limits have always been stated in terms of the *means* required to sustain life. But if the implications of this middle position are unpacked a bit, they will allow us, perhaps, to adapt to the type of quality-of-life judgment we are now called on to make without tumbling into vitalism or a utilitarian pessimism.

A beginning can be made with a statement of Pope Pius XII[4] in an allocution to physicians delivered Nov. 24, 1957. After noting that we are normally obliged to use only ordinary means to preserve life, the Pontiff stated: "A more strict obligation would be too burdensome for most men and would render the attainment of the higher, more important good too difficult. Life, death, all temporal activities are in fact subordinated to spiritual ends." Here it would be helpful to ask two questions. First, what are these spiritual ends, this "higher, more important good"? Second, how is its attainment rendered too difficult by insisting on the use of extraordinary means to preserve life?

The first question must be answered in terms of love of God and neighbor. This sums up briefly the meaning, substance, and consummation of life from a Judeo-Christian perspective. What is or can easily be missed is that these two loves are not separable. . . . It is in others that God demands to be recognized and loved. If this is true, it means that, in Judeo-Christian perspective, the meaning, substance, and consummation of life is found in human *relationships*, and the qualities of justice, respect, concern, compassion, and support that surround them.

Second, how is the attainment of this "higher, more important (than life) good" rendered "too difficult" by life-supports that are gravely burdensome? One who must support his life with disproportionate effort focuses the time, attention, energy, and resources of himself and others not precisely on relationships, but on maintaining the condition of relationships. Such concentration easily becomes overconcentration and distorts one's view of and weakens one's pursuit of the very relational goods that define our growth and flourishing. The importance of relationships gets lost in the struggle for survival. The very Judeo-Christian meaning of life is seriously jeopardized when undue and unending effort must go into its maintenance.

I believe an analysis similar to this is implied in traditional treatises on preserving life. The illustrations of grave hardship (rendering the means to preserve life extraordinary and nonobligatory) are instructive, even if they are outdated in some of their particulars. Older moralists often referred to the hardship of moving to another climate or country. As the late Gerald Kelly[5] noted of this instance: "They (the classical moral theologians) spoke of other inconveniences, too: e.g., of moving to another climate or another country to preserve one's life. For people whose lives were, so to speak, rooted in the land, and whose native town or village was as dear as life itself, and for whom, moreover, travel was always difficult and often dangerous—for such people, moving to another country or climate was a truly great hardship, and more than God would demand as a 'reasonable' means of preserving one's health and life."

Similarly, if the financial cost of life-preserving care was crushing, that is, if it would create grave hardships for oneself or one's family, it was considered extraordinary and nonobligatory. Or again, the grave inconvenience of living with a badly mutilated body was viewed, along with other factors (such as pain in preanesthetic days, uncertainty of success), as constituting the means extraordinary. Even now, the contemporary moralist, M. Zalba,[6] states that no one is obliged to preserve his life when the cost is "a most oppressive convalescence" (*molestissima convalescentia*).

The Quality of Life

In all of these instances—instances where the life could be saved—the discussion is couched in terms of the means necessary to preserve life. But often enough it is the kind of, the quality of the life thus saved (painful, poverty-stricken and deprived, away from home and friends, oppressive) that establishes the means as extraordinary. *That* type of life would be an excessive hardship for the individual. It would distort and jeopardize his grasp on the overall meaning of life. Why? Because, it can be argued, human relationships—which are the very possibility of growth in love of God and neighbor—would be so threatened, strained, or submerged that they would no longer function as the heart and meaning of the individual's life as they should. Something other than the "higher, more important good" would occupy first place. Life, the condition of other values and achievements, would usurp the place of these and become itself the ultimate value. When that happens, the value of human life has been distorted out of context. . . .

Can these reflections be brought to bear on the grossly malformed infant? I believe so. Obviously there is a difference between having a terribly mutilated body as the result of surgery and having a terribly mutilated body from birth. There is also a difference between a long, painful, oppressive convalescence resulting from surgery and a life that is from birth one long, painful, oppressive convalescence. Similarly, there is a difference between being plunged into poverty by medical expenses and being poor without ever incurring such expenses. However, is there not also a similarity? Can not these conditions, whether caused by medical intervention or not, equally absorb attention and energies to the point where the "higher, more important good" is simply too difficult to attain? It would appear so. Indeed, is this not precisely why abject poverty (and the systems that support it) is such an enormous moral challenge to us? It simply dehumanizes.

Life's potentiality for other values is dependent on two factors, those external to the individual, and the very condition of the individual. The former we can and must change to maximize individual potential. That is what social justice is all about. The latter we sometimes cannot alter. It is neither inhuman nor unchristian to say that there comes a point where an individual's condition itself represents the negation of any truly human—i.e., relational—potential. When that point is reached, is not the best treatment no treatment? . . .

Human Relationships

If these reflections are valid, they point in the direction of a guideline that may help in decisions about sustaining the lives of grossly deformed and deprived infants. That guideline is the potential for human relationships associated with the infant's condition. If that potential is simply nonexistent or would be utterly submerged and undeveloped in the mere struggle to survive, that life has achieved its potential. There are those who will want to continue to say that some terribly deformed infants may be allowed to die *because* no extraordinary means need be used. Fair enough. But they should realize that the term "extraordinary" has been so relativized to the condition of the patient that it is this condition that is decisive. The means is extraordinary because the infant's condition is extraordinary. And if that is so, we must face this fact head-on—and discover the substantive standard that allows us to say this of some infants, but not of others.

Here several caveats are in order. First, this guideline is not a detailed rule that preempts decisions; for relational capacity is not subject to mathematical analysis but to human judgment. However, it is the task of physicians to provide some more concrete categories or presumptive biological symptoms for this human judgment. For instance, nearly all would very likely agree that the anencephalic infant is without relational potential. On the other hand, the same cannot be said of the mongoloid infant. The task ahead is to attach relational potential to presumptive biological symptoms for the gray area between such extremes. In other words, individual decisions will remain the anguishing onus of parents in consultation with physicians.

Second, because this guideline is precisely that, mistakes will be made. Some infants will be judged in all sincerity to be devoid of any meaningful relational potential when that is actually not quite the case. This risk of error should not lead to abandonment of decisions; for that is to walk away from the human scene. Risk of error means only that we must proceed with great humility, caution, and tentativeness. Concretely, it means that if err we must at times, it is better to err on the side of life—and therefore to tilt in that direction.

Third, it must be emphasized that allowing some infants to die does not imply that "some lives are valuable, others not" or that "there is such a thing as a life not worth living." Every human being, regardless of age or condition, is of incalculable worth. The point is not, therefore, whether this or that individual has value. Of course he has, or rather *is* a value. The only point is whether this undoubted value has any potential at all, in continuing physical survival, for attaining a share, even if reduced, in the "higher, more important good." This is not a question about the inherent value of the individual. It is a question about whether this worldly existence will offer such a valued individual any hope of sharing those values for which physical life is the fundamental condition. Is not the only alternative an attitude that supports mere physical life as long as possible with every means?

Fourth, this whole matter is further complicated by the fact that this decision is being made for someone else. Should not the decision on whether life is to be supported or not be left to the individual? Obviously, wherever possible. But there is nothing inherently objectionable in the fact that parents with physicians must make this decision at some point for infants. Parents must make many crucial decisions for children. The only concern is that the decision not be shaped out of the utilitarian perspectives so deeply sunk into the consciousness of the contemporary world. In a highly technological culture, an individual is always in danger of being valued for his function, what he can do, rather than for who he is. . . .

Were not those who disagreed with the Hopkins decision saying, in effect, that for the infant, involved human relationships were still within reach and would not be totally submerged by survival? If that is the case, it is potential for relationships that is at the heart of these agonizing decisions.

References

1. Duff, S., Campbell, A. G. M., Moral and ethical dilemmas in the special-care nursery. *N. Engl. J. Med.* 289:890–894, 1973.
2. Gustafson, J. M., Mongolism, parental desires, and the right to life. *Perspect. Biol. Med.* 16:529–559, 1973.

3. Shaw, A., Dilemmas of "informed" consent in children. *N. Engl. J. Med.* 289:885–890, 1973.
4. Pope Pius XII, *Acta Apostolicae Sedis.* 49:1031–1032, 1957.
5. Kelly, G., *Medico-Moral Problems*. St. Louis, Catholic Hospital Association of the United Stales and Canada, 1957, p. 132.
6. Zalba, M., *Theologiae Moralis Summa*. Madrid, La Editorial Catolica. 1957, vol. 2, p. 71.

Active and Passive Euthanasia

James Rachels

James Rachels, a professor of moral philosophy at University of Alabama, Birmingham until his death in 2003, offered a radical critique of a distinction that was common among medical ethicists, namely, that passive euthanasia, or allowing to die, was morally acceptable, whereas active euthanasia, equivalent to killing, was not. This article did not notably change the prevalence of the received distinction at the time of its publication, but it did set in motion a skepticism about its viability as a useful ethical distinction that eventually gave credibility to the arguments for assisted suicide that appeared in the 1990s.

The distinction between active and passive euthanasia is thought to be crucial for medical ethics. The idea is that it is permissible, at least in some cases, to withhold treatment and allow a patient to die, but it is never permissible to take any direct action designed to kill the patient. This doctrine seems to be accepted by most doctors, and it is endorsed in a statement adopted by the House of Delegates of the American Medical Association on December 4, 1973:

> The intentional termination of the life of one human being by another—mercy killing—is contrary to that for which the medical profession stands and is contrary to the policy of the American Medical Association.
>
> The cessation of the employment of extraordinary means to prolong the life of the body when there is irrefutable evidence that biological death is imminent is the decision of the patient and/or his immediate family. The advice and judgment of the physician should be freely available to the patient and/or his immediate family.

However, a strong case can be made against this doctrine. In what follows I will set out some of the relevant arguments, and urge doctors to reconsider their views on this matter.

To begin with a familiar type of situation, a patient who is dying of incurable cancer of the throat is in terrible pain, which can no longer be satisfactorily alleviated. He is certain to die within a few days, even if present treatment is continued, but he does not want to go on living for those days since the pain is

Reprinted from James Rachels, "Active and Passive Euthanasia," *The New England Journal of Medicine*, Volume 292, 1975, pp. 78–80. Reprinted by permission of *The New England Journal of Medicine*. Copyright © 1975, Massachusetts Medical Society.

unbearable. So he asks the doctor for an end to it, and his family joins in the request.

Suppose the doctor agrees to withhold treatment, as the conventional doctrine says he may. The justification for his doing so is that the patient is in terrible agony, and since he is going to die anyway, it would be wrong to prolong his suffering needlessly. But now notice this. If one simply withholds treatment, it may take the patient longer to die, and so he may suffer more than he would if more direct action were taken and a lethal injection given. This fact provides strong reason for thinking that, once the initial decision not to prolong his agony has been made, active euthanasia is actually preferable to passive euthanasia, rather than the reverse. To say otherwise is to endorse the option that leads to more suffering rather than less, and is contrary to the humanitarian impulse that prompts the decision not to prolong his life in the first place.

Part of my point is that the process of being "allowed to die" can be relatively slow and painful, whereas being given a lethal injection is relatively quick and painless. Let me give a different sort of example. In the United States about one in 600 babies is born with Down's syndrome. Most of these babies are otherwise healthy—that is, with only the usual pediatric care, they will proceed to an otherwise normal infancy. Some, however, are born with congenital defects such as intestinal obstructions that require operations if they are to live. Sometimes, the parents and the doctor will decide not to operate, and let the infant die. Anthony Shaw describes what happens then:

> . . . When surgery is denied [the doctor] must try to keep the infant from suffering while natural forces sap the baby's life away. As a surgeon whose natural inclination is to use the scalpel to fight off death, standing by and watching a salvageable baby die is the most emotionally exhausting experience I know. It is easy at a conference, in a theoretical discussion, to decide that such infants should be allowed to die. It is altogether different to stand by in the nursery and watch as dehydration and infection wither a tiny being over hours and days. This is a terrible ordeal for me and the hospital staff—much more so than for the parents who never set foot in the nursery.*

I can understand why some people are opposed to all euthanasia, and insist that such infants must be allowed to live. I think I can also understand why other people favor destroying these babies quickly and painlessly. But why should anyone favor letting "dehydration and infection wither a tiny being over hours and days"? The doctrine that says that a baby may be allowed to dehydrate and wither, but may not be given an injection that would end its life without suffering, seems so patently cruel as to require no further refutation. The strong language is not intended to offend, but only to put the point in the clearest possible way.

---

* Shaw A: "Doctor, Do We Have a Choice?" *The New York Times Magazine*, January 30, 1972, p. 54.

My second argument is that the conventional doctrine leads to decisions concerning life and death made on irrelevant grounds.

Consider again the case of the infants with Down's syndrome who need operations for congenital defects unrelated to the syndrome to live. Sometimes, there is no operation, and the baby dies, but when there is no such defect, the baby lives on. Now, an operation such as that to remove an intestinal obstruction is not prohibitively difficult. The reason why such operations are not performed in these cases is, clearly, that the child has Down's syndrome and the parents and doctor judge that because of that fact it is better for the child to die.

But notice that this situation is absurd, no matter what view one takes of the lives and potentials of such babies. If the life of such an infant is worth preserving, what does it matter if it needs a simple operation? Or, if one thinks it better that such a baby should not live on, what difference does it make that it happens to have an unobstructed intestinal tract? In either case, the matter of life and death is being decided on irrelevant grounds. It is the Down's syndrome, and not the intestines, that is the issue. The matter should be decided, if at all, on that basis, and not be allowed to depend on the essentially irrelevant question of whether the intestinal tract is blocked.

What makes this situation possible, of course, is the idea that when there is an intestinal blockage, one can "let the baby die," but when there is no such defect there is nothing that can be done, for one must not "kill" it. The fact that this idea leads to such results as deciding life or death on irrelevant grounds is another good reason why the doctrine should be rejected.

One reason why so many people think that there is an important moral difference between active and passive euthanasia is that they think killing someone is morally worse than letting someone die. But is it? Is killing, in itself, worse than letting die? To investigate this issue, two cases may be considered that are exactly alike except that one involves killing whereas the other involves letting someone die. Then, it can be asked whether this difference makes any difference to the moral assessments. It is important that the cases be exactly alike, except for this one difference, since otherwise one cannot be confident that it is this difference and not some other that accounts for any variation in the assessments of the two cases. So, let us consider this pair of cases:

In the first, Smith stands to gain a large inheritance if anything should happen to his six-year-old cousin. One evening while the child is taking his bath, Smith sneaks into the bathroom and drowns the child, and then arranges things so that it will look like an accident.

In the second, Jones also stands to gain if anything should happen to his six-year-old cousin. Like Smith, Jones sneaks in planning to drown the child in his bath. However, just as he enters the bathroom Jones sees the child slip and hit his head, and fall face down in the water. Jones is delighted; he stands by, ready to push the child's head back under if it is necessary, but it is not necessary. With only a little thrashing about, the child drowns all by himself, "accidentally," as Jones watches and does nothing.

Now Smith killed the child, whereas Jones "merely" let the child die. That is the only difference between them. Did either man behave better, from a moral point of view? If the difference between killing and letting die were in itself a morally important matter, one should say that Jones's behavior was less reprehensible than Smith's. But does one really want to say that? I think not. In the first place, both men acted from the same motive, personal gain, and both had exactly the same end in view when they acted. It may be inferred from Smith's conduct that he is a bad man, although that judgment may be withdrawn or modified if certain further facts are learned about him—for example, that he is mentally deranged. But would not the very same thing be inferred about Jones from his conduct? And would not the same further considerations also be relevant to any modification of this judgment? Moreover, suppose Jones pleaded, in his own defense, "After all, I didn't do anything except just stand there and watch the child drown. I didn't kill him; I only let him die." Again, if letting die were in itself less bad than killing, this defense should have at least some weight. But it does not. Such a "defense" can only be regarded as a grotesque perversion of moral reasoning. Morally speaking, it is no defense at all.

Now, it may be pointed out, quite properly, that the cases of euthanasia with which doctors are concerned are not like this at all. They do not involve personal gain or the destruction of normal healthy children. Doctors are concerned only with cases in which the patient's life is of no further use to him, or in which the patient's life has become or will soon become a terrible burden. However, the point is the same in these cases: the bare difference between killing and letting die does not, in itself, make a moral difference. If a doctor lets a patient die, for humane reasons, he is in the same moral position as if he had given the patient a lethal injection for humane reasons. If his decision was wrong—if, for example, the patient's illness was in fact curable—the decision would be equally regrettable no matter which method was used to carry it out. And if the doctor's decision was the right one, the method used is not in itself important.

The AMA policy statement isolates the crucial issue very well; the crucial issue is "the intentional termination of the life of one human being by another." But after identifying this issue, and forbidding "mercy killing," the statement goes on to deny that the cessation of treatment is the intentional termination of a life. This is where the mistake comes in, for what is the cessation of treatment, in these circumstances, if it is not "the intentional termination of the life of one human being by another"? Of course it is exactly that, and if it were not, there would be no point to it.

Many people will find this judgment hard to accept. One reason, I think, is that it is very easy to conflate the question of whether killing is, in itself, worse than letting die, with the very different question of whether most actual cases of killing are more reprehensible than most actual cases of letting die. Most actual cases of killing are clearly terrible (think, for example, of all the murders reported in the newspapers), and one hears of such cases every day. On the other hand, one hardly ever hears of a case of letting die, except for the actions of doctors who are motivated by humanitarian reasons. So one learns to think of killing in a much worse light than of letting die. But this does not mean that there is something about killing that makes it in itself worse than letting die, for it is not the bare difference between killing and letting die that makes the difference in these cases. Rather, the other factors—the murderer's motive of personal gain, for example, contrasted with the doctor's humanitarian motivation—account for different reactions to the different cases.

I have argued that killing is not in itself any worse than letting die; if my contention is right, it follows that active euthanasia is not any worse than passive euthanasia. What arguments can be given on the other side? The most common, I believe, is the following:

"The important difference between active and passive euthanasia is that, in passive euthanasia, the doctor does not do anything to bring about the patient's death. The doctor does nothing, and the patient dies of whatever ills already afflict him. In active euthanasia, however, the doctor does something to bring about the patient's death: he kills him. The doctor who gives the patient with cancer a lethal injection has himself caused his patient's death; whereas if he merely ceases treatment, the cancer is the cause of the death."

A number of points need to be made here. The first is that it is not exactly correct to say that in passive euthanasia the doctor does nothing, for he does do one thing that is very important: he lets the patient die. "Letting someone die" is certainly different, in some respects, from other types of action—mainly in that it is a kind of action that one may perform by way of not performing certain other actions. For example, one may let a patient die by way of not giving medication, just as one may insult someone by way of not shaking his hand. But for any purpose of moral assessment, it is a type of action nonetheless. The decision to let a patient die is subject to moral appraisal in the same way that a decision to kill him would be subject to moral appraisal: it may be assessed as wise or unwise, compassionate or sadistic, right or wrong. If a doctor deliberately let a patient die who was suffering from a routinely curable illness, the doctor would certainly be to blame for what he had done, just as he would be to blame if he had needlessly killed the patient. Charges against him would then be appropriate. If so, it would be no defense at all for him to insist that he didn't "do anything." He would have done something very serious indeed, for he let his patient die.

Fixing the cause of death may be very important from a legal point of view, for it may determine whether criminal charges are brought against the doctor. But I do not think that this notion can be used to show a moral difference between active and passive euthanasia. The reason why it is considered bad to be the cause of someone's death is that death is regarded as a great evil—and so it is. However, if it has been decided that euthanasia—even passive euthanasia—is desirable in a given case, it has also been decided that in this instance death is no greater an evil than the patient's continued existence. And if this is true, the usual reason for not wanting to be the cause of someone's death simply does not apply.

Finally, doctors may think that all of this is only of academic interest—the sort of thing that philosophers may worry about but that has no practical bearing on their own work. After all, doctors must be concerned about the legal consequences of what they do, and active euthanasia is clearly forbidden by the law. But even so, doctors should also be concerned with the fact that the law is forcing upon them a moral doctrine that may well be indefensible, and has a considerable effect on their practices. Of course, most doctors are not now in the position of being coerced in this matter, for they do not regard themselves as merely going along with what the law requires. Rather, in statements such as the AMA policy statement that I have quoted, they are endorsing this doctrine as a central point of medical ethics. In that statement, active euthanasia is condemned not merely as illegal but as "contrary to that for which the medical profession stands," whereas passive euthanasia is approved. However, the preceding considerations suggest that there is really no moral difference between the two, considered in themselves (there may be important moral differences in some cases in their *consequences*, but, as I pointed out, these differences may make active euthanasia, and not passive euthanasia, the morally preferable option). So, whereas doctors may have to discriminate between active and passive euthanasia to satisfy the law, they should not do any more than that. In particular, they should not give the distinction any added authority and weight by writing it into official statements of medical ethics.

In the Matter of Karen Quinlan

The Supreme Court, State of New Jersey

The case of Karen Ann Quinlan thrust bioethics into public attention. It is also a landmark legal case about the major bioethical question of forgoing life-supporting medical treatment. In 1975, Karen Ann Quinlan, age 21, was brought to the emergency department of a local hospital. She was in a coma, induced apparently by alcohol and barbiturates. Her breathing was supported by a respirator. During the next five months, she slipped from coma to a persistent vegetative state. Her parents requested that the respirator be withdrawn, allowing Karen to die. The doctors and the hospital objected, forcing the Quinlan family to take the matter to the courts. The public followed the case with close attention. On March 31, 1976, the Supreme Court of New Jersey rendered the opinion excerpted below. This was the first case of its sort to reach a judicial decision in the United States. The words of this opinion both reflected and shaped bioethical and legal discourse about life support.

IN THE MATTER OF KAREN QUINLAN: THE SUPREME COURT, STATE OF NEW JERSEY (1976)

The Factual Base

An understanding of the issues in their basic perspective suggests a brief review of the factual base developed in the testimony and documented in greater detail in the opinion of the trial Judge. [*In re Quinlan*, 137 N.J.Super. 227, 348 A.2d 801 (Ch.Div.1975).]

On the night of April 15, 1975, for reasons still unclear, Karen Quinlan ceased breathing for at least two 15 minute periods. She received some ineffectual mouth-to-mouth resuscitation from friends. She was taken by ambulance to Newton Memorial Hospital. There she had a temperature of 100 degrees, her pupils were unreactive and she was unresponsive even to deep pain. The history at the time of her admission to that hospital was essentially incomplete and uninformative.

Three days later, Dr. Morse examined Karen at the request of the Newton admitting physician, Dr. McGee. He found her comatose with evidence of decortication, a condition relating to derangement of the cortex of the brain causing a physical posture in which the upper extremities are flexed and the lower extremities are extended. She required a respirator to assist her breathing. Dr. Morse was unable to obtain an adequate account of the circumstances and

---

Condensed from 344 Atlantic Reporter 2d Series. 647.

events leading up to Karen's admission to the Newton Hospital. Such initial history or etiology is crucial in neurological diagnosis. Relying as he did upon the Newton Memorial records and his own examination, he concluded that prolonged lack of oxygen in the bloodstream, anoxia, was identified with her condition as he saw it upon first observation. When she was later transferred to Saint Clare's Hospital she was still unconscious, still on a respirator and a tracheotomy had been performed. On her arrival Dr. Morse conducted extensive and detailed examinations. An electroencephalogram (EEG) measuring electrical rhythm of the brain was performed and Dr. Morse characterized the result as "abnormal but it showed some activity and was consistent with her clinical state." Other significant neurological tests, including a brain scan, an angiogram, and a lumbar puncture were normal in result. Dr. Morse testified that Karen has been in a state of coma, lack of consciousness, since he began treating her. He explained that there are basically two types of coma, sleep-like unresponsiveness and awake unresponsiveness. Karen was originally in a sleep-like unresponsive condition but soon developed "sleep-wake" cycles, apparently a normal improvement for comatose patients occurring within three to four weeks. In the awake cycle she blinks, cries out and does things of that sort but is still totally unaware of anyone or anything around her.

Dr. Morse and other expert physicians who examined her characterized Karen as being in a "chronic persistent vegetative state." Dr. Fred Plum, one of such expert witnesses, defined this as a "subject who remains with the capacity to maintain the vegetative parts of neurological function but who . . . no longer has any cognitive function."

Dr. Morse, as well as the several other medical and neurological experts who testified in this case, believed with certainty that Karen Quinlan is not "brain dead." They identified the Ad Hoc Committee of Harvard Medical School report (*infra*) as the ordinary medical standard for determining brain death, and all of them were satisfied that Karen met none of the criteria specified in that report and was therefore not "brain dead" within its contemplation.

In this respect it was indicated by Dr. Plum that the brain works in essentially two ways, the vegetative and the sapient. He testified:

> We have an internal vegetative regulation which controls body temperature which controls breathing, which controls to a considerable degree blood pressure, which controls to some degree heart rate, which controls chewing, swallowing and which controls sleeping and waking. We have a more highly developed brain which is uniquely human which controls our relation to the outside world, our capacity to talk, to see, to feel, to sing, to think. Brain death necessarily must mean the death of both of these functions of the brain, vegetative and the sapient. Therefore, the presence of any function which is regulated or governed or controlled by the deeper parts of the brain which in laymen's terms might be considered purely vegetative would mean that the brain is not biologically dead.

Because Karen's neurological condition affects her respiratory ability (the respiratory system being a brain stem function) she requires a respirator to assist her breathing. From the time of her admission to Saint Clare's Hospital Karen has been assisted by an MA-1 respirator, a sophisticated machine which delivers a given volume of air at a certain rate and periodically provides a "sigh" volume, a relatively large measured volume of air designed to purge the lungs of excretions. Attempts to "wean" her from the respirator were unsuccessful and have been abandoned.

The experts believe that Karen cannot now survive without the assistance of the respirator; that exactly how long she would live without it is unknown; that the strong likelihood is that death would follow soon after its removal, and that removal would also risk further brain damage and would curtail the assistance the respirator presently provides in warding off infection.

It seemed to be the consensus not only of the treating physicians but also of the several qualified experts who testified in the case, that removal from the respirator would not conform to medical practices, standards and traditions.

The further medical consensus was that Karen, in addition to being comatose, is in a chronic and persistent "vegetative" state, having no awareness of anything or anyone around her and existing at a primitive reflex level. Although she does have some brain stem function (ineffective for respiration) and has other reactions one normally associates with being alive, such as moving, reacting to light, sound and noxious stimuli, blinking her eyes, and the like, the quality of her feeling impulses is unknown. She grimaces, makes stereotyped cries and sounds and has chewing motions. Her blood pressure is normal.

Karen remains in the intensive care unit at Saint Clare's Hospital, receiving 24-hour care by a team of four nurses characterized, as was the medical attention, as "excellent." She is nourished by feeding by way of a nasal-gastro tube and is routinely examined for infection, which under these circumstances is a serious life threat. The result is that her condition is considered remarkable under the unhappy circumstances involved.

Karen is described as emaciated, having suffered a weight loss of at least 40 pounds, and undergoing a continuing deteriorative process. Her posture is described as fetal-like and grotesque; there is extreme flexion-rigidity of the arms, legs and related muscles, and her joints are severely rigid and deformed.

From all of this evidence, and including the whole testimonial record, several basic findings in the physical area are mandated. Severe brain and associated damage, albeit of uncertain etiology, has left Karen in a chronic and persistent vegetative state. No form of treatment which can cure or improve that condition is known or available. As nearly as may be determined, considering the guarded area of remote uncertainties characteristic of most medical science predictions, she can never be restored to cognitive or sapient life. Even with regard to the vegetative level and improvement therein (if such it may be called) the prognosis is extremely poor and the extent unknown if it should in fact occur.

She is debilitated and moribund, and although fairly stable at the time of argument before us (no new information having been filed in the meanwhile in expansion of the record), no physician risked the opinion that she could live more than a year and indeed she may die much earlier. Excellent medical and nursing care so far has been able to ward off the constant threat of infection, to which she is peculiarly susceptible because of the respirator, the tracheal tube and other incidents of care in her vulnerable condition. Her life accordingly is sustained by the respirator and tubal feeding, and removal from the respirator would cause her death soon, although the time cannot be stated with more precision.

The determination of the fact and time of death in past years of medical science was keyed to the action of the heart and blood circulation, in turn dependent upon pulmonary activity, and hence cessation of these functions spelled out the reality of death [note omitted].

Developments in medical technology have obfuscated the use of the traditional definition of death. Efforts have then been made to define irreversible coma as a new criterion for death, such as by the 1968 report of the Ad Hoc Committee of the Harvard Medical School (the Committee comprising ten physicians, an historian, a lawyer and a theologian), which asserted that:

> From ancient times down to the recent past it was clear that, when the respiration and heart stopped, the brain would die in a few minutes; so the obvious criterion of no heart beat as synonymous with death was sufficiently accurate. In those times the heart was considered to be the central organ of the body; it is not surprising that its failure marked the onset of death. This is no longer valid when modern resuscitative and supportive measures are used. These improved activities can now restore "life" as judged by the ancient standards of persistent respiration and continuing heart beat. This can be the case even when there is not the remotest possibility of an individual recovering consciousness following massive brain damage. ["A Definition of Irreversible Coma," 205 J.A.M.A. 337, 339 (1968)].

The Ad Hoc standards, carefully delineated, included absence of response to pain or other stimuli, pupillary reflexes, corneal, pharyngeal and other reflexes, blood pressure, spontaneous respiration, as well as "flat" or isoelectric electroencephalograms and the like, with all tests repeated "at least 24 hours later with no change." In such circumstances, where all of such criteria have been met as showing "brain death," the Committee recommends with regard to the respirator:

> The patient's condition can be determined only by a physician. When the patient is hopelessly damaged as defined above, the family and all colleagues who have participated in major decisions concerning the patient, and all nurses involved, should be so informed. Death is to be declared and then the respirator turned off. The decision to do this and the responsibility for it are to be taken by the physician-in-charge, in consultation with one or more physicians who have been directly involved in the case. It is unsound and undesirable to force the family to make the decision. [205 J.A.M.A., supra at 338].

But, as indicated, it was the consensus of medical testimony in the instant case that Karen, for all her disability, met none of these criteria, nor indeed any comparable criteria extant in the medical world and representing, as does the Ad Hoc Committee report, according to the testimony in this case, prevailing and accepted medical standards.

We have adverted to the "brain death" concept and Karen's disassociation with any of its criteria to emphasize the basis of the medical decision made by Dr. Morse. When plaintiff and his family, finally reconciled to the certainty of Karen's impending death, requested the withdrawal of life support mechanisms, he demurred. His refusal was based upon his conception of medical standards, practice and ethics described in the medical testimony, such as in the evidence given by another neurologist, Dr. Sidney Diamond, a witness for the State. Dr. Diamond asserted that no physician would have failed to provide respirator support at the outset, and none would interrupt its life-saving course thereafter, except in the case of cerebral death. In the latter case, he thought the respirator would in effect be disconnected from one already dead, entitling the physician under medical standards and, he thought, legal concepts, to terminate the supportive measures. We note Dr. Diamond's distinction of major surgical or transfusion procedures in a terminal case not involving cerebral death, such as here:

> The subject has lost human qualities. It would be incredible, and I think unlikely, that any physician would respond to a sudden hemorrhage, massive hemorrhage or a loss of all her defensive blood cells, by giving her large quantities of blood. I think that . . . major surgical procedures would be out of the question even if they were known to be essential for continued physical existence.

This distinction is adverted to also in the testimony of Dr. Julius Korein, a neurologist called by plaintiff. Dr. Korein described a medical practice concept of "judicious neglect" under which the physician will say:

> Don't treat this patient anymore, . . . it does not serve either the patient, the family, or society in any meaningful way to continue treatment with this patient.

Dr. Korein also told of the unwritten and unspoken standard of medical practice implied in the foreboding initials DNR (do not resuscitate), as applied to the extraordinary terminal case:

> Cancer, metastatic cancer, involving the lungs, the liver, the brain, multiple involvements, the physician may or may not write: Do not resuscitate. . . . [I]t could be said to the nurse: if this man stops breathing don't resuscitate him. . . . No physician that I know personally is going to try and resuscitate a man riddled with cancer and in agony and he stops breathing. They are not going to put him on a respirator. . . . I think that would be the height of misuse of technology.

While the thread of logic in such distinctions may be elusive to the non-medical lay mind, in relation to the supposed imperative to sustain life at all costs, they nevertheless relate to medical decisions, such as the decision of Dr. Morse in the present case. We agree with the trial court that that decision was in accord with Dr. Morse's conception of medical standards and practice.

It is from this factual base that the Court confronts and responds to three basic issues:

1. Was the trial court correct in denying the specific relief requested by plaintiff, *i.e.*, authorization for termination of the life-supporting appa-

ratus, on the case presented to him? Our determination on that question is in the affirmative.

2. Was the court correct in withholding letters of guardianship from the plaintiff and appointing in his stead a stranger? On that issue our determination is in the negative.

3. Should this Court, in the light of the foregoing conclusions, grant declaratory relief to the plaintiff? On that question our Court's determination is in the affirmative.

Constitutional and Legal Issues . . .

The Right of Privacy. [Note omitted.] It is the issue of the constitutional right of privacy that has given us most concern, in the exceptional circumstances of this case. Here a loving parent, qua parent and raising the rights of his incompetent and profoundly damaged daughter, probably irreversibly doomed to no more than a biologically vegetative remnant of life, is before the court. He seeks authorization to abandon specialized technological procedures which can only maintain for a time a body having no potential for resumption or continuance of other than a "vegetative" existence.

We have no doubt, in these unhappy circumstances, that if Karen were herself miraculously lucid for an interval (not altering the existing prognosis of the condition to which she would soon return) and perceptive of her irreversible condition, she could effectively decide upon discontinuance of the life-support apparatus, even if it meant the prospect of natural death. To this extent we may distinguish *Heston, supra*, which concerned a severely injured young woman (Delores Heston), whose life depended on surgery and blood transfusion; and who was in such extreme shock that she was unable to express an informed choice (although the Court apparently considered the case as if the patient's own religious decision to resist transfusion were at stake), but most importantly a patient apparently salvable to long life and vibrant health—a situation not at all like the present case.

We have no hesitancy in deciding, in the instant diametrically opposite case, that no external compelling interest of the State could compel Karen to endure the unendurable, only to vegetate a few measurable months with no realistic possibility of returning to any semblance of cognitive or sapient life. We perceive no thread of logic distinguishing between such a choice on Karen's part and a similar choice which, under the evidence in this case, could be made by a competent patient terminally ill, riddled by cancer and suffering great pain: such a patient would not be resuscitated or put on a respirator in the example described by Dr. Korein, and a fortiori would not be kept against his will on a respirator. . . .

The claimed interests of the State in this case are essentially the preservation and sanctity of human life and defense of the right of the physician to administer medical treatment according to his best judgment. In this case the doctors say that removing Karen from the respirator will conflict with their professional judgment. The plaintiff answers that Karen's present treatment serves only a maintenance function; that the respirator cannot cure or improve her condition but at best can only prolong her inevitable slow deterioration and death; and that the interests of the patient, as seen by her surrogate, the guardian, must be evaluated by the court as predominant, even in the face of an opinion contra by the present attending physicians. Plaintiff's distinction is significant. The nature of Karen's care and the realistic chances of her recovery are quite unlike those of the patients discussed in many of the cases where treatments were ordered. In many of those cases the medical procedure required (usually a transfusion) constituted a minimal bodily invasion and the chances of recovery and return to functioning life were very good. We think that the State's interest contra weakens and the individual's right to privacy grows as the degree of bodily invasion increases and the prognosis dims. Ultimately there comes a point at which the individual's rights overcome the State interest. It is for that reason that we believe Karen's choice, if she were competent to make it, would be vindicated by the law. Her prognosis is extremely poor—she will never resume cognitive life. And the bodily invasion is very great—she requires 24-hour intensive nursing care, antibiotics, the assistance of a respirator, a catheter and feeding tube.

Our affirmation of Karen's independent right of choice, however, would ordinarily be based upon her competency to assert it. The sad truth, however, is that she is grossly incompetent and we cannot discern her supposed choice based on the testimony of her previous conversations with friends, where such testimony is without sufficient probative weight. [137 N.J. Super. at 260, 348 A.2d 801.] Nevertheless we have concluded that Karen's right of privacy may be asserted on her behalf by her guardian under the peculiar circumstances here present.

If a putative decision by Karen to permit this non-cognitive, vegetative existence to terminate by natural forces is regarded as a valuable incident of her right of privacy, as we believe it to be, then it should not be discarded solely on the basis that her condition prevents her conscious exercise of the choice. The only practical way to prevent destruction of the right is to permit the guardian and family of Karen to render their best judgment subject to the qualifications hereinafter stated, as to whether she would exercise it in these circumstances. If their conclusion is in the affirmative, this decision should be accepted by a society the overwhelming majority of whose members would, we think, in similar circumstances, exercise such a choice in the same way for themselves or for those closest to them. It is for this reason that we determine that Karen's right of privacy may be asserted in her behalf, in this respect, by her guardian and family under the particular circumstances presented by this record.

The Medical Factor

The medical obligation is related to standards and practice prevailing in the profession. The physicians in charge of the case, as noted above, declined to withdraw the respirator. That decision was consistent with the proofs below as to the then existing medical standards and practices.

Under the law as it then stood, Judge Muir was correct in declining to authorize withdrawal of the respirator.

However, in relation to the matter of the declaratory relief sought by plaintiff as representative of Karen's interests, we are required to reevaluate the applicability of the medical standards projected in the court below. The question is whether there is such internal consistency and rationality in the application of such standards as should warrant their constituting an ineluctable bar to the effectuation of substantive relief for plaintiff at the hands of the court. We have concluded not.

In regard to the foregoing it is pertinent that we consider the impact on the standards both of the civil and criminal law as to medical liability and the new technological means of sustaining life irreversibly damaged.

The modern proliferation of substantial malpractice litigation and the less frequent but even more unnerving possibility of criminal sanctions would seem, for it is beyond human nature to suppose otherwise, to have bearing on the practice and standards as they exist. The brooding presence of such possible liability, it was testified here, had no part in the decision of the treating physicians. As did judge Muir, we afford this testimony full credence. But we cannot believe that the stated factor has not had a strong influence on the standards, as the literature on the subject plainly reveals. . . . Moreover our attention is drawn not so much to the recognition by Drs. Morse and Javed of the extant practice and standards but to the widening ambiguity of those standards themselves in their application to the medical problems we are discussing.

The agitation of the medical community in the face of modern life prolongation technology and its search for definitive policy are demonstrated in the large volume of relevant professional commentary [note omitted].

The wide debate thus reflected contrasts with the relative paucity of legislative and judicial guides and standards in the same field. The medical profession has sought to devise guidelines such as the "brain death" concept of the Harvard Ad Hoc Committee mentioned above. But it is perfectly apparent from the testimony we have quoted of Dr. Korein, and indeed so clear as almost to be judicially noticeable, that humane decisions against resuscitative or maintenance therapy are frequently a recognized de facto response in the medical world to the irreversible, terminal, pain-ridden patient, especially with familial consent. And these cases, of course, are far short of "brain death."

We glean from the record here that physicians distinguish between curing the ill and comforting and easing the dying; that they refuse to treat the curable as if they were dying or ought to die, and that they have sometimes refused to treat the hopeless and dying as if they were curable. In this sense, as we were reminded by the testimony of Drs. Korein and Diamond, many of them have refused to inflict an undesired prolongation of the process of dying on a patient in irreversible condition when it is clear that such "therapy" offers neither human nor humane benefit. We think these attitudes represent a balanced implementation of a profoundly realistic perspective on the meaning of life and death and that they respect the whole Judeo-Christian tradition of regard for human life. No less would they seem consistent with the moral matrix of medicine, "to heal," very much in the sense of the endless mission of the law, "to do justice."

Yet this balance, we feel, is particularly difficult to perceive and apply in the context of the development by advanced technology of sophisticated and artificial life-sustaining devices. For those possibly curable, such devices are of great value, and, as ordinary medical procedures, are essential. Consequently, as pointed out by Dr. Diamond, they are necessary because of the ethic of medical practice. But in light of the situation in the present case (while the record here is somewhat hazy in distinguishing between "ordinary" and "extraordinary" measures), one would have to think that the use of the same respirator or like support could be considered "ordinary" in the context of the possibly curable patient but "extraordinary" in the context of the forced sustaining by cardio-respiratory processes of an irreversibly doomed patient. And this dilemma is sharpened in the face of the malpractice and criminal action threat which we have mentioned.

We would hesitate, in this imperfect world, to propose as to physicians that type of immunity which from the early common law has surrounded judges and grand jurors; see e.g., *Grove v. Van Duyn*, 44 N.J.L. 654, 656-57 (E. & A.1882); *O'Regan v. Schermerhorn*, 25 N.J.Misc. 1, 19-20, 50 A.2d 10 (Sup.Ct.1940), so that they might without fear of personal retaliation perform their judicial duties with independent objectivity. In *Bradley v. Fisher*, 80 U.S. (13 Wall.) 335, 347, 20 L.Ed. 646, 649 (1872), the Supreme Court held:

> [I]t is a general principle of the highest importance to the proper administration of justice that a judicial officer, in exercising the authority vested in him, shall be free to act upon his own convictions, without apprehension of personal consequences to himself.

Lord Coke said of judges that "they are only to make an account to God and the King [the State]." [12 Coke Rep. 23, 25, 77 Eng.Rep. 1305, 1307 (S.C.1608).]

Nevertheless, there must be a way to free physicians, in the pursuit of their healing vocation, from possible contamination by self-interest or self-protection concerns which would inhibit their independent medical judgments for the well-being of their dying patients. We would hope that this opinion might be serviceable to some degree in ameliorating the professional problems under discussion.

A technique aimed at the underlying difficulty (though in a somewhat broader context) is described by Dr. Karen Teel, a pediatrician and a director of Pediatric Education, who writes in the *Baylor Law Review* under the title "The Physician's Dilemma: A Doctor's View: What The Law Should Be." Dr. Teel recalls:

> Physicians, by virtue of their responsibility for medical judgments are, partly by choice and partly by default, charged with the responsibility of making ethical judg-

ments which we are sometimes ill-equipped to make. We are not always morally and legally authorized to make them. The physician is thereby assuming a civil and criminal liability that, as often as not, he does not even realize as a factor in his decision. There is little or no dialogue in this whole process. The physician assumes that his judgment is called for and, in good faith, he acts. Someone must and it has been the physician who has assumed the responsibility and the risk.

I suggest that it would be more appropriate to provide a regular forum for more input and dialogue in individual situations and to allow the responsibility of these judgments to be shared. Many hospitals have established an Ethics Committee composed of physicians, social workers, attorneys, and theologians. . . . which serves to review the individual circumstances of ethical dilemma and which has provided much in the way of assistance and safeguards for patients and their medical caretakers. Generally, the authority of these committees is primarily restricted to the hospital setting and their official status is more that of an advisory body than of an enforcing body.

The concept of an Ethics Committee which has this kind of organization and is readily accessible to those persons rendering medical care to patients, would be, I think, the most promising direction for further study at this point. . . .

. . . [This would allow] some much needed dialogue regarding these issues and [force] the point of exploring all of the options for a particular patient. It diffuses the responsibility for making these judgments. Many physicians, in many circumstances, would welcome this sharing of responsibility. I believe that such an entity could lend itself well to an assumption of a legal status which would allow courses of action not now undertaken because of the concern for liability. [27 Baylor L.Rev. 6, 8-9 (1975).]

The most appealing factor in the technique suggested by Dr. Teel seems to us to be the diffusion of professional responsibility for decision, comparable in a way to the value of multi-judge courts in finally resolving on appeal difficult questions of law. Moreover, such a system would be protective to the hospital as well as the doctor in screening out, so to speak, a case which might be contaminated by less than worthy motivations of family or physician. In the real world and in relationship to the momentous decision contemplated, the value of additional views and diverse knowledge is apparent.

Conclusion

We therefore remand this record to the trial court to implement (without further testimonial hearing) the following decisions:

1. To discharge, with the thanks of the Court for his service, the present guardian of the person of Karen Quinlan, Thomas R. Curtin, Esquire, a member of the Bar and an officer of the court.
2. To appoint Joseph Quinlan as guardian of the person of Karen Quinlan with full power to make decisions with regard to the identity of her treating physicians.

We repeat for the sake of emphasis and clarity that upon the concurrence of the guardian and family of Karen, should the responsible attending physicians conclude that there is no reasonable possibility of Karen's ever emerging from her present comatose condition to a cognitive, sapient state and that the life-support apparatus now being administered to Karen should be discontinued, they shall consult with the hospital "Ethics Committee" or like body of the institution in which Karen is then hospitalized. If that consultative body agrees that there is no reasonable possibility of Karen's ever emerging from her present comatose condition to a cognitive, sapient state, the present life-support system may be withdrawn and said action shall be without any civil or criminal liability therefore on the part of any participant, whether guardian, physician, hospital or others.

By the above ruling we do not intend to be understood as implying that a proceeding for judicial declaratory relief is necessarily required for the implementation of comparable decisions in the field of medical practice.

Modified and Remanded

For modification and remandment: Chief Justice Hughes, Justices Mountain, Sullivan, Pashman, Clifford and Schreiber and Judge Conford-7.
Opposed: None.

QUESTIONING THE HISTORY OF BIOETHICS

ह&

Medical Morality Is Not Bioethics:
Medical Ethics in China and the United States

Renée C. Fox and Judith P. Swazey

Renée C. Fox received a doctorate in sociology from Radcliffe College in 1954 and spent her research and teaching career at the University of Pennsylvania's Departments of Sociology, Psychiatry, and Medicine, where she is currently Professor Emeritus. Judith P. Swazey received her Ph.D. in the History of Science from Harvard University. These authors compare the tenets of Chinese medical morality and American bioethics to illustrate that Western values are not embraced by all societies. They go on to argue that American bioethicists tend to be unaware of their cultural assumptions and of the "American-ness" of the value concerns they raise.

Introduction to Chinese Medical Morality

In the summer of 1981, we spent 6 weeks doing medical sociology fieldwork in the People's Republic of China, primarily in the city of Tianjin. Our trip was arranged by the program of scientific exchanges created by the American Association for the Advancement of Science in Washington, D.C., and the China Association of Science and Technology (CAST) in Beijing. The focal point of our work was a mini-ethnographic study of a profoundly Chinese urban hospital that is energetically committed to modern scientific and technological medicine. . . .

Abridged from Renée C. Fox and Judith P. Swazey, "Medical Morality Is Not Bioethics—Medical Ethics in China and the United States," *Perspectives in Biology and Medicine*, Volume 27, 1984, pp. 336–360. Copyright © 1984 by University of Chicago Press.

The hospital that our Chinese colleagues chose as the base for our research and teaching proved to be the center of medical modernization that we had asked to study—and more. It contained the only free-standing Critical Care Unit in China [1], conducted hemodialysis for acute renal failure, included a bioengineering-oriented absorbent artificial kidney group, and had made some forays into the transplantation of human organs. It was also highly active in matters pertaining to our work in medical ethics—an interest which we had not thought of pursuing in a Chinese setting.

In retrospect, it seems far from accidental that the hospital we were sent to turned out to be as notable for its leadership in what the Chinese term *medical morality* as in the "fourth modernization" of (medical) science and technology. We soon learned that the First Central Hospital's intensive involvement in medical ethics was partly due to the influence of its vice-director, Madame She Yunzhu, a remarkable 75-year-old woman who is one of the pioneers of modern nursing in China. . . .

But even without the dynamic presence of Madame She, we probably would have been introduced to medical morality as representatives of what our Chinese colleagues conceived to be medical sociology and expected from it—albeit in a somewhat more tentatively conceptualized and less vigorously uplifting way. For, in a number of medical and nursing schools that we visited or about which we were told, first steps were being taken to develop courses that were alternately called "Medical Sociology," "Medical Psychology," and/or "Medicine, Morals, and Society." As the last course title suggests, in China, medical sociology and medical ethics are not only interrelated—they are virtually synonymous. Social relationships and a conception of what one of our hosts termed "the individual as a social community" are at the heart of what the Chinese have always defined as ethics. And ethics is the center of the Chinese world view—its very core and essence in Chinese society today as it has been for thousands of years. What is more, participant observation—which the Chinese recognized as a guiding principle as well as a major technique of our research—is also inherent to their own inductive, humanistic approach to ethics. For them, thinking in an entirely abstract or speculative way about moral or social questions runs the risk of what Chinese scholars historically have called "playing with emptiness." What seems to them more "practical" and "right," as well as comfortably familiar, is to work from everyday, empirically observable human reality, focusing particularly on the relationship between specific, identifiable persons, and on their "lived-in," reciprocal existence.

It was both surprising and satisfying to learn in a firsthand way that, despite the thousands of geographical miles and historical years that separate Chinese society and our own, and their very different cosmic outlooks, these aspects of Chinese thought are compatible with the conceptual and methodological framework in which we observe, analyze, interpret, and evaluate as sociologists. . . .

Medical morality is "the kind of morality that doctors and nurses should have." It is concerned with three sets of interconnected goals and with the obligation of "medical workers," individually and collectively, to do everything possible—"sparing no effort"—to attain these goals. Repairing the moral and intellectual as well as the economic and political damage of the Cultural Revolution (1966–1976) is one of the primary objectives of medical morality. Supreme importance is given to restoring the basic "order" that was "smashed" by the Cultural Revolution (and its personification in the Gang of Four): the "task of straightening things out in every field of [medical] work" that must be accomplished, especially the reteaching of "what is right and wrong." A second basic aim of medical morality is to "scale the heights" of modern medicine and thereby achieve the "golden-dream" benefits that come from applying advanced science and technology to problems of health, illness, and the care of patients. This is the medical facet of the national policy of "Four Modernizations" (agriculture, industry, defense, science and technology) that currently prevails in China. In turn, "order" and "modernization" are part of the third general goal of medical morality: the dynamic and creative continuation of the "Great Liberation," the revolution that established the People's Republic of China in 1949.

As this implies, medical morality fits into a larger societal frame. "Work ethics" and "civic virtues" and their relationship to the integrity and development of the whole society are constantly stressed, ideologically and politically, in every sphere of Chinese life. Nationwide campaigns like the civic virtues month and the *Wujiang Simei* ("Five Efforts" and "Four Beauties") movement have been organized around these themes. Workers of all kinds are continually reminded that they are expected to pay attention to morality. Medical workers are among those who have special ethical responsibilities because their job is to care for patients, "relieve them from pain," help them to recover from their illnesses, and "save them from death." Leading nurses and doctors, in particular, are exhorted to demonstrate "the highest level of ethics" in their own behavior—to be "the first to observe the principles and disciplines" entailed—and thereby to set an example that is "a silent order" to those who work with them.

Medical morality is rooted in a conception of the individual in relation to statuses and roles, enmeshed in the network of human relationships that this involves. In this conception, the individual steadfastly strives to meet his responsibilities and carry out his duties ever more totally and perfectly, guided by certain principles, inspired by particular maxims and exemplars, and in conformity with concrete rules. At the vital center of this morality is the continuous effort that each person is expected to make to perfect his at once individual and social self through relationships to significant others and the fulfillment of obligations to these persons. The relationships encompassed by medical morality include those of physicians, nurses, other medical workers, and hospital administrators with each other and with patients and their families. It also concerns the relations between the unit or *danwei* [2] to which medical workers belong and the local bureaucrats and Communist party officials associated with their professional activities. The bedrock and point of departure of medical morality lie in the quality of these human relationships: in how correct, respectful, harmonious, complementary, and reciprocal they are. . . .

. . . But medical morality and its attainment require more. In the words of Dr. Wang Chin-ta (director of the Critical Care Unit of Tianjin First Central Hospital): "No matter how good doctors and nurses are technically, if they do not have noble thinking, they cannot serve the patient, the people, and the country."

"Noble thinking" is an epigrammatic way of referring to the moral virtues that good medical professionals are ideally expected to demonstrate in their work and work relations. Foremost among these medically relevant virtues are the following:

Humanity, compassion, kindness, helpfulness to others;
Trust in others;
A spirit of self-sacrifice;
A high sense of responsibility;
A good sense of discipline, good order;
Hard, conscientious work that is also systematic, careful, precise, punctual, and prudent;
Devotion, dynamic commitment;
Courage to think, act, innovate, blaze new trails, overcome difficulties;
Alertness, high spirits, optimism, a positive attitude;
Patience;
Modesty;
Self-control, a sense of balance and equilibrium;
Politeness, good manners, proper behavior;
Cleanliness, tidiness, good hygiene, keeping healthy;
Lucidity, clarity, intelligence, wisdom;
Honesty, integrity;
Self-knowledge, self-examination, self-criticism, self-cultivation, self-improvement;
Frankness about difficulties, limitations, shortcomings, and mistakes—admitting them, and working to overcome them.

Seen as a whole, these virtues have a number of patterned characteristics. They are relatively concrete ethical qualities, close to the empirical reality of medical practice and patient care. They are formulated as responsibilities and duties, generally stated as positive "musts" and "can do's," rather than as admonitory "must nots" and "do nots." They are punctuated by aphorisms and proverb-like political slogans. A "we shall overcome" moralistic optimism pervades the outlook that they represent. But seen in closer detail, the dynamic nature of these moral virtues is a product of the balancing and blending of the active and passive, traditional and innovative, intellectual and emotional, personal and interpersonal, individual and collective qualities that their fulfillment requires. They are as neo-Confucian as they are Maoist, and more of both than they are Marxist or Leninist.

Particular individuals are singled out because they personify the virtues of medical morality. They are considered to be models whose "example will inspire and encourage others to follow them. . . ."

In the particular settings where nurses, physicians, and their co-workers carry out their medical duties, the principles and virtues of medical morality are translated into specific sets of rules. The ethical importance of these rules is aesthetically expressed through the high quality of their calligraphy, the care with which they are framed, and the prominence with which they are displayed. Here, for example, are the rules of the Critical Care Unit of Tianjin First Central Hospital. Composed by the members of the unit out of their shared experiences, they are written in elegant black script under the title, "Regulations of the Critical Care Unit," which is written in contrasting red ink. Enclosed in an ornate green frame, they hang alone on a wall adjacent to the unit's doorway. The nine sets of regulations are explicit and detailed statements of the unit's work norms, and also of the problematic attitudes and behaviors—the persistent "shortcomings" at work—that have been identified as needing improvement:

1. Patients in this room need critical care. When they are better, they should be transferred to the recovery ward.
2. Care should be given to these patients day and night.
3. The work should be done strictly by the medical workers. They should cooperate. In these ways, they will be able to serve the people wholeheartedly.
4. Medical workers must check the equipment, drugs, and machines on every shift, e.g., ventilator, EKG, tracheal tubes, IV, catheters, abdominal dialyzer, suction, extension cord. Everything must be kept in its place. Do not lend things out.
5. Adhere strictly to regulations. Visiting doctors should make rounds twice a day, in the morning and in the afternoon. Doctors on duty should carefully observe the patients. They should make careful observations at the bedside in order to discover any developments, and be prepared to give emergency treatment if necessary.
6. The staff on one shift should tell the staff on the next shift about any changes that have taken place with the patients. Explain things clearly for continuing with the next shift.
7. For emergency treatment, use Western and Chinese medicine together. There are four principles to follow in administering Chinese medicine— four things that must be done in time:
   a. Prescription;
   b. Fetching of drugs;
   c. Cooking of prescription;
   d. Administering of prescription.
8. Check regulations, and carry them out strictly, in time. Carry out the doctor's orders, to do well and avoid complications.
9. Perform a case history in time.

The First Central Hospital's various sets of rules, regulations, and requirements are now being organized into a centralized system of total quality control (called TQC) under the direction of the hospital's medical administration office and the Party dialectician who is attached to it. The TQC is an elaborate moral accountancy system, designed principally to apply to nurses and doctors and to raise the overall level of medical care. One of its major features is a "shortcomings control" classification scheme which identifies and categorizes various kinds of "technical" and "responsibility" errors and mistakes that physicians and nurses can make in giving care. It then attaches quantitative weights to them according to how major or minor they are considered to be. "Responsibility shortcomings" are viewed largely as moral errors. They are therefore defined as more grave than are shortcomings judged to be primarily technical in nature and are correspondingly subject to more severe penalties and punishments. Eventually, First Central Hospital hopes to translate the variables of its TQC system into a computer program for calculating individual and group "medical quality scores" in a sophisticated, modern way. The hospital regards the computer not only as an important technological tool in this effort but also as a powerful empirical and symbolic expression of the medical modernization toward which it strives. . . .

The "Chinese-ness" of Medical Morality

. . . The central status that medical morality attaches to the primary ethical importance of fulfilling one's duties to specified others in commonplace settings and in everyday acts, and to the continual improvement and perfecting of self in and through these relations and duties, is . . . anciently and quintessentially Chinese. These are core ideas in the traditional morality: of how striving to perfect one's individual and social being expresses the principle of immanent order in the cosmos and contributes to a society that embodies it.

The "noble" ethical virtues emphasized by medical morality are also closely related to the Confucian virtues that structured the ongoing ethical effort required by traditional Chinese morality (*ren* [humanity], *li* [sense of rites], *yi* [sense of duties], *zhi* [wisdom]). Rules were one of the principal forms in which these ethical obligations were made explicit, as is the case with medical morality. In China's past, they were often developed out of the group experience of a guild or a clan and were posted in temples, clan halls, and schools in a manner comparable to the way that medical morality rules are currently displayed in First Central Hospital.

The stylistic features and ambience of medical morality—with its emphasis on orthodoxy, righteousness, and propriety; its concern about "moral sympathy"; its stress on the power of didacticism; its ritualization and bureaucratization; and its preeminently public nature—all have their counterparts and origins in Chinese tradition.

Tianjin First Central Hospital's TQC system has significant antecedents in Chinese history and tradition. It could be said that ancestral versions of TQC existed in the "morality books" (*shanshu*) and the "ledgers of merit and demerit" (*gongguoge*) that were kept by individuals and families in the sixteenth- and early seventeenth-century years of China's Ming dynasty [3, pp. 193–196, 197, 4]. These were moral account books, based on the self-examination of conscience, the written confession of wrongdoing, and the recording of both good and bad thoughts and acts. Not only were thoughts and deeds morally classified in this positive and negative way, but they were sometimes weighted by a point system. In some ledgers, good points were recorded in red ink, bad ones in black ink. (Since at least the time of the Han dynasty, red had been the color of positive numbers and black of negative ones.) At periodic intervals, the person keeping this moral audit added up the points, thereby arriving at a quantitative score of his current state of ethicality and his cumulated "moral capital." Sinologists identify Taoist and Buddhist influences in these morality books, including concepts of judgments meted out by a "cosmic bureaucracy" that involved one's present life, life span, and rebirth [5].

This is not to imply that medical morality is a pure emanation of Confucianism, Taoism, and Chinese Buddhism, or that Marxism-Leninism-Maoism has played a negligible role in shaping its form and content. Certain precepts of medical morality constitute radical departures from the concepts on which the traditional system of Chinese ethics was built. Most notable are the ways in which the egalitarian and universalistic principles of Marxism have strongly influenced the tenets of medical morality. Doctors and nurses are urged to unite, to work closely together professionally and "transprofessionally," and to treat colleagues equally. This egalitarian view runs counter to the traditional Chinese thesis that morality and public order consist of, and depend on, a series of hierarchically structured relationships and the fulfillment of the duties associated with them. The archetype and the ethical keystone of these superordinate-subordinate relationships was that between father and son; filial piety (*xiao*) was regarded as the model and the source of all other virtues. . . .

. . . In contradistinction to American bioethics, as already indicated, Chinese medical morality is not preoccupied with social and ethical problems associated with the advancement of medical science and technology. At the present time, medical modernization—enriched by its incorporation of traditional Chinese medicine—is viewed as morally good, socially desirable, economically necessary, and politically obligatory. Substantive issues that *do* fall under the aegis of medical morality include the population's response to the new, one-child-per-family policy; the wisdom of telling or not telling seriously ill patients (particularly those with cancer) about the gravity of their illness; the role that the family ought and ought not to play in the care of ill relatives who are hospitalized; the causes, prevention, and treatment of suicide attempts; and the problem of obtaining blood donations. All are issues that are encountered by nurses and doctors in the various medical milieus where they currently work. Prominent bioethical concerns, such as human experimentation, the "gift of life" and "quality of life," the termination of treatment, and questions about the allocation of scarce resources associated with therapeutic innovations like organ transplantation and hemodialysis in our own society, are not yet considered to be problems in China. Chinese nurses, physicians, and relevant officials are aware that, as the process of medical modernization goes forward, they may face comparable difficulties. But, in accord with Chinese pragmatism, they are disinclined to engage in abstract speculation about hypothetical problems that may (or may not) develop, and about how they should be handled if they do. It is not until the Chinese face such issues in a firsthand way, and can meet and analyze them as "lived-in experiences," that these matters will become part of their medical morality.

From all the foregoing, it is amply clear that medical morality is not bioethics. It is as Chinese as bioethics is American. We now turn to bioethics, an arena in which we have worked and observed since the mid-1960s. In the final analysis, as we shall see, American bioethics and Chinese medical morality are so culturally dissimilar that they are not sufficiently related to form a yin-yang (opposing, but complementary) pair.

American Bioethics

In contrast to medical morality, the phenomena with which bioethics is primarily concerned are related to some of the ways in which modern, Western, American medicine has already *succeeded* in what the Chinese call "scaling the high peaks" of science and technology. Bioethics is focused on what we consider serious *problems* associated with these advances, rather than on the achievements they represent or the "golden dream" promises that they hold forth [6]:

> Actual and anticipated developments in genetic engineering and counseling, life support systems, birth technology, population control, the implantation of human, animal, and artificial organs, as well as in the modification and control of human thought are principal [areas] of concern. Within this framework, special attention is concentrated on the implications of amniocentesis, abortion, in vitro fertilization, the prospect of cloning, organ transplantation, the use of the artificial kidney machine, the development of an artificial heart, the modalities of the intensive care unit, the practice of psychosurgery, and the introduction of psychotropic drugs. Cross-cutting the consideration . . . given to these general and concrete [spheres] of biomedical development, there is marked preoccupation with the ethicality of human experimentation under various conditions. . . .

Bioethics has also been concerned with the proper definition of life and death and personhood and with the humane treatment of "emerging life and life that is passing away" [7]—especially with the justifiability of forgoing life-sustaining forms of medical therapy. One of the most significant general characteristics of this ensemble of bioethical concerns is the degree to which they cluster around problems of natality and mortality, at the beginning and at the end of the human life cycle.

The chief intellectual and professional participants in American bioethics are philosophers (above all, those who are called "ethicists" in the United States, and "moral philosophers" in Europe), theologians (predominantly Catholic and Protestant), jurists, physicians, and biologists. Lately, the thought and presence of economists have been strongly felt in the field; but relatively few other social scientists are actively involved or notably influential in bioethical discussion, research, writing, and action. The limited participation of anthropologists, sociologists, and political scientists in bioethics is a complex phenomenon, caused as much by the prevailing intellectual orientations and the weltanschauung of present day American social science as by the framework of bioethics [8].

The disciplinary backgrounds of bioethicists contrast sharply with those of the key participants in medical morality because of historic differences in the American and Chinese cultures as well as current differences in our respective political and economic systems. For example, although one could argue that there are functional parallels between the role of a Chinese dialectician and that of an American theologian, one would hardly expect a theologian trained in the Judeo-Christian tradition to define value and belief issues and make decisions in the same way as someone whose world view is shaped by Confucian, Taoist, and Buddhist thought. Nor does the pivotal place of lawyers and judges in bioethics have its counterpart in medical morality. The status and role of jurists in bioethics are integrally connected with the singular importance that Americans attach to the principle as well as to the fact of being "a society under law, rather than under men."

On the other hand, the control in degree and kind over medical morality exercised by the central government in Beijing is not only an emanation of the Chinese Communist party, its present-day leadership, and its current doctrine. It is thoroughly compatible with the at once profane and sacred power over the order and organization of the entire society accorded to the emperor and his imperial bureaucracy throughout all the dynasties of Chinese history. In the United States, quite to the contrary, the fact that bioethical questions, with their moral and religious connotations, have been appearing more frequently and prominently in national and local political arenas—in our legislatures, courts, and in specially created commissions—constitutes a societal dilemma. Although ours is a "society under law," it is also a nation founded on the separation of church and state as one of its sacredly secular principles. What ought we to do, then, about the fact that bioethics (like Mr. Smith) has gone to Washington? In the light of the religious and even metaphysical, as well as moral, nature of bioethical issues, is it legitimate or wise for our government to deal with them? If so, at what level, through what branch, using what mechanisms? If not, are there other means through which we can try to resolve such matters of our collective conscience, on behalf of the whole society? These are distinctly American questions that are decidedly not Chinese.

The pluralism of American society and its voluntarism have contributed to the development of the numerous centers, institutes, and associations of varying orientations that have been organized around bioethical activities in the United States over the course of the past 20 years, both inside and outside university settings. Most of the persons who are professionally active in the field of bioethics belong to one or several of such groups and participate in their interconnected and to some extent overlapping activities (discussions, research, teaching, consultations, meetings, publications, etc.). In this sense, they form a sort of "invisible college," although not a unified school of thought.

Such a plethora of voluntary associations, organized around common interests, but with somewhat different origins, auspices, memberships, and outlook, is a very American configuration. It embodies a set of culture patterns and social traits, not confined to bioethics, that always have been strikingly characteristic of American society and its conception of democracy. As early as the 1830s, Alexis de Tocqueville, the astute French observer-analyst of our new nation-state, identified our tendency to form and join voluntary associations as one of our most notable societal attributes. Again, for reasons broader, deeper, and older than the particular contemporaneous circumstances that have given rise to bioethics and medical morality, this is not a pattern that exists in China or that one would expect to find there.

But above all, it is in the values and beliefs emphasized and deemphasized by bioethics, and in its cognitive framework and style, that its Western and American orientation is both most evident and most fully articulated.

To begin with, as already indicated, individualism is the primary value-complex on which the intellectual and moral edifice of bioethics rests. Individualism, in this connection, starts with a belief in the importance, uniqueness, dignity, and sovereignty of the individual, and in the sanctity of each individual life. From this flows the assumption that every person, singularly and respectfully defined, is entitled to certain individual rights. Autonomy of self, self-determination, and privacy are regarded as fundamental among these rights. They are also considered to be necessary preconditions for another value-precept of individualism: the opportunity for persons to "find," develop, and realize themselves and their self-interests to the fullest—to achieve and enjoy individual well-being. In this view, "individuals are entitled to be and do as they see fit, so long as they do not violate the comparable rights of others" [9]. "Paternalism" is defined as interfering with and limiting a person's freedom and liberty of action for the sake of his or her own good or welfare. It is regarded as ethically dubious because, however beneficent its intentions or outcome, it restricts autonomy, involves coercion, implies that someone else knows better what is best for a given individual, and may insidiously impair that individual's ability to decide and act independently.

The notion of contract plays a major role in the way relations between autonomous individuals are conceived in bioethics. Self-conscious, rational, specific agreements by persons involved in interaction with one another, that explicitly delineate the scope, content, and conditions of their joint activities, are presented as ethical models. They are considered to be exemplary expressions of the way that moral relationships, protective of individual rights, can be structured. The archetype of such contractual relations is the kind of informed, voluntary consent agreement between subjects and investigators in medical research which the field of bioethics helped to formulate and that is now required by all federal and most private agencies funding this research. The informed consent contract, though mutual, is asymmetric. It is principally concerned with the rights and welfare of one of the two partners—the human subject, often a patient—because he is the most vulnerable, disadvantaged, and least powerful of the pair. The special contractual obligation to watch over and safeguard the rights of the person(s) most susceptible to exploitation or harm in this type of exposed and unequal situation is a part of the bioethics conception of individualism and of moral relations between individuals. But little mention is made by bioethicists of what sociologist Emile Durkheim termed the "noncontractual aspects of contract": that is, the more implicit and informal commitment, fidelity, and trust aspects of social relationships that reciprocally bind persons to live up to their promises and their responsibilities to one another.

Veracity and truth-telling, the "faithfulness" dimension of relationships on which bioethics fixes its attention, is more specific and circumscribed than the Durkheimian concept. In keeping with the overall orientation of bioethics, what is stressed is the right of patients or research subjects to "know the truth" about the discomforts, hazards, uncertainties, and "bad news" that may be associated with medical diagnosis, prognosis, treatment, and experimentation. The physician's obligation to communicate the truth to the patient or subject is derived from and based on the latter's presumed right to know. Discerning what is the truth and what is a lie is seen as relatively unproblematic. And there is a decided tendency to look on the use of denial by the patient as an undesirable defense, because it complicates truth-telling and blocks truth-receiving. Here, the affirmation that patients have the right to know the truth veers toward insistence that they ideally ought to face the truth consciously and deal with it rationally (in keeping with the particular definitions of "truth" and of "rationality" inherent to bioethics).

Another major value preoccupation of bioethics—and one that it has increasingly emphasized since the mid-1970s—concerns the allocation of scarce, expensive resources for advanced medical care, research, and development. What proportion of our national and local resources should be designated for these purposes, in what ways, and according to what principles and criteria? The resources with which bioethics is chiefly concerned are material ones, mainly economic and technological in nature. The allocation of nonmaterial resources such as personnel, talent, skill, time, energy, caring, and compassion is rarely mentioned. Bioethics situates its allocation questions within a rather abstract, individual rights-oriented notion of the general or common good, assigning greater importance to equity than to equality. The ideally moral distribution of goods is defined as one that all rational, self-interested persons are willing to accept as just and fair, even if goods are allotted unequally. "Cost containment" is also an essential value-component of this view of rightful distribution. In the bioethical calculus, it is not just a practical or necessary response to an empirical situation of economic scarcity. It has become a more categorical moral imperative.

Finally, what is usually referred to as "the principle of beneficence" or "benevolence" is also a key value of bioethics. This enjoinder to "do good" and to "avoid harm" is structured and limited by the supremacy of individualism. The benefiting of others advocated in bioethical thought is circumscribed and constrained by the obligation to respect individual rights, interests, and autonomy. Furthermore, rather than being seen as an independent virtue, doing good is generally conceived to be part of a "benefit-harm ratio" in which, ideally, benefits should outweigh costs. "Minimization of harm" rather than "maximization of good" is more strongly emphasized in this bioethical equation.

These values are predominant in American bioethics and are considered to be the most fundamental. They are accorded the highest intellectual and moral significance and are set forth with the greatest certainty and the least qualification. Other values and virtues and principles and beliefs that are part of the ethos of bioethics occupy a more secondary and less secure status. They are less frequently invoked and when introduced into ethical discussion and analysis are likely to elicit debate or require special justification. . . .

The emphasis that bioethics places on individualism and on contractual relations freely entered into by voluntarily consenting adults tends to minimize and obscure the interconnectedness of persons and the social and moral importance of their interrelatedness. Particularly when compared with Chinese medical morality, it is striking how little attention bioethics pays to the web of human relationships of which the individual is a part and to the mutual obligations and interdependence that these relations involve. Concepts like reciprocity, solidarity, and community, which are rooted in a social perspective on our moral life and our humanity, are not often employed. Characteristically, bioethics deals with the "more-than-individual" in terms of the "general good," the "common good," or the "public interest." In the bioethical use of these concepts, the "collective good" tends to be seen atomistically and arithmetically as the sum total of the rights and interests, desires and demands of an aggregate of self-contained individuals. The fair and just distribution of limited collective resources is the major dimension of commonality that is stressed, often to the exclusion of other aspects, and usually with a propensity to define resources as material (primarily economic), and quantitative. In this view, private and public morality are sharply distinguished from one another in keeping with the underlying essential dichotomy between individual and social. Social and cultural factors are largely seen as external constraints that limit individuals. They are rarely presented as enabling and empowering forces, *inside* as well as outside of individuals, that are constituent, dynamic elements in making them human persons. . . .

These assumptions about what is and is not purely moral are integrally related to the major cognitive characteristics of bioethical thought: to how participants in bioethics actually *do* think, and especially, to what they define as ideal standards of ethical thinking. A high value is placed on logical reasoning— preferably based on a general moral theory and concepts derived from it—that is systematically developed according to codified methodological rules and techniques around select, analytically designated variables and problems. Rigor, precision, clarity, consistency, parsimony, and objectivity are regarded as earmarks of the intellectually and ethically "best" kind of moral thought. Flawed logical and conceptual analysis is considered to be not only a concomitant of moral error but also, to a significant degree, responsible for producing it. This way of thought also tends toward dichotomous distinctions and bipolar choices. Self versus others, body versus mind, individual versus group, public versus private, objective versus subjective, rational versus nonrational, lie versus truth, benefit versus harm, rights versus responsibilities, independence versus dependence, autonomy versus paternalism, liberty versus justice are among the primary ones. Even the field's own self-defining conception of what is and is not a moral problem is formulated in a bipolar, either/or fashion.

Bioethics is an applied field that brings its theory, methods, and knowledge to bear on phenomena and situations deemed ethically problematic. It seeks to identify and illuminate points of moral consideration and provide a way of thinking about them that can contribute to their practical moral resolution through concrete choices and specific acts. Bioethics attempts this by proceeding in a largely deductive manner to impose its mode of reasoning on the phenomenological reality addressed. The amount of detailed investigation of the actual situations in which the ethical problems occur varies. But what philosophers call "thought experiments" are more often conducted in bioethics than is empirical, in situ research. . . .

Within its rigorously stripped-down analytic and methodological framework, bioethics is prone to reify its own logic and to formulate absolutist, self-confirming principles and insights. These tendencies are associated with the disinclination of bioethics to critically examine its own moral epistemology: to searchingly identify and evaluate the presuppositions and assumptions on which it rests. In a scholastic sense, the field of bioethics is knowledgeably aware of the traditions of Western thought on which it draws (e.g., act and rule utilitarianism and various theories of justice). But there is a more latent level on which it nevertheless considers its principles, its style of reasoning, and its perceptions to be objective, unbiased, and reasonable to a degree that not only makes them socially and culturally neutral but also endows them with a kind of universality. Paradoxically, these very suppositions of bioethical thought contribute to its inadvertent propensity to reflect and systematically support conventional, relatively conservative American concepts, values, and beliefs.*

These value, belief, and thought patterns of bioethics have developed within an interdisciplinary matrix. But particularly since the mid-1970s, when philosophers began "arriving by the score" in bioethics (and "in applied ethics more broadly") [10], moral philosophy has had the greatest molding influence on the field. It is principally American analytic philosophy—with its emphasis on

---

*Several major critiques of the emergence of the new philosophical subdiscipline of applied ethics on the American scene have been published in the *Hastings Center Report* which coincide in numerous respects with our characterization of bioethics, particularly its "seeming indifference to history, social context, and cultural analysis" [10–12].

theory, methodology, and technique, and its utilitarian, Kantian, and "contractarian" outlooks—in which most of the philosophers who have entered bioethics were trained. Defined as "ethicists" who are specialized experts in moral problems associated with biomedicine, they have established themselves, and their approach to matters of right and wrong, as the "dominant force" [12] in the field.

This is not to say that all analytic philosophers who actively participate in bioethics think and write in a uniform way, or that every philosopher-bioethicist is grounded in this analytic tradition. Major contributors to bioethics, for example, also include a number of highly esteemed philosopher-scholars whose work incorporates more phenomenological, social, and religious dimensions rooted in the traditions of moral theology and American social ethics. The respect that such individuals are accorded notwithstanding, the perspective that they represent has had far less influence on the predominant ethos of bioethics than has analytic philosophy. . . .

Bioethics Is Not Just Bioethics

In our sociological view, the paradigm of values and beliefs, and of reflections on them, that has developed and been institutionalized in American bioethics is an impoverished and skewed expression of our society's cultural tradition. In a highly intellectualized but essentially fundamentalistic way, it thins out the fullness of that tradition and bends it away from some of the deepest sources of its meaning and vitality. . . .

Bioethics has "sprung loose from that broader [religious] framework" [13] in which the values of our cultural tradition are historically embedded. In turn, the particular forms that the secularism, the rationality, and the individualism of bioethics take, and the ways in which they interact with each other, contribute to another of the field's constricting features: its provincialism. Bioethics is sealed into itself in such a way that it tends to take its own characteristics and assumptions for granted. It is relatively uncritical of its premises and unaware of its cultural specificity. It is this sort of parochialism, with its mix of naiveté and arrogance, that makes it difficult for bioethicists not only to recognize medical morality and its Chinese-ness when they encounter it but also to perceive the "American-ness" of their particular value-concerns and of how they approach them. . . .

References

1. Fox, R. C., and Swazey, J. P. Critical care at Tianjin's First Central Hospital and the fourth modernization. *Science* 217:700–705, 1982.
2. Henderson, G. E. Danwei: The Chinese work unit: a participant observation study of a hospital. Dissertation submitted in partial fulfillment of the requirements of the Ph.D. (Sociology) at the University of Michigan, Ann Arbor, 1982.
3. Gernet, J. *Chine et Christianisme*. Paris: Gallimard, 1982.

4. Berling, J. Religion and popular culture: the management of moral capital in *The Romance of the Three Teachings*. In *Popular Culture in Late Imperial China* edited by A. Nathan, D. Johnson, and E. Rawski. Berkeley and Los Angeles: Univ. California Press, 1985.

5. Sivin, N. Ailment and cure in traditional China (unpublished manuscript).

6. Fox, R. C. Ethical and existential developments in contemporaneous American medicine: their implications for culture and society. In *Essays in Medical Sociology*. New York: Wiley, 1979.

7. Bok, S. In discussion at the Conference on the Problem of Personhood, organized by Medicine in the Public Interest (MIPI). New York City, April 1–2, 1982.

8. Fox, R. C. Advanced medical technology—social and ethical implications. In *Essays in Medical Sociology*. New York: Wiley, 1979.

9. Gorovitz, S. *Doctors' Dilemmas: Moral Conflicts and Medical Care*. New York: Macmillan, 1982.

10. Callahan, D. At the center: from "wisdom" to "smarts." *Hastings Cent. Rep.* 12:4, 1982.

11. Noble, C. N. Ethics and experts. *Hastings Cent. Rep.* 12:7–9, 15, 1982.

12. Callahan, D. Minimalistic ethics. *Hastings Cent. Rep.* 11:19–25, 1981.

13. De Craemer, W. See [7].

The New Conservatives in Bioethics:
Who Are They and What Do They Seek?

Ruth Macklin

Ruth Macklin, Ph.D. is professor of bioethics at Albert Einstein College of Medicine. She received her doctorate in philosophy from Case Western University and has been a leading contributor to the field of bioethics since its beginnings. In this article she comments on the tendency to see bioethical issues in light of the politically tinged terms *liberal* and *conservative*. The former, it is claimed, would look more favorably on topics as abortion, stem cell research, and euthanasia; the latter would be much more negative toward these topics. Liberals would foster the notions of personal autonomy and technological progress, whereas conservatives would advocate for moral values that would stem the onslaught of technology against human dignity. Macklin challenges the validity of this distinction and criticizes the concepts that conservative bioethicists espouse.

The ever-changing landscape of bioethics has a new feature: a "conservative" movement, whose spokespersons attack what they see as the liberal cast of "mainstream bioethics,"[1] a field that, since its inception, has mainly comprised academic scholars . . . These new contributors to the field have applied the label "conservative" to themselves as they mount a critique of the assumedly "liberal" tradition of mainstream bioethics. Yet I question whether these labels are meaningful, or at least consistent . . .

What has led to the introduction into bioethics of labels once reserved for a stance on political issues? A possible explanation is that the most conservative wing of the Republican Party in American politics, now dominating both the executive and legislative branches of government, is obsessed with matters related to procreation, prenatal life, and extracorporeal embryos, and has put those issues high on its political and legislative agendas. Some academics, but also journalists and other public intellectuals who find common cause with political conservatives on these issues, are dissatisfied with what they see as the approach of mainstream bioethics. Once the label "conservative" is adopted by those who are critical of "permissive" positions on procreation and the status of embryos, bioethicists who take an opposing view are virtually by default labeled "liberals."

But to apply that label to everyone in bioethics who does not explicitly identify with the conservative movement is misleading, at best. To characterize mainstream bioethics as "liberal" is to lump together—uncritically and irresponsibly—an array of widely divergent and often nuanced positions. Even the question of what constitutes a conservative position in bioethics has received different answers.

Who Are the Liberals and Conservatives in Bioethics?

Back in 1973, when Daniel Callahan began his three-decade critique of runaway science and technology, arguing for the need for limits in the use of technology,[2] he did not identify himself as a "conservative" or label the position he criticized as "liberal." Today, the conservative bioethicists could consider Callahan one of them, as he has opposed the use of embryos in research, is critical of developing life-extending technologies, and has questioned the reigning paradigm of autonomy in bioethics. Yet Callahan's views on social justice in health care and the breadth of his writings in the field make it impossible to shoehorn him into a neat category, notwithstanding his views on embryo research and life-extending technologies. Indeed, liberals could also count Callahan as one of their own since he is in favor of universal health care, has been pro-choice on abortion since before *Roe vs. Wade*, and has been an outspoken critic of the pharmaceutical industry.

There have always been contributors to the bioethics literature whose views could uncontroversially be characterized as conservative. Probably the leading example is Leon Kass, today the chief public spokesperson for conservative bioethicists in the United States and, until recently, a political appointee of George W. Bush to head the President's Council on Bioethics (PCB). Kass has been a consistent opponent of medical technology that diverges from a "more natural science" (part of the title of his 1988 book).[3] He was an opponent of in vitro fertilization years before the birth of the first infant by that technique,[4] and he has continued to criticize the use of IVF long after that event. By any measure, Kass . . . has demonstrated his unswerving commitment to that position in numerous books and articles up to the present.

Aside from topics related to human reproduction (abortion, research on embryos, assisted reproduction, fetal tissue research) and certain genetic manipulations, few controversial issues in mainstream bioethics have given rise to characterizations of opponents in those debates as "liberals" and "conservatives." Unless the term "liberal" is used to mean something rather different from its meaning in the political context, it may be difficult to determine which advocates are "liberals" and which "conservatives." If by "liberals" one means "those who seek to liberalize" existing laws or policies, or to keep existing laws and policies from being made more restrictive and thereby constrain individual choices, then one can examine the writings of bioethicists and see where they stand on prominent policy issues. Beyond that, the term has a much less unambiguous application.

But the new conservatives in bioethics have identified themselves and promoted their agenda, so they leave no doubt as to who they are. They are opponents of biotechnology and its use in all sorts of interventions they term "artificial": artificial reproduction,[5] artificial life extension,[6] artificial intelligence and artificial life,[7] and, in general, making ourselves "artificially better." They are opponents of reproductive freedom, including the right to abortion,[8] new reproductive technologies and practices,[9] stem cell research that involves destruction of human embryos,[10] and biomedical efforts of any sort to enhance physical or mental capabilities.[11]

While the new conservatives themselves may harbor no doubts about their identity, at least some of their stated positions do not really distinguish them from others that appeared much earlier, promoted by groups that were anything but conservative. Interestingly and perhaps paradoxically, contemporary conservatives in bioethics were preceded in their opposition to assisted reproduction and even some forms of birth control by a group that can hardly be considered conservative or even liberal—the radical feminists. The organization known as the Feminist International Network of Resistance to Reproductive and Genetic Engineering (FINRRAGE) has long opposed the use of assisted reproductive technologies, hormonal methods of contraception, the medical abortifacient RU-486 and research on immunological contraceptives. Does this make the radical feminists "conservatives"?

Writing in the *Hastings Center Report* in 1971, Robert Veatch said, "we are trying to outgrow the simplistic ethical notion that if something has been artificially processed it is intrinsically evil." Veatch presciently warned against "pitting *naturo-centrism vs. techno-centrism*" because it "escalates the battle to one of two opposing world views."[12] Ironically, perhaps, the main critics of the technology to which Veatch was referring were members of that generation's counterculture—here again, not a group one would characterize as "conservative." Indeed, they were the "radicals" of yesteryear. Like FINRRAGE, the counterculture of the 1960s that sought more "natural" approaches had common cause with today's conservative critics of biotechnology. Perhaps the new conservatives would not refer to themselves as "naturo-centrics," of course, even though Kass titled one of his books *Toward a More Natural Science* and recently pointed to "the danger of violating or deforming the deep structure of *natural* human activity" (emphasis mine).[13]

If the examples cited here are not sufficient to conclude that the labels "conservative" and "liberal" are confusing, if not altogether confused, we have only to look at M. L. Tina Stevens's book attacking bioethics.[14] Stevens argues that, despite its early history of questioning and even criticizing advances in medical technology, bioethics soon became a comfortable ally of the medical establishment and the developers of biomedical technology. Those who support the medical establishment, the pharmaceutical industry, and the business of biotechnology are, according to Stevens, conservatives.

Stevens writes, "the movement from the sixties to the seventies was a shift from critique to management."[15] In a chapter devoted to The Hastings Center from 1969 to the present, Stevens says that "the conservative position into which the center fell was not foreseen at its inception."[16] But just what is that "conservative position"? According to Stevens, it is "absorption into existing medical and scientific institutional frameworks." It is avoidance of an "intense, sweeping examination of political values undergirding medicine and science."[17] In what Stevens depicts as a march toward conservatism, she describes a meeting of The Hastings Center's board of directors in 1972 in which the members "reached an informal conclusion: the institute should make an effort to elect more members with a pro-technology bias."[18] For Stevens, this marked an early step in a shift from skepticism about technology to an embrace of the medical and scientific establishment—a "conservative" stance.

Herein lies the confusion of the labels. The very position Stevens calls conservative is termed "liberal" by the recently emerged conservative movement in bioethics. The conservatives in this latter group are part of the broader neo-conservative political movement.[19] In contrast to the academic field of bioethics, now almost four decades old, a significant number of those who are aligned with the conservative views of the new bioethics movement are journalists (such as *Washington Post* columnist Charles Krauthammer), editors (such as William Kristol of *The Weekly Standard* and Eric Cohen of *The New Atlantis*), or contributors to publications such as *Commentary*, *First Things* (a conservative religious publication), and *The Public Interest*. Many are members of politically conservative think tanks such as the American Enterprise Institute, American Heritage Foundation, and the Ethics and Public Policy Center. In an article discussing the members and staff of the President's Council on Bioethics, R. Alta Charo observes that "Overall, there is a sense of an interlocking network of conservative and neo-conservative figures, appearing together at fora, writing for and editing the same journals, and working with one another as members or staff within a small circle that orbits the chair, Leon Kass."[20] . . .

Whether mainstream bioethics has become "conservative," as Stevens charges, or has always been predominantly "liberal," as the neoconservatives would have it, those who characterize bioethicists as unquestioning allies of medical science and technology make a sweeping generalization that does not stand up to scrutiny. An example is the following passage in an article by Gilbert Meilaender, a member of the PCB, contributor to the conservative publications *The Weekly Standard* and *The New Atlantis* and a member of the editorial advisory board of *First Things*:

> At a time in the early 1980s—just as bioethics was emerging as a profession recognized by universities, hospitals, and even the courts—the scientific research community was worried about the establishment of regulatory commissions, which might have placed limits on developing possibilities for human genetic engineering. And so the researchers joined hands with the newly emerging profession of bioethics in order to prevent a wide-ranging public examination of where biotechnology might be taking us. In order to accomplish this—in order to marginalize participation by those eager to discuss larger questions, and in order to keep elected officials from concerning themselves with these questions—scientists were willing to give an advisory, though clearly secondary, role to some bioethicists. And those bioethicists, in exchange for their place at the table (their place in clinical and research centers, on institutional review boards, and on national bioethics commissions) were on the whole happy to support the cause of advancing science.[21]

Charo questions the accuracy of this characterization of the field, pointing out that since the 1970s, bioethics has promoted a comprehensive regulatory scheme for research with human subjects, urged greater regulation of the practice of organ donation, and contributed to the development of other model legislation.[22]

Aside from evidence showing that individual bioethicists and bioethics commissions actually sought to place limits on existing practices in research and medicine, it is a mistake to think that promoting scientific advances in medicine is somehow incompatible with reining in unacceptable practices or putting brakes on the so-called "technological imperative." Few, if any, bioethicists have ever argued that all attempts to advance medical science and technology should proceed apace, without scrutiny, without regulation, and without oversight. Meilaender's comment impugns the motives of bioethicists, implying that their main aim has been to secure a "place at the table." This ignores the positive contributions of those individuals and commissions, especially when their efforts have led to legislation or regulation that provides greater protection of the rights and welfare of individuals.

The Mission of Conservative Bioethics

In contrast to the large majority of scholars who have contributed to the bioethics literature and served on public policy bodies over the past thirty-odd years, the new conservatives have an explicit "mission." In 2003, in the first issue of *The New Atlantis*, Yuval Levin, one of the senior editors who also became acting executive director of the President's Council on Bioethics, identified the "mission" of conservative bioethics: "to prevent our transformation into a culture without awe filled with people without souls."[23] What could be the literal meaning of this mission statement? The poetic nature of Levin's description of the aim of conservative bioethics is typical of the style of writing found in many articles in the journal, as discussed further below. The implication of Levin's characterization of the conservatives' mission is that without their effort, our culture is somehow on the brink of being transformed into something terrible, and we, its citizens, will change from being Cartesian creatures into . . . what? Robotic entities? The literal meaning of Levin's warning is far from clear.

Somewhat less mystical is Levin's claim that "the present task of a conservative bioethics . . . must be to develop and articulate a coherent world view—to put meat on the bones of loosely defined terms like 'human dignity.' "[24] So far, appeals to dignity continue to abound in the writings of contributors to *The New Atlantis* and members of the PCB without any attempt to define or analyze the concept. Although one can applaud the intention to clarify a notion whose meaning has remained vague and largely unanalyzed, attempts to do what Levin promises are so far absent in the writings of the new conservatives in bioethics. In contrast, several scholars in the mainstream tradition in bioethics have recently contributed thoughtful essays and clear-minded analyses of "dignity" in a journal issue devoted entirely to the topic.[25]

Where do these conservatives think mainstream bioethics has failed? One charge is that "it has too often disregarded the moral and religious views that shape the outlook of many citizens. . . . And it has too often abandoned the deeper questions of human dignity and human nature."[26] In addition, as Charo notes, "conservatives seem to feel that liberal bioethics has been an obstacle to regulating medicine and scientific research. For some, regulation to guard against concrete harms to identifiable persons represents an impoverished, consequentialist approach to policy-making. Instead, they yearn for a more activist government that regulates or bans whole technologies merely to ensure the endurance of current social structures."[27]

Although contributors to the large body of literature in bioethics often write with conviction, sometimes even with passion, they do not claim to be speaking immutable truths. In contrast, some conservatives seem to think they have a handle on the "one, true bioethics," akin to religious believers who assert that theirs is the "one, true religion." Meilaender strives for "a truly human bioethics." He claims that "a truly human bioethics will recognize not only the creative but also the destructive possibilities in the exercise of our freedom."[28] What distinguishes a "truly human bioethics" from the academic field that emerged in the mid-to-late 1960s and is still thriving today? How would Meilaender characterize what I have chosen to call "mainstream bioethics": Nonhuman? Inhuman? Or simply, not *truly* human?

Eric Cohen speaks of "bioethics, rightly understood."[29] Cohen refers also to "a true bioethics"—one that "wrestles truthfully with the meaning of our biological humanity."[30] What is the contrast with "a true bioethics"—a false bioethics? An activity that is not bioethics at all? In a sweeping indictment of the field, Cohen charges bioethics with being "largely adrift in recent years," with a focus on autonomy and social justice.[31] It is true that there has been a focus on autonomy—perhaps too much focus—in the bioethics literature, a point noted by many in mainstream bioethics over the years and by thoughtful critics such as Renee Fox. However, the emphasis on autonomy has rarely been one of "maximizing personal choice," as Cohen contends, but rather, a reaction against the several-thousand-year-old tradition of paternalism in the medical profession, more recently supplanted by the authority of managed care organizations making decisions that take away the autonomy even of doctors.

As for distributing medical resources more equitably (social justice), what could possibly be wrong with that? Cohen does acknowledge that autonomy and social justice "are both important concerns, but they are not the only concerns, and they do not finally help us understand how biotechnology will change how we live, what we value, and who we are."[32] There is no further defense of social justice in Cohen's articles or, for that matter, in any of the others in the journal he edits. One can only surmise that like their neoconservative counterparts in the political arena, conservatives in bioethics have little interest in justice in the realm of health and disease, viewing it as a concern of old-fashioned liberals and neither an obligation of government nor a main focus of a "true bioethics." Although it is true that with a few notable exceptions,[33] mainstream bioethicists have not, until recently, paid a great deal of attention to justice in health care, health policy has always been a topic of concern.

Do liberals have a "mission" in bioethics? That is an impossible question to answer, since there are no spokespersons nor is there any group in mainstream bioethics that has applied the term "liberal" to itself. Surely, the so-called liberals (conservatives never identify them by name and never cite specific writings of mainstream bioethicists) would not characterize their mission as *promoting* "our transformation into a culture without awe filled with people without souls." Nor would they uniformly and collectively proclaim that biotechnology should march ever forward, without limits or governmental regulation. Bioethicists who have contributed to the literature in the past thirty-five years do not have a single approach, perspective, or ideology in the broad array of topics that they address. In fact, many have been critical of—or at least skeptical about—scientific advances and technologies, evincing many of the same concerns as these recent conservative critics.

How the Conservatives Address Topics in Bioethics

Several characteristics of the writings of conservatives exhibit sharp differences from those found in books and articles by mainstream bioethicists. These are: metaphors and slogans as substitutes for empirical evidence and reasoned arguments; patently offensive analogies; deliberately misleading terminology; and an almost total absence of quotations from and citations to the people they are criticizing. I discuss each of these in turn.

The Use of Poetic and Metaphoric Language

I have already mentioned Yuval Levin's statement of the mission of conservative bioethics—"to prevent our transformation into a culture without awe filled with people without souls." A more pervasive example is the portrayal of "children as a gift." Conservative writers use this phrase in various criticisms leveled against the prospect of human reproductive cloning and the possibility of genetic manipulations of future offspring. Stated by a theologian, the phrase might even be intended literally (not metaphorically) to mean that children are a gift of God. In that context, one could reasonably expect an acknowledgment by the speaker of the intended meaning. Without that acknowledgment or further elucidation, however, the phrase remains metaphorical.

At a conference on genetic selection and enhancement in which Leon Kass made a presentation, he repeatedly stated—without accompanying arguments— that it is wrong for parents to be able to choose the traits of their children. In response to a question that followed his talk, asking why he so adamantly rejects the role parents might have in choosing characteristics of their future children, Kass replied: "Children are a gift." Kass's answer may have been understood and accepted by members of a homogenous, religious audience, especially if he had made reference to the giver of the gift (is it even meaningful to speak of a "giverless gift"?). But as a response to a question at a secular conference, and without further explication, Kass's reply was a conversation stopper, not a stimulant to dialogue.

When theologians or scholars in religious studies write on topics in bioethics, their commitment to a religious perspective is open, acknowledged, and understood. Meilaender, for one, also speaks of "the child as a gift."[34] Meilaender is a professor of theological ethics, and so readers understand such references as coming from the beliefs and tradition of religious scholars. Meilaender and others who explicitly derive their moral beliefs from a religious tradition are justified in their use of terminology, including metaphors, that are accepted by their coreligionists. But in the context of a general discussion of the relation between generations, Meilaender does not specify that children are a gift of God. Instead, he draws on a poem in which "we are given a captivating image of the child as gift." That is a use of metaphor, not an intended literal truth of theology.

Appeals to Emotion, Sentiment, and Intuition

Another rhetorical device differs from the use of metaphor and other poetic language, but is similarly designed to appeal to readers' emotions or intuitions. Kass often uses this kind of device in his essays. In a well-known article, his response to the prospect of reproductive cloning was "the wisdom of repugnance": the emotion of revulsion (his own) as sufficient grounds for rejecting any public policy short of total prohibition.[35] That is not an argument. As Yuval Levin correctly observes, an "important reason not to rely on moral intuitions is that they may simply be wrong or unjust." Levin notes that interracial marriage "turned the stomachs of many in white America . . . but that gut reaction could not stand up to scrutiny, and should not have been allowed to determine government policy."[36] Well, how can we know which emotions or intuitions cannot stand up to scrutiny and which can?

Yet despite his apparent recognition of the limits of moral intuition, Levin is equally suspicious of the use of rational argument. He claims that "the democratic transformation of sentiments into arguments creates a deep and serious problem."[37] What is that problem? It is the same concern that Kass expresses in defending the wisdom of repugnance. Levin writes:

> By transforming a deep moral sentiment into an argument, we abandon and likely lose forever its power as a sentiment. In appealing to a clear and explicit rational argument, we begin to overcome our deep repugnance and may diminish it in others. We create an argument that rests, as arguments do, upon premises and postulates, rather than a deep taboo resting on some profound common moral foundation that animates us forcefully but that we cannot adequately put into words.[38]

The trouble is, of course, that in pluralistic societies, people do not all share the same repugnance or the same moral foundations. Racists and anti-Semites retain their repugnance regarding interracial or interreligious marriage. Many Christians find abortion repugnant and destroying embryos equally so, whereas a large number of other Christians, many Jews, agnostics, and atheists do not share either that repugnance or the religious moral foundation that prohibits those practices. In the end, Levin accepts even the epistemological method of intuition of which he appeared to be skeptical: "The very fact that everything must be laid out in the open in the democratic age is destructive of the reverence that gives moral intuition its authority."[39] The alternative to rational debate and argumentation is to rely on intuition. But whose intuition, especially when it comes to making public policy?

There is another problem with a rejection of clear and explicit rational arguments. The new conservatives are seeking changes in public policy, and they believe bioethics should be engaged in that. But in a democratic society, how can public policy be established or altered without reasoned debate? No one in any branch of government will be persuaded of any proposed policy by opponents who cite their profound moral sentiments or deep repugnance or heartfelt intuitions. Legislators debate the issues before them, and members of the judiciary have to determine the merits of arguments that lawyers put forth.

Levin poses a false dichotomy. There is no inconsistency in holding deep moral sentiments and formulating rational arguments in their defense. Indeed, rational arguments can often be mounted in a "passionate" defense of a position—a description that is not self-contradictory.[40] Mary Midgley argues for the compatibility of attention to both emotions and reason. She writes: "Feeling is an essential part of our moral life, though of course not the whole of it," and "we must spell out the message of our emotions and see what they are trying to tell us."[41] The difference between Midgley's approach and that of conservatives writing on this topic is that Midgley provides an ethical analysis of the differences between a consequentialist mode of reasoning and one based on intrinsic objections, while conservatives rely on readers' readiness to accept their position without a supporting analysis.

Because of his eloquence and his wide acquaintance with literary and philosophical writings, Kass could hardly be called "anti-intellectual." Yet in his essays he comes across as opposed to reason and rational arguments, embracing instead the same intuitionist epistemology that Levin endorses. (It is probably more accurate to say the Levin endorses the same epistemology as Kass.) The following passage is illustrative:

> What if anything can we say to justify our disquiet over the individual uses of performance-enhancing genetic engineering or mood-brightening drugs? . . . It is difficult to put this disquiet into words. We are in an area where initial repugnances are hard to translate into sound moral arguments . . . But is there wisdom in this repugnance? . . . it is going to be hard to say what is wrong with any biotechnological intervention that could give us (more) ageless bodies or happier souls. If there is a case to be made against these activities—for individuals—we sense that it may have something to do with what is natural, or what is humanly dignified, or with the attitude that is properly respectful of what is naturally and dignifiedly human.[42]

Kass's appeals to what is "natural," what is "dignified," and what is distinctively "human" pervade his writings without any apparent need to analyze such terms or to justify their use. It is a style of writing alien to that found in mainstream bioethics. On that account it precludes the type of reasoned response characteristic of debates among bioethicists who take sides on issues such as physician-assisted suicide, monetary payments to research subjects or to egg donors, the limits of medical confidentiality, the risks and benefits of germ-line alteration, and the myriad other controversies that populate the field.

Mean-Spirited Rhetoric

A feature of the writings of some conservatives is their use of emotionally charged, high-flown rhetoric. Granted, rhetorical devices can enhance rational arguments and have a respectable place in reasoned discourse. But when the purpose is to attack opponents by mislabeling their positions or using analogies likely to be highly offensive to some people, rhetoric becomes mean-spirited. One prominent conservative has used the term "eugenics" in a way that clearly distorts its original meaning (whether the distortion is deliberate or done out of ignorance is not evident). In a piece written in 1988, before the recent consolidation of a conservative movement in bioethics, Richard John Neuhaus asserted that "the question of euthanasia is . . . an integral part of the progress of the eugenics project." In this article, Neuhaus broadened the concept of eugenics to mean "new ways of using and terminating undesired human life."[43] This misconstrues the meaning of "eugenics," which has nothing to do with euthanasia and which bears a highly negative connotation from historical practices in the United States as well as in Nazi Germany. Proponents of euthanasia or physician-assisted suicide (PAS) seek to end the suffering of terminally ill patients, not to alter the human gene pool. The idea that potential candidates for euthanasia or PAS are likely or even able to procreate and thereby "contaminate" the gene pool is absurd. Moreover, Neuhaus's statement implies that people who favor euthanasia are engaged in some sort of "eugenics project"; but there is no such "project" among the bioethicists who have argued for assisted suicide or voluntary euthanasia.

Neuhaus's use of the term "eugenics" in this context is either a deliberate distortion or else exhibits woeful ignorance of concepts commonly discussed in bioethics. But there is something worse than this misleading linkage between euthanasia and eugenics, two concepts that already bring to mind abhorrent practices during the Third Reich. It is Neuhaus's explicit use of an analogy with the Nazis. Neuhaus likened the practice of legal abortion to the killing of Jews, gypsies, homosexuals, and Slavs by the Nazis. He wrote: "If one believes that 20 million abortions are equivalent to 20 million instances of the taking of innocent human life, does not the analogy with the Holocaust become more appropriate? Perhaps even inevitable?"[44] It is hard to think of an analogy more likely to offend and even outrage large numbers of people.

In an article written at about the same time, Charles Krauthammer took aim at techniques of assisted reproduction, characterizing as "manufacture" those techniques that involve "artificial babies," "artificial families," and "artificial sex." This use of language replaces reasoned argument with labels that bear a distinctly negative connotation: "artificial" is definitely bad when it comes to babies, families, and sex.[45] However, the babies are real, not artificial. The families (parents and children) that result from assisted reproduction are real, not artificial. And while there is *no* sex—real or artificial—when the techniques of assisted reproduction are used, the couple who contributed their gametes can nevertheless enjoy real sex, though not for the purpose of procreation.

Although contributors to mainstream bioethics may have come across the articles by Neuhaus and Krauthammer, they were unlikely to pay very much attention to them as harbingers of a new conservative movement in bioethics, probably because the articles appeared in magazines (*Commentary* and *The New Republic*) rather than in scholarly journals. At the time he wrote his article, Neuhaus was a Lutheran theologian. In 1990 he converted to Catholicism, and today he is editor-in-chief of *First Things*, which describes itself as "the Journal of Religion, Culture, and Public Life." Krauthammer writes a syndicated column for *The Washington Post*, contributes to the conservative publication *The Weekly Standard*, and is a member of the President's Council on Bioethics. If there is now a mainstream *conservative* bioethics, Neuhaus and Krauthammer were forerunners and are still engaged, despite the fact that both write more frequently on other topics.

Proliferation of "Projects"

Virtually all the conservative bioethicists speak of "projects" of various sorts, and in contexts in which there appears to be nothing that fits the commonly held notion of a "project." According to its usual meaning, a project is a concerted effort, by one or more individuals, to bring about an intended result. When we speak of a group as engaged in a project of some sort, the implication is that there is a common, agreed upon goal that members of the group, working together, are striving to achieve. If any groups writing or speaking on topics in bioethics have what can legitimately be termed a "project," it is these conservatives themselves.

What are these projects the conservatives claim to identify? There is Neuhaus's reference to "the eugenics project," noted above. Then there are the "biotechnology project" and the "American project." Eric Cohen writes that "the biotechnology project cuts to the very foundation of the American project."[46] There may well be lots of activities in biotechnology among profit-making companies and scientists working at the NIH and in academia. But these various efforts hardly deserve to be identified as a single "project," a vision and coherent plan of one or more individuals or groups.

The "American project"? According to Cohen, this has something to do with "our idea of equality,"[47] the idea that we should strive to promote and ensure the equality of all members of society. "The Declaration of Independence says that 'all men are created equal,' but in some unfortunate ways this is clearly not so."[48] One way in which it is "clearly not so" is the unfortunate circumstance of children born with serious diseases and disabilities. In numerous writings, conservative bioethicists denounce proposals or efforts to use the knowledge and techniques of biotechnology to enhance "normal" human traits or attributes. But Cohen takes a much more extreme position, rejecting the aspiration "to make ourselves biologically more equal" by conducting research on stem cells that may someday help these children with grave diseases approach a physiological norm. Cohen refers to this as the "guiding sentiment of liberal humanitarians, both scientists and politicians, [who] want children born with grave diseases to live full lives—like everybody else."[49]

Is it a mere "sentiment" that leads scientists, politicians, bioethicists, and many citizens in the United States, the United Kingdom, and other countries to advocate a line of research that may relieve suffering for many sick and disabled people, children, and adults? It is easier to understand opposition to stem cell research on the grounds that it involves the destruction of human embryos than it is to comprehend Cohen's apparent opposition to helping suffering children lead more normal lives by making them "biologically more equal." Why is it acceptable (or is it?) to alter the physical environment to benefit individuals with disabilities (such as public accommodations for wheelchairs) but not their biological attributes? Does Cohen reject a conception of justice that would seek to reduce disparities between the less fortunate who have congenital or acquired disabilities and people without such physical or mental impairments? . . .

In the writings of bioethicists in the mainstream bioethics literature, one rarely finds such sweeping generalizations without citations or attribution. Nor do mainstream bioethicists employ the kinds of rhetorical devices characteristic of the writings of conservatives cited in this article. Contributors to mainstream bioethics most certainly do not all agree with one another. If asked, some might consider themselves liberals, but others conservatives; and most would choose these labels with specific topics in bioethics in mind. What characterizes their approach to any topic is analysis of the evidence and arguments of their opponents, an approach that lends itself to reasoned rebuttal.

Given the contrast between the modes of argument and methodology of mainstream bioethics and the style the new conservatives have adopted, it might be tempting to say that the latter are not bioethicists at all. They are something else—social critics, perhaps—who rely on dramatic impact and rhetorical persuasion rather than rational argument to convince their readers.[50] But to exclude these conservative writers from the ranks of bioethicists would be a mistake. It would play into the rules of the game they have devised, stipulating what is "bioethics, rightly understood" or "a true bioethics."[51] It is possible to be a bioethicist and also a social critic, a characterization that could apply as well to some prominent contributors to mainstream bioethics.[52]

Bioethics and Justice

So-called liberals in bioethics have been paying increasing attention to matters of justice in access to health care and the gap in health status between rich and poor, as well as broader issues of global justice in medical research and health disparities between industrialized and developing countries. These concerns appear to be entirely absent from the conservatives' "project," to borrow their term. It is possible even to detect a hostility toward justice, as evidenced by Eric Cohen's criticism of "liberal humanitarians" who promote stem cell research in the hope that it can help children born with grave diseases: "They want justice where fate or genes or both has denied it."[53]

In railing against biotechnology, conservatives ignore what many take to be much larger ethical concerns, both domestic and international. Biotechnology holds out the only promise for an effective preventive vaccine for HIV/AIDS, which in turn is the only hope for gaining control of the pandemic; but researchers are a long way from achieving that goal. When avian influenza comes to our shores—and it surely will—are the conservatives in bioethics going to refuse to be inoculated because the vaccine is manufactured by a biotechnology company? Or because vaccines are a form of "enhancement" of normal immunologic functioning, and therefore an unacceptable—because "unnatural"—use of biotechnology?

Opponents in such ongoing debates as abortion rights, therapeutic cloning, stem cell research, and biological attempts to enhance human traits will almost certainly continue to disagree in their substantive positions. It is not clear, however, whether a dialogue between mainstream bioethicists and the new conservatives is possible. As long as the methodology and style of writing of these two camps remain so different, it is not likely that the issues can be joined. If the conservatives reject an appeal "to a clear and explicit rational argument," then there is no way that allegedly liberal bioethicists can respond to what they write and say. If the conservatives can describe their mission only in terms that refer to preventing "a culture without awe" and "people without souls," then mainstream bioethicists seeking to examine the conservatives' mission will be left in the dark.

As a liberal, humanitarian bioethicist, I acknowledge that my chief concerns lie in striving for greater social justice within and among societies, and reducing disparities in health, wealth, and other resources among populations in the world. Unless the conservative bioethicists begin to address those topics, I for one will not find common cause with their main worries about where we are headed.

Acknowledgment

I thank Daniel Callahan for his helpful suggestions, and also two anonymous reviewers for their comments on earlier versions of this article. I am especially grateful to the reviewer whose suggestions for specific wording I have adopted in several places.

References

1. I am using this term to refer to the field whose more prominent members are nationally and internationally recognized as contributors to the literature, or as educators and consultants. These include physicians, other health professionals, biological scientists, philosophers, lawyers, social scientists, theologians, scholars in religious studies, political scientists, health economists, and

any others who may not fall into these categories. In using the adjective "main-stream" I do not intend to marginalize any individuals or groups.

2. D. Callahan, "Science: Limits and Prohibitions," *Hastings Center Report* 3, no. 5 (1973): 5–7.

3. L. Kass, *Toward a More Natural Science: Biology and Human Affairs* (New York: The Free Press, 1985).

4. L. Kass, "Making Babies—the New Biology and the 'Old' Morality," *The Public Interest*, no. 26 (Winter 1972). Quoted in L. Kass, " 'Making Babies' Revisited," *The Public Interest*, no. 54 (Winter 1979): 32–60.

5. C. Krauthammer, "The Ethics of Human Manufacture," *The New Republic*, May 4, 1987, 17–21. Krauthammer is a member of the President's Council on Bioethics.

6. L. R. Kass, "Ageless Bodies, Happy Souls: Biotechnology and the Pursuit of Perfection," *The New Atlantis*, no. 1 (Spring 2003): 9–28.

7. C. T. Rubin, "Artificial Intelligence and Human Nature," *The New Atlantis*, no. 1 (Spring 2003): 88–100.

8. G. Meilaender, "Abortion: The Right to an Argument," *Hastings Center Report* 19, no. 6 (1989): 13–16; R. J. Neuhaus, "The Way They Were, The Way We Are," in A. L. Caplan, ed., *Bioethics and the Holocaust* (Totowa, N.J.: Humana Press, 1992): 211–30.

9. C. Krauthammer, "The Ethics of Human Manufacture."

10. E. Cohen and W. Kristol, "The Politics of Bioethics: Playing Defense Is Not Enough," *The Weekly Standard* 9, no. 33 (May 10, 2004). Although Cohen and Kristol do not state clearly and unequivocally their opposition to the destruction of human embryos in stem cell research, their use of the terms "innocent human life" and "nascent human life" place them presumptively in the pro-life camp on the morality of destroying embryos.

11. President's Council on Bioethics, *Beyond Therapy: Biotechnology and the Pursuit of Happiness* (Washington, D.C., 2003); L. Kass, "Ageless Bodies, Happy Souls," 22.

12. R. Veatch, "Doing What Comes Naturally: The Dangers of Anti-Technological Romanticism," *Hastings Center Report* 1, no. 2 (1971): 1–2.

13. Kass, "Ageless Bodies, Happy Souls," 22.

14. M. L. T. Stevens, *Bioethics in America: Origins and Cultural Politics* (Baltimore, Md.: Johns Hopkins University Press, 2000).

15. Ibid., xiii.

16. Ibid., 47.

17. Ibid.

18. Ibid, 61.

19. "Neoconservative" is a term that has proven difficult to define. As one writer notes: "Although neo-cons profess devotion to liberal democracy, they have never hesitated to assail 'liberalism', or what they sometimes call with their Christian Right allies 'secular humanism', whose relativism, in their view, can lead to 'a culture of appeasement', nihilism or worse. So, even while suppos-edly defending 'liberal' and democratic ideals, their attitude is at best ambiva-

lent." J. Lobe, "What is a Neo-Conservative, Anyway?" Inter Press Service News Agency, http://www.ipsnews.net/interna. asp?idnews=19618, accessed July 20, 2004. See also R. Alta Charo, "Passing on the Right: Conservative Bioethics is Closer Than it Appears," *Journal of Law, Medicine & Ethics* 32 (2004); 307–314.

20. Charo, "Passing on the Right," at 310.
21. G. Meilaender, "The Politics of Bioethics: In Defense of the Kass Council," The Weekly Standard (April 12, 2004–April 19, 2004). In the quoted passage, Meilaender is summarizing an account by the sociologist John H. Evans without representing it as his own view. Yet Meilaender appears to accept the account uncritically as he does not question Evans's characterization of what bioethicists were up to.
22. Charo, "Passing on the Right," at 310.
23. Y. Levin, "The Paradox of Conservative Bioethics," *The New Atlantis*, no. 1 (Spring 2003): 53–65, at 65.
24. Ibid., at 64.
25. *Journal of Palliative Care* 20, no. 3 (2004).
26. E. Cohen, "The New Politics of Technology," *The New Atlantis*, no. 1 (Spring 2003): 3–8, at 5.
27. Charo, "Passing on the Right," at 310.
28. G. Meilaender, "Bioethics and the Character of Human Life," *The New Atlantis*, no. 1, (Spring 2003): 67–78, at 72.
29. E. Cohen, "Bioethics in Wartime," *The New Atlantis*, no. 3 (Fall 2003): 23–33, at 25.
30. Ibid., at 32.
31. Cohen, "The New Politics of Technology," p. 5.
32. Cohen, "The New Politics of Technology," at 5.
33. In addition to Callahan, other prominent contributors to the literature over the years have been Norman Daniels, Steven Miles, Daniel Wikler, Dan W. Brock, and Frances Kamm.
34. G. Meilaender, "Bioethics and the Character of Human Life," at 72–73.
35. L. Kass, "The Wisdom of Repugnance," *The New Republic*, June 2, 1997, 17–26.
36. Levin, "The Paradox of Conservative Bioethics," at 57.
37. Ibid.
38. Ibid., at 58.
39. Ibid., at 59.
40. An anonymous reviewer of an earlier version of this article characterized my criticism of Levin's and Kass's appeal to emotions and intuitions as an endorsement of "pure, olympian rationality in ethical reflection." That is another example of a false dichotomy.
41. M. Midgley, *Hastings Center Report* 30, no. 5 (2000): 7–15.
42. Kass, "Ageless Bodies, Happy Souls," at 17.

43. R. J. Neuhaus, "The Return of Eugenics," *Commentary* 85, no. 4 (1988), at 15.
44. Neuhaus, "The Way They Were, The Way We Are," at 222.
45. C. Krauthammer, "The Ethics of Human Manufacture," *The New Republic* (May 4, 1987): 17–21.
46. E. Cohen, "Bioethics in Wartime," at 29.
47. Ibid.
48. Ibid., at 30.
49. Ibid.
50. I owe this point to Jeffrey Blustein.
51. Cohen, "Bioethics in Wartime."
52. To name only a few: Peter Singer, George Annas, Ronald Bayer, Solomon Benatar, Udo Schüklenk, Susan Sherwin and other feminist bioethicists, among others.
53. Cohen, "Bioethics in Wartime," at 30.

Conservative Bioethics
and the Search for Wisdom

Eric Cohen

Eric Cohen is adjunct fellow of the Ethics and Public Policy Center. He served as a consultant
to the President's Council on Bioethics, 2001–2007. In this article, Cohen proposes that contemporary bioethics reflects a strongly liberal trend as it analyses its standard topics, that it
works almost exclusively with a broad notion of personal autonomy, and that it uncritically
supports technological progress. He sketches the vision of a conservative bioethics that would
foster a wisdom based on a conception of human dignity and normative critique of technological manipulation of nature.

. . . [D]espite the grave limitations and internal contradictions of the terms "conservative" and "liberal," they still offer us an introductory (if imperfect) understanding of distinct approaches to the major issue of bioethics, which is the focus
of what follows. Liberals tend to believe that many abortions are morally justified; that embryo research is morally good; that hastening death is sometimes
the best way to end suffering; and that the government has a moral responsibility to ensure that every citizen has access to quality health care. Conservatives
tend to believe that abortion is morally justified only when the life or physical health of the mother is in danger; that embryo research exploits some lives
to help others; that caregiving means never seeking death as an aim; and that
universalizing the role of government in health care has the potential to make
medicine worse for nearly everyone.

These political divides are often rooted in differing understandings of certain shared ideals—like human equality—and different understandings of fundamental human experiences—like birth and death. In this essay, I attempt to
explore the moral anthropology and governing philosophy that inform conservative bioethics, and perhaps to make the sharpest conservative-liberal divisions
seem more like disagreements among friends: that is, disagreements about the
meaning of principles we all hold dear, or about the most prudent way to advance
shared ideals in an imperfect world filled with imperfect people.

The People and the State

Let us begin with a crucial distinction between the "philosophy of the person" and the "philosophy of the state." The philosophy of the person deals with existential questions—"what shall we do and how shall we live?" as Tolstoy once put it—that all individuals face at various points in their lives.[1] I suspect every professor of bioethics has had students or colleagues come to their office seeking advice: the freshman deciding whether to have an abortion; the research scientist deciding whether to destroy human embryos for research; the young woman deciding whether to get tested for mutations that would elevate her risk of breast cancer; the doctor deciding whether to prescribe beta blockers to numb the sharp edges of a rape victim's memory; the two sons deciding whether to treat their mother's pneumonia in a case of advanced Alzheimer's.

These life decisions are always complex and often puzzling. Defending the individual's "right to choose" without government interference offers little guidance about which choices are better or worse. And saying that individuals should "rely upon their own values" ignores the responsibility of wise elders to help shape those values. The autonomous individual, after all, never starts from moral scratch. He evaluates moral alternatives that previous generations set before him, and stands as a moral alternative to the generations that follow. Bioethics as a vocation has a responsibility to offer normative guidance on normative choices, and to search for wisdom in those puzzling human situations where wisdom is most needed.[2] Of course, different cultures and traditions often have different values, and there is much to learn universally by understanding why particular groups live the way they do. But it is also possible—and sometimes necessary—to judge cultures from the outside, and to make arguments against deeply ingrained cultural practices (like the forced circumcision of women) that violate the dignity of all human beings.

Liberals sometimes assume that the conservative idea of human dignity is entirely biblically grounded, and thus unavailable to those who are not already religious. As a sociological matter, it is surely the case that most citizens who hold conservative views on bioethics are traditional Christians, Jews, and Muslims. But as a philosophical matter, the idea of the human person (or ethical animal) that informs conservative bioethics does not require any particular faith in any particular God, even if living in accordance with its ideals is often bolstered by faith. Personally, I hope God exists, but I am hardly certain. And while I have great reverence for the teachings of Judaism—my own religious tradition—there are cases when I believe that deferring to prevailing rabbinical opinion is morally wrong, including on the issue of embryonic stem cell research. But what is most unfortunate in bioethics today is that defining a position as "religious" is sometimes used as a tactic by nonreligious intellectuals to avoid confronting the rational arguments of people who happen to believe in God. It is a secular form of philosophical laziness, no less commendable than believing something simply "because Jesus says so."

While moral philosophy or moral anthropology is the essential grounding for thinking about the good life, it is hardly sufficient. We also need a serious "philosophy of the state" that grapples with how imperfect human beings live together in community and with the proper relationship among the many layers of authority (the federal government, state governments, private institutions, free individuals) where moral decisions are made and moral obligations are met. Not everything worth doing should b[e] done by the state; not everything immoral should be unlawful; and not everything that is legally permissible is ethically sound. The philosophical challenge is discerning when the state should remain neutral and uninvolved and when the state should set certain boundaries or promote certain goods—from health insurance for the poor to federal funding of embryo research to regulations on the use of psychotropic drugs in children. Taken together, the philosophy of the person and the philosophy of the state provide the ground for both private and public bioethics, beginning with an account of what makes us dignified animals, and moving to an account of what this dignity requires both existentially and politically . . .

Toward a Moral Anthropology

The conservative idea of the person might be explored in five parts: (1) an understanding of human beings as ethical animals; (2) an understanding of human equality from conception to natural death; (3) a sensibility about the meaning of mortality; (4) an account of the nature of marriage, family, and procreation; and (5) an understanding of the character of human experience and human flourishing. Without question, many of the ideas and insights that follow are not uniquely conservative; they seek to address what is universally human. Nor do these ideas have uniquely conservative origins, which would give our current political categories far more credit than they deserve. Indeed, it is my hope that much here could be embraced by those who see themselves as liberals, independents, or just non-ideological human beings seeking wisdom about the good life and good society.

The Human Animal

*Bio*ethics begins with biology and specifically human biology: What does it mean to have a biological *life*, and what does it mean to have a distinctly *human* biological life?[3] As biological beings, we are not alone in the world but animals among animals. We are conceived and born; we depend on food and water to live; we move about and feel pain; we perpetuate ourselves sexually; we are vulnerable and resilient; we live with death as a possibility at any moment and an inevitability in due time. This continuity with other biological animals partially defines who and what we are; it defines what it means to have a *life*.

But equally significant is our radical discontinuity with other animals—our distinctly *human* life. We are the only beings with complex language; the only beings who marry; the only beings with courts of law; the only beings who keep Kosher or observe the Sabbath; the only beings with theories of our own evolution. We are special animals, separated by our distinct powers of reason and by our moral aspirations and moral failures. The other animals live outside good and evil—we would never say that a bear that attacks a hiker in the woods is immoral. Human beings live within good and evil—the hiker who tortures a bear acts immorally.

Taken together, these two dimensions of our humanity define what it means to have a human life and to be a human person: We have a life the way all animals have life—as living organisms with mortal beginnings and mortal endings; as beings who are conceived, develop, ripen, and die. And we have a human life by being members of the human family and human species. To be sure, infants do not yet manifest all the characteristics that are distinctly human, adults with dementia have lost some of them, and the severely disabled may never manifest them at all.[4] But all members of the human family—all living human bodies— have a human life, and therefore deserve the respect that such membership commands. This is the egalitarian definition of human dignity—the dignity that all human beings possess regardless of size, age, wealth, stage of development, cognitive powers, or level of dependence upon others. This dignity is not merely *experiential*—contingent, for example, on the capacity to feel pain or pleasure. It is *ontological*, dependent merely on being here as a human being.

Equality at the Edges of Life

This egalitarian understanding of human dignity shapes how conservatives understand the many ethical issues "at the edges of life,"[5] such as embryo research, prenatal screening and abortion, and caregiving for persons with dementia. It shapes how conservatives understand the moral standing of human beings at the embryonic stage, when the neurological capacity and human form are unfolding in unison; and how they understand the moral standing of human beings at the geriatric stage, when the mind and body are winding down, often not in unison until the very moment of death.

The embryo question is obviously central to the public bioethics debate, and perhaps paradigmatic of the broader philosophical divides within bioethics about the worth of human life. To discern that embryos are equal, in the minimal sense of not being available for our use, runs against the grain of our moral feelings. Most people looking under a microscope could not tell the difference between a human embryo, a monkey embryo, and a clump of adult human cells. We feel no special emotional attachment to human embryos simply by virtue of sight; we experience no visceral repugnance if we do not know in advance that a human embryo is being destroyed, and even then our moral feelings may slumber.[6] But feelings are often poor guides to moral action, and to understand what we owe embryonic human life, we must engage in the hard work of ethical reason in light of the newly discovered facts of modern embryology.

From the moment of conception, a human embryo is a new human life in process.[7] The very first cellular divisions are purposeful and orderly, involving the unfolding of an inner-directed plan of development.[8] If we trace an individual human life backwards biologically—from the adult stage to the infant stage to the fetal stage to the embryonic stage—the ontological line that separates being from nonbeing is fertilization. Before fertilization, we have an egg and many sperm; we have many possibilities and no organism. After fertilization, we have an individual life unfolding. Every reader of this essay was once a zygote, but never a sperm or an egg, since the gametes that produced them could have supplied the genetic material for a never-existing sibling.

To be sure, there are other key moments in embryological development, moments that some people believe are more significant than fertilization for conferring human worth. For example, there is the moment when certain powers—like primordial neurological activity or the capacity to feel pain—first become manifest. There is the moment when the discernible human form first becomes visible. But these moments do not mark the arrival of a *new* person; they mark the arrival of certain attributes in an *existing* person. To believe that crossing these hurdles is the prerequisite for human dignity contradicts the first principle of modern democracy that all human individuals are created equal. It makes our equality conditional on the judgment of others rather than intrinsic to who we are. Moreover, if dignity requires possessing the very powers that embryos lack—like physical independence or the capacity for language—then many nonembryonic human beings will surely not pass muster. This includes those with advanced dementia whom we seek to help with embryonic stem cell research, those who need the caring hand of the strong in their final moments of weakness, those who lack permanently the very powers that embryos are in the process of developing.

Of course, we can never prove rationally that all human beings possess equal dignity or that human beings possess any dignity at all. Equality is an ideal we uphold, not an obvious fact about the human condition. But if we abandon this democratic ideal, then the moral foundations of all caregiving will be eroded. We will undermine, in principle, not only the dignity of embryos in the laboratory or disabled fetuses in the womb, but the dignity of the uninsured child or disabled parent—two classes of persons that modern liberals rightly seek to protect. And if we seek to live by the principle of equality for all, we must reason carefully about what that means in practice, and not succumb to moral feelings based on the small size of an embryo or the cognitive incapacities of an aging parent.

In the end, equality is never free, especially when it requires loving those who cannot love us back or accepting death if the therapeutic alternative is using some (weaker) lives as tools to extend our own. Equality often requires heroism—the heroic sacrifice of the caregiver and the heroic courage of the patient, who see the other as equal to themselves, even when the equality of the other is hardly obvious and the suffering of the self is very great.

The Meaning of Mortality

While equality is not an obvious dimension of being human, death and dying surely are. To be human is to be mortal, and living well with mortality is central to many of the dilemmas of bioethics.[9] This surely does not mean accepting death without a fight, and part of living well with mortality is using human ingenuity to resist and conquer discrete causes of death, especially those that rob the young of a full and flourishing life. Technology—especially medical technology—is one of the distinguishing marks of our humanity, and surely a great blessing for everyone fortunate enough to live in the modern age. But there is also a danger that the morally sound project of defeating particular causes of death will deform into the desire to conquer death itself by human will—with both the hubris and discontent that comes with embracing the illusion of man-made immortality. Already, a group of technologically sophisticated "transhumanists" speak of downloading the cognitive self into machines so they can outlive their bodies, and leaders in the biotechnology industry speak about conquering death through regenerative medicine.[10] At the same time, the revolt against mortality takes a different turn: the final assertion of autonomy through suicide, the only way to die a fully autonomous death.

Conservative bioethics rejects these two extremes—the quest for immortality and the embrace of suicide—and seeks a moral framework to live well with what Hans Jonas called the "burdens and blessings of mortality."[11] The burden is that death may strike at any moment, and that death often strikes with little rhyme or reason—killing the young, killing the virtuous before their time, killing the parent before he reconciles with his child. But mortality, rightly understood, can also lead us to live more urgently and to savor the sweet things of life that will not last forever, sweeter still because they are always for us brief moments in time.[12]

While death is a permanent dimension of being human, how we die changes at the hands of new technologies and new cultures. In his essay on the meaning of mortality, Jonas quotes the following passage from the Psalms: "Teach us to number our days, that we may get a heart of wisdom."[13] In the age of genetic testing, the instruction to "number our days" takes on new meaning, since these tests might allow us to number them with increasing precision.[14] In those situations like Huntington's, where the diagnosis is clear and there is no cure, genetic self-knowledge seems like both a blessing and a curse. It is a blessing because it might lead us to an uncommon wisdom about the preciousness of life, or move us to live without wasting time because we know every day how short time really is. And yet such foreknowledge must also seem like a curse: the permanent presence of looming death might make living seem worthless, with too many projects we can never finish and too many ambitions we can never fulfill. Our genetic death sentence may come to feel like a living death, with no escape except pharmacology or suicide.

In the end, conservatives embrace the "culture of life" as a limit on the willful negation of life entailed in abortion or euthanasia; and they accept the human reality of mortality, knowing that resisting death is not the highest human good—since accepting death is always preferable to betraying our neighbors, our family, or our nation. But the "good death" remains—and always will remain—a puzzling concept.[15] Death is never good in itself, even if it provides a welcome end to great physical suffering . . . And death will always be a burden we must accept if we are to live courageously and morally with life's many uncertainties.

The Nature of the Family

While death looms large within human life, we are not impotent in the face of our mortality, and we need not see the future as a period of oblivion after our own personal death. As Hans Jonas, Leon Kass, and Yuval Levin have all argued—each of them drawing heavily on Jewish sources of wisdom—children are one human answer to mortality; procreation is a way of believing in and securing a human future that is more reliable than life-extending and life-improving technology alone.[16] It is indeed ironic that many of the most technologically advanced regions of the world—especially Europe and Japan—have stopped having children at a rate sufficient to sustain themselves.[17] For all their sophistication, both scientific and cultural, they seem to lack a compelling answer to the most fundamental human question: Why have children at all?

Many people have children because they believe that they have something essential to pass down or something vital to preserve—like a particular culture, tradition, or family name. In this way, the future depends in part on reverence for the past, on the belief that what was given to us must be perpetuated by those who replace us, in the bris and in the baptism. This means seeing ourselves as more than free individuals pursuing our own happiness—children, after all, are a great limit on personal freedom—and seeing procreation as not only a choice but also a duty. It requires an acceptance of one's own limited but essential place in the nexus of generations, and a willingness to see oneself replaced by those who follow.

Children, of course, are typically born and reared in families, and the conservative idea of the family attempts to connect the sexual character of human procreation with the moral activity of raising the young. Children have always been the fruit of one biological father and one biological mother, connected to them as flesh of their flesh but independent of them with a biological identity of their own. This is also true of other sexual animals. But human sexuality is clearly different: We marry and divorce; we use birth control and make pornography; we pass laws against prostitution; we circumcise our young. In various ways, human beings seek to transcend the merely animalistic character of sex and embrace its deeply moral character.[18] And so far in human history, the family seems like the best institution to serve this moral purpose. It binds biological parents together in ties of fidelity to one another and to their children, and it grants husbands and wives the exclusive right to have children with and through one another.[19] (Adoption, of course, is the great exception, but the model for loving and raising an adoptive child is to love her and raise her *as if* she were a biological child. Biological love becomes the standard for forming an adoptive love that transcends the biological.) Within families—if not only within families—human beings learn what it means to keep a covenant: as spouses, as parents, and as children who eventually care for the parents who once cared for them, and who come to understand the sacrifices of their mothers and fathers as they rear young children of their own.

This idea of the family shapes the conservative approach to certain bioethical dilemmas, especially about the responsible uses of technological intervention in human reproduction. Because being a parent means accepting offspring unconditionally—certainly when the offspring themselves are morally innocent—conservatives reject the practice of aborting fetuses *because* they are genetically disabled. Of course, every parent hopes for a healthy child, and no one should belittle the misery that often accompanies being disabled or the sacrifice that is required to raise a disabled child. But selective abortion is a form of eugenics antithetical to the spirit of parenthood and the ideal of human equality, even if performed for supposedly compassionate reasons or in the name of pursuing equality by eliminating the unfit.

Conservatives also reject various technological possibilities—like human cloning, gamete engineering, and the creation of man-animal hybrids—that would exert novel parental control over the genetic makeup of new life; that would confound the relations between the generations by making our twins into our children; that would produce orphans by design by procuring gametes from aborted fetuses or embryonic stem cells; or that would blur the line between human and nonhuman procreation by seeking to produce humans with animal traits or animals with human traits (both unlikely possibilities).

To be sure, many conservatives disagree about the moral meaning of certain reproductive biotechnologies, such as in vitro fertilization (IVF). Some conservatives defend IVF as a proper use of human ingenuity, a proper answer to the pathos of infertility, and a moral way to procreate within marriage. Other conservatives oppose IVF because it separates the "unitive" and "procreative" purposes of human sexuality, because it turns the mysterious birth of new life into a technological project, because it paves the way for the age of human cloning and genetic engineering, and because it destroys thousands of embryos as "byproducts" and abandons thousands more as "spares."[20] But in the end, such differences over particular technologies exist within a shared conservative understanding of the dignity of marriage, family, and procreation—as activities that reveal the truth of being human, and as institutions central to a decent society that believes in the future. This sentiment was captured powerfully by Jonas decades ago: "Youth is our hope, the eternal promise of life's retaining its spontaneity. With their ever new beginning, with all their foolishness and fumbling, it is the young that ever renew and thus keep alive the sense of wonder, of relevance, of the unconditional, of ultimate commitment, which (let us be frank) goes to sleep in us as we grow older and tired. With young life pressing after us, we can grow old and, sated with days, resign ourselves to death—giving youth and therewith life a new chance."[21]

Human Experience and Human Flourishing

And this leaves one final set of questions: What is the meaning of human flourishing for the new life that follows us? What does it mean to live better and do better, to pursue happiness and be happy? One of the central concerns of contemporary bioethics is the so-called problem of "enhancement"—the worry that novel ways of biologically engineering our offspring or reengineering ourselves might be unjust, unwise, or unethical. For decades, there has been endless speculation about genetic engineering. But in reality, the prospect of picking and choosing the attributes of our children *de novo* is very unlikely.[22] The powers of biological enhancement seem limited by the complexity of human biology, and by the fact that aiming to improve one set of human attributes risks undermining the human whole that makes us function well in the first place. At most, we may discover certain genetic patterns that correlate with certain desirable traits—like high levels of intelligence, athletic ability, or musical talent—and use this knowledge to pick and choose the "best" embryos that nature herself has created. But the traits we most care about are usually the most biologically complex and thus the least prone to mastery.[23]

That said, it is also clear that we have developed significant new ways to alter the functioning of the human body and human psyche—such as interventions that radically expand the human capacity to develop physical strength or interventions that remake our emotional life by altering the levels of serotonin in the brain. These forms of human intervention—"enhancement" seems like the wrong term, since it is not clear that these interventions are genuine enhancements at all—have many advocates, many opponents, and many who worry about the challenge of using them wisely. At stake is the very meaning and nature of human excellence and human happiness—the meaning of what we do at our best, and the connection between our real experiences and our inner understanding of the world.

In its report *Beyond Therapy*, the President's Council on Bioethics explored the meaning of human excellence through a discussion of performance-enhancing drugs in sport.[24] The Olympic athlete on steroids will certainly run more quickly. He will perform better in every quantifiable category of measurement. But is he truly a better human *athlete*, who runs in a fittingly *human* way? Even if steroids were safe and legal, would the Olympian want to be seen shooting up in public before the race—showing the world his *dependence* on chemicals right before demonstrating his supposed *excellence* on the racetrack?[25] At what point does he become more like a thoroughbred passively bred for the race than a man who actively prepares for it—that is, not *super*human but *sub*human?

There is obviously a spectrum of physiological interventions—from eating a well-balanced diet to taking daily vitamins to using steroids to engaging in "gene doping." And even the most sophisticated techniques of enhancement still require the activity of the willing self in the form of training; no one becomes excellent just by popping a pill. Moreover, many average people might use performance-enhancing drugs just to compete with those whose bodies are naturally more predisposed to athleticism. Why not permit steroids (or some safe equivalent) to make competition more just by correcting the inequities of nature? Why are nature's endowments more praiseworthy than those of the biological engineer?

These are legitimate questions with no easy answers. Unlike embryo research or selective abortion—which involve the mistreatment of weak, disabled, or dependent forms of human life—the perils of "enhancement" are more ambiguous. Athletic bodies and musical minds are never simply the creation of those who possess them, but these cultivated bodies and cultivated minds must remain enough "our doing" if they are to be worthy of genuine admiration. Otherwise, we reduce every human activity to a form of mass production, making our greatest exemplars too common to revere and too similar to machines designed from scratch to work every time.

The possible interventions in the human psyche are even more ethically puzzling. In many cases, such interventions are medically necessary. They aim to restore the connection between lived experience and emotional effect, to correct chemical imbalances that lead to chronic misery and self-destructive behavior, or to give individuals the raw neurological ingredients necessary to feel happy in response to genuinely happy things. But these interventions in the human psyche can also sever our inner life from our lived experience—by making people feel happy for no good reason, by allowing people to live through miserable experiences without feeling miserable, or by replacing real experience with a neurological simulation (such as the bliss of being on Ecstasy rather than the feelings of real love).[26] Mind- and mood-altering drugs might make us apathetic in a world where apathy is hardly a fitting response.

In *Beyond Therapy*, the President's Council on Bioethics studied a class of drugs with the potential to numb the sharp edges of bad memories: from the horrible recollection of being raped to the death and destruction of war, from the terrible things that we would prevent if we could only turn back the clock to the reprehensible things we are about to do but desire not to remember too strongly or too well.[27] It is perhaps an irony of human life that many of the things most worth remembering are things we wish never happened at all. But wishes cannot change realities, and the moral dilemma we face is whether we possess a duty to remember painful events as they really were when we possess the biotechnical capacity to alter our perception of the past. Beta-blockers—or the more advanced memory-numbing drugs that might follow—do not erase bad memories altogether. But they potentially allow those who take them to remember falsely, if more comfortably, by making flat what is jagged and bland what is momentous.

When we remember the planes crashing into the World Trade Center towers, or see people flinging themselves to their deaths out of skyscraper windows, we should shudder. We should ache. We should hunger for justice, spurred by the enraging memory of being attacked. If we sought to ease our psychic pain with memory-anesthetizing drugs, we might make the shudder go away, and with it the insight that only bad memories, deeply felt, can truly provide. We might feel better, but we would not necessarily behave better in the future. We might still remember the past, but without the emotional power that provokes us not to repeat it. We might still mourn the dead, but without the heartache that our lost loved ones deserve.

Of course, we should not belittle the great difficulty of living well with bad memories or living with the psychic despair that is often due mostly or entirely to chemical imbalances in the brain. Some situations are so desperate—some people are so sick with the diseases of the brain—that psychotropic interventions are a blessing. But we must also ensure that we do not make a trouble-free life the moral aim of biotechnology. Life without troubles, after all, lacks the hatred of injustice, the honest regrets, and the misery of loss that are defining marks of our humanity. A well-balanced brain should not mean an untroubled soul— since there is much in the world to be troubled about.

Bioethics in the Public Square

The troubles of the world lead us from the realm of anthropology to the realm of politics. The controversies of bioethics often present themselves as public questions: questions not only for individuals, but for citizens. This is certainly the case for questions at the beginning and ending of human life, and most recently the case in the congressional debate over the use of performance-enhancing drugs in sports. I surely cannot do justice here to the many complex issues that arise at the intersection of bioethics and public policy, so I will limit my analysis to three core subjects: (1) the limits of moral neutrality in the public square, with the embryo research debate as a prime example; (2) the need for certain minimum ethical boundaries to govern decisions at the beginning and end of human life; and (3) the politics of equality and especially the different ways conservatives and liberals pursue this common ideal.

The Limits of Neutrality

. . . [W]hile neutrality often prevails in public policy, it rarely satisfies the moral aspirations of most citizens. On the stem cell question, liberals believe that advancing medicine and promoting scientific freedom are such important values that public silence is irresponsible. And conservatives believe that defending innocent human life from willful destruction is such an important value that state neutrality is irresponsible. On this particular issue, I believe conservatives are the truer friends of democratic justice. If America never made a single new medical discovery, we could still be a moral nation; if we spent the NIH budget on providing existing therapies to those who do not now have them, we might even be a more just nation. Of course, I am not advocating replacing the National Institutes of Health with the National Agency for Medical Access: scientific research is a civic good worthy of our national support, and new scientific discoveries often make existing therapies more available for more people. But scientific progress is not nearly as essential for civic morality as defending and as promoting human equality; medical progress is less crucial for democracy than securing the bedrock principle that even the most vulnerable human lives are more than mere things for our use, even when our aim is compassionate and our motives are pure.

The Need for Moral Boundaries

In general, both conservatives and liberals believe that the state is ill-equipped to make hard existential decisions, and that those moral issues that involve prudential trade-offs between competing goods—such as using or not using an experimental therapy—are best made by individuals and families with minimum government interference. But prudence must operate within certain shared legal protections—such as laws that ensure informed consent of research subjects—and within certain shared moral boundaries—such as laws that prevent harvesting organs from minimally conscious patients before they die.

Of course, not everything that raises moral concerns should be illegal, just as not everything legal is morally sound. But there are certain moral limits that many conservatives believe should be enacted in law, and that many nonconservatives might be willing to support.[28] Advancing this public policy agenda was the purpose of a series of meetings I helped organize at the Ethics and Public Policy Center in late 2004 and early 2005 to discuss the conservative bioethics agenda for the next several years.[29] This agenda deals mostly with the beginning and the end of human life—not because these are the only important issues in bioethics, but because they are the issues where certain inviolable boundaries are most needed.

The agenda discussed at these meetings comprised six key areas. First, the federal government should ban the creation of human embryos solely for research. Second, the federal government should prohibit certain radical new ways of making babies—including (1) human cloning, (2) the production of children using eggs procured from aborted fetuses or gametes produced using embryonic stem cells, and (3) the production of children by fusing the blastomeres from two or more human embryos. Third, the federal government should prohibit certain experiments that blur the line between human being and animal—including the implantation of a human embryo into a nonhuman uterus, or the fusion of animal sperm and human egg or human sperm and animal egg in the effort to produce a hybrid embryo. Fourth, the federal government should maintain the existing prohibition on the buying and selling of human organs and pass an additional prohibition on the buying and selling of human embryos. Fifth, it should be unlawful to initiate a pregnancy solely to conduct research on the developing fetus or to harvest fetal organs for transplant. Finally, individual states should pass laws prohibiting assisted suicide, euthanasia, or other practices that involve doctors and caregivers in the morally misguided project of deliberately hastening a loved one's death.

Such policies, if enacted, would leave most bioethics decisions in the hands of individuals, families, scientists, and doctors. And they would leave many areas of grave bioethical concern untouched—such as the use of preimplantation genetic diagnosis for sex selection, the abortion of fetuses with Down syndrome, the buying and selling of gametes, the deliberate omission of medical care so that patients will die, or the use of embryos left over in fertility clinics for research purposes. But conservatives recognize that public policy must always begin with those areas where there is the broadest moral consensus, and in those areas (unlike abortion in the post–*Roe v. Wade* era) where self-governing citizens still have the freedom to govern themselves.

The Politics of Equality

Building such a consensus is not only a political challenge but an intellectual one, and perhaps the most promising avenue for finding consensus is for conservatives and liberals to reexamine the ideal of equality that they mutually hold dear. Surely, there are some on the political fringes who deny that the vulnerable have any special claim on our care: libertarians who believe that the downtrodden deserve their misery or that an excessive obsession with the weak weighs down the strong; utilitarians who believe that people with disabilities or dementia are burdens on the rest of society with no moral claim on our protection. But equality is the founding faith of modern democratic societies, rooted in our common sense of vulnerability to experiential and biological misfortune. We are always potentially dependent persons. We are always progressing toward the loss of independence that comes with growing old, and toward the final loss of autonomy that comes with being mortal. This recognition of universal neediness awakens us to the universal reality of human equality.

While equality is our democratic faith, however, it is not the only or most obvious way to understand the human condition. As the conservative political theorist Harvey Mansfield once quipped, the idea that "all men are created equal" is the "self-evident half-truth" of the American Founding.[30] Some men and women are better than others—better mothers and fathers, better athletes and musicians, more generous to the needy, more productive in their work. And some types of human excellence are made possible or impossible by our biological predispositions—by how we are created in the first place. At the same time, some individuals are born with disabilities or diseases that threaten the equal pursuit of happiness, or born into such impoverished economic or cultural circumstances that rising above them requires real heroism. To see the genuine equality of human beings sometimes requires seeing beyond the genuine inequality of human beings. This means seeing both the possibility of change (equality as an aspiration) and seeing the dignity of all human beings regardless of their circumstances (equality as a commandment). Conservatives tend to emphasize equality as a commandment; liberals tend to emphasize equality as an aspiration. This difference lies at the root of many of our deepest political divisions.

Equality as an aspiration is both inspiring and dangerous. It inspires us to try to make things better for those who suffer—by curing terrible diseases, by providing drugs for people in poor countries, by passing laws that protect people with disabilities. It challenges the self-satisfied to remember those less fortunate, and challenges the cynical to believe in the possibility of progress. But the prophecy of equality, like all political prophecies, is also dangerous. As Paul Ramsey once put it, "any person, or any society or age, expecting ultimate success where ultimate success is not to be reached, is peculiarly apt to devise extreme and morally illegitimate means for getting there."[31] This is why equality as a commandment is both less utopian and more fundamental: it obliges us to treat everyone with at least a minimum level of respect, rather than making equality contingent on some hoped for improvement. It provides a floor of dignity for all persons in a world where perfect equality is impossible, both because the human body is frail and the human character is imperfect.[32]

Of course, it would be wrong to see these two ideas of equality as simply in tension. For it is precisely the belief in equality as a commandment—the belief that all human beings are created equal, deserving of equal protection and equal rights—that often moves people to devote their lives to equality as an aspiration. In America, this unity of commandment and aspiration was best displayed in the civil rights movement—a movement heroically led by many liberals in the 1950s and 1960s, but whose principles and achievements are now rightly embraced by the vast majority of Americans. But the tension between equality as a commandment and equality as an aspiration persists in those dimensions of human life where equality cannot be achieved—that is, when the problem is not society's sins but the mysteries, frailties, and imperfections of human nature . . .

[M]ost liberals seem to reject the very ideal of human equality when it comes to certain classes of human beings; the disagreement is not one of prudence but one of principle. In pursuit of equality for the sick who suffer the inequities of disease, many liberals seem willing to destroy human life at its earliest stages. In an effort to remedy the inequities of disability, many liberals seem willing to screen and abort the "genetically unfit" using amniocentesis, or to transform reproduction into a process of division and exclusion using preimplantation genetic diagnosis. No one should doubt the laudable motives of those who defend embryo research or eugenic abortion; they seek cures for terrible diseases and respite from the genuine burdens of living with disability. But, as Ramsey feared, they also "devise extreme and morally illegitimate means for getting there." They deny life to the developing fetus *because* she has Down syndrome; they deny food and water to the person in a persistent vegetative state *because* she is cognitively disabled. Disability becomes the basis for lethal discrimination. Those who engage in such discrimination are often moved by the egalitarian desire to give everyone the best "genetic equipment," or by the desire to divert scarce resources from "futile cases" to those who can still be helped. But lethal discrimination is a dangerous ethical game, one that undermines the equality project by eroding its foundations.

My great hope—perhaps futile—is that a renewed appreciation of human equality and what it demands, among both conservatives and liberals, will serve as the basis for renewed conversation and even political common ground—including providing increased public assistance for uninsured persons who need it, and increased protection for human embryos who always need it. This is not to say that everything good about being human is egalitarian—surely it is not. But equality is America's defining ideal and the best foundation for a decent society. It is also an ideal that both conservatives and liberals can enthusiastically endorse, even as we continue to disagree about the details.

Disclaimer

I have served part-time on the Council staff as a senior research consultant since March 2002, and before that as a senior research analyst. All views expressed here are entirely my own; in no way do I speak for the Council as a body.

References

1. Tolstoy's question is famously quoted in Max Weber's essay on "Science as a Vocation," where Weber discusses the limits of modern science to answer questions of human meaning and value. M. Weber, "*Wissenschaft als Beruf* [Science as a Vocation]," *Gesammelte Aufsaetze zur Wissenschaftslehre* (Tubingen, 1922), 524–55. Originally a speech at Munich University, 1918, published in 1919 by Duncker & Humblot, Munich. Available at www2.pfeiffer.edu/~lridener/DSS/Weber/scivoc.html.

2. This moral purpose and moral framework are the starting point for all bioethics—left, right, center, and otherwise. If good and evil are an illusion, then the plight of the uninsured is a matter of indifference. It is no more ethical to help them than it is to ignore them. If nothing is morally sacred, then the health of the environment is a matter of indifference. It is no more ethical to protect nature's treasures than it is to destroy them. If all cultures are morally equal, then the political equality of women is a matter of indifference. It is no more ethical to extend voting rights in patriarchal societies than it is to deny them. Both liberals and conservatives believe in right and wrong, justice and injustice, and both believe that the state has a role in promoting good and restraining evil—whether the issue is health care for the poor or embryo research. Almost no one seeks neutrality alone. And inasmuch as bioethics as a vocation has some obligation to guide those who face hard bioethical dilemmas in their own lives, neutrality is not an option.

3. This section is influenced heavily by G. Meilaender, "*Terra es Animata:* On Having a Life," *Hastings Center Report* 23, no. 4 (1993), 25–32, and by Kass, *Toward a More Natural Science*, 249–345.

4. The issue of human equality and human worth is explored in some detail in *Taking Care: Ethical Caregiving in Our Aging Society* (Washington, D.C.: President's Council on Bioethics, 2005), which I had the privilege to help draft in my role as senior consultant. See especially 103–107.

5. The phrase "at the edges of life" is taken from P. Ramsey, *Ethics at the Edges of Life* (New Haven, Conn.: Yale University Press, 1980).

6. For a fuller discussion of the meaning of sight as it relates to the embryo question, see E. Cohen, "Of Embryos and Empire," *The New Atlantis*, no. 2 (Summer 2003), 10–13.

7. The literature on the moral standing of human embryos is vast, including discussions of the ethical significance and biological nature of continuous embryological development from fertilization forward. Of special importance are the personal statements by William Hurlbut and Robert George/ Alfonso Gomez-Lobo in President's Council on Bioethics, *Human Cloning and Human Dignity;* chapter 6 of *Human Cloning and Human Dignity* (especially 152–59); and R. George and P. Lee, "Acorns and Embryos," *The New Atlantis*, no. 7 (Fall 2004/Winter 2005), 90–100. Both Hurlbut's personal statement and chapter 6 in *Human Cloning and Human Dignity* address the

issue of twinning, and how this biological possibility does not undermine the biological fact that the embryo is, from conception, an individual life in process. The discussion here also draws heavily upon my own essay, E. Cohen, "The Tragedy of Equality," *The New Atlantis*, no. 7 (Fall 2004/ Winter 2005): 101–109.

8. See, for example, the important work in mammalian embryology being conducted by the Zernicka-Goetz Group at the University of Cambridge: K. Piotrowska-Nitsche et al., "Four-Cell Stage Mouse Blastomeres Have Different Developmental Properties," *Development* 132(3): 479–91; B. Plusa et. al, "The First Cleavage of the Mouse Zygote Predicts the Blastocyst Axis," *Nature* 434: 391–95.

9. This section is influenced heavily by L. Kass, "L'Chaim and Its Limits: Why Not Immortality?" *First Things* 113 (May 2001), 17–24; H. Jonas, "The Burden and Blessing of Mortality," *Hastings Center Report* 22, no. 1 (1992), 34–40; *Beyond Therapy: Biotechnology and the Pursuit of Happiness* (Washington, D.C.: President's Council on Bioethics, 2003), 159–204.

10. For a critical review and survey of transhumanist thinking, see C. Rubin, "Artificial Intelligence and Human Nature," *The New Atlantis*, no. 1 (Spring 2003), 88–100; see also the quote from William Haseltine on making death "a series of preventable diseases" in L. M. Fisher, "The Race to Cash in on the Genetic Code," *The New York Times*, August 29, 1999, sec. 3, p. 1.

11. Jonas, "The Burden and Blessing of Mortality."

12. Kass, "L'Chaim and Its Limits."

13. Jonas, "The Burden and Blessing of Mortality."

14. These themes are developed further in E. Cohen, "The Real Meaning of Genetics," *The New Atlantis*, no. 9 (Summer 2005): 29–41.

15. For an extended discussion of "the good death," see President's Council on Bioethics, *Taking Care*, 108–113.

16. H. Jonas, *Philosophical Essays*, 182; Kass, "L'Chaim and Its Limits," 23–24; Y. Levin, "Imagining the Future," *The New Atlantis*, no. 4 (Winter 2004): 55–58.

17. For a detailed discussion of the demographic issues facing Europe and Japan, see P. Longman, *The Empty Cradle* (New York: Basic Books, 2004), 47–67.

18. For a discussion of the character and significance of human procreation, see *Reproduction and Responsibility: The Regulation of New Biotechnologies* (Washington, D.C.: President's Council on Bioethics, 2004), 13–19, and L. Kass, *The Beginning of Wisdom* (New York: Free Press, 2003), 111–22.

19. I am indebted to Father Tad Pacholczyk for this account of the exclusive right retained by husband and wife to have children only through one another.

20. For a fuller discussion of disagreements about IVF, especially among traditional Catholics and traditional Jews, see E. Cohen, "A Jewish-Catholic Bioethics?" *First Things* 154 (2005), 7–10.

21. Jonas, *Philosophical Essays*, 182.

22. President's Council on Bioethics, *Beyond Therapy*, 37–40.

23. Cohen, "The Real Meaning of Genetics," 32–33, 34–37.
24. President's Council on Bioethics, *Beyond Therapy*, 101–157. In my role as a senior consultant to the Council, I had the privilege of helping draft the chapter on "superior performance." The discussion that follows in this section on human excellence and human performance draws heavily on the themes and examples contained in that chapter. See also L. Kass and E. Cohen, "The Price of Winning at Any Cost," *The Washington Post*, February 1, 2004.
25. This example is taken from President's Council on Bioethics, *Beyond Therapy*, 128.
26. Ibid., 252–55.
27. Ibid., 214–34. In my role as a senior consultant to the Council, I had the privilege of helping draft the section on "memory and happiness." The discussion that follows draws heavily on the themes and examples contained in that section. It also draws upon my own previously published essay, E. Cohen, "Our Psychotropic Memory," *SEED*, no. 8, Fall 2003, 42; see also G. Meilaender, "Why Remember?" *First Things* 135 (2003), 20–24.
28. Many of the items in the agenda described below were recommended unanimously by the President's Council on Bioethics in *Reproduction and Responsibility*, 220–27.
29. These meetings were reported in R. Weiss, "Conservatives Draft a 'Bioethics Agenda' for President," *The Washington Post*, March 8, 2005. See also E. Cohen, "The Bioethics Agenda and the Bush Second Term," *The New Atlantis*, no. 7 (Fall 2004/Winter 2005), 11–18.
30. H. Mansfield, "Returning to the Founders," *The New Criterion*, September 1993.
31. P. Ramsey, *Fabricated Man*, 30–31.
32. This issue is explored in some detail in President's Council on Bioethics, *Taking Care*, especially 106–107.

PART II

The Methods of Ethical Analysis

Applying Ethical Reasoning:
Philosophical, Clinical, and Cultural Challenges

Nancy S. Jecker

In its early days, the field of bioethics focused primarily on ethical concerns that arose in the clinical setting of medicine. Yet as the field matured, it encompassed a broader range of inquiry, both practical and theoretical. Today, bioethicists investigate conceptual questions related to the meaning of central terms, such as "medical futility" (see Part III, Section 2) or "privacy" (see Part III, Sections 1 and 2), and they study the scope and limits of ethical principles and rules such as equality and respect for persons (see Part II, Section 2). Empirical contributions to bioethics are also noteworthy and include, for example, studies determining the actual moral beliefs and practices of providers and patients; an example (included in Part III) is the empirical report of physician and patient participation in Oregon's Death with Dignity, the 1997 law legalizing physician-assisted suicide for Oregon residents. Bioethics scholarship reflects an increasing diversity. Whereas bioethicists were once primarily theologians and philosophers, today they come from a variety of educational backgrounds, including law, medicine, nursing, literature, sociology, anthropology, and other fields.

Not only has the field of bioethics matured and broadened, its gaze has also become more self-reflective, with bioethicists turning a critical eye on the very tools of ethical analysis they themselves use. It is this critical reflection about the methods of ethical analysis that furnishes the focus of Part II of this book. This introductory chapter will acquaint you with three important methods bioethicists use to analyze the ethical features of a case: principlism, casuistry, and narrative ethics. It will also invite you to consider some of the clinical challenges that arise when these methods are applied in clinical settings. Before introducing these specific methods, however, we must first consider some preliminary questions: To what extent can there be a "method" for making ethical decisions? Are the recommendations of bioethicists backed by ethical reasoning, or do they merely reflect what is accepted as the norm by a given community? More broadly, do ethical judgments *ever* have an independent foundation, or are they instead determined by what a particular group or culture believes at a particular time?

Philosophical and Cultural Challenges

Even if we are persuaded of the truth of our own moral beliefs, in the health care setting we must come to terms with the fact that we care for patients in multi-cultural contexts. This has not always been the case. During the first half of the 20th century, most immigrants arriving on U.S. shores were Caucasians with European ancestry who shared the Judeo-Christian religion and traditions that were dominant in the U.S. at that time. However, during the second half of the 20th century, and continuing today, many more immigrants come from the Far East and Middle East. They bring with them moral traditions and religious values, such as Islam, Shintoism, and Buddhism, that are dissimilar from the Judeo-Christian tradition. No longer can we assume that we all share the same ethical values and principles. We must instead tackle the question of whether we are ever justified in applying one group's values to another group.

Even in the absence of immigrant populations arriving here from distant lands, questions of this sort arise in medicine. After all, many Native American populations embrace religious and cultural traditions that are very different from the Judeo-Christian tradition. Thus, a Western physician caring for a traditional Navajo patient may face the question of whether or not to attempt to obtain the patient's informed consent for treatment, which requires reviewing the risks and benefits of treatment, or to instead set aside this practice out of respect for the patient's traditional beliefs, which hold that talking about negative risks may cause them to occur (see the article by Carrese and Rhodes, Part III, Section 3).

Second, even in the absence of immigrant populations bringing diverse cultural and religious beliefs to our shores, we face questions of applying Western values to non-Western groups in the context of international research. In these situations, it is Western physicians who travel to non-Western countries, bringing with them "different" ideas and values than their host country's. Should Western physicians doing research in developing countries assimilate to the cultural values of the country they are visiting? Should they simply adopt the norms governing research with human subjects that prevail in the research subjects' country? Or should they instead appeal to their own Western values? For example, must a group of Western physicians studying the spread of HIV in Uganda offer treatment to subjects with HIV, even if the subjects would not have access to treatment without the study (see Farmer and Campos, Part II, Section 3)?

These questions lead to further questions. Are one group's moral beliefs better or more valid than another group's? Or is it more accurate to regard our own values as no better and no worse than the values and norms of other societies? To begin to address these questions, we need to look directly at the philosophical position known as "ethical relativism."

Ethical relativism is the view that the truth or falsity of an ethical claim is relative to a particular culture. According to the ethical relativist, whether an ethical claim is true or false depends upon the culture of the person who is making the claim. If the ethical claim is accepted as true by the speaker's culture, then the claim is true; otherwise not. According to this position, each culture has its own moral code, and all moral codes have an equal claim to validity. Expressed differently, there is no standard external to all cultures, by which the ethical codes of particular cultures can be judged. This view offers us one kind of answer to the questions we have been considering. It suggests that physicians and researchers are justified in adopting whatever values their patients hold, because these values are just as good or "true" as the moral beliefs and practices they themselves hold. Whereas adopting the patient's values may *appear* "wrong" from the point of view of the doctor's community, according to the ethical relativist, the values are not *really* wrong, because every culture's moral code is equally valid.

People have a variety of motives for adopting ethical relativism:

1. The practical experience of trying to resolve very difficult ethical problems may lead some people to feel that there must not *be* any answer, if one cannot be found or agreed upon. Generalizing this, one might conclude that it is simply the "nature" of ethical problems that they do not have answers. In other words, one might conclude that ethical positions have no justification outside of what a particular person or group believes.
2. Some believe that toleration of diverse views is a virtue, and that ethical relativism will enable this virtue to flourish. Rather than viewing other culture's moral beliefs as "wrong," the relativist regards the beliefs as equally valid.
3. Some believe that moral beliefs are culturally determined. In other words, we simply adopt the values of the culture in which we happen to be born. If this is how we acquire moral beliefs, it might be thought that we do not have any culturally independent basis for our moral beliefs.
4. The fact that we have not yet found a generally agreed-upon foundation on which a universal moral code can be based may lead people to conclude that there is no universally binding moral code.

Yet these motivations are suspect for a variety of reasons. First, the claim that one cannot determine or agree about the answer to a particular question does not entail that there is no right answer, or that all answers are equally valid. After all, a group of 5-year-olds may not be able to figure out how to divide 10 pieces of cake equally among nine children, because they are not acquainted with the idea of fractions, and hence the idea of dividing the 10th piece into nine parts has not occurred to them. But it hardly follows that there is no way to divide the 10 pieces equally. Nor does it follow that all attempts to solve this problem are equally valid.

Second, although toleration of diverse views may well be a virtue, this virtue has limits. Critics of ethical relativism point to egregious abuses of human rights, and argue that toleration of such actions is unconscionable. For example, many culturally based practices aim to control women or render them servile to men's desires and interests.[1] When a non-Western culture resists assimilation to Western liberal values and asserts group rights as a means of holding on to patriarchal values, critics of ethical relativism argue that this should not be tolerated. Instead, critics argue, our response should be to assert that women and men are moral equals and sex discrimination is wrong.

Finally, it is fallacious to argue that a proposition is false on the grounds that it has not yet been proved to be true. Philosophers have a name for this fallacy: argumentum ad ignorantiam (argument from ignorance). It asserts the principle that "Nothing follows from nothing." For example, the fact that we have not yet found a basis for a universal moral code does not prove anything. It does not establish that there is no basis for a universal moral code.

Motivations aside, what are the *arguments* typically offered in support of ethical relativism? One of the most popular defenses of ethical relativism is the "cultural differences argument" (for a discussion of this argument, see Rachels, Part II, Section 3). This argument begins with empirical observations of cultures at different times and places exhibiting different moral behaviors. For example, careful observations of the moral behaviors of different cultural groups establish that attitudes toward homosexuality have varied widely from time to time and place to place (see Benedict, Part II, Section 3). In some societies (the ancient Greek civilization and certain Native American tribes) homosexuality was practiced openly and accepted as a chief means to the good life. In other times and places, homosexuality has been hidden and considered aberrant or shameful. Based on such observations, it might be thought that different cultures have different *moral beliefs*. The conclusion of this argument is that one culture's moral code is no better than any other culture's moral code. In other words, the belief in culturally independent moral truths is a myth.

In response to this argument, it can be noted that even if different cultures practice different moral behaviors, they may nonetheless share the same underlying moral values. Yet even if we accept the claim that different societies have radically different moral beliefs, this does not suffice to show that all moral codes are equally valid. To return to the example noted above, a class of children may have different beliefs about how to divide 10 pieces of cake equally among nine people, but that does not yet show that all answers to this problem are equally good. The child who says, "Give each child one and one-ninth pieces" has a better answer than the child who says, "Give it all to me." Or to take another example, throughout history different societies have held different scientific beliefs. Yet it hardly follows that all scientific beliefs are equally valid. For example, we do not think that the belief that the earth is flat is no better than the belief that the earth is round. Differences of opinion can indicate many things, and do not necessarily establish that opposing views are equally valid. Nor does disagreement establish that there is not truth about a particular matter.

These difficulties aside, the central objection to the cultural differences argument is that it moves from a purely descriptive claim, in its premises, to an evaluative claim, in its conclusion. Yet this reasoning commits an important fallacy. The fallacy, which moral philosophers refer to as the "naturalistic fallacy"[2] consists of moving from factual premises to evaluative conclusions. Consider an analogous argument. I observe hundreds of white swans, and conclude that "Swans should be white." The problem with my reasoning is that the empirical observation that hundreds of swans are white can, at best, lend inductive support to a more general empirical statement, such as "All swans are white"; it does not establish the very different, non-empirical claim that "Swans should be white" or "White swans are better than other colors."

Three Methods of Ethical Reasoning

If the above reasoning is sound, then some of the most common motives and reasons for favoring ethical relativism do not withstand scrutiny. Clearly, more could be said about the question of ethical relativism and about the broader question of how ethical claims are ultimately justified. For now, however, let us set these questions aside, and move forward with the assumption that it is possible to justify ethical claims. Let us assume further that the justification of ethical claims is not based solely on the beliefs or attitudes of the culture of the person making the claim. For example, we cannot simply consult the standards of our society to determine what is right and what is wrong. Rather than applying this simple test, let us assume that reason provides the necessary tools for moral justification, and let us consider how one would go about appealing to reason to justify moral claims in particular cases.

This brings us to our next task, which is to examine some of the central methods of ethical reasoning used to support ethical judgments in particular cases.

Principlism

How can reason illuminate practical ethical problems? It will be instructive to consider the trajectories of two quite different responses one might give. The first regards answers to practical ethical problems as based on philosophical principles and theories of ethics. This approach, known as "principlism," calls upon ethical principles to lend support to practical decisions. This view, which is associated with philosophers Tom Beauchamp and James Childress (see the selection from Beauchamp and Childress, Part II, Section 1), typically draws upon four principles that are claimed to derive authority from considered judgments in common morality and medical traditions. The four principles can be stated as follows.

1. Principle of Respect for Autonomy. One ought to respect the autonomous choices of persons.
2. Principle of Beneficence. One ought to do good, and prevent or remove harm.
3. Principle of Nonmaleficence. One ought to refrain from inflicting harm.
4. Principle of Justice. One ought to treat equals equally.

Principlism holds that particular ethical judgments are justified by showing that they follow from one of these four ethical principles. The principles themselves are said to match our considered judgments in a wide range of situations. The idea of "considered judgments" refers to

> judgments in which our moral capacities are most likely to be displayed without distortion ... Considered judgments are simply those rendered ... where the more common excuses and explanations for making a mistake do not obtain ... [T]he

criteria . . . are not arbitrary. They are, in fact, similar to those that single out considered judgments of any kind. . . . involving the exercise of thought . . . those given under
conditions favorable for deliberation and judgment in general.[3]

The suggestion is that we discard moral judgments made with hesitation, or in
which we have little confidence, as well as those given when we are upset or
frightened, or when we stand to gain one way or the other. Such judgments are
likely to be erroneous, or to be biased in favor of our own interests.

To illustrate the method of principlism, consider the case of Terri Schiavo, a
woman in a persistent vegetative state whose husband requested removal of artificial nutrition and hydration, but whose parents insisted that everything possible
be done to keep their daughter alive (see the article by Perry, Churchill, and
Kirshner, Part III, Section 2). According to principlism, we can justify a particular
claim, such as the claim that nutrition and hydration should be removed from
Ms. Schiavo, by (1) appealing to an ethical principle, such as the principle of respect
for patient autonomy and (2) showing that the patient left convincing evidence
that she would not want life-sustaining treatment, including nutrition and hydration, if she were in a persistent vegetative state. The principle of respect for autonomy is itself justified by showing that we continue to accept this principle in many
situations and under conditions favorable to moral deliberation and judgment.

It is important to note that principlism allows for the possibility of revising
either our considered judgments about a particular case or our general moral
beliefs or principles. According to principlism, the ultimate support for ethical
claims is not principles per se, but considered judgments. In other words, ethics
is ultimately grounded on those moral judgments about which we have the highest degree of confidence. So understood, one objection to principlism is that it
does not provide clear guidance about the crucial process of balancing specific
and general moral judgments. For example, Beauchamp and Childress (Part II,
Section 1) do not offer moral deliberators guidance about how to decide whether
to sacrifice a concrete moral judgment or a general moral conviction in particular cases. In the absence of such instruction, the tendency may be for moral rea-
soners to make arbitrary decisions that merely reflect their pre-reflective leanings
toward a judgment or principle.

Casuistry

Whereas principlism directs our attention away from the specific circumstances
of cases and toward more general moral ideas, the approaches we explore next,
casuistry and narrative ethics, direct our attention to the concrete aspects of the
case. Both casuistry and narrative approaches hold that experience and observation, rather than philosophical principles and theories, provide the premises of
ethical argument. A first step for both is to pay attention to the particular details
of the case. This is the first step, because without the ability to respond to the
particular features of a case, one cannot begin to decipher what more general
principles and obligations are operative. Next, one determines what ethical prin-

ciples or general guides are suitable to the case at hand. Finally, without ever resorting to a philosophical theory, one attempts to balance general guides and obligations with particular facts and insights. This may require, for example, imagining what might be done in analogous situations.

Let us look in more detail at the approach of casuistry espoused by scholars such as Albert Jonsen (see the selection from Jonsen, Part II, Section 1). According to casuistry, the heart of ethical understanding is not mastery of philosophical theories, but moral experience and judgment. The casuistic method of ethical analysis requires us to

1. Pay attention to the morally relevant features of the case;
2. Find analogous cases or "paradigms" that share these features; and
3. Consider the maxim associated with the paradigm and use it to shed light on the case at hand.

In rare cases, a fourth step may be required. This occurs primarily in situations in which the validity of a particular maxim is persuasively challenged. Such a challenge may spring from the recognition that certain assumptions animate the principle or obligation and are no longer considered tenable. For example, the ethical commentaries on marriage that medieval casuists wrote were eventually challenged on the grounds that they were based on outmoded beliefs about sexuality and about equality between the sexes. When a general principle or obligation invoked in ethical analysis is called into question, a fourth step requires us to reevaluate the general claim directly. Thus, we might

4. Reject or modify a general ethical principle after identifying the assumptions underlying it.

Unlike principlism, which prompts us to stand back from the details of the case and think more broadly about the more general values or principles at stake, casuistry directs our attention to the case. Casuistry prompts us to examine the case until we find its most important details, and then to examine other cases until we find ones that share these features. To facilitate case comparison, the casuist organizes cases on the basis of "paradigms." A "paradigm" is a case that offers a relatively clear and simple moral situation. Cases that move away from a paradigm introduce specific circumstances that make the moral situation more obscure or complex. Cases can thus be organized into a kind of taxonomy, beginning with those that present the most clear and simple moral dilemmas (the paradigms), and then introducing specific circumstances that depart from the paradigm in specific ways.

Associated with each paradigm is a maxim, or well-established moral principle, that provides warrant or justification for the expert opinion about the case. Maxims have a variety of sources. Al Jonsen and Stephen Toulmin, for example, cite maxims such as "force may be repulsed by force" and "defense measured to the need of the occasion," which are based on Roman law, as well as maxims such as "don't kick a man when he is down," which reflect what ordinary people might say when arguing about a moral issue.[4]

Consider again the case of Terri Schiavo. A casuistic analysis would begin with the details of the patient's situation, comparing it to analogous cases. Several analogous cases suggest themselves. Karen Ann Quinlan (see "In the Matter of Karen Quinlan," Part I, Section 1) was also a young woman in a persistent vegetative state, whose family members requested removal of life-sustaining treatment. In *Quinlan*, the Supreme Court of New Jersey held that Karen Ann Quinlan's family had the authority to withdraw life support. However, the details of Terri Schiavo's case differ from those of Karen Ann Quinlan in significant respects, rendering the Schiavo case less clear and convincing than the paradigm. In *Schiavo* the family was divided about what to do, whereas in *Quinlan* the family was united in wanting removal of life-sustaining treatment. Moreover, in *Schiavo* the question was whether or not to remove nutrition and hydration, but in *Quinlan* the question dealt with removal of a respirator.

In the years since *Quinlan*, the court's opinion has been tested and confirmed in a series of analogous cases. In these cases, courts have granted families authority to withdraw life support, including artificial nutrition and hydration, from patients in a persistent vegetative state provided there is evidence that this is what the incapacitated patient would have preferred if able to speak on his or her behalf.[5] By drawing on these precedent cases, casuistic reasoning builds a cumulative argument for the maxim that treatment may be withdrawn from a patient in a persistent vegetative state if this is what the patient would have wanted. On this basis, a casuistic analysis could conclude that withdrawing nutrition and hydration in the case of Terri Schiavo is ethically supported. The dispute over which family member had authority to make decisions on Terri Schiavo's behalf is well established in clinical practice and Florida law. The spouse, not the parents, of an incapacitated adult patient have decisional authority. In conclusion, although principlism and casuistry represent different approaches to ethical reasoning, both can be used to reach a similar conclusion in the case of Terri Schiavo.

Narrative Ethics

A final approach to ethical reasoning, narrative ethics, bears a close resemblance to casuistry. Like casuistry, narrative ethics invites us to think about a case by considering analogous situations, but it finds these analogies in fictional stories rather than paradigm cases. Narrative ethics uses stories in a variety of ways. Kathryn Montgomery Hunter, for example, appeals to stories to teach health professionals about the narrative context of patients' lives (see the selection from Hunter, Part II, Section 1). She urges health professionals to view their patients as more than merely "cases" and associated medical diagnoses, but as persons with narrative histories who must try to fit disease and disability into their life story. So understood, narrative techniques assist health providers to understand and respond to patients in richer, more meaningful ways. A related interpretation of narrative ethics, proposed by Rita Charon, encourages health professionals to develop "narrative competence" to improve their understanding of and connection with patients (see the article from Charon, Part II, Section 1).

Narrative competence refers to the techniques of literary analysis, such as techniques for analyzing the characters, plots, and meanings of stories. Charon maintains that applying these techniques to patients' stories enhances empathy and understanding in the doctor–patient relationship. A somewhat different interpretation of narrative ethics encourages using fictional stories like a laboratory, where ethical principles can be "experimented with" or applied to moral issues under controlled conditions.[6] For example, we can find out how a rule, such as "tell the truth," plays out over the course of a story in the lives of fictional characters. By seeing moral rules and principles in narrative context, rather than in abstraction, we can potentially appreciate better what it means to follow certain rules or principles.

Narrative ethics renders moral judgments in a variety of ways. Narrative approaches may ask us to

1. Think about how moral decisions play out in the lives of fictional characters;
2. Judge moral rules by using stories to teach and question them;
3. Judge other people's choices more empathically by reading stories that enable us to understand their point of view; or
4. Regard the patient's medical situation in narrative terms, as a story that is itself situated within the broader narrative of the patient's life.

By drawing on works of fiction in these ways, narrative ethics fosters critical thinking and helps us form more considered judgments. Narrative approaches also encourage us to view discrete medical events in a broader narrative context, where they are imbued with specific meanings for patients and families.

Ultimately, the question of how useful these methods of ethical reasoning are is an empirical and practical one. It requires considering whether or not these methods help us to think more deeply about practical ethical situations we face. In Part III of this book, the methods of ethical analysis discussed here will be used to address particular ethical problems that arise at the beginning and end of life. For example, Judith Jarvis Thomson applies narrative techniques when she uses analogical reasoning about fictional cases to address the problem of abortion (see her paper in Part III, Section 1). Ronald Dworkin and colleagues apply casuistic methods of analysis to defend the idea that patients have the right to exercise control over the timing and manner of their death; they appeal to analogous cases and associated expert opinion to build a cumulative argument in support of their position (see Dworkin, Nagel, Nozick, Rawls, Scanlon, and Jarvis Thomson, Part III, Section 2). Finally, John Robertson illustrates the method of principlism when he invokes the ethical principle of autonomy to address ethical issues related to assisted reproductive technologies (see his paper in Part III, Section 1). When you notice these methods of ethical analysis being used, ask yourself whether this helps you to reach a more carefully considered judgment about the ethical situation.

To summarize, various methods of ethical reasoning are available to help us think more critically about medical cases. Principlism directs our attention to general moral principles and applies these principles, or their derivative rules, to the case at hand. Both casuistry and narrative approaches ask us to compare the case before us to similar situations and apply what we have learned in these analogous cases to our present predicament. To the extent that no single method of ethical analysis is generally regarded as having an exclusive authoritative status, it could be argued that the strategy of employing multiple methods affords the most comprehensive and systematic analysis of an issue.

Clearly the methods of ethical reasoning described here are limited in scope. They are intended to apply to individual cases, and are of limited value in addressing other kinds of ethical issues. For example, none of the methods presented here offers a consistent analytic process for dealing with the kinds of human rights questions that arise in connection with global health topics, such as health inequities between rich and poor nations or ethical standards for conducting transnational research. (For a discussion of these issues, see Farmer and Campos, Part II, Section 3.)

The Challenge of Using Methods in Clinical Settings

Despite their limitations, the methods of principlism, casuistry, and narrative ethics can help us to think more critically about case-based ethical problems. However, further challenges await us when these methods are brought to bear in clinical settings. The most common means of introducing ethical reasoning in the clinical setting is through ethics committees and ethics consultation services. Ethics committees are groups of individuals, usually including a mix of health professionals and others, who assist their institution with ethics consultation, ethics education, and policies that raise ethical issues. Ethics consultants are individuals or teams of individuals who respond to specific requests by health professionals, or by patients or family members, for assistance with ethical issues that arise in the care of an individual patient. In some institutions, ethics consultants are members of the institution's ethics committee who serve as consultants on a rotating basis. Like ethics committee members, ethics consultants may be health professionals, such as physicians or nurses, or individuals with expertise outside health care, in areas such as philosophy or religious studies. The diverse membership of ethics committees and ethics consultation services offers the opportunity for broader representation of diverse viewpoints, yet it also creates specific challenges. As La Puma and Schiedermayer note in Part II, Section 2, physicians who wish to become ethics consultants require training in moral reasoning and ethical decision making. They need opportunities to reflect on common clinical ethical dilemmas, to discover and discuss multidisciplinary perspectives, and to learn and apply techniques of facilitation and negotiation. Non-clinicians require a different set of skills and experiences: routine participation with medical teams in clinics, hospital rooms, and special care units, and firsthand experience dealing with diverse medical situations.

Ethics committees and ethics consultants may introduce ethical reasoning to the clinical setting directly, through teaching ethical principles or methods of ethical analysis, or, more often, indirectly, through modeling ethical reasoning in clinical situations. Examples include

1. An ethics committee sponsoring a lecture series, panel discussion, or other educational forum to stimulate discussion among house staff about the application of a particular ethical principle to cases within the institution;
2. An ethics consultant regularly using a particular method of ethical reasoning when consulting on cases and noting this method in the patient's chart, thus making the method familiar to health professionals; and
3. An ethics committee critiquing an institutional policy by appealing to ethical principles that undergird alternative policy proposals, and in the process clarifying the meaning and implications of ethical principles.

To be effective in clinical settings, ethics committees and ethics consultants must meet various challenges. These include

- Addressing diverse perspectives on an issue;
- Applying a fair procedure in dealing with cases, and treating similar cases similarly;
- Giving minority viewpoints a fair hearing;
- Being attuned to interpersonal and communication issues;
- Being sensitive to cultural differences;
- Distinguishing between situations in which there is a genuine ethical dilemma and situations in which conflicts are the result of other problems, such as a failure to communicate effectively about a plan of care;
- Gaining the confidence and trust of health care professionals in the institution; and
- Avoiding "group think," or the tendency to pressure inadvertently the members of a group to think alike (see the paper by Lo, Part II, Section 2).

In addition, ethics committees and consultants face the challenges of educating their own members and developing strategies to evaluate the quality of their own services and gain feedback from the various constituencies they serve.

Ultimately, the question of whether or not ethics committees and ethics consultants are useful is an empiric one; it requires seeing how effective or ineffective the committees and consultants actually are in improving patient care and the quality of thinking applied to ethical decisions. Assessments of ethics committees and ethics consultation to date suggest that several of the challenges noted above deserve further attention. As Fox, Myers, and Pearlman point out (Part II, Section 2), although ethics consultations services are widespread throughout U.S. hospitals, most consult services are not provided by individuals with expertise in ethics. Only 41% of individuals performing ethics consultation have undergone formal, supervised training in ethics consultation, and only 5% have completed a fellowship or graduate school program in bioethics. Nor can the lack of ethics training be dealt with by pairing, e.g., health professionals with philosophers, because less than 1% of individuals performing ethical consultations are philosophers.

Another difficulty noted by Fox and colleagues is that the majority of ethics consultants are white and non-Hispanic. This raises the question of whether there is sufficient sensitivity to cultural dimensions of patient care, as well as whether there is sufficient inclusion of diverse perspectives on an issue.

Finally, as Robert Veatch points out (see Part II, Section 2), the role of ethics committees and consultants is evolving. To respond effectively to new clinical challenges, ethics committees and consultants must be aware of emerging ethical, legal, social, and cultural concerns. Just as ethics committees have evolved from "prognosis committees" and "patient selection committees" in their early days, to their contemporary function of consultation, education, and policy review, the committees undoubtedly will undergo further changes as the demands of clinical medicine change. Among the many future challenges facing ethics consultants is the problem of how to assure the competence and quality of ethics consultation services. As noted by Scofield (Part II, Section 2), the field of bioethics has not yet determined training or other requirements for its practitioners, set standards of accreditation or licensure for clinical ethics consultants, or developed a professional code of conduct.

Conclusion

This chapter explored the practical problem of caring for patients in multicultural settings and the related philosophical problem of ethical relativism. The chapter argued that some of the most common motivations and arguments favoring ethical relativism do not survive scrutiny. The chapter has shown various techniques for appealing to reason to justify moral claims, and noted some of their advantages and limitations. Finally, we saw that applying these methods in the clinical setting is not easy. Further work is needed to assure the quality of ethics committees and ethics consultation services.

Acknowledgments

This chapter draws on my paper, "Reasoning in Ethics," which appeared in D. Micah Hester, ed., *Ethics By Committee: A Textbook on Consultation, Organization, and Education for Hospital Ethics Committees* (New York: Rowman and Littlefield, 2008): 27–47.

References

1. Okin, Susan. "Is Multiculturalism Bad for Women?" In Susan Moller Okin, ed., *Is Multiculturalism Bad for Women?* (Princeton, NJ: Princeton University Press, 1999), 9–24.
2. Moore, George Edward. *Principia Ethica* (New York: Cambridge University Press, 1903).
3. Rawls, John. *A Theory of Justice*. Cambridge, MA: Harvard University Press, 1971.

4. Jonsen, Albert R. and Toulmin, Stephen. *The Abuse of Casuistry* (Berkeley: University of California Press, 1988).
5. Cruzan, V. Director, Missouri Department of Health: U.S. Supreme Court, reprinted in Albert R. Jonsen, Robert M. Veatch, LeRoy Walters, eds., *Sourcebook in Bioethics: A Documentary History* (Washington, DC: Georgetown University Press, 1998): 229–237.
6. Rosenstand, Nina. "Introduction," in Nina Rosenstand, ed., *The Moral of the Story, 4th Edition* (Boston: McGraw-Hill, 2003).

THE METHODS OF PHILOSOPHY, CASUISTRY, AND NARRATIVE

PHILOSOPHICAL THEORIES AND PRINCIPLES

෮෧

Equality and Its Implications

Peter Singer

Peter Singer studied moral and social philosophy at Oxford University during the 1970s under the direction of philosopher R. M. Hare. Singer is currently Laureate Professor, University of Melbourne, Centre for Applied Philosophy and Public Ethics. In this essay, excerpted from the book *Practical Ethics*, Singer espouses a principle requiring the equal consideration of each person's interests. He specifically rejects the possibility of taking into account any facts about persons other than their interests. One implication of this idea—which Singer does not address—is that it provides no ethical justification for showing partiality toward persons with whom we stand in special relationships (e.g., friends, parents, siblings).

The principle that all humans are equal is now part of the prevailing political and ethical orthodoxy. But what, exactly, does it mean and why do we accept it? . . .

John Rawls has suggested, in his influential book *A Theory of Justice*, that equality can be founded on the natural characteristics of human beings, provided we select what he calls a "range property." Suppose we draw a circle on a piece of paper. Then all points within the circle—this is the "range"—have the property of being within the circle, and they have this property equally. Some points may be closer to the centre and others nearer the edge, but all are, equally,

Abridged from Peter Singer, "Equality and Its Implications," *Practical Ethics*, 80, 1993, pp. 14–23. Reprinted by permission of Cambridge University Press.

points inside the circle. Similarly, Rawls suggests, the property of "moral personality" is a property which virtually all humans possess, and all humans who possess this property possess it equally. By "moral personality" Rawls does not mean "morally good personality"; he is using "moral" in contrast to "amoral." A moral person, Rawls says, must have a sense of justice. More broadly, one might say that to be a moral person is to be the kind of person to whom one can make moral appeals, with some prospect that the appeal will be heeded. . . .

There are problems with using moral personality as the basis of equality. One objection is that moral personality is, unlike being inside a circle, a matter of degree. Some people are highly sensitive to issues of justice and ethics generally; others, for a variety of reasons, have only a very limited awareness of such principles. The suggestion that being a moral person is the minimum necessary for coming within the scope of the principle of equality still leaves it open just where this minimal line is to be drawn. Nor is it intuitively obvious why, if moral personality is so important, we should not have grades of moral status, with rights and duties corresponding to the degree of refinement of one's sense of justice.

Still more serious is the objection that it is not true that all humans are moral persons, even in the most minimal sense. Infants and small children, along with some mentally defective humans, lack the required sense of justice. Shall we then say that all humans are equal, except for very young or mentally defective ones? This is certainly not what we ordinarily understand by the principle of equality. If this revised principle implies that we may disregard the interests of very young or mentally defective humans in ways that would be wrong if they were older or more intelligent, we would need far stronger arguments to induce us to accept it. (Rawls deals with infants and children by including potential moral persons along with actual ones within the scope of the principle of equality. But this is an *ad hoc* device, confessedly designed to square his theory with our ordinary moral intuitions, rather than something for which independent arguments can be produced. Moreover, although Rawls admits that those with irreparable mental defects "may present a difficulty" he offers no suggestions towards the solution of this difficulty.)

So the possession of "moral personality" does not provide a satisfactory basis for the principle that all humans are equal. I doubt that any natural characteristic, whether a "range property" or not, can fulfill this function, for I doubt that there is any morally significant property which all humans possess equally.

There is another possible line of defence for the belief that there is a factual basis for a principle of equality. . . . We can admit that humans differ as individuals, and yet insist that there are no morally significant differences. . . . Knowing that someone is black or white, female or male, does not enable us to draw conclusions about her or his intelligence, sense of justice, depth of feelings, or anything else that would entitle us to treat her or him as less than equal. . . .

The fact that humans differ as individuals, not as races or sexes, is important . . . yet it provides neither a satisfactory principle of equality, nor an adequate defence against a more sophisticated opponent of equality. . . .

... the claim to equality does not rest on intelligence, moral personality, rationality or similar matters of fact. There is no logically compelling reason for assuming that a difference in ability between two people justifies any difference in the amount of consideration we give to their interests. Equality is a basic ethical principle, not an assertion of fact. . . .

... when I make an ethical judgment I must go beyond a personal or sectional point of view and take into account the interests of all those affected. This means that we weigh up interests, considered simply as interests and not as my interests, or the interests of Australians, or of whites. This provides us with a basic principle of equality: the principle of equal consideration of interests.

The essence of the principle of equal consideration of interests is that we give equal weight in our moral deliberations to the like interests of all those affected by our actions. This means that if only X and Y would be affected by a possible act, and if X stands to lose more than Y stands to gain, it is better not to do the act. We cannot, if we accept the principle of equal consideration of interests, say that doing the act is better, despite the facts described, because we are more concerned about Y than we are about X. What the principle really amounts to is: an interest is an interest, whoever's interest it may be.

We can make this more concrete by considering a particular interest, say the interest we have in the relief of pain. Then the principle says that the ultimate moral reason for relieving pain is simply the undesirability of pain as such, and not the undesirability of X's pain, which might be different from the undesirability of Y's pain. Of course, X's pain might be more undesirable than Y's pain because it is more painful, and then the principle of equal consideration would give greater weight to the relief of X's pain. Again, even where the pains are equal, other factors might be relevant, especially if others are affected. If there has been an earthquake we might give priority to the relief of a doctor's pain so she can treat other victims. But the doctor's pain itself counts only once, and with no added weighting. The principle of equal consideration of interests acts like a pair of scales, weighing interests impartially. True scales favour the side where the interest is stronger or where several interests combine to outweigh a smaller number of similar interests; but they take no account of whose interests they are weighing. . . .

The principle of equal consideration of interests prohibits making our readiness to consider the interests of others depend on their abilities or other characteristics, apart from the characteristic of having interests. It is true that we cannot know where equal consideration of interests will lead us until we know what interests people have, and this may vary according to their abilities or other characteristics. Consideration of the interests of mathematically gifted children may lead us to teach them advanced mathematics at an early age, which for different children might be entirely pointless or positively harmful. But the basic element, the taking into account of the person's interests, whatever they may be, must apply to everyone, irrespective of race, sex, or scores on an intelligence test. . . .

Equal consideration of interests is a minimal principle of equality in the sense that it does not dictate equal treatment. Take a relatively straightforward example of an interest, the interest in having physical pain relieved. Imagine that after an earthquake I come across two victims, one with a crushed leg, in agony, and one with a gashed thigh, in slight pain. I have only two shots of morphine left. Equal treatment would suggest that I give one to each injured person, but one shot would not do much to relieve the pain of the person with the crushed leg. She would still be in much more pain than the other victim, and even after I have given her one shot, giving her the second shot would bring greater relief than giving a shot to the person in slight pain. Hence equal consideration of interests in this situation leads to what some may consider an inegalitarian result: two shots of morphine for one person, and none for the other.

There is a still more controversial inegalitarian implication of the principle of equal consideration of interests. In the case above, although equal consideration of interests leads to unequal treatment, this unequal treatment is an attempt to produce a more egalitarian result. By giving the double dose to the more seriously injured person, we bring about a situation in which there is less difference in the degree of suffering felt by the two victims than there would be if we gave one dose to each. Instead of ending up with one person in considerable pain and one in no pain, we end up with two people in slight pain. This is in line with the principle of declining marginal utility, a principle well-known to economists, which states that for a given individual, a set amount of something is more useful when the individual has little of it than when he has a lot. If I am struggling to survive on 200 grammes of rice a day, and you provide me with an extra fifty grammes per day, you have improved my position significantly; but if I already have a kilo of rice per day, I probably couldn't care less about the extra fifty grammes. When marginal utility is taken into account the principle of equal consideration of interests inclines us towards an equal distribution of income, and to that extent the egalitarian will endorse its conclusions. What is likely to trouble the egalitarian about the principle of equal consideration of interests is that there are circumstances in which the principle of declining marginal utility does not hold or is overridden by countervailing factors.

We can vary the example of the earthquake victims to illustrate this. Let us say, again, that there are two victims, one more severely injured than the other, but this time we shall say that the more severely injured victim, A, has lost a leg and is in danger of losing a toe from her remaining leg; while the less severely injured victim, B, has an injury to her leg, but the limb can be saved. We have medical supplies for only one person. If we use them on the more severely injured victim the most we can do is save her toe, whereas if we use them on the less severely injured victim we can save her leg. In other words, we assume that the situation is: without medical treatment, A loses a leg and a toe, while B loses only a leg; if we give the treatment to A, A loses a leg and B loses a leg; if we give the treatment to B, A loses a leg and a toe, while B loses nothing.

Assuming that it is worse to lose a leg than it is to lose a toe (even when that toe is on one's sole remaining foot) the principle of declining marginal utility does not hold in this situation. We will do more to further the interests, impartially considered, of those affected by our actions if we use our limited resources on the less seriously injured victim than on the more seriously injured one. Therefore, this is what the principle of equal consideration of interests leads us to do. . . .

Kantian Ethics

Fred Feldman

Fred Feldman received his formal training in philosophy at Brown University during the late 1960s. He is a faculty member of the Department of Philosophy at the University of Massachusetts, Amherst. This essay, which is taken from a larger work entitled *Introductory Ethics*, analyzes Kant's formulation of the categorical imperative. According to Feldman, Kantian philosophy regards the requirements of morality as derived from the requirements of rationality, such that the person who wills an immoral act wills inconsistently. This approach holds that the universality of moral imperatives is as self-evident as the universality of reason itself.

Sometimes our moral thinking takes a decidedly nonutilitarian turn. That is, we often seem to appeal to a principle that is inconsistent with the whole utilitarian standpoint. One case in which this occurs clearly enough is the familiar tax-cheat case. A person decides to cheat on his income tax, rationalizing his misbehavior as follows: "The government will not be injured by the absence of my tax money. After all, compared with the enormous total they take in my share is really a negligible sum. On the other hand, I will be happier if I have the use of the money. Hence, no one will be injured by my cheating, and one person will be better off. Thus, it is better for me to cheat than it is for me to pay."

In response to this sort of reasoning, we may be inclined to say something like this: "Perhaps you are right in thinking that you will be better off if you cheat. And perhaps you are right in thinking that the government won't even know the difference. Nevertheless, your act would be wrong. For if everyone were to cheat on his income taxes, the government would soon go broke. Surely you can see that you wouldn't want others to act in the way you propose to act. So you shouldn't act in that way." While it may not be clear that this sort of response would be decisive, it should be clear that this is an example of a sort of response that is often given.

There are several things to notice about this response. For one, it is not based on the view that the example of the tax cheat will provoke everyone else to cheat too. If that were the point of the response, then the response might be explained on the basis of utilitarian considerations. We could understand the responder to be saying that the tax cheater has miscalculated his utilities. Whereas he thinks his act of cheating has high utility, in fact it has low utility because it

---

Feldman, Fred. "Kantian Ethics." *Introductory Ethics*, © 1978, pp. 97–99, 101–117. Adapted by permission of Prentice Hall, Upper Saddle River, New Jersey.

145

will eventually result in the collapse of the government. It is important to recognize that the response presented above is not based upon any such utilitarian considerations. This can be seen by reflecting on the fact that the point could just as easily have been made in this way. "Of course, very few other people will know about your cheating, and so your behavior will not constitute an example to others. Thus, it will not provoke others to cheat. Nevertheless, your act is wrong. For if everyone were to cheat as you propose to do, then the government would collapse. Since you wouldn't want others to behave in the way you propose to behave, you should not behave in that way. It would be wrong to cheat."

Another thing to notice about the response in this case is that the responder has not simply said, "What you propose to do would be cheating; hence, it is wrong." The principle in question is not simply the principle that cheating is wrong. Rather, the responder has appealed to a much more general principle, which seems to be something like this: If you wouldn't want everyone else to act in a certain way, then you shouldn't act in that way yourself.

This sort of general principle is in fact used quite widely in our moral reasoning . . . [It is] used against the person who refrains from giving to charity; the person who evades the draft in time of national emergency; the person who tells a lie in order to get out of a bad spot; and even the person who walks across a patch of newly seeded grass. In all such cases, we feel that the person acts wrongly not because his actions will have bad results, but because he wouldn't want others to behave in the way he behaves.

A highly refined version of this nonutilitarian principle is the heart of the moral theory of Immanuel Kant.[1] In his *Groundwork of the Metaphysic of Morals*,[2] Kant presents, develops, and defends the thesis that something like this principle is the "supreme principle of morality." . . .

Kant formulates his main principle in a variety of different ways.

> I ought never to act except in such a way that my maxim should become a universal law.[3]
>
> Act only on that maxim through which you can at the same time will that it should become a universal law.[4]
>
> Act as if the maxim of your action were to become through your will a universal law of nature.[5]
>
> We must be able to will that a maxim of our action should become a universal law—this is the general canon for all moral judgment of action.[6]

Before we can evaluate this principle, which Kant calls the *categorical imperative*, we have to devote some attention to figuring out what it is supposed to mean. To do this, we must answer a variety of questions. What is a maxim? What is meant by "universal law"? What does Kant mean by "will"? Let us consider these questions in turn.

Maxims

. . . Kant defines *maxim* as "a subjective principle of volition. . . ."[7]

Kant apparently believes that when a person engages in genuine action, he always acts on some sort of general principle. The general principle will explain what the person takes himself to be doing and the circumstances in which he takes himself to be doing it. For example, if I need money, and can get some only by borrowing it, even though I know I won't be able to repay it, I might proceed to borrow some from a friend My maxim in performing this act might be, "Whenever I need money and can get it by borrowing it, then I will borrow it, even if I know I won't be able to repay it."

Notice that this maxim is *general*. If I adopt it, I commit myself to behaving in the described way *whenever* I need money and the other conditions are satisfied. In this respect, the maxim serves to formulate a general principle of action rather than just some narrow reason applicable in just one case.[8] So a maxim must describe some general sort of situation, and then propose some form of action for the situation. To adopt a maxim is to commit yourself to acting in the described way whenever the situation in question arises. . . .

It would be implausible to maintain that before we act, we always consciously formulate the maxim of our action. Most of the time we simply go ahead and perform the action without giving any conscious thought to what we're doing, or what our situation is. We're usually too intent on getting the job done. Nevertheless, if we are asked after the fact, we often recognize that we actually were acting on a general policy, or maxim. . . .

For our purposes, it will be useful to introduce a concept that Kant does not employ. This is the concept of the *generalized form* of a maxim. Suppose I decide to go to sleep one night and my maxim in performing this act is this:

M: Whenever I am tired, I shall sleep.

My maxim is stated in such a way as to contain explicit references to me. It contains two occurrences of the word "I". The generalized form of my maxim is the principle we would get if we were to revise my maxim so as to make it applicable to everyone. Thus, the generalized form of my maxim is this:

GM: Whenever anyone is tired, he will sleep.

In general, then, we can represent the form of a maxim in this way:

M: Whenever I am——, I shall——.

Actual maxims have descriptions of situations in the first blank and descriptions of actions in the second blank. The generalized form of a maxim can be represented in this way:

GM: Whenever anyone is——, she will——.

So much, then, for maxims. Let us turn to our second question, "What is meant by universal law?"

When, in the formulation of the categorical imperative, Kant speaks of "universal law," . . . sometimes he seems to be thinking of a *universal law of nature*. . . .

A *law of nature* is a fully general statement that describes not only how things are, but how things always *must* be. Consider this example: If the temperature of a gas in an enclosed container is increased, then the pressure will increase too. This statement accurately describes the behavior of gases in enclosed containers. Beyond this, however, it describes behavior that is, in a certain sense, necessary. The pressure not only *does* increase, but it *must* increase if the volume remains the same and the temperature is increased. This "must" expresses not logical or moral necessity, but "physical necessity." Thus, a law of nature is a fully general statement that expresses a physical necessity. . . .

Willing

. . . The Kantian concept of willing is a bit more complicated . . .

Some states of affairs are impossible. They simply cannot occur. For example, consider the state of affairs of your jumping up and down while remaining perfectly motionless. It simply cannot be done. Yet a sufficiently foolish or irrational person might will that such a state of affairs occur. That would be as absurd as commanding someone else to jump up and down while remaining motionless. Kant would say of a person who has willed in this way that his will has "contradicted itself." We can also put the point by saying that the person has willed inconsistently.

Inconsistency in willing can arise in another, somewhat less obvious way. Suppose a person has already willed that he remain motionless. He does not change this volition, but persists in willing that he remain motionless. At the same time, however, he begins to will that he jump up and down. Although each volition is self-consistent, it is inconsistent to will both of them at the same time. This is a second way in which inconsistency in willing can arise.

It may be the case that there are certain things that everyone must always will. For example, we may have to will that we avoid intense pain. Anyone who wills something that is inconsistent with something everyone must will, thereby wills inconsistently.

Some of Kant's examples suggest that he held that inconsistency in willing can arise in a third way. . . . Suppose a person wills to be in Boston on Monday and also wills to be in San Francisco on Tuesday. Suppose, furthermore, that because of certain foul-ups at the airport it will be impossible for her to get from Boston to San Francisco on Tuesday. In this case, Kant would perhaps say that the person has willed inconsistently.

In general, we can say that a person wills inconsistently if he wills that p be the case and he wills that q be the case and it is impossible for p and q to be the case together.

With all this as background, we may be in a position to interpret the first version of Kant's categorical imperative. Our interpretation is this:

CI_1: An act is morally right if and only if the agent of the act can consistently will that the generalized form of the maxim of the act be a law of nature.

We can simplify our formulation slightly by introducing a widely used technical term. We can say that a maxim is *universalizable* if and only if the agent who acts upon it can consistently will that its generalized form be a law of nature. Making use of this new term, we can restate our first version of the categorical imperative as follows:

CI₁′: An act is morally right if and only if its maxim is universalizable.

As formulated here, the categorical imperative is a statement of necessary and sufficient conditions for the moral rightness of actions. Some commentators have claimed that Kant did not intend his principle to be understood in this way. They have suggested that Kant meant it to be understood merely as a necessary but not sufficient condition for morally right action. Thus, they would prefer to formulate the imperative in some way such as this:

CI₁″: An act is morally right only if its maxim is universalizable.

Understood in this way, the categorical imperative points out one thing to avoid in action. That is, it tells us to avoid actions whose maxims cannot be universalized. But it does not tell us the distinguishing feature of the actions we should perform. Thus, it does not provide us with a criterion of morally right action. Since Kant explicitly affirms that his principle is "the supreme principle of morality," it is reasonable to suppose that he intended it to be taken as a statement of necessary and sufficient conditions for morally right action. In any case, we will take the first version of the categorical imperative to be CI, rather than CI₁. . . .

In a very famous passage in Chapter II of the *Groundwork*, Kant presents four illustrations of the application of the categorical imperative.[9] In each case, in Kant's opinion, the act is morally wrong and the maxim is not universalizable. Thus, Kant holds that his theory implies that each of these acts is wrong. If Kant is right about this, then he has given us four positive instances of his theory. That is, he has given us four cases in which his theory yields correct results. . . .

Kant distinguishes between "duties to self" and "duties to others." He also distinguishes between "perfect" and "imperfect" duties. This gives him four categories of duty: "perfect to self," "perfect to others," "imperfect to self," and "imperfect to others." Kant gives one example of each type of duty. By "perfect duty," Kant says he means a duty "which admits of no exception in the interests of inclination."[10] Kant seems to have in mind something like this: If a person has a perfect duty to perform a certain kind of action, then he must *always* do that kind of action when the opportunity arises. For example, Kant apparently holds that we must always perform the (negative) action of refraining from committing suicide. This would be a perfect duty. On the other hand, if a person has an imperfect duty to do a kind of action, then he must at least *sometimes* perform an action of that kind when the opportunity arises. For example, Kant maintains that we have an imperfect duty to help others in distress. We should devote at least some of our time to charitable activities, but we are under no obligation to give all of our time to such work. . . .

Kant's first example illustrates the application of CI₁ to a case of perfect duty to oneself—the alleged duty to refrain from committing suicide. Kant describes the miserable state of the person contemplating suicide, and tries to show that his categorical imperative entails that the person should not take his own life. In order to simplify our discussion, let us use the abbreviation "a_1" to refer to the act of suicide the man would commit, if he were to commit suicide. According to Kant, every act must have a maxim. Kant tells us the maxim of a_1: "From self-love I make it my principle to shorten my life if its continuance threatens more evil than it promises pleasure."[11] Let us simplify and clarify this maxim, understanding it as follows:

M (a_1): When continuing to live will bring me more pain than pleasure, I shall commit suicide out of self-love.

The generalized form of this maxim is as follows:

GM(a_1): Whenever continuing to live will bring anyone more pain than pleasure, he will commit suicide out of self-love.

Since Kant believes that suicide is wrong, he attempts to show that his moral principle, the categorical imperative, entails that a_1 is wrong. To do this, of course, he needs to show that the agent of a_1 cannot consistently will that GM (a_1) be a law of nature. Kant tries to show this in the following passage:

. . . a system of nature by whose law the very same feeling whose function is to stimulate the furtherance of life should actually destroy life would contradict itself and consequently could not subsist as a system of nature. Hence this maxim cannot possibly hold as a universal law of nature and is therefore entirely opposed to the supreme principle of all duty.[12]

The general outline of Kant's argument is clear enough:

Suicide Example

1. GM (a_1) cannot be a law of nature.
2. If GM (a_1) cannot be a law of nature, then the agent of a_1 cannot consistently will that GM (a_1) be a law of nature.
3. a_1 is morally right if and only if the agent of a_1 can consistently will that GM (a_1) be a law of nature.
4. Therefore, a_1 is not morally right. . . .

Let us turn now to the second illustration. Suppose I find myself hard-pressed financially and I decide that the only way in which I can get some money is by borrowing it from a friend. I realize that I will have to promise to repay the money, even though I won't in fact be able to do so. For I foresee that my financial situation will be even worse later on than it is at present. If I perform this action, a_2, of borrowing money on a false promise, I will perform it on this maxim:

M(a_2): When I need money and can get some by borrowing it on a false promise, then I shall borrow the money and promise to repay, even though I know that I won't be able to repay.

The generalized form of my maxim is this:

> GM(a$_2$): Whenever anyone needs money and can get some by borrowing it on a false promise, then he will borrow the money and promise to repay, even though he knows that he won't be able to repay.

Kant's view is that I cannot consistently will that GM(a$_2$) be a law of nature. This view emerges clearly in the following passage:

> . . . I can by no means will a universal law of lying; for by such a law there could properly be no promises at all, since it would be futile to profess will for future action to others who would not believe my profession or who, if they did so over-hastily, would pay me back in like coin; and consequently my maxim, as soon as it was made a universal law, would be bound to annul itself.[13]

It is important to be clear about what Kant is saying here. He is not arguing against lying on the grounds that if I lie, others will soon lose confidence in me and eventually won't believe my promises. Nor is he arguing against lying on the grounds that my lie will contribute to a general practice of lying, which in turn will lead to a breakdown of trust and the destruction of the practice of promising. These considerations are basically utilitarian. Kant's point is more subtle. He is saying that there is something covertly self-contradictory about the state of affairs in which, as a law of nature, everyone makes a false promise when in need of a loan. Perhaps Kant's point is this: Such a state of affairs is self-contradictory because, on the one hand, in such a state of affairs everyone in need would borrow money on a false promise, and yet, on the other hand, in that state of affairs no one could borrow money on a false promise—for if promises were always violated, who would be silly enough to loan any money?

Since the state of affairs in which everyone in need borrows money on a false promise is covertly self-contradictory, it is irrational to will it to occur. No one can consistently will that this state of affairs should occur. But for me to will that GM(a$_2$) be a law of nature is just for me to will that this impossible state of affairs occur. Hence, I cannot consistently will that the generalized form of my maxim be a law of nature. According to CI$_1$, my act is not right unless I can consistently will that the generalized form of its maxim be a law of nature. Hence, according to CI$_1$, my act of borrowing the money on the false promise is not morally right.

We can restate the essentials of this argument much more succinctly:

Lying-Promise Example

1. GM(a$_2$) cannot be a law of nature.
2. If GM(a$_2$) cannot be a law of nature, then I cannot consistently will that GM(a$_2$) be a law of nature.
3. a$_2$ is morally right if and only if I can consistently will that GM(a$_2$) be a law of nature.
4. Therefore, a$_2$ is not morally right. . . .

Let us turn, then, to the third example. Kant now illustrates the application of the categorical imperative to a case of imperfect duty to oneself. The action in question is the "neglect of natural talents." Kant apparently holds that it is wrong for a person to let all of his natural talents go to waste. Of course, if a person has several natural talents, he is not required to develop all of them. Perhaps Kant considers this to be an imperfect duty partly because a person has the freedom to select which talents he will develop and which he will allow to rust.

Kant imagines the case of someone who is comfortable as he is and who, out of laziness, contemplates performing the act, a_3, of letting all his talents rust. His maxim in doing this would be:

$M(a_3)$: When I am comfortable as I am, I shall let my talents rust.

When generalized, the maxim becomes:

$GM(a_3)$: Whenever anyone is comfortable as he is, he will let his talents rust.

Kant admits that $GM(a_3)$ could be a law of nature. Thus, his argument in this case differs from the arguments he produced in the first two cases. Kant proceeds to outline the reasoning by which the agent would come to see that it would be wrong to perform a_3:

> He then sees that a system of nature could indeed always subsist under such a universal law, although (like the South Sea Islanders) every man should let his talents rust and should be bent on devoting his life solely to idleness, indulgence, procreation, and, in a word, to enjoyment. Only he cannot possibly *will* that this should become a universal law of nature or should be implanted in us as such a law by a natural instinct. For as a rational being he necessarily wills that all his powers should be developed, since they serve him, and are given him, for all sorts of possible ends.[14]

Once again, Kant's argument seems to be based on a rather dubious appeal to natural purposes. Allegedly, nature implanted our talents in us for all sorts of purposes. Hence, we necessarily will to develop them. If we also will to let them rust, we are willing both to develop them (as we must) and to refrain from developing them. Anyone who wills both of these things obviously wills inconsistently. Hence, the agent cannot consistently will that his talents rust. This, together with the categorical imperative, implies that it would be wrong to perform the act, a_3, of letting one's talents rust.

The argument can be put as follows:

Rusting-Talents Example

1. Everyone necessarily wills that all his talents be developed.
2. If everyone necessarily wills that all his talents be developed, then the agent of a_3 cannot consistently will that $GM(a_3)$ be a law of nature.
3. a_3 is morally right if and only if the agent of a_3 can consistently will that $GM(a_3)$ be a law of nature.
4. Therefore a_3 is not morally right. . . .

In Kant's fourth illustration the categorical imperative is applied to an imperfect duty to others—the duty to help others who are in distress. Kant describes a man who is flourishing and who contemplates performing the act, a_4, of giving nothing to charity. His maxim is not stated by Kant in this passage, but it can probably be formulated as follows:

M(a_4): When I'm flourishing and others are in distress, I shall give nothing to charity.

When generalized, this maxim becomes:

GM(a_4): Whenever anyone is flourishing and others are in distress, he will give nothing to charity.

As in the other example of imperfect duty, Kant acknowledges that GM (a_4) could be a law of nature. Yet he claims once again that the agent cannot consistently will that it be a law of nature. He explains this by arguing as follows:

For a will which decided in this way would be in conflict with itself, since many a situation might arise in which the man needed love and sympathy from others, and in which, by such a law of nature sprung from his own will, he would rob himself of all hope of the help he wants for himself.[15]

Kant's point here seems to be this: The day may come when the agent is no longer flourishing. He may need charity from others. If that day does come, then he will find that he wills that others give him such aid. However, in willing that GM(a_4) be a law of nature, he has already willed that no one should give charitable aid to anyone. Hence, on that dark day, his will will contradict itself. Thus, he cannot consistently will that GM(a_4) be a law of nature. This being so, the categorical imperative entails that a_4 is not right. . . .

References

1. Immanuel Kant (1724–1804) is one of the greatest Continental philosophers. He produced quite a few philosophical works of major importance. The *Critique of Pure Reason* (1781) is perhaps his most famous work.
2. Kant's *Grundlegung zur Metaphysik der Sitten* (1785) has been translated into English many times. All references here are to Immanuel Kant, *Groundwork of the Metaphysic of Morals*, translated and analysed by H. J. Paton (New York: Harper & Row, 1964).
3. Kant, *Groundwork*, p. 70.
4. *Ibid.*, p. 88.
5. *Ibid.*, p. 89.
6. *Ibid.*, p. 91.
7. *Ibid.*, p. 69n.
8. In some unusual cases, it may accidentally happen that the situation to which the maxim applies can occur only once, as, for example, in the case of successful suicide. Nevertheless, the maxim is general in form.

9. *Ibid.*, pp. 89–91.
10. *Ibid.*, p. 89n.
11. *Ibid.*, p. 89.
12. *Ibid.*
13. *Ibid.*, p. 71.
14. *Ibid.*
15. *Ibid.*, p. 91.

From *Principles of Biomedical Ethics*

Tom L. Beauchamp and James F. Childress

Philosopher Tom Beauchamp was a member of the Kennedy Institute of Ethics at George-town University when the first edition of *Principles of Biomedical Ethics* appeared in 1977. He continues to serve in this capacity and is also Professor of Philosophy at Georgetown. James Childress is Professor of Religious Studies at the University of Virginia, Charlottesville. Now in its fifth edition, the excerpt from *Principles of Biomedical Ethics* reprinted below espouses a "common-morality theory," which takes its basic premises from the shared morality of society's members. Elsewhere in the book, Beauchamp and Childress set forth the principles of autonomy, beneficence, nonmaleficence, and justice as the central ethical considerations that constitute a common-morality theory in health care.

Principle-Based, Common-Morality Theories

. . . A common-morality theory takes its basic premises directly from the morality shared in common by the members of a society—that is, unphilosophical common sense and tradition. Such a theory need not be principle-based, but we treat these two types of theories together in order to develop the tradition of ethics in which our account should be situated. This [discussion], then, is best understood overall as a statement of the type of ethical theory that we accept and utilize. . . .

Principle-based theories share with utilitarian and Kantian theories an emphasis on principles of *obligation*, but these theories share little else. Two main differences distinguish them. First, utilitarianism and Kantianism are *monistic* theories. One supreme, absolute principle supports all other action-guides in the system. Common-morality theories, as we here stipulatively define them, are *pluralistic*. Two or more nonabsolute (prima facie) principles form the general level of normative statement. Second, common-morality ethics relies heavily on ordinary shared moral beliefs for its content, rather than relying on pure reason, natural law, a special moral sense, and the like. The principles embedded in these shared moral beliefs are also usually accepted by rival ethical theories. Although not the most general principles in many normative theories, the principles are nonetheless accepted in most types of ethical theory. . . .

Any theory that eventuates in moral judgments that cannot be brought into reflective equilibrium with pretheoretical commonsense judgments will be

---

Abridged from Tom L. Beauchamp and James F. Childress, *Principles of Biomedical Ethics*, 1994, pp. 100–106. Reprinted by permission of Oxford University Press.

considered seriously flawed. However, this is not to maintain *either* that ... a common-morality theory is merely a systematizing of commonsense judgments *or* that all *customary moralities* qualify as part of the common morality. An important function of the standards in the common morality (from which the principles we defend and their correlative rights are developed) is to provide a basis for the evaluation and criticism of actions in countries and communities whose customary moral viewpoints fail to acknowledge basic principles. A customary morality, then, is not synonymous with the common morality. The latter is a pretheoretic moral point of view that transcends merely local customs and attitudes. Analogous to beliefs in the universality of basic human rights, the principles of the common morality are universal standards.

Our method ... is to unite principle-based, common-morality ethics with the coherence model of justification. ... This strategy allows us to rely on the authority of the indispensable principles in the common morality, while incorporating tools to refine and correct its weaknesses and unclarities and to allow for additional specification. Because our strategy accepts the goal of reflective equilibrium and, in part, *constructs* principles and rules from considered judgments in the common morality, while also *specifying* principles and rules, we will not end with the identical content with which we began. ...

The Common Morality as Primary Source

As a rough generalization, what Henry Sidgwick called the commonsense morality (morality's core principles and assorted rules of veracity, fidelity, and the like) is the source of the initial moral content for this type of theory. Ethical theory augments this sparse content by a method (1) to clarify and interpret the content, (2) to make the various strands coherent, and (3) to further specify and balance the requirements of norms. ...

Consider why the common morality should play an essential role in ethical theory. If we could be confident that some abstract moral theory was a better source for codes and policies than the common morality, we could work constructively on practical and policy questions by progressive specification of the norms in that theory. But fully analyzed norms in ethical theories are invariably more contestable than the norms in the common morality. We cannot reasonably expect that a contested moral theory will be better for practical decisionmaking and policy development than the morality that serves as our common denominator. Far more social consensus exists about principles and rules drawn from the common morality (for example, our four principles) than about theories. This is not surprising, given the central social role of the common morality and the fact that its principles are, at least in schematic form, usually embraced in some form by all major theories. Theories are rivals over matters of justification, rationality, and method, but they often converge on mid-level principles. ...

Common-morality ethics does not preclude the possibility of reform, which often occurs through interpretation, specification, and balancing. ... Interpretation and innovation are almost always carried out by appeal to justifications *within* rather than *beyond* norms already shared in the community. For example, if our policies on AIDS are so uncompassionate that we need to alter our conception of how therapeutic drugs are brought to the market, purchased, and distributed, this reevaluation will invoke available conceptions of compassion, fair funding, and distribution, rather than totally new principles of justice. Moreover, social agreements, traditions, and norms are inherently indeterminate, thereby failing to anticipate adequately the full range of moral problems and solutions. Interpretation and specification of norms, reconstruction of traditional beliefs, balancing different values, and negotiation are essential. This approach to construction in theory invites evolutionary change while insisting that the common morality provides the starting point and the constraining framework.

Two Examples of Principle-Based Theories

Commonsense convictions played only a minor role in ethical theory prior to the eighteenth century, when philosophers such as Francis Hutcheson, Jean-Jacques Rousseau, and Joseph Butler argued that a native moral sense or an intuitive conscience possessed by all persons is far more important in the moral life than the more complicated systems of philosophers. Their moral psychology did not survive, but their commonsense emphasis did, and Hume, Kant, Hegel, and other leading moral theorists were deeply affected by it. Two twentieth-century writers in ethical theory will serve here to illustrate how a principle-based, common-morality theory is still alive and well.

Frankena's theory. An elegant and simple example of a common-morality theory that resembles ours is William Frankena's version of Hume's postulate that the two major "principles of morals" are beneficence and justice. Frankena appeals to what Bishop Butler called "the moral institution of life," together with what Frankena calls "the moral point of view," meaning a dispassionate attitude of sympathy in which moral decisions are reached by appeal to principled good reasons. For Frankena, the principle of beneficence . . . resembles, but is not identical to, the utilitarian demand that we maximize good over evil, whereas the principle of justice (primarily an egalitarian principle) guides "our distribution of good and evil" independently of judgments about maximizing and balancing good outcomes. Frankena's theory comprises these two general principles, together with an argument that they capture the essence of the moral point of view.

Ross's theory. A second example is the ethics of W. D. Ross, who has had a particularly imposing influence on twentieth-century ethical theory, and more influence on the present authors than any recent writer in ethical theory. He is best known for his intuitionism and his scholarship on Aristotle, but we will largely ignore these dimensions of his work. Ross's starting point is Aristotelian. The moral convictions of thoughtful persons are "the data of ethics just as sense-perceptions are the data of a natural science. Just as some of the latter have to be rejected as illusory, so have some of the former." The "plain" person is, for Ross, the beginning rather than the end of the matter. Using this data base of ordinary standards, Ross thinks *acts* are properly categorized as right and wrong, whereas *motivation* and character are good and bad. This allows him to say that a right act can be done from a bad motive and that a good motive may eventuate in a wrong act.

Ross defends several basic and irreducible moral principles that express prima facie obligations. For example, promises create obligations of fidelity, wrongful actions and debts create obligations of reparation, and the generous services or gifts of others create obligations of gratitude. In addition to fidelity, reparation, and gratitude, Ross lists obligations of self-improvement, justice, beneficence, and nonmaleficence. He holds that the principle of nonmaleficence (noninfliction of harm) takes precedence over the principle of beneficence (production of benefit) when the two come into conflict, but he assigns no priorities among the other principles. This list of obligations is not grounded in any overarching principle.

In a noteworthy methodological statement, Ross maintains that principles are "recognized by intuitive induction as being implied in the judgments already passed on particular acts." His studies of Greek philosophy also led him to distinguish knowledge from opinion. We know principles in the same way the plain person knows the main lines of moral obligation. Here we have *knowledge*, not *opinion*. However, when two or more obligations conflict and balancing, overriding, and judgment are necessary, Ross says we must examine the situation carefully until we form a "considered opinion (it is never more)" that one obligation is more incumbent in the circumstances than any other. These judgments are about the *weight* of principles. They are not judgments that straightforwardly *apply* principles.

The Centrality of Principles and Rules

We can now develop the perspectives and assumptions that make [our theory] a form of common-morality ethics.

The source of the principles.   To say that principles have their origins in the common morality is not to suggest that the final form in which they greet a reader of this book is identical to their appearance in the common morality. Conceptual clarification and methods to introduce coherence are needed to give shape and substance to our moral commitments, much as grammarians, lexicographers, and stylists investigate the nature of our commitments in using words, punctuation, forms of citation, and the like. If unacceptable content is discovered in formulations of principles (for example, if a vigorous strong paternalism in clinical medicine is uncovered) or if incoherence is located, an attempt is made to find acceptable content and achieve coherence. This is work *in* ethical theory, even if its product should not be spoken of as *an* ethical theory. The objective is to give each principle a precise, plausible, thorough, and independent

statement, without presupposing that our familiar ways of formulating principles are necessarily the best or the most coherent ways. After the principles are
so formulated, they will still have to be further interpreted, specified, and balanced to produce an ethics for biomedicine. This is the heart of our strategy.

The prima facie and specifiable nature of the principles.   Like Ross, we
construe principles as prima facie binding. Some theories recognize rules, but
treat them as expendable rules of thumb that summarize past experience by expressing better and worse ways to handle recurrent problems. Other theories
contain absolute principles. Still other theories give a hierarchical (or lexical) or-
dering to moral norms. We reject all three interpretations as inadequate to capture the nature of moral norms and moral reasoning. Rules of thumb permit too
much discretion, as if principles or rules were not binding; absolute principles
and rules disallow all discretion for moral agents and also encounter unresolv-
able moral conflicts; and a hierarchy of rules and principles suffers from damaging counterexamples whose force depends on our reservoir of considered
judgments. (Unlike Ross, we assign no form of priority weighing or hierarchical
ranking to our principles.)

By contrast, we treat principles as both prima facie binding and subject to
revision. So understood, a prima facie principle is a normative guideline stating
conditions of the permissibility, obligatoriness, rightness, or wrongness of
actions that fall within the scope of the principle. The latitude to balance principles in cases of conflict leaves room for compromise, mediation, and negotiation. The account is thereby rescued from the charge that principles cannot be
compromised and so become tyrannical. In stubborn cases of conflict there may
be no single right action, because two or more morally acceptable actions are
unavoidably in conflict and yet have equal weight in the circumstances. Here we
can give good but not decisive reasons for more than one action.

For instance, although murder is absolutely prohibited because of the normative content in the word *murder*, it is not plausible to hold that killing is absolutely prohibited. Killing persons is *prima facie* wrong, but killing to prevent a
person's further extreme pain or suffering is not wrong in every circumstance.
Killing may be the only way to meet some obligations, even though it is prima
facie wrong. However, when a prima facie obligation is outweighed or overridden, it does not simply disappear or evaporate. It leaves what Nozick calls
"moral traces," which should be reflected in the agent's attitudes and actions.

A disadvantage of this account, some say, is that it moves relentlessly to the
paradoxical conclusion that, as Hume put it, "the principles upon which men
reason in morals are always the same; though the conclusions which they draw
are often very different." True, a relativity of judgment is inevitable, but a relativity of the principles embedded in the common morality is not. When people
reach different conclusions, their moral judgments are still subject to justification by good reasons. They are not purely arbitrary or subjective judgments. A
judgment can be proposed for consideration on any basis a person chooses—
random selection, emotional reaction, mystical intuition, etc.—but to propose is

not to justify, and one part of justification is to test judgments and norms by their coherence with the other norms in the moral life.

We conclude that although flexibility and diversity in judgment are ineliminable, judgment generally should be constrained by the demands of moral justification, which typically involves appeal to principles. Our presentation of principles—together with arguments to show the coherence of these principles with other aspects of the moral life, such as the moral emotions, virtues, and rights [will] *constitutes* [our] theory. This web of norms and arguments is the theory. There is no single unifying principle or concept, no description of the highest good, and the like.

A Critique of Principlism

K. Danner Clouser and Bernard Gert

K. Danner Clouser was one of the first American philosophers to join the faculty at a major medical school. He is Emeritus Professor of Humanities at the Pennsylvania State University College of Medicine. Bernard Gert is Stone Professor of Intellectual and Moral Philosophy at Dartmouth College. This essay rejects the use of principles to replace both moral theory and particular moral rules, arguing that such an approach lacks logical consistency and fails to provide a workable method of resolving practical ethical problems.

I. Introduction and Overview

Throughout the land, arising from the throngs of converts to bioethics awareness, there can be heard a mantra ". . . beneficence . . . autonomy . . . justice . . ." It is this ritual incantation in the face of biomedical dilemmas that beckons our inquiry.

In the last twenty years the field of biomedical ethics has expanded in an unprecedented way. The numbers of persons involved, its acceptance as an important field, the myriad university courses, the ubiquitous workshops and conferences, and the plethora of articles, books, and journals have exceeded all expectations. In response to this enormous demand for training in ethics, there have appeared countless books, workshops, and courses that package the theories and methods of ethics, making them readily available to more people in a shorter time.

The major strategy in the most influential of these responses is the deployment of "principles" of biomedical ethics. Conceptually, as diagrammed for example by Beauchamp and Childress (1983), the principles are located just below theories and just above rules. The general notion is that principles follow from moral theories and, in turn, generate particular rules that are then used to make moral judgments. Brandishing these several principles, adherents to the "principle approach" go forth to confront the quandaries of biomedical ethics.

We believe that the "principles of biomedical ethics" approach (hereinafter referred to as "principlism") is mistaken and misleading. Principlism is mistaken about the nature of morality and is misleading as to the foundations of ethics. It misconceives both theory and practice. By no means do we wish to impugn the

Abridged from K. Danner Clouser and Bernard Gert, "A Critique of Principlism," *The Journal of Medicine and Philosophy*, Volume 15, 1990, pp. 219–236. Reprinted by permission of Kluwer Academic Publishers.

many significant moral insights of the proponents of principlism. Our quarrel is not so much with the content of the various "principles" as it is with the use of "principles" at all. We consider this to be crucial and not just a matter of philosophical style. Our focus is on [one] philosophical point: the conceptual or systematic status of "principles" as used in principlism.

Our bottom line, starkly put, is that "principle," as conceived by the proponents of principlism, is a misnomer and that "principles" so conceived cannot function as they are in fact claimed to be functioning by those who purport to employ them. At best, "principles" operate primarily as checklists naming issues worth remembering when considering a biomedical moral issue. At worst "principles" obscure and confuse moral reasoning by their failure to be guidelines and by their eclectic and unsystematic use of moral theory. . . .

A. Our General Claim

Our general contention is that the so-called "principles" function neither as adequate surrogates for moral theories nor as directives or guides for determining the morally correct action. Rather they are primarily chapter headings for a discussion of some concepts which are often only superficially related to each other. When, for example, we are told that a particular case calls for the application of the principle of beneficence, this can mean that the case involves either (1) the utilitarian ideal of promoting some good, or (2) the moral ideal of preventing some harm or removing some harm, or (3) some duty which is morally required. This use of "principles" bears no similarity to principles that "summarize" theories, e.g., as used by Rawls and Mill. Rawls's principle of justice and Mill's principle of utility or principle of liberty are directives toward a moral resolution of particular cases. The principles of Rawls and Mill are effective summaries of their theories; they are shorthand for the theories that generated them. However, this is not the case with principlism, because principlism often has two, three, or even four competing "principles" involved in a given case, for example, principles of autonomy, justice, beneficence, and nonmaleficence. This is tantamount to using two, three, or four conflicting moral theories to decide a case. Indeed some of the "principles"—for example, the "principle" of justice— contain within themselves several competing theories. . . .

Why do we make so much of the fact that in principlism the "principles" provide no systematic guidance? After all, the proponents of principlism would simply say, "Principles are complicated directives. When we say 'apply the principle of beneficence,' we mean consider those points that we discuss in our chapter on the principle of beneficence." In other words, they would say that "the principle of beneficence" is shorthand for their discussion of beneficence. But in that case there is really nothing to be "applied." In effect the agent is being told "think about beneficence and here's thirty pages of distinctions and deliberations to get you started," and that is very different from being told, e.g., "Do that act which will create the greatest good for the greatest number." At best the agent may be reflecting on the relevance of beneficence to the current problem, but he is only deceiving himself if he believes that he has some useful guideline to apply.

There are two problems with an agent's being deceived about whether or not he has a principle that can be applied. One is that the principles are assumed to be firmly established and justified. A person feels secure in applying or in presuming to apply them. The other problem is that an agent will not be aware of the real grounds for his moral decision. If the principle is not a clear, direct imperative at all, but simply a collection of suggestions and observations, occasionally conflicting, then he will not know what is really guiding his action nor what facts to regard as relevant nor how to justify his action. The language of principlism suggests that he has applied a principle which is morally well-established and hence *prima facie* correct. But a closer look at the situation shows that in fact he has looked at and weighed many diverse moral considerations, which are superficially interrelated and herded under a chapter heading named for the "principle" in question.

The agent meanwhile may have "applied" other competing "principles" as well, e.g., autonomy and justice, to the same case. This actually amounts simply to thinking about the case from diverse and conflicting points of view. By "applying" the "principles" of autonomy, beneficence, and justice, the agent is unwittingly using several diverse and conflicting accounts rather than simply applying a well-developed unified theory. It is risky to be doing the former while believing one is doing the latter. . . .

B. Our Thesis Illustrated

It is necessary to see some real examples of principlism. . . .

. . . what is surely the most popular of all biomedical ethics textbooks [is] Beauchamp and Childress's *Principles of Biomedical Ethics* (1983). The authors enunciate four basic principles, each of which illustrates the problems that we have been delineating. Consider their principle of beneficence. For Beauchamp and Childress beneficence is a duty "to help others further their important and legitimate interests" (1983, p. 149); it is morally required (p. 148). The "principle" explicitly prescribes at least two very different kinds of action: (1) to prevent and remove harm, and (2) to confer benefits. These are both included in the general duty of beneficence. Additionally, there seem to be other subprinciples buried in the general "principle." Some are genuine duties to help, which accrue by virtue of special relationships and roles, whereas others are triggered by needs and one's ability to meet those needs, though without clear limitations on the scope of such obligations. All these are included in "*the* principle of beneficence." Clearly, this "principle" is simply a chapter heading under which many superficially related topics are discussed; it is primarily a label for a general concern with consequences. . . .

The Beauchamp and Childress "principle of justice" manifests our point even more than their other "principles." There is not even a glimmer of a usable guide to action. There is a discussion of the concept of justice and about various well-known and conflicting accounts of justice, yet there is no specific action-guide stated. Nevertheless, they refer to a principle of justice as though it is something we ought to apply to moral situations. It is clearly not a guide to action, but rather a checklist of considerations that should be kept in mind when reflecting on moral problems. Not being the kind of classical principle that summarizes a theory and yields specific action-guides, it is deceptive in purporting to have conceptual status and systematic validity. Their "principle" is neither derived from a theory nor does it provide a usable guide to action. . . .

III. Principlism: Systematic Considerations

Relativism: The Anthology Syndrome

Beauchamp and Childress accompany their account of moral reasoning with [the following] diagram:

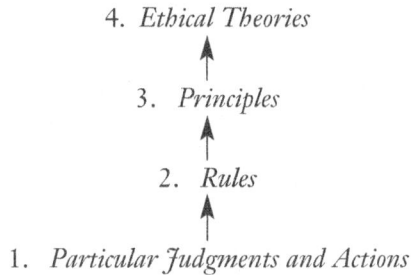

4. *Ethical Theories*

↑

3. *Principles*

↑

2. *Rules*

↑

1. *Particular Judgments and Actions*

"According to this diagram, judgments about what ought to be done in particular situations are justified by moral rules, which in turn are justified by principles, which ultimately are justified by ethical theories" (p. 5). Admitting that their diagram "may be oversimplified," they nevertheless claim that "its design indicates that in moral reasoning we appeal to different reasons of varying degrees of abstraction and systematization" (p. 5).

The authors give no argument for this account of moral reasoning. We suspect that they give no argument because none exists to support the role of principles in the hierarchy they propose. We believe that giving principles a significant role in moral reasoning is not only mistaken, but it also has unfortunate practical and theoretical consequences.

We had earlier seen a kind of relativism embodied by their "principles." Each principle seemed to have a life and logic of its own, as well as a number of internal conflicts. This relativism seems to be endorsed by their diagram having *theories* at the top of the hierarchy rather than a single unified ethical theory. This same kind of ethical relativism is endorsed by almost all anthologies in medical ethics, as well as in all other areas of applied and professional ethics. These anthologies (as well as most courses) almost invariably start by providing brief summaries of some standard ethical theories, e.g., utilitarianism, Kantianism, and contractualism. Next, the inadequacies of each of these theories are pointed out. There is no attempt to repair or remedy these defects, nor to present readers with a theory that they can actually use in solving the problems that are presented in the main body of the book (or course). Rather, the theories are either completely ignored and each problem is dealt with on an *ad hoc* basis, or the student is told to apply whatever inadequate theory he thinks is most useful in dealing with the problem at hand. Often he is told to apply several different, inadequate theories to a given problem, using whatever part of each theory seems most appropriate. This is an extraordinary way to proceed. It is difficult to imagine any respectable discipline proceeding in a similar fashion. Having acknowledged that all of the standard theories are inadequate, one is then told to apply them anyway, and even to apply competing theories, without any attempt to show how the theories can be reconciled.

In effect, the "anthology" approach is that of principlism. The proponents of principlism claim to derive principles from several different theories, none of which they judge to be adequate, and then they urge the student or health care professional to apply one or more of these competing principles to a given case. There is no attempt to show how or even whether these different principles can be reconciled. There is no attempt to show that the different theories, from which the principles are presumably derived, can be reconciled, or that any one of the theories can be revised so as to remove its defects and inadequacies.

Reference

Beauchamp, T.L., and Childress, J.F.: 1983, *Principles of Biomedical Ethics*, second edition, Oxford University Press, New York.

Feminism and Moral Theory

Virginia Held

Virginia Held received her Ph.D. from Columbia University in the late 1960s and is Professor Emeritus at Hunter College of the City University of New York. She argues here that
women's moral experience has generally been discounted in the construction of ethical theories and principles. To remedy this omission, Held analyzes the experience of women who engage in mothering. She concludes that the practice of mothering has important perspectives
to contribute to ethics; these perspectives emphasize an ethic of caring, place a premium on
sensitivity to the particularity of context, and are suspicious of general impartial rules.

The tasks of moral inquiry and moral practice are such that different moral
approaches may be appropriate for different domains of human activity. I
have argued in a recent book that we need a division of moral labor.[1] In *Rights
and Goods*, I suggest that we ought to try to develop moral inquiries that will be
as satisfactory as possible for the actual contexts in which we live and in which
our experience is located. Such a division of moral labor can be expected to yield
different moral theories for different contexts of human activity, at least for the
foreseeable future. In my view, the moral approaches most suitable for the
courtroom are not those most suitable for political bargaining; the moral approaches suitable for economic activity are not those suitable for relations within
the family, and so on. The task of achieving a unified moral field theory covering all domains is one we may do well to postpone, while we do our best to devise and to "test" various moral theories in actual contexts and in light of our
actual moral experience.

What are the implications of such a view for women? Traditionally, the experience of women has been located to a large extent in the context of the family. In recent centuries, the family has been thought of as a "private" domain
distinct not only from that of the "public" domain of the polis, but also from the
domain of production and of the marketplace. Women (and men) certainly need
to develop moral inquiries appropriate to the context of mothering and of family relations, rather than accepting the application to this context of theories developed for the marketplace or the polis. We can certainly show that the moral
guidelines appropriate to mothering are different from those that now seem

Abridged from Virginia Held, "Feminism and Moral Theory," *Women and Moral Theory*,
1987, pp. 112–127. Reprinted by permission of Rowman and Littlefield.

suitable for various other domains of activity as presently constituted. But we need to do more as well: we need to consider whether distinctively feminist moral theories, suitable for the contexts in which the experience of women has or will continue to be located, are better moral theories than those already available, and better for other domains as well.

The Experience of Women

We need a theory about how to count the experience of women. It is not obvious that it should count equally in the construction or validation of moral theory. To merely survey the moral views of women will not necessarily lead to better moral theories. In the Greek thought that developed into the Western philosophical tradition,[2] reason was associated with the public domain from which women were largely excluded. If the development of adequate moral theory is best based on experience in the public domain, the experience of women so far is less relevant. But that the public domain is the appropriate locus for the development of moral theory is among the tacit assumptions of existing moral theory being effectively challenged by feminist scholars. We cannot escape the need for theory in confronting these issues.

We need to take a stand on what moral experience is. As I see it, moral experience is "the experience of consciously choosing, of voluntarily accepting or rejecting, of willingly approving or disapproving, of living with these choices, and above all of acting and of living with these actions and their outcomes . . . . Action is as much a part of experience as is perception."[3] Then we need to take a stand on whether the moral experience of women is as valid a source or test of moral theory as is the experience of men, or on whether it is more valid.

Certainly, engaging in the process of moral inquiry is as open to women as it is to men, although the domains in which the process has occurred have been open to men and women in different ways. Women have had fewer occasions to experience for themselves the moral problems of governing, leading, exercising power over others (except children), and engaging in physically violent conflict. Men, on the other hand, have had fewer occasions to experience the moral problems of family life and the relations between adults and children. Although vast amounts of moral experience are open to all human beings who make the effort to become conscientious moral inquirers, the contexts in which experience is obtained may make a difference. It is essential that we avoid taking a given moral theory, such as a Kantian one, and deciding that those who fail to develop toward it are deficient, for this procedure imposes a theory on experience, rather than letting experience determine the fate of theories, moral and otherwise.

We can assert that as long as women and men experience different problems, moral theory ought to reflect the experience of women as fully as it reflects the experience of men. The insights and judgments and decisions of women as they engage in the process of moral inquiry should be presumed to be as valid as those of men. In the development of moral theory, men ought to have no privileged position to have their experience count for more. If anything, their privileged position in society should make their experience more suspect rather than more worthy of being counted, for they have good reasons to rationalize their privileged positions by moral arguments that will obscure or purport to justify these privileges.[4] . . .

Mothering and Markets

When we bring women's experience fully into the domain of moral consciousness, we can see how questionable it is to imagine contractual relationships as central or fundamental to society and morality. . . .

The most central and fundamental social relationship seems to be that between mother or mothering person and child. It is this relationship that creates and recreates society. It is the activity of mothering which transforms biological entities into human social beings. Mothers and mothering persons produce children and empower them with language and symbolic representations. Mothers and mothering persons thus produce and create human culture.

Despite its implausibility, the assumption is often made that human mothering is like the mothering of other animals rather than being distinctively human. In accordance with the traditional distinction between the family and the polis, and the assumption that what occurs in the public sphere of the polis is distinctively human, it is assumed that what human mothers do within the family belongs to the "natural" rather than to the "distinctively human" domain. Or, if it is recognized that the activities of human mothers do not resemble the activities of the mothers of other mammals, it is assumed that, at least, the difference is far narrower than the difference between what animals do and what humans who take part in government and industry and art do. But, in fact, mothering is among the most human of human activities.

Consider the reality. A human birth is thoroughly different from the birth of other animals, because a human mother can choose not to give birth. However extreme the alternative, even when abortion is not a possibility, a woman can choose suicide early enough in her pregnancy to consciously prevent the birth. A human mother comprehends that she brings about the birth of another human being. A human mother is then responsible, at least in an existentialist sense, for the creation of a new human life. The event is essentially different from what is possible for other animals.

Human mothering is utterly different from the mothering of animals without language. The human mother or nurturing person constructs with and for the child a human social reality. The child's understanding of language and of symbols, and of all that they create and make real, occurs in interactions between child and caretakers. Nothing seems more distinctively human than this. In comparison, government can be thought to resemble the governing of ant colonies, industrial production to be similar to the building of beaver dams, a market exchange to be like the relation between a large fish that protects and a small fish that grooms, and the conquest by force of arms that characterizes so much of human history to be like the aggression of packs of animals. But the imparting of language and the creation within and for each individual of a human social reality, and often a new human social reality, seems utterly human.

An argument is often made that art and industry and government create new human reality, while mothering merely "reproduces" human beings, their cultures, and social structures. But consider a more accurate view: in bringing up children, those who mother create new human *persons*. They change persons, the culture, and the social structures that depend on them, by creating the kinds of persons who can continue to transform themselves and their surroundings. Creating new and better persons is surely as "creative" as creating new and better objects or institutions. It is not only bodies that do not spring into being unaided and fully formed; neither do imaginations, personalities, and minds.

Perhaps morality should make room first for the human experience reflected in the social bond between mothering person and child, and for the human projects of nurturing and of growth apparent for both persons in the relationship. In comparison, the transactions of the marketplace seem peripheral; the authority of weapons and the laws they uphold, beside the point. . . .

Between the Self and the Universal

Perhaps the most important legacy of the new insights will be the recognition that more attention must be paid to the domain *between* the self—the ego, the self-interested individual—on the one hand, and the universal—everyone, others in general—on the other hand. . . .

. . . Moral theory has neglected the intermediate region of family relations and relations of friendship, and has neglected the sympathy and concern people actually feel for particular others. . . .

Standard moral philosophy has construed personal relationships as aspects of the self-interested feelings of individuals, as when a person might favor those he loves over those distant because it satisfies his own desires to do so. Or it has let those close others stand in for the universal "other," as when an analysis might be offered of how the conflict between self and others is to be resolved in something like "enlightened self-interest" or "acting out of respect for the moral law," and seeing this as what should guide us in our relations with those close, particular others with whom we interact. . . .

. . . What feminist moral theory will emphasize, in contrast, will be the domain of particular others in relations with one another. . . .

Moral theories must pay attention to the neglected realm of particular others in actual contexts. In doing so, problems of egoism vs. the universal moral point of view appear very different, and may recede to the region of background insolubility or relative unimportance. The important problems may then be seen to be how we ought to guide or maintain or reshape the relationships, both close and more distant, that we have or might have with actual human beings.

Particular others can, I think, be actual starving children in Africa with whom one feels empathy or even the anticipated children of future generations, not just those we are close to in any traditional context of family, neighbors, or friends. But particular others are still not "all rational beings" or "the greatest number." . . .

In recognizing the component of feeling and relatedness between self and particular others, motivation is addressed as an inherent part of moral inquiry. Caring between parent and child is a good example.[5] . . . When the relationship between "mother" and child is as it should be, the caretaker does not care for the child (nor the child for the caretaker) because of universal moral rules. The love and concern one feels for the child already motivate much of what one does. This is not to say that morality is irrelevant. One must still decide what one ought to do. But the process of addressing the moral questions in mothering and of trying to arrive at answers one can find acceptable involves motivated acting, not just thinking. . . .

Principles and Particulars

When we take the context of mothering as central, rather than peripheral, for moral theory, we run the risk of excessively discounting other contexts. It is a commendable risk, given the enormously more prevalent one of excessively discounting mothering. But I think that the attack on principles has sometimes been carried too far by critics of traditional moral theory. . . .

We should not forget that an absence of principles can be an invitation to capriciousness. Caring may be a weak defense against arbitrary decisions, and the person cared for may find the relation more satisfactory if both persons, but especially the person caring, are guided, to some extent, by principles concerning obligations and rights. To argue that no two cases are ever alike is to invite moral chaos. Furthermore, for one person to be in a position of caretaker means that that person has the power to withhold care, to leave the other without it. The person cared for is usually in a position of vulnerability. The moral significance of this needs to be addressed along with other aspects of the caring relationship. Principles may remind a giver of care to avoid being capricious or domineering. While most of the moral problems involved in mothering contexts may deal with issues above and beyond the moral minimums that can be covered by principles concerning rights and obligations, that does not mean that these minimums can be dispensed with. . . .

That aspect of the attack on principles which seems entirely correct is the view that not all ethical problems can be solved by appeal to one or a very few simple principles. It is often argued that all more particular moral rules or principles can be derived from such underlying ones as the Categorical Imperative or the Principle of Utility, and that these can be applied to all moral problems. The call for an ethic of care may be a call, which I share, for a more pluralistic view of ethics, recognizing that we need a division of moral labor employing different moral approaches for different domains, at least for the time being.[6] Satisfactory intermediate principles for areas such as those of international affairs, or family relations, cannot be derived from simple universal principles, but must be arrived at in conjunction with experience within the domains in question.

Attention to particular others will always require that we respect the particularity of the context, and arrive at solutions to moral problems that will not give moral principles more weight than their due. But their due may remain considerable. And we will need principles concerning relationships, not only concerning the actions of individuals, as we will need evaluations of kinds of relationships, not only of the character traits of individuals.

Notes

1. See Virginia Held, *Rights and Goods: Justifying Social Action* (New York: Free Press, Macmillan, 1984).
2. See Genevieve Lloyd, *The Man of Reason: "Male" and "Female" in Western Philosophy* (Minneapolis: University of Minnesota Press, 1984).
3. Virginia Held, *Rights and Goods*, p. 272. See also V. Held, "The Political 'Testing' of Moral Theories," *Midwest Studies in Philosophy* 7 (1982):343–63.
4. For discussion, see especially Nancy Hartsock, *Money, Sex, and Power* (New York: Longman, 1983), chaps. 10, 11.
5. See, e.g., Nell Noddings, *Caring: A Feminine Approach to Ethics and Moral Education* (Berkeley: University of California Press, 1984), pp. 91–94.
6. Participants in the conference on Women and Moral Theory offered the helpful term "domain relativism" for the version of this view that I defended.

CASUISTRY

The Four Topics:
Case Analysis in Clinical Ethics

Albert R. Jonsen, Mark Siegler, and William J. Winslade

Albert R. Jonsen, who trained at the Yale Divinity School, is Professor Emeritus, University of Washington, School of Medicine, where he was Chairman of the Department of Medical History and Ethics from 1987 to 1999. He is also Co-director and Senior Ethics Scholar at the Program in Medicine and Human Values, California Pacific Medical Center, San Francisco. Mark Siegler is a general internist and Professor of Medicine, University of Chicago, Division of Biological Sciences, Department of Medicine, where he directs the MacLean Center for Clinical Medical Ethics. William J. Winslade, a philosopher, attorney and psychoanalyst, is Professor, University of Texas Medical Branch, Institute for Medical Humanities. The following excerpt explains a method of ethical reasoning about cases that has come to be known as "the four box method."

Clinical ethics is a practical discipline that provides a structured approach for identifying, analyzing, and resolving ethical issues in clinical medicine. . . . Clinical situations are complex, because they involve a wide range of medical facts, a multitude of circumstances, and a variety of values. Often, decisions must be reached quickly. The authors believe that clinicians need a straightforward method of sorting out the pertinent facts and values of any case into an orderly pattern that facilitates the discussion and resolution of ethical problems. . . .

We suggest that every clinical case, especially those raising an ethical problem, should be analyzed by means of the following four topics: (1) medical indications, (2) patient preferences, (3) quality of life, and (4) contextual features, defined as the social, economic, legal, and administrative context in which the case occurs. Although the facts of each case can differ, these four topics are always relevant. The topics organize the various facts of the particular case and, at the same time, call attention to the ethical principles appropriate to the case. It is our intent to show readers how these four topics provide a systematic method of identifying and analyzing the ethical problems occurring in clinical medicine.

Clinicians will recall the method of case presentation that they learned at the beginning of their professional training. They were taught to "present" a patient by stating in order (1) the chief complaint, (2) history of the present illness, (3) past medical history, (4) family and social history, (5) physical findings, and (6) laboratory data. These are the topics that an experienced clinician uses to reach a diagnosis and to formulate a plan for management of the case. Although the particular details under each of these topics can differ from patient to patient, the topics themselves are constant and are always relevant to the task of arriving at a management plan. Sometimes one topic, for example, the patient's family history or the physical examination, may be particularly important or, conversely, may not be relevant to the present problem. Still, clinicians are expected to review all of the topics in every case.

Our four topics help clinicians understand how the ethical principles connect with the circumstances of the clinical case . . .

Dax's Case. We can illustrate our method by a brief summary of a case familiar to many who have studied medical ethics, namely, the case of Donald "Dax" Cowart, the burn patient who related his experience in the videotape *Please Let Me Die* and the documentary *Dax's Case.*

In 1973, "Dax" Cowart, aged 25 years, was severely burned in a propane gas explosion. Rushed to the Burn Treatment Unit of Parkland Hospital in Dallas, he was found to have severe burns over 65% of his body; his face and hands suffered third-degree burns, and his eyes were severely damaged. Full-burn therapy was instituted. After an initial period during which his survival was in doubt, he was stabilized and underwent amputation of several fingers and removal of his right eye. During much of his 232-day hospitalization at Parkland, his few weeks at The Texas Institute of Rehabilitation and Research at Houston, and his subsequent 6-month stay at University of Texas Medical Branch in Galveston, he repeatedly insisted that treatment be discontinued and that he be allowed to die. Despite this demand, wound care was continued, skin grafts were performed, and nutritional and fluid support were provided. He was discharged totally blind, with minimal use of his hands, badly scarred, and dependent on others to assist in personal functions.

Discussion of the ethics involved in a case like this can begin by asking any number of questions. Did Dax have the moral or the legal right to refuse care? Was Dax competent to make a decision? Were the physicians paternalistic? What was Dax's prognosis? All these questions, and many others, are relevant and can result in vigorous debate. However, we suggest that the ethical analysis should begin by an orderly review of the four basic topics. We recommend that the same order be followed in all cases; that is, (1) medical indications, (2) patient preferences, (3) quality of life, and (4) contextual features. The use of this procedure will lay out the ethically relevant facts of the case (or show where further information is needed) before debate begins. This order of review does not constitute an order of ethical priority. The determination of relative importance of these topics will be explained in the four chapters.

Medical Indications. This topic includes the usual content of a clinical discussion: the diagnosis, prognosis, and treatment of the patient's medical problem. "Indications" refers to the diagnostic and therapeutic interventions that are appropriate to evaluate and treat the problem. Although this is the usual material covered in the presentation of any patient's clinical problems, the ethical discussion reviews the medical facts and evaluates them in the light of the fundamental ethical features of the case, such as the possibility of benefiting the patient and respecting the patient's preferences.

In Dax's case, the medical indications include the clinical facts necessary to diagnose the extent and seriousness of his burns, to make a prognosis for survival or restoration of function, and to determine the options for treatment, including the risks, benefits, and probable outcomes of each treatment modality. For example, certain prognoses are associated with burns of given severity and extent. Various forms of treatment, such as fluid replacement, skin grafting, and antibiotics are associated with certain probabilities of outcome and risk. After initial emergency treatment, Dax's prognosis for survival was approximately 20%, but the quality of life after his survival was likely to be greatly diminished by blindness, disability, and deformity. After 6 months of intensive care, his prognosis for survival improved to almost 100%. If his request to stop wound care and grafting during that first hospitalization had been respected, he would almost certainly have died. A clear view of the possible benefits of intervention is the first step in assessing the ethical aspects of a case.

Patient Preferences. In all medical treatment, the patient's preferences that are based on the patient's own values and personal assessment of benefits and burdens are ethically relevant. In every clinical case, certain questions must be asked: What does the patient want? and What are the patient's goals? The systematic review of this topic requires the following additional questions: Has the patient been provided sufficient information? Does the patient comprehend? Does the patient understand the uncertainty inherent in any medical recommendation and the range of reasonable options that exist? Is the patient consenting voluntarily? and Is the patient coerced? In some cases, an answer to these questions might be: "We don't know because the patient is incapable of formulating a preference or expressing one." If the patient is mentally incapacitated at the time a decision must be made, we must ask: Who has the authority to decide on behalf of this patient? What are the ethical and legal limits of that authority? and What is to be done if no one can be identified as surrogate?

In Dax's case, his mental capacity was questioned in the early days of his refusal of care. Had the physical and emotional shock of the accident undermined his ability to decide for himself? Initially, it was assumed that he lacked the capacity to make his own decisions, at least about refusing life-saving therapy. The doctors accepted the consent of Dax's mother in favor of treatment over his refusal of treatment. Later, when Dax was rehospitalized in the Galveston Burn Unit, a psychiatric consultation was requested, which affirmed his capacity to make decisions. Once that capacity was established, the ethical implications of his desire to refuse care became central. The following ethical questions immediately had to be considered: Should his preference be respected? Did Dax appreciate sufficiently the prospects for his rehabilitation? Are physicians obliged to pursue therapies they believe have promise over the objections of a patient? Would they be cooperating in a suicide if they assented to Dax's wishes? Any case involving the ethics of patient preferences relies on clarification of these questions.

Quality of Life. Any injury or illness threatens persons with actual or potentially reduced quality of life, manifested in the signs and symptoms of their disease. One goal of medical intervention is to restore, maintain, or improve quality of life. Thus, in all medical situations, the topic of quality of life must be considered. Many questions surround this topic: What does this phrase "quality of life" mean in general? How should it be understood in particular cases? How do persons other than the patient perceive the patient's quality of life, and of what ethical relevance are their perceptions? Above all, what is the relevance of quality of life to ethical judgment? This topic, important as it is in clinical judgment, opens the door for bias and prejudice. Still, it must be confronted in the analysis of clinical ethical problems.

In Dax's case, we note the quality of his life before the accident. He was a popular, athletic young man, just discharged from the Air Force, after serving as a fighter pilot in Vietnam. He worked in a real estate business with his father (who was also injured in the explosion and died on the way to the hospital). Before his accident, Dax's quality of life was excellent. During the course of medical care, he endured excruciating pain and profound depression. After the accident, even with the best of care, he was confronted with significant physical deficits, including notable disfigurement, blindness, and limitation of activity. During most of his hospital course, Dax had the capacity to determine what quality of life he wished for himself. However, in the early weeks of his hospitalization, he may have suffered serious deficits in mental capacity at the time critical decisions had to be made. Others would have to make quality-of-life decisions on his behalf. Was the prospect for return to a normal or even acceptable life so poor that no reasonable person would choose to live, or is any life worth living regardless of its quality? Who should make such decisions? What values should guide the decision makers? The ethical controversy occurred because Dax believed, even though his mother and physicians did not, that he had the capacity and the right to make his own quality-of-life decisions, including the right to refuse all treatment. The meaning and import of such considerations must be clarified in any clinical ethical analysis.

Contextual Features. Preferences and quality of life bring out the most common features of the medical encounter. However, every medical case is embedded in a larger context of persons, institutions, and financial and social arrangements. The possibilities and the constraints of that context influence patient care, positively or negatively. At the same time, the context itself is affected by the decisions made by or about the patient: these decisions may have psychological, emotional, financial, legal, scientific, educational, or religious impact on others. In every case, the relevance of the contextual features must be determined and assessed. These contextual features may be crucially important to the understanding and resolution of the case.

In Dax's case, several of these contextual features were significant. Dax's mother was opposed to termination of his medical care for religious reasons. The legal implications of honoring Dax's demand were unclear at the time. The costs of 16 months of intensive burn therapy were substantial. Dax's refusal to cooperate with treatment may have influenced the attitudes of physicians and nurses toward him. These and other contextual factors must be made explicit and assessed for their relevance.

Rules and Principles. These four topics are relevant to any clinical case, whatever the actual circumstances. They serve as a useful organizing device for teaching and discussion. Some clinicians have even found them useful for organizing a plan for patient management. A review of these topics can also help to move the discussion of an ethical problem toward a resolution. Any serious discussion of an ethical problem must go beyond merely talking about it in an orderly way: it must push through to a reasonable and practical resolution. Ethical problems, no less than medical problems, cannot be left hanging. Thus, after presenting a case, the task of seeking a resolution must begin.

The discussion of each topic includes certain standards of behavior that are ethically appropriate to the topic. These can be called ethical principles or ethical rules. For example, one version of the principle of beneficence states, "There is an obligation to assist others in the furthering of their legitimate interests." The ethical rule, "Physicians have a duty to treat patients, even at risk to themselves," is a specific expression of that broad principle, suited to a particular sphere of professional activity, namely, medical care. The topic of medical indications, in addition to the clinical data that must be discussed, includes the additional questions, "How much can we do to help this patient?" and "What risks of adverse effects can be tolerated in the attempt to treat the patient?" Answers to these questions, arising so naturally in the discussion of medical indications, can be guided by familiar historical rules of medical ethics such as, "Be of benefit and do no harm" or "Risks should be balanced by benefits." Rules such as these reflect in a specific way the broad principle that the philosophers have named beneficence. Similarly, the topic of patient preferences contains rules that instruct clinicians to tell patients the truth, to respect their deliberate preferences, and to honor their values. Rules such as these fall under the general scope of the principles of autonomy and respect for persons.

Our method of analysis begins, not with the principles and rules, as do many other ethics treatises, but with the factual features of the case. We refer to principles and rules as they become relevant to the discussion of the topics. In this way, abstract discussion of principles is avoided, as is also the tendency to think of only one principle, such as autonomy or beneficence, as the sole guide in the case. Ethical rules and principles are best appreciated in the specific context of the actual circumstances of a case. For example, a key issue in Dax's case is the autonomy of the patient. However, the significance of autonomy in Dax's case is derived not simply from the principle that requires we respect it, but from the confluence of considerations about preferences, medical indications for treatment, quality of life, decisional capacity, and the role of his mother, the doctors, the lawyers, and the hospitals. Only when all these are seen and evaluated in relation to each other will the meaning of the principle of autonomy be appreciated in this case.

. . . [A] chart [on this and the following page] depicts the four topics in quadrants. This chart can serve as a convenient way to record the facts of a case in an orderly way. However, it has a much more important purpose as a guide for ethical deliberation about a case. The many facts do not remain isolated in their respective quadrants. Rather, once they are displayed, the ethical task begins: to evaluate the facts in relation to each other and in light of the principles. In some cases, once the facts are clear, it also becomes clear that the issue is easily resolved: confusion about the facts or failures in communication may have obscured the obvious priority of one or another principle. In other cases, reflective balancing of the principles is required, so as to reveal which principle should take priority. Finally, there are cases where the facts and the principles may be clear, but a genuine ethical dilemma may remain. We are aware that there are such dilemmas in clinical medicine, but we are convinced that many ethical

MEDICAL INDICATIONS	PATIENT PREFERENCES
The Principles of Beneficence and Nonmaleficence	The Principle of Respect for Autonomy
1. What is the patient's medical problem? history? diagnosis? prognosis?	1. Is the patient mentally capable and legally competent? Is there evidence of incapacity?
2. Is the problem acute? chronic? critical? emergent? reversible?	2. If competent, what is the patient stating about preferences for treatment?
3. What are the goals of treatment?	3. Has the patient been informed of benefits and risks, understood this information, and given consent?
4. What are the probabilities of success?	
5. What are the plans in case of therapeutic failure?	4. If incapacitated, who is the appropriate surrogate? Is the surrogate using appropriate standards for decision making?
6. In sum, how can this patient be benefited by medical and nursing care, and how can harm be avoided?	5. Has the patient expressed prior preferences, e.g., Advance Directives?
	6. Is the patient unwilling or unable to cooperate with medical treatment? If so, why?
	7. In sum, is the patient's right to choose being respected to the extent possible in ethics and law?

QUALITY OF LIFE	CONTEXTUAL FEATURES
The Principles of Beneficence and Nonmaleficence and Respect for Autonomy 1. What are the prospects, with or without treatment, for a return to normal life? 2. What physical, mental, and social deficits is the patient likely to experience if treatment succeeds? 3. Are there biases that might prejudice the provider's evaluation of the patient's quality of life? 4. Is the patient's present or future condition such that his or her continued life might be judged undesirable? 5. Is there any plan and rationale to forgo treatment? 6. Are there plans for comfort and palliative care?	The Principles of Loyalty and Fairness 1. Are there family issues that might influence treatment decisions? 2. Are there provider (physicians and nurses) issues that might influence treatment decisions? 3. Are there financial and economic factors? 4. Are there religious or cultural factors? 5. Are there limits on confidentiality? 6. Are there problems of allocation of resources? 7. How does the law affect treatment decisions? 8. Is clinical research or teaching involved? 9. Is there any conflict of interest on the part of the providers or the institution?

problems can be reasonably resolved. When value conflicts are encountered, reasonable persons should make choices only after careful, honest consideration of the ethical aspects and the facts of the situation. . .

Reference

1. Kliever L.D., ed. *Dax's Case. Essays in Medical Ethics and Human Meaning.* Dallas: Southern Methodist University Press, 1989.

Casuistry and Clinical Ethics

Albert R. Jonsen

Albert Jonsen is an Emeritus Professor in the Department of Medical History and Ethics at the University of Washington, School of Medicine, where he served as chair from 1987 to 1999. He is also Co-Director and Senior Ethics Scholar at the Program in Medicine and Human Values, California Pacific Medical Center, San Francisco. He served as Commissioner on the National Commission for the Protection of Human Subjects of Biomedical and Behavioral Research (1974–1978) and on The President's Commission for the Study of Ethical Problems in Medicine (1979–1982). The essay that appears here reviews the method of casuistry and defends its use for contemporary bioethics.

The remoteness of academic ethics from moral problems was dramatically revealed in the 1960s. The war in Southeast Asia and the civil rights activities, which had fired the nation, excited college students. Moral philosophy courses and their teachers were expected to speak to these issues. Yet the finest elaboration of rule utilitarianism or ideal observer theory seemed weak instruments when brought to bear on anxious questions about civil disobedience and draft avoidance. I was a graduate student in ethics during these years; we were struggling to make our ethical theories match the reality of moral perplexities. Our professors were questioning the "relevance" of their theories. They began to write, at first hesitantly then with greater assurance, about conscientious objection and divestment in South African firms. They began to look at cases as seriously as they had looked at theories.

At about the same time, the interest in medical ethics was stimulated by the advent of cardiac transplantation and the rationing of renal dialysis. These interests brought the ethicists closer to the case than had any other issue. As they read about medical care and talked with physicians, they realized that the "case" was the unit of health care and the unwaning center of attention of its providers. They learned that cases were filled with details, obscured by many unknowns, described differently by different observers. They came to recognize stable elements about which generalizations could be competently made and idiosyncratic features that were irreducibly singular. Agents A and B were now this young Dr. Smith and that elderly, very sick Mrs. Jones. Act M became "turning off the respirator," or was it "allowing to die" or was it "killing"? . . .

Abridged from Albert Jonsen, "Casuistry and Clinical Ethics," *Theoretical Medicine*, Volume 7, 1986, pp. 65–74. Reprinted by permission of Kluwer Academic Publishers.

Those ethicists who did venture into the hospital discovered quickly that their philosophical skills, while certainly useful, needed supplementation. They needed to learn the added skill of *interpreting the case*. Many of these ethicists devised their own approaches to the cases, their methods, systems, paradigms and matrices. Common to all these was the need to comprehend the complexity of detail and the variety of values and principles that seemed embedded in the case. These ethicists often described their work as "helping to clarify thinking" about these difficult cases. But unquestionably, their "clarification" was far different than the programs of clarification proposed by the modern moral philosophers. One of the most distinguished of these, R. M. Hare, has recently expressed this in his book *Moral Thinking*. It is in the modern tradition, concerned with helping us think more clearly about moral issues, but, unlike the preponderant tradition, Hare affirms "we can get a long way by logic alone but, in the selection of principles for use in this world of ours, facts about the world and the people in it are relevant" ([3] p. 5).

A case is filled with facts about the world and the people in it. Ethicists' methods must find a way to take them into account, to think about principles in their presence and to draw conclusions that seem to fit them. This sort of ethicist then becomes a casuist *malgre lui*. Without knowing much about casuistry and casuists, they act as did those shadowy forebearers who counseled kings and conquistadors, bankers and bishops.

The Casuistical Method

What was the casuistical method? Strangely enough, it is difficult to discover. Although they produced a vast literature (600 works, many in multiple editions between 1550 and 1650), casuists rarely exposed their method in explicit terms. Indeed, they may not have thought much about method, as we understand the idea. Yet, several features of their style emerge after much reading of their dusty tomes.

The first feature is surprising: they did not apply principles to cases in any carefully deductive or inferential fashion. True, they universally acknowledged a doctrine of "natural law," but that doctrine, although elaborated in considerable intricacy by the theologians of the era, remained at a high level of abstraction and served more as a source of maxims than a system of premises. . . .

The "system" they were familiar with was not a philosophical structure of linked ideas as in the Cartesian or better, Spinozan fashion—an *ethica modo geometrica demonstrata*—but the system of the ancient discipline of rhetoric. It is hardly known today that rhetoric and moral philosophy were closely associated in the ancient and medieval education. Aristotle's *Rhetoric* (hardly opened by modern philosophers) begins with a synopsis of his *Nicomachean Ethics*, then goes on to set out the sorts of reasoning appropriate to recommendations about "acts to be pursued or avoided, with justification or condemnation of actions performed" [1]. The forms of reasoning, namely examples, the enthymeme, maxims, and commonplaces are, for Aristotle, the probabilistic counterparts of formal reasoning based on induction and deductions. Examples are real or fictitious presentations of cases; the enthymeme is a syllogism starting from a general proposition true only for the most part; maxims are familiar, widely accepted, but inconclusive statements about the moral life, such as "know thyself" or "nothing in excess." Certain basic and common understandings about the world and about people give structure to certain forms of argument which the rhetoricians called "topics" or "commonplace." In the most general form, these might be cause and effect, before and after, possible and impossible, greater or less. Cicero, also writing about rhetoric, adds to the Aristotelian framework the element of the particular occasion or the circumstances of the case. These elements came to be summarized as "Who did what? Where? Why? In what manner? With what help?" [2]

These rhetorical categories informed the minds of the casuists. The categories, more than the categories of formal logic, were used to formulate a case and devise a recommendation about "acts to be pursued or avoided, justification or condemnation of acts performed." The broad concepts of moral philosophy, whether Aristotle's virtues, Plato's ideals, Augustine's charity or Aquinas' natural law, were an inexhaustible source of examples and maxims. Casuists were not adherents of one or another master or school. All the ancients provided wisdom about the mystery of life. Also, the Church Law, Roman Law and Jurisprudence were dipped into when convenient. . . .

The "new" casuistry must be more than talking about cases. It must be an articulated art, that is, it must be able to discuss the singular and unique in terms that can be generally understood and appreciated. It must have the quality of moral discourse, that is, its judgments about particulars must reflect the features of universality and prescriptivity now commonly appreciated as essential to moral thought [3]. . . .

Modern casuists may avail themselves of certain features of classical casuistry, even without delving into the actual history of that arcane endeavor. . . . However, the features that can be profitably copied are first, reliance on paradigm cases, second, reference to broad consensus, and, finally, acceptance of "probable certitude."

The casuists' thinking moved from the clear and obvious cases toward the more problematic ones. They had a base, the manifest relevance of a strong principle to a certain case, and moved away from that base by the moral judgments, but the questions do not, in themselves, rule out the prudent assertion of "probable certitudes."

Ethicists associated with the provision of medical care can be casuists in ways that their counterparts in other fields do not enjoy. They daily meet cases in urgent need of practical resolution. They are aware that elegant theory and critical questions do not lead to answers for demanding clinical problems. They have at hand a series of paradigm cases which they understand and which can be readily communicated to their listeners. They have an ample body of carefully assessed opinions about these cases that reveal both areas of consensus and points of disagreement. They can assure their hearers, on the grounds of the paradigms, of considered opinions and of reasoning by analogy, that this or that resolution lies within the realm of "probable certitude" [4]. In this way, the spirits of the antique casuists inhabit the corridors of modern hospitals.

References

[1] Aristotle: 1941, *The Rhetoric*, R. McKeon (trans.), Random House, New York.

[2] Cicero: 1949, *De Inventione*, H. M. Hubbell (trans.), Harvard University Press, Cambridge, I, 33–44.

[3] Hare, R. M.: 1981, *Moral Thinking: Its Levels, Method and Point*, Clarendon Press, Oxford, p. 5.

[4] Jonsen, A. R., Siegler, M., Winslade, W.: 1982, *Clinical Ethics*, Macmillan Publishing Co., New York.

Getting Down to Cases:
The Revival of Casuistry in Bioethics

John Arras

A philosopher by training, John Arras is Porterfield Professor of Biomedical Ethics and Director of the Bioethics Minor Program at the University of Virginia, Department of Philosophy, Charlottesville, Virginia. This article critically evaluates the method of casuistry espoused by Al Jonsen and others. After identifying important shortcomings, Arras recommends ways of improving casuistic analysis in health care.

The Revival of Casuistry

Developed in the early Middle Ages as a method of bringing abstract and universal ethico-religious precepts to bear on particular moral situations, casuistry has had a checkered history (Jonsen and Toulmin, 1988). In the hands of expert practitioners during its salad days in the 16th and 17th centuries, casuistry generated a rich and morally sensitive literature devoted to numerous real-life ethical problems, such as truth-telling, usury, and the limits of revenge. By the late 17th century, however, casuistical reasoning had degenerated into a notoriously sordid form of logic-chopping in the service of personal expediency. To this day, the very term "casuistry" conjures up pejorative images of disingenuous argument and moral laxity.

In spite of casuistry's tarnished reputation, some philosophers have claimed that casuistry, shorn of its unfortunate excesses, has much to teach us about the resolution of moral problems in medicine. Indeed, through the work of Albert Jonsen (1980, 1986a, 1986b, 1988) and Stephen Toulmin (1981; Jonsen and Toulmin, 1988) this "new casuistry" has emerged as a definite alternative to the hegemony of the so-called "applied ethics" method of moral analysis that has dominated most bioethical scholarship and teaching since the early 1970s (Beauchamp and Childress, 1989). In stark contrast to methods that begin from "on high" with the working out of a moral theory and culminate in the deductivistic application of norms to particular factual situations, this new casuistry works from the "bottom up," emphasizing practical problem-solving by means of nuanced interpretations of individual cases.

Abridged from John Arras, "Getting Down to Cases: The Revival of Casuistry in Bioethics," *The Journal of Medicine and Philosophy*, Volume 16, 1991, pp. 29–51. Reprinted by permission of Kluwer Academic Publishers.

This paper will assess the promise of this reborn casuistry for bioethics education. . . .

Problems with the Casuistical Method

Since the new casuistry attempts to define itself by turning applied ethics on its head, working from cases to principles rather than vice-versa, it should come as no surprise to find that its strengths correlate perfectly with the weaknesses of applied ethics. Thus, whereas applied ethics, and especially deductivism, are often criticized for their remoteness from clinical realities and for their consequent irrelevance (Fox *et al.*, 1984; Noble, 1982) casuistry prides itself on its concreteness and on its ability to render useful advice to caregivers in the medical trenches. Likewise, if the applied ethics model appears rather narrow in its single-minded emphasis on the application of principles and in its corresponding neglect of moral interpretation and practical discernment, the new casuistry can be viewed as a defense of the Aristotelian virtue of *phronesis* (or sound, practical judgment).

Conversely, it should not be surprising to find certain problems with the casuistical method that correspond to strengths of the applied ethics model. I shall devote the second half of this essay to an inventory of some of these problems. It should be stressed, however, that not all of these problems are unique to casuistry, nor does applied ethics fare much better with regard to some of them.

What Is "a Case"?

For all of their emphasis upon the interpretation of particular cases, casuists have not said much, if anything, about how to select problems for moral interpretation. What, in other words, gets placed on the "moral agenda" in the first place, and why? This is a problem because it is quite possible that the current method of selecting agenda items, whatever that may be, systematically ignores genuine issues equally worthy of discussion and debate (O'Neil, 1988).

I think it safe to say that problems currently make it onto the bioethical agenda largely because health practitioners and policy makers put them there. While there is usually nothing problematic in this, and while it always pays to be scrupulously attentive to the expressed concerns of people working in the trenches, practitioners may be bound to conventional ways of thinking and of conceiving problems that tend to filter out other, equally valid experiences and problems. As feminists have recently argued, for example, much of the current bioethics agenda reflects an excessively narrow, professionally driven, and male outlook on the nature of ethics (Carse, 1989). As a result, a whole range of important ethical problems—including the unequal treatment of women in health care settings, sexist occupational roles, personal relationships, and strategies of *avoiding* crisis situations—have been either downplayed or ignored completely (Warren, 1989, pp. 77–82). It is not enough, then, for casuistry to tell us *how* to interpret cases; rather than simply carrying out the agenda dictated by health professionals, all of us (casuists and applied ethicists alike) must begin to think more about the problem of *which* cases ought to be selected for moral scrutiny.

An additional problem, which I can only flag here, concerns not the identification of "a case"—i.e., what gets placed on the public agenda—but rather the specification of "the case"—i.e., what description of a case shall count as an adequate and sufficiently complete account of the issues, the participants and the context. One of the problems with many case presentations, especially in the clinical context, is their relative neglect of alternative perspectives on the case held by other participants. Quite often, we get the attending's (or the house officer's) point of view on what constitutes "the case," while missing out on the perspectives of nurses, social workers and others. Since most cases are complicated and enriched by such alternative medical, psychological and social interpretations, our casuistical analyses will remain incomplete without them. Thus, in addition to being long, the cases that we employ should reflect the usually complementary (but often conflicting) perspectives of all the involved participants.

Is Casuistry Really Theory-Free?

The casuists claim that they make moral progress by moving from one class of cases to another without the benefit of any ethical principles or theoretical apparatus. Solutions generated for obvious or easy categories of cases adumbrate solutions for the more difficult cases. In a manner somewhat reminiscent of pre-Kuhnian philosophers of science clinging to the possibility of "theory free" factual observations, to a belief in a kind of epistemological "immaculate perception," the casuists appear to be claiming that the cases simply speak for themselves.

. . . One problem with this suggestion is that it does not acknowledge or account for the way in which different theoretical preconceptions help determine which cases and problems get selected for study in the first place. Another problem is that it does not explain what allows us to group different cases into distinct categories or to proceed from one category to another. In other words, the casuists' account of case analysis fails to supply us with principles of relevance that explain what binds the cases together and how the meaning of one case points beyond itself toward the resolution of subsequent cases. The casuists obviously cannot do without such principles of relevance; they are a necessary condition of any kind of moral taxonomy. Without principles of relevance, the cases would fly apart in all directions, rendering coherent speech, thought, and action about them impossible.

But if the casuists rise to this challenge and convert their implicit principles of relevance into explicit principles, it is certainly reasonable to expect that these will be heavily "theory laden." Take, for example, the novel suggestion that anencephalic infants should be used as organ donors for children born with fatal heart defects. What is the relevant line of cases in our developed "morisprudence" for analyzing this problem? To the proponents of this suggestion, the brain death debates provide the appropriate context of discussion. According to this line of argument, anencephalic infants most closely resemble the brain dead; and since we already harvest vital organs from the latter category, we have a moral warrant for harvesting organs from anencephalics (Harrison, 1986). But to some of those opposed to any change in the status quo, the most relevant line of cases is provided by the literature on fetal experimentation. Our treatment of the anencephalic newborn should, they claim, reflect our practices regarding nonviable fetuses. If we agree with the judgment of the National Commission that research which would shorten the already doomed child's life should not be permitted, then we should oppose the use of equally doomed anencephalic infants as heart donors (Meilaender, 1986).

How ought the casuist to triangulate the moral problem of the anencephalic newborn as organ donor? What principles of relevance will lead him to opt for one line of cases instead of another? Whatever principles he might eventually articulate, they will undoubtedly have something definite to say about such matters as the concept of death, the moral status of fetuses, the meaning and scope of respect, the nature of personhood, and the relative importance of achieving good consequences in the world versus treating other human beings as ends in themselves. Although one's position on such issues perhaps need not implicate any full-blown ethical theory in the strictest sense of the term, they are sufficiently theory-laden to cast grave doubt on the new casuists' ability to move from case to case without recourse to mediating ethical principles or other theoretical notions.

Although the early work of Jonsen and Toulmin can easily be read as advocating a theory-free methodology comprised of mere "summary principles," their recent work appears to acknowledge the point of the above criticism. Indeed, it would be fair to say that they now seek to articulate a method that is, if not "theory free," then at least "theory modest." Drawing on the approach of the classical casuists, they now concede an indisputably normative role for principles and maxims drawn from a variety of sources, including theology, common law, historical tradition, and ethical theories. Rather than viewing ethical theories as mutually exclusive, reductionistic attempts to provide an apodictic *foundation* for ethical thought, Jonsen and Toulmin now view theories as limited and complementary *perspectives* that might enrich a more pragmatic and pluralistic approach to the ethical life (1988, Chapter 15). They thus appear reconciled to the usefulness, both in research and education, of a severely chastened conception of moral principles and theories.

One lesson of all this for bioethics education is that casuistry, for all its usefulness as a method, is nothing more (and nothing less) than an "engine of thought" that must receive *direction* from values, concepts and theories outside of itself. Given the important role such "external" sources of moral direction must play even in the most case-bound approaches, teachers and students need to be self-conscious about which traditions and theories are in effect driving their casuistical interpretations. This means that they need to devote time and energy to studying and criticizing the values, concepts and rank-orderings implicitly or explicitly conveyed by the various traditions and theories from which they derive their overall direction and tools of moral analysis. In short, it means that adopting the casuistical method will not absolve teachers and students from studying and evaluating either ethical theories or the history of ethics.

Indeterminacy and Consensus

One need not believe in the existence of uniquely correct answers to all moral questions to be concerned about the casuistical method's capacity to yield determinate answers to problematical moral questions. Indeed, anyone familiar with Alastair MacIntyre's (1981) disturbing diagnosis of our contemporary moral culture might well tend to greet the casuists' announcement of moral consensus with a good deal of skepticism. According to MacIntyre, our moral culture is in a grave state of disorder: lacking any comprehensive and coherent understanding of morality and human nature, we subsist on scattered shards and remnants of past moral frameworks. It is no wonder, then, according to MacIntyre, that our moral debates and disagreements are often marked by the clash of incommensurable premises derived from disparate moral cultures. Nor is it any wonder that our debates over highly controversial issues such as abortion and affirmative action take the form of a tedious, interminable cycle of assertion and counter-assertion. In this disordered and contentious moral setting, which MacIntyre claims is *our* moral predicament, the casuists' goal of consensus based upon intuitive responses to cases might well appear to be a Panglossian dream.

One need not endorse MacIntyre's pessimistic diagnosis in its entirety to notice that many of our moral practices and policies bear a multiplicity of meanings; they often embody a variety of different, and sometimes conflicting, values. An ethical methodology based exclusively on the casuistical analysis of these practices can reasonably be expected to express these different values in the form of conflicting ethical conclusions. . . .

Conventionalism and Critique

. . . Eschewing any theoretical derivation of principles and insisting that the locus of moral certainty is the particular, the casuist asks "What principles best organize and account for what we have already decided?" Viewed from this angle, the casuistic project amounts to nothing more than an elaborate refinement of our intuitions regarding cases. As such, it begins to resemble the kind of relativistic conventionalism recently articulated by Richard Rorty (Rorty, 1989).

Obviously, one problem with this is that our intuitions have often been shown to be wildly wrong, if not downright prejudicial and superstitious. To the extent that this is true of *our own* intuitions about ethical matters, then casuistry will merely refine our prejudices. Any casuistry that modestly restricts itself to interpreting and cataloguing the flickering shadows on the cave wall can easily be accused of lacking a critical edge. If applied ethics might rightly be said to have purchased critical leverage at the expense of the concrete moral situation, then casuistry might be charged with having purchased concreteness and relevance at the expense of philosophical criticism. This charge might take either of two forms. First, one could claim that the casuist is a mere expositor of *established* social meanings and thus lacks the requisite critical distance to formulate telling critiques of regnant social understandings. Second, casuistry could be accused of ignoring the power relations that shape and inform the social meanings that its practitioners interpret. . . .

Reinforcing the Individualism of Bioethics

Analytical philosophers working as applied ethicists have often been criticized for the ahistorical, reductionist, and excessively individualistic character of their work in bioethics (Fox *et al.*, 1984; Noble, 1982; MacIntyre, 1982). While the casuistical method cannot thus be justly accused of importing a short-sighted individualism into the field of bioethics—that honor already belonging to analytical philosophy—it cannot be said either that casuistry offers anything like a promising remedy for this deficiency. On the contrary, it seems that the casuists' method of reasoning by analogy only promises to exacerbate the individualism and reductionism already characteristic of much bioethical scholarship.

Consider, for example, how a casuist might address the problem of heart transplants. He or she might reason like this: Our society is already deeply committed to paying for all kinds of "half-way technologies" for those in need. We already pay for renal dialysis and transplantation, chronic ventilatory support for children and adults, expensive open-heart surgery, and many other "high tech" therapies, some of which might well be even more expensive than heart transplants. Therefore, so long as heart transplants qualify medically as a proven therapy, there is no reason why Medicaid and Medicare should not fund them (Overcast *et al.*, 1985).

Notwithstanding the evident fruitfulness of such analogical reasoning in many contexts of bioethics, and notwithstanding the possibility that these particular examples of it might well prevail against the competing arguments on heart transplantation, it remains true that such contested practices raise troubling questions that tend not to be asked, let alone illuminated, by casuistical reasoning by analogy. The extent of our willingness to fund heart transplantation has great bearing on the kind of society in which we wish to live and on our priorities for spending within (and without) the health care budget. Even if we already fund many high technology procedures that cost as much or more than heart transplants, it is possible that this new round of transplantation could threaten other forms of care that provide greater benefits to more people; and we might therefore wish to draw the line here (Massachusetts Task Force, 1984; Annas, 1985).

The point is that, no matter where we stand on the particular issue of heart transplants, we *might* think it important to raise such "big questions," depending on the nature of the problem at hand. We might want to ask, to borrow from a recent title, "What kind of life?" (Callahan, 1990). But the kind of reasoning by analogy championed by the new casuists tends to reduce our field of ethical vision down to the proximate moral precedents, and thereby suppresses the important global questions bearing on who we are and what kind of society we want. The result is likely to be a method of moral reasoning that graciously accommodates us to any and all technological innovations, no matter what their potential long-term threat to fundamental and cherished institutions and values.

Conclusions

. . . It remains to be seen whether casuistry, as a program in practical ethics, will be able to marshall sufficient internal resources to respond to these criticisms. Whatever the outcome of that attempt, however, an equally promising approach might be to incorporate the insights and tools of casuistry into the methodological approach known as "reflective equilibrium" (Rawls, 1971; Daniels, 1979). According to this method, the casuistical interpretation of cases, on the one hand, and moral theories, principles and maxims, on the other, exist in a symbiotic relationship. Our intuitions on cases will thus be guided, and perhaps criticized, by theory; while our theories and moral principles will themselves be shaped, and perhaps reformulated, by our responses to paradigmatic moral situations. Whether we attempt to flesh out this method of reflective equilibrium or further develop the casuistical program, it should be clear by now that the methodological issue between theory and cases is not a dichotomous "either/or" but rather an encompassing "both-and."

In closing I would like to gather together my various recommendations, strewn throughout this paper, for the use of casuistry in bioethics education:

1. Use real cases rather than hypotheticals whenever possible.
2. Avoid schematic case presentations. Make them long, richly detailed, messy, and comprehensive. Make sure that the perspectives of all the major players (including nurses and social workers) are represented.
3. Present complex sequences of cases that sharpen students' analogical reasoning skills.
4. Engage students in the process of "moral diagnosis."
5. Be mindful of the limits of casuistical analysis. As a mere engine of moral argument, casuistry must be supplemented and guided by appeals to ethical theory, the history of ethics, and moral norms embedded in our traditions and social practices. It must also be supplemented by critical social analyses that unmask the power behind much social consensus and raise larger questions about the kind of society we want and the kind of people we want to be.

Bibliography

Annas, G.: 1985, "Regulating heart and liver transplants in Massachusetts," *Law, Medicine and Health Care* 13(1), 4–7.

Beauchamp, T.L. and Childress, J.F.: 1989, *Principles of Biomedical Ethics*, 3rd edition, Oxford University Press, New York, New York.

Callahan, D.: 1990, *What Kind of Life?*, Simon and Schuster, New York, New York.

Carse, A.L.: 1991, "The 'voice of care': Implications for bioethics education," *Journal of Philosophy and Medicine* 16, 5–28.

Daniels, N.: 1979, "Wide reflective equilibrium and theory acceptance in ethics," *The Journal of Philosophy* 76, 256–82.

Fox, R.C. and Swazey, J.P.: 1984, "Medical morality is not bioethics—medical ethics in China and the United States," *Perspectives in Biology and Medicine* 27, 336–360.

Jonsen, A.R.: 1980, "Can an ethicist be a consultant?," in V. Abernethy (ed.), *Frontiers in Medical Ethics*, Ballinger Publishing Company, Cambridge, Massachusetts, pp. 157–171.

Jonsen, A.R.: 1986a, "Casuistry and clinical ethics," *Theoretical Medicine* 7, 65–74.

Jonsen, A.R.: 1986b, "Casuistry," in J.F. Childress and J. Macgvarrie (eds.), *Westminster Dictionary of Christian Ethics*, Westminster Press, Philadelphia, Pennsylvania, pp. 78–80.

Jonsen, A.R. and Toulmin, S.: 1988, *The Abuse of Casuistry*, University of California Press, Berkeley, California.

MacIntyre, A.: 1981, *After Virtue*, University of Notre Dame Press, Notre Dame, Indiana.

Massachusetts Task Force on Organ Transplantation: 1984, *Report of the Massachusetts Task Force on Organ Transplantation*, Boston, Massachusetts.

Meilaender, G.: 1986, "The anencephalic newborn as organ donor: Commentary," *Hastings Center Report* 16, 22–23.

Noble, C.: 1982, "Ethics and experts," *Hastings Center Report* 12, 7–9.

O'Neill, O.: 1988, "How can we individuate moral problems?" in D.M. Rosenthal and F. Shehadi (eds.), *Applied Ethics and Ethical Theory*, University of Utah Press, Salt Lake City, pp. 84–99.

Overcast, D. *et al.*: 1985, "Technology assessment, public policy and transplantation," *Law, Medicine and Health Care* 13 (3), 106–111.

Pascal, B.: 1981, *Lettres écrites à un provincial*, A. Adam (ed.), Flammarion, Paris.

Rawls, J.: 1971, *A Theory of Justice*, Harvard University Press, Cambridge Massachusetts.

Rorty, R.: 1989, *Contingency, Irony, and Solidarity*, Cambridge University Press, Cambridge, England.

Toulmin, S.: 1981, "The tyranny of principles," *Hastings Center Report* 11, 31–39.

Warren, V.: 1989, "Feminist directions in medical ethics," *Hypatia* 4, 73–87.

Feminist and Medical Ethics:
Two Different Approaches to Contextual Ethics

Susan Sherwin

Trained in philosophy at Stanford University during the early 1970s, Susan Sherwin currently teaches philosophy and women's studies at Dalhousie University. Her essay distinguishes feminist ethics from other methods of analysis. Unlike casuistry and narrative ethics, feminist ethics sees its task as calling attention to the particular features of context that oppress women. In health care, this means exposing patriarchal features of medical practice.

Introduction

. . . In this paper, I shall explore the ways in which context is appealed to in both feminist and medical ethics, and I shall argue that particular sorts of details should be included in the recommended narrative approaches to ethical problems. In particular, I claim that incorporating an explicitly feminist political analysis in our discussion of context is critical to our ethical deliberations. . . .

The Role of Context in Feminist Ethics

Turning first to examine the role of context in feminist ethics, we must acknowledge the important influence of Carol Gilligan (1982). In identifying a female tendency to approach ethical problems in a personalized, contextual manner, Gilligan helped articulate the sense of alienation many women have experienced in trying to work within the structures of contemporary moral theory.[1] She identified distinct masculine and feminine voices in ethical reasoning, allowing us to recognize that mainstream ethical theory has been carried on in a voice that is overwhelmingly masculine—the voices of women have been largely excluded or ignored. Feminists stress that it is important in ethics, as in all fields, to include women's moral experiences and reasoning in the deliberations. Hence, most theorists seeking to develop a feminist approach to ethics have given serious consideration to the gender map which Gilligan has provided and have tried to incorporate many of her observations into their approach to ethics.[2]

Abridged from Susan Sherwin, "Feminist and Medical Ethics: Two Different Approaches to Contextual Ethics," *Hypatia*, Volume 4, Summer 1989, pp. 57–71. Reprinted by permission of the author.

In her research, Gilligan found that girls and women tend to approach ethical dilemmas in a contextualized, narrative way that looks for resolution in particular details of a problem situation; in contrast, boys and men seem inclined to try to apply some general abstract principle without attention to the unique circumstances of the case. For instance, in Kohlberg's famous Heinz case, Gilligan found that males tended to answer in terms of the logical implications of a general rule, such as that stealing is wrong or that the duty to save a life outweighs other moral rules. In contrast, she found that female subjects tried to preserve relationships and to find new options through better communication and a presumption of co-operation; they tended to respond by seeking more information or by trying to re-conceive the terms set by the example. Gilligan recognized two different patterns of reasoning here: one which pursues universal rules in an endeavour to ensure fairness, and one which is focused on the actual feelings and interactions of those involved. The first approach, which she found to be associated with male moral thinking, she labelled an ethic of justice; the latter, which she found to be more commonly exercised by female subjects, she identified as an ethic of care.

The gender difference she describes is two-fold, characterized by differences in both scope and values: men seem to be preoccupied with developing comprehensive, generalizable, abstract ethical systems which are based on rights, while women seem to be concerned with understanding the specific human dynamics of a situation and, hence, concentrate on particular narrative details with the aim of avoiding hurt and providing care. As a result, we can identify distinct methodological differences between men and women in their approaches to morally troubling questions. But we should be cautious in interpreting the significance of the gender correlations of these differences; much of the discussion in feminist ethics has been occupied with evaluating the implications of developing what might be called a feminine ethics or a woman-centered ethics.[3] Since, in our sexist society, gender is inseparable from oppression, we should be sensitive to the fact that characteristics associated with gender are also likely to be associated with oppression. Obviously, it is important that women's distinctive moral reasoning be (at last) acknowledged as worthy of respect, but many feminists—including Gilligan herself—have expressed caution in interpreting the gender patterns her research reveals in the context of a society that systematically oppresses women. In particular, many feminists are wary of enthusiasm for virtues like caring which are associated with both gender and oppression.[4] Gilligan recommends that we work towards an androgynous ethics that could combine elements of both approaches, but other feminists have pointed out that notions of androgyny seem, themselves, to perpetuate the old gender system. . . .

The Role of Context in Medical Ethics

The theme of seeking a practical, context-specific approach to ethics is not restricted to feminist literature, however. The literature of medical ethics also contains frequent discussions about the inadequacy of abstract moral reasoning for resolving real moral dilemmas; there, too, we can find evidence of a widespread recognition that we must go beyond "mere theory." Further, there is frequent mention of the need to engage considerations of caring in medical ethics, usually couched in the language of the beneficence which is owed to patients. When placed in context (even if hypothetically), medical dilemmas are often discussed in terms that appear to rank sensitivity and caring ahead of applications of principle.[5]

In the "early days" of philosophical medical ethics (i.e., the 1970s), there was an attempt to try to fit responses to moral dilemmas into the general framework offered by standard moral theories, especially utilitarianism and Kantian deontology. It became apparent quite early on, however, that the simple appeal to theory and principle did not offer satisfying analyses of the sorts of dilemmas that arise in medical ethics. Case studies became a central element in influential journals, in many textbooks, and in individual articles. The texture and the details of cases have become important in trying to decide about perennial issues such as confidentiality, truth-telling, and euthanasia. Clear answers deduced from precise principles are not at hand for most of the topics addressed; many authors now accept the assumption that universal principles cannot be found which will govern such issues in all cases. . . .

Some philosophers still entrenched in mainstream moral theory have difficulty in seeing the distinction being cited here, since surely all moral theories are context sensitive to some degree. Kantian theory, for example, demands an interpretation of context in order to determine which maxim applies in a given case. But Kantian theory does assume that the maxims, once identified, will be universal and our policy on suicide, truth-telling, or confidentiality will be consistent across the full spectrum of relevant cases. It does not direct us to make our ethical assessments in terms of particular details of the lives of the individuals.

Utilitarianism is often espoused precisely as an antidote to such a rigid ethics. It certainly seems to be extremely sensitive to contextual features, in that it recommends we calculate relevant utilities for all possible options in a given set of circumstances. Nonetheless, it discounts some important features which medical ethics and feminist ethics consider important. Utilitarianism requires that we calculate the relevant utility values for all persons (or beings) affected by an action or practice and proceed according to a calculation of the relevant balances. In contrast, those engaged in doing feminist or medical ethics often reflect a desire to take account of the details of specific relationships and to give added weight to some particular utility-related qualities like caring and responsibility. Many of those engaged in feminist ethics diverge even further from standard utilitarianism, for they argue that the preferences of the oppressed ought to be counted differently from those of the dominant group. (Feminist objections to pornography, for instance, do not rely merely on the weighing of harms done against pleasure produced but reflect concern about the dehumanizing effect of the message of pornography whatever the utilities involved turn out to be.) In feminist and medical ethics, it is important to consider factors that do not carry any special weight in utilitarianism. There is need to look at the nature of the persons and the relationships involved in our analysis and not merely to record such values as preference satisfaction or pleasure or pain; while the latter values are specifically held, their importance comes from some abstract sum and not from their attachment to any particular persons in particular situations. Hence, neither Kantian nor utility theory satisfies the requirement of particularity as it is conceived in feminist and medical ethics. . . .

Some Requirements for a Feminist Ethics

For medical ethics to be thought feminist, it must also reflect a political dimension, but this is mostly lacking in the literature to date. Although there are currently many diverse attempts to characterize feminist ethics, all share some political analysis of the unequal power of women and men, of white people and people of colour, of first world and third world people, of rich and poor, of healthy and disabled, etc.[6] Ours is a world structured by hierarchies and a sense of supremacy on the part of the powerful; there are numerous social patterns which shape the people we are and the sorts of relationships we will have with one another. In attending to the quality of actual interactions among people in ethics, we need to account for the influence of social and political factors on the nature of those relationships. From either the caring or the justice perspective (to use Gilligan's language), we can see that empowerment of people who are currently victims of oppression is an ethical as well as a political issue, and ethical investigations of particular problem areas should reflect these dimensions. Many feminist critics have observed that current medical practice constitutes a powerful social institution which contributes to the oppression of women. They have demonstrated that the practice of medicine serves as an important instrument in the continuing disempowerment of women (and members of other oppressed groups) in society and thrives on hierarchical power structures. By medicating socially induced depression and anxiety, medicine helps to perpetuate unjust social arrangements. With its authority to define what is normal and what is pathological and to coerce compliance to its norms, medicine tends to strengthen patterns of stereotyping and reinforce existing power inequalities. It serves to legitimize practices such as woman battering or male sexual aggression that might otherwise be evaluated in moral and political terms.[7]

Nonetheless, the discussion in medical ethics to date has been largely myopic, failing to comment on this important political role of medicine. That is, the institution of medicine is usually accepted as given in discussions of medical ethics, and debate has focused on certain practices within that structure: for example, truth-telling, obtaining consent, preserving confidentiality, the limits of paternalism, allocation of resources, dealing with incurable illness, and matters of reproduction. The effect is to provide an ethical legitimization of the institution overall, with acceptance of its general structures and patterns. With the occasional exception of certain discussions of resource allocation, it would appear from much of the medical ethics literature that all that is needed to make medical interactions ethically acceptable is a bit of fine-tuning in specific problem cases.

A good indication of the legitimizing function of medical ethics can be seen by noting its gradual acceptance among those who are influential within the medical profession. Increasingly, medical practitioners seem to be recognizing the value of incorporating discussions of medical ethics within their own work, for they can thereby demonstrate their serious interest in moral matters. Such serious professional concern in matters of medical ethics serves to encourage the public to place even greater trust in their judgment. Keeping the scope of medical ethics narrowed to specific problems of interaction helps physicians maintain their supportive stance towards it.

Feminists must be critical of the fact that medical ethics has remained largely silent about the patriarchal practice of medicine. Few authors writing on medical ethics have been critical of practices and institutions that contribute to the oppression of women. The deep questions about the structure of medical practice and its role in a patriarchal society are largely inaccessible within the framework; they are not considered part of the standard curriculum in textbooks of medical ethics. Consequently, medical ethics, as it is mostly practiced to date, does not amount to a feminist approach to ethics. . . .

Notes

1. For an apt description of the "moral madness" women commonly experience when confronted with patriarchal ethical demands, see Morgan (1987).
2. See, for instance, the various discussions in the collection, *Women and Moral Theory*, Kittay and Meyers, 1987.
3. Nel Noddings (1984) has spelled out the full implications of pursuing the feminine approach to ethics exclusively in *Caring*. Though I would not classify her as a feminist, Noddings has given theoretical voice to the ethic of care described by Gilligan, rejecting all aspects of an ethics based in abstract principles in favour of an ethics concerned only with particular relationships based on caring.
4. See, for instance, the arguments put forward by Houston (1987) and Wilson (1988).
5. This tendency seems to me to be especially common in the contributions of physicians to medical ethics; it is less apparent in the philosophical discussions in the field.
6. I am well aware of the rich diversity of views clustered under the label of "feminist analysis," but I think that there are some core views that transcend the differences which divide feminists in their internal debates. For the purposes of this paper, I will focus only on the common themes which include a recognition that women are in a subordinate position in society, that oppression is a form of injustice and hence is intolerable, that there are further forms of oppression in addition to gender oppression (and that there are

women victimized by each of these forms of oppression), that it is possible to change society in ways that could eliminate oppression, and that it is a goal of feminism to pursue the changes necessary to accomplish this. I believe that the argument presented in this paper is unchanged however we explain the cause of women's oppression and whatever we imagine is best for bringing about the desired changes. Therefore, I shall speak of a feminist analysis without being specific here about which particular variation I have in mind.

7. See Stark, Flitcraft, and Frazier (1983). For a more far ranging discussion, see the powerful indictment of medicine's contribution to the oppression of women in the survey by Ehrenreich and English (1979).

References

Ehrenreich, Barbara and Dierdre English. 1979. *For her own good: 150 years of the experts' advice to women.* Garden City, NY: Anchor Press, Doubleday.

Houston, Barbara. 1987. Reclaiming moral virtues: Some dangers of moral reclamation. In *Science, morality, and feminist theory.* Marsha Hanen and Kai Nielsen, eds. *Canadian Journal of Philosophy* (supplementary volume) 13.

Kittay, Eva Feder and Diana T. Meyers, eds., 1987. *Women and moral theory.* Totowa, N.J.: Rowman and Littlefield.

Morgan, Kathryn. 1987. Women and moral madness. In *Science, morality, and feminist theory.* Marsha Hanen and Kai Nielsen, eds. *Canadian Journal of Philosophy* (supplementary volume) 13.

Noddings, Nel. 1984. *Caring: A feminine approach to ethics and moral education.* Berkeley: University of California Press.

Stark, Evan, Anne Flitcraft, and William Frazier. 1983. Medicine and patriarchal violence: The social construction of a "private" event. In *Women and health: The politics of sex in medicine.* Elizabeth Fee, ed. Farmingdale, N.Y.: Baywood Publishing Company.

Wilson, Leslie. 1988. Is a "feminine" ethics enough? *Atlantis* 5 (17):15–23.

NARRATIVE APPROACHES

From *Doctors' Stories*

Kathryn M. Hunter

Kathryn Montgomery (formerly Kathryn M. Hunter) is on the faculty of Northwestern University School of Medicine and serves as Director of Northwestern's Medical Humanities and Bioethics Program. The essay below, excerpted from her 1991 book, *Doctors' Stories*, explores how narrative shapes clinical judgment. Hunter defends a case-based empirical way of making ethical decisions that requires health professionals to become familiar with patients' stories.

Early in his illness, Tolstoy's Ivan Ilych recognizes in his physician the "new method" he has perfected for himself as an examining magistrate. In that office he "acquired a method of eliminating all considerations irrelevant to the legal aspect of the case, and reducing even the most complicated case to a form in which it would be presented on paper only in its externals, completely excluding his personal opinion of the matter, while above all observing every prescribed formality." Now, with his physician, there is

> the sounding and listening, and the questions which called for answers that were foregone conclusions and were evidently unnecessary, and the look of importance which implied that "if only you put yourself in our hands we will arrange everything—we know indubitably how it has to be done, always in the same way for everybody alike." It was all just as it was in the law courts. The doctor put on just the same air towards him as he himself put on towards an accused person.[1]

Although impartiality and restraint in the new bureaucratic administration of power are preferable to bribe-taking and favoritism, we are meant to see the deficits of the "new method" for Ivan Ilych—first as a person who practices it and then, once he is ill, as the victim of such cool professionalism in its medical form. Patients need more than diagnoses, and to supply their need physicians must have richer case narratives than the traditional medical case history.

Narrative shapes clinical judgment. In medical practice, the vast body of knowledge about human biology is applied to the patient analogically through

---

Abridged from Kathryn M. Hunter, *Doctors' Stories.* 1991, pp. 148–160. Reprinted by permission of Princeton University Press.

narratives of the experience of comparable instances. The capacity to provide good medical care depends upon both the physician's stock of clinical stories and an understanding of how they are (or are not) relevant to this particular case. Maxims and rules are absorbed during clinical training, just as the principles of the biomedical sciences are memorized in the first two years of medical school. But despite the increased specificity of practice-based maxims, their applicability is still often uncertain. As in other case-based inquiry—law, moral theology, criminal detection—judgment in medicine is shaped (and the relativism of the individual interpretation of principle controlled) by comparing the narrated circumstances of the present case with others of more or less the same kind.

The prevalence of narrative in medicine suggests that this case-based, experiential way of knowing is well accepted in clinical practice, even if its implications are seldom acknowledged. Ethical decisions are made in much the same case-based way, and as a consequence, philosophers fresh from the classroom in the early days of the bioethics movement were frustrated at what seemed to be physicians' ignorance or unconsciousness of the overarching principles that guided their ethical decision making. They soon learned the strength of case-based deliberations. Not surprisingly, the recent defense of the philosophical position of casuistry has been undertaken by philosophers working in medicine and influenced by the methods used in the work of their clinical colleagues.[2] Medicine, as Leon Kass has observed, is a fertile ground for understanding "the moral relation between knowledge or expertise and the concerns of life."[3]

The construction of the case history is an integral part of medical thinking, essential to clinical education and to making decisions about the care of an individual patient. But good decisions about patient care beyond the diagnosis call for a richer narrative than the traditional medical case. In an era dominated by chronic disease, a physician's narrative stock should include not only clinical cases, which traditional medical education provides, but also a practical knowledge of human character and life patterns for both the well and the ill. In addition to an encyclopedic, Sherlock Holmesian knowledge of pathological cases, physicians need a literary sense of the lives in which illness and medical care take place. In the past twenty-five years medical thinkers have sought to broaden the practitioner's understanding of the individual case. Whether by increasing the number and kind of "facts" regarded as relevant to the grasp of a case, by organizing what is observed into a more coherent chronicle, or by attending to narrative shape and subtleties of representation, these critics have expanded and enriched the concept of the "case" itself. Taken together, they are working toward reshaping the medical narrative. Their arguments have met with only mixed success at a time when medical practice has been altered by technology and economic constraints. Yet these pressures only make more necessary a richer sense of patients and their life choices. Physicians feel strongly the danger of becoming mere technicians. Whether they treat patients in brief, almost anonymous encounters or take care of the chronically ill, intellectual and moral support for patient care can be found in the "color and life" of enriched case narrative. A larger sense of the patient's story enlivens the everyday practice of medicine and improves the quality of attention given to the person who is ill.

The Shield of Achilles

In the midst of the *Iliad* Homer interrupts the progress of the war to describe the making of a new shield for Achilles. It is a narrative digression that (among other things) is itself about narrative art. Because the besieged Trojans have driven the Greeks back to their ships, Achilles has allowed his friend Patroclus to go into battle disguised in the armor Achilles inherited from his father. Patroclus has been slain by Hector and his body and the armor have been taken captive. Grief-stricken and enraged, Achilles is ready at last to return to battle, and his mother, the nymph Thetis, persuades Hephaistos, "smith of the strong arms," to forge new armor for her son. At this turning point in the war Homer gives us a long digressive account of the wonders of the god's creation. The shield is marvelous. Homer describes it as "fivefold," five layers thick, but it is easy to imagine that the phrase might mean five layers folded like pages, for the whole world is represented there. First is the physical universe:

> earth, heaven, and sea,
> unwearied sun, moon waxing, all the stars
> that heaven bears for garland.
> (XVIII, 557–560, trans. Robert Fitzgerald)

Next are two cities, one at peace, celebrating weddings and adjudicating a blood quarrel; the other at war, besieged like Troy, refusing a treaty, attempting an ambush, breaking into open battle. As Homer describes them, these are not still scenes but moving pictures, full of action. They are eventful slices of narratable life. Only with difficulty can we anchor such descriptions to static representation on the surface of a shield, and as readers immersed in the narrative we do not try. It seems right that this visual representation convey action over time; it is, after all, a god's handiwork.

Cosmology and politics are not the shield's only themes nor, evidently, has Hephaistos begun to fill the available space. He adds a field being plowed, the gathering of a king's wheat, and the preparation by the people of a celebratory feast. Elsewhere, an irrigated vineyard is harvested by singing children; cattle in a pasture are surprised by a pair of lions; shepherds not far away tend their sheep in a quiet valley. Last, there is a dancing floor where young men and women link arms, moving effortlessly in a circle, and all around the shield's rim runs the circle of the mighty ocean stream.

Here in the midst of the *Iliad*'s account of bravery, death, and loss we find a full, richly detailed representation of human life, and, moreover, it is a part of what we might have expected to be a simple piece of military equipment. The digression's unexpected length and its intrusion on the principal action of the epic suggest the importance of the shield and its art. Despite Achille's renown and his connection with the gods, the scenes represent small, ordinary events. They are the life out of which the story of the *Iliad* arises. Their representation of the whole of human experience places war and the epic poem itself in a larger context of human activity, and we are led to feel that this representation is a part of the protection offered to the hero returning to battle.

The shield is an epitome of narrative and its representation of human life. Works of art generally give us intellectual pleasure by representing to us the things that are not. These are not lies, of course, but fiction: true accounts of the way things would be, if only we had experienced them in just this way. In the ninth year of the Trojan War, Hephaistos paints enameled pictures of a besieged, chaotic city and a peaceful, orderly one. We recognize Troy in the one and are reminded, perhaps reassured, that, outside the *Iliad*, the other still exists. Beyond is the countryside at peace where life is marked by (and stories are part of) the regular events of the seasons—plowing, harvest, festival. There nature and art provide all that is narratable: lions rampage; one boy among the happy grape harvesters sings "a summer dirge"; two tumblers handspring across the circle of the "magical" dance.

The representative wholeness of Hephaistos's creation, which might be a metaphor for all art for all of us, seems to have a special relevance for physicians. Literature in particular constitutes a source of knowledge. For those to whom the experiences are familiar, narrative is confirming; for those reading about something new, the view of human possibility is enlarged. This is especially useful for physicians and for medical students. Their education often proceeds as if the practice of medicine were only a science and not also a social enterprise subject to cultural and emotional variants. Illness is assumed to be an inhuman evil, death is wrong, and medicine's task is simply described: to restore the sick to health, preventing or at least forestalling death. There are few sick people and no physicians at all in medical textbooks and none of the human experience that gives illness and the practice of medicine their meaning. "Cure sometimes, relieve often, comfort always." Only the old maxim, more often read than heard these days, exists as a reminder of medicine's role in the face of the entropic reality of human life.

Literature and Medicine

Where does medicine look for an understanding of its activities and its values? What is the source of its ideas about its place in the sum of human activity? In its confidence in its knowledge and the significance of that knowledge, the medical profession emulates early twentieth-century positivist physical scientists rather than the social scientists and the humanists whose work physicians' work more nearly resembles. The human sciences, by contrast, have begun to study themselves obsessively. Anthropology and sociology puzzle over how other human beings can be reliably understood, and historians are absorbed by the impossibility of separating the event from its telling. Social science generally has been gripped by debate over its suspension, like quantum mechanics, between the knower and what is known. Literary scholars approach their texts armed with theories of reading, writing, and knowing, and philosophers debate not only the foundations of knowledge but their very possibility. Medicine, however, has remained turned resolutely outward toward the "real world."

Medicine cannot of itself address questions of its meaning or the meaning of illness. This is not a part of its province as it is currently conceived, but is rather the province of religion, the humanities, and the values-oriented social sciences. Philosophers, historians, and sociologists have studied medicine as they have studied science, and these studies include valuable descriptions of what medicine is and does. Although many literary critics find inadequate the view that literature provides an unmediated record of social reality, it is nevertheless a vivid means of understanding the physician's often quite lonely job, the hard work of nursing the desperately ill, the patient's experience of illness, the process of dying. Fiction, poetry, and drama all offer medicine their visions of human experience. Within its representation of the full range of human possibility, we may see how doctoring, being sick, and learning to heal fit into the whole.

Literature and medicine are distinguished from most other studies of humankind by the particular account they give of individual experience. Just as literary criticism can be abstract, so can medical knowledge, especially in its textbook form; but as medicine, it has meaning only in its applicability to the individual case. The medical case history, like literary narrative, can be about only one set of circumstances at a time. Medicine has its origin—in several senses—in the patient's presentation to medical attention, and it has its end—also in more than one sense—in the treatment of that sick person. Like literature, the medical case history embodies the attention medicine accords the individual. The case history concerns instances that at once test our generalizations about human beings and embody the aggregate of human experience.

Reading about Patients

This particularization of widespread experience is easily seen in contemporary stories about illness. In recent years, the patient's story of "my operation" has gathered force in a flood of books and articles. There are a few traditional plays and stories in which illness plays an essential part—the Book of Job, Sophocles's *Philoctetes* come to mind. But only recently, as we have begun to live long enough to fall victim to chronic disease, have very many stories been written to examine illness as the individual experiences it. Thomas Mann's *The Magic Mountain* is in some ways the model of the genre, yet in that novel the hero is distanced from the narrator and both are distanced from the author; its story becomes a metaphor for a sick society. The contemporary narrative of illness in the United States is instead a midlife version of the growing-up novel. Hero-narrator and author are collapsed into one figure in an autobiographical (or autobiographically fictional) account of an individual's growth in circumstances not of his or her choosing.

The lack of choice compels our attention. As medicine has become central to our understanding of what it means to be human in an unreligious, techno-cratic time, a voracious public appetite has developed for medical narrative: fic-tion, autobiography, reportage, played out in drama. These stories are about the failure of control and the threat of extinction. They enable us to think about the value society places on a human life, about the meaning of pain, the definition of the person, the limits of even an American autonomy, and our attitudes to-ward authority, choice, chance, and the death of the individual. Much of our public discourse, even before the AIDS epidemic, concerned disease and med-ical care. Fleshed out with "human interest" detail, case narrative has become the stuff of the well-publicized adventures of transplantation teams. It is a main-stay of television's daily dose of hospital crisis and intrigue. Pathographies of the seriously ill have become a subgenre of contemporary biography.[4] Issues of health-care policy are debated and necessarily have their legal manifestation not in the abstract realm of public policy but as individual cases. Names of individ-ual people come to stand for issues—and now and then for their resolution: Dax Cowart, Karen Ann Quinlan, Baby Jane Doe, Elizabeth Bouvier, Rock Hudson, Baby M., Nancy Cruzan. Medical cases are often the germ of fiction and, espe-cially, drama. Brian Clark's *Whose Life Is It Anyway?*,[5] which stirred public dis-cussion of unwanted life-sustaining medical treatment, was very probably inspired by the 1975 discussion of Cowart's case by Robert White and H. Tris-tram Engelhardt, Jr., in the *Hastings Center Report*.[6] In 1985 Larry Kramer's *The Normal Heart*[7] and William Hoffman's *As Is*[8] gave their newly alarmed off-Broadway audiences a glimpse of the human and political costs of ignoring AIDS. We look to such accounts of illness to tell us who we are as a society. Their crises concern the exercise of individual rights, the problem of balancing conflicting interests in decision making, and the allocation of resources that our political choices have made scarce. In their imaginative exploration of the ordi-nary person's opportunity for self-definition and even for heroism, they depict for us a boundary condition of our humanity.

These stories and plays and autobiographies are based on what are, at other times and places, with other narrators, plain, stripped-down medical case histo-ries. They go beyond the conventions of medical storytelling to supply plots and themes missing from medical narrative. Indeed, pathographies, especially first-person narratives of fatal illness, seem to have been written by patients precisely to supply those things that the case history rigorously excludes. Many of them are quite forthright about this motive. They write not simply to supply that lack for us as readers (although, having few other ways to think about the unthink-able, we are morbidly curious about fatal diagnoses) but primarily to repair the loss to themselves. As Eric Cassell has pointed out, suffering is quite distinct from pain and even from dying, and, although the goal of medicine is the relief of suffering, many patients suffer not from their disease but from their medical treatment.[9] Pathographies address and sometimes seek to avenge the damage done by medical care to their bodies, care that took no care of them. They tackle complex questions of human suffering and response to illness, the difficulties and hopes of the doctor-patient relationship, the acceptance or rejection of medical therapy, and the meaning of illness in the life of the person who is ill. In so doing, they bring to bear on the medical "facts" of the case assumptions broader than rigorously scientific ones about causality and consequence in the course of human illness.

Physicians' familiarity with patients' stories may ultimately enrich their understanding of medical cases, but pathography cannot substitute for case narrative. However valuable pathography may be, its focus is on the patient rather than the disease. The case, by contrast, serves an essential diagnostic purpose, and for this purpose its narrowness makes sense. It orders the messy and confusing details of experience and filters out clinical "irrelevancies." It promotes medicine's focus on—indeed, its obsession with—the particular in the care of patients, for it serves as a constant reminder that not all cases of the same kind are actually the same. It clarifies medical intervention by preserving the physician's awareness of the problematic relation between general laws and a particular circumstance. It encourages a tolerance of uncertainty by providing the means of recording and memorializing exceptions to the rule. For all these reasons, case narrative is central to the epistemology as well as to the practice of medicine. It is a construct of that epistemology, necessary to rational investigation in a domain where subjective experience (and subjective accounts of that experience by another person) are the original and grounding data of clinical care. As patient history, narrative makes possible the communication of one human being's experience to another and underlines its status as mediated fact: it is often the best we have to go on. As the account of a patient's course of illness, the case narrative's representation of change through time is an essential tool of clinical reasoning, facilitating the comparison and contrast of developing patterns. Narrative accommodates the uncontrolled, uncontrollable variables of the individual circumstance, making possible a clinical flexibility that dare not ossify into inalterable rules. Scientifically it fosters a fidelity to the phenomena and a recognition that they are the final arbiters of the explanatory power of the principles themselves. Pedagogically, narrative encourages and improves clinical judgment by making possible a kind of practical, clinical knowledge that mediates biological principles and the facts of the particular clinical case.

Nevertheless, in any situation but acute emergency care—and sometimes even there—the traditional medical case is restrictive, limiting the practice of medicine and the care of patients to diagnosis and prescription. Just as that narrative hones clinical judgment, so fiction and pathography enlarge the physician's awareness of the human problems that are the context of disease and injury. Knowledge of cases sharpens the awareness of clinical possibilities; knowledge of life stories helps cultivate attention to patients, an interest in their oddities and their ordinariness—and a tolerance of both. Especially in primary-care practice, this interest and attention can be vital. John Berger in his portrait of a physician, *A Fortunate Man*, describes the daily rounds of a general practitioner in the north of England. By the standards of a resident in a North American tertiary-care hospital, the array of illness this village physician sees is grindingly "uninteresting."[10] Its only distinction from a general practice anywhere at all lies in the recent decline in the region's standard of living as industry has departed. What fascinates the physician (and thus the narrator and his readers) is the role he plays in his patients' lives: for them he is the "requested clerk of their records."[11] It is a historical, even a literary task. He is not simply a bystander in the unfolding story of their lives. He knows the hard truths. He understands something of what goes wrong in their lives. William Carlos Williams, fascinated by his own quite ordinary patients, admitted, "my 'medicine' was the thing which gained me entrance to these secret gardens of the self."[12] But fewer physicians these days have such a practice, and, although "continuity of care" has been a medical buzzword off and on for more than two decades, few schools set out to prepare their students to understand its pleasures.

Is reading autobiography and fiction necessary to the formation of this narrative sensibility? The wisdom and sensitivity of experienced physicians may have been gathered without benefit of literature through a lifetime of careful attention to patients. But what are young practitioners to do before they have acquired a store of cases on which to found a sustaining overview of life? And what are middle-aged physicians to do when the physical consequences of bad luck and folly and human evil lead them to harden themselves to the life stories of their patients? Physicians turn to professional journals for accounts of difficult or unusual cases and new developments that offer hope of altering the plots in old stories of disease. Likewise, in fiction, autobiography, and drama they can broaden their knowledge of human beings not only beyond the textbooks in human behavior but beyond the ethnic and chronological limits of their own experience. The physician who has read Tolstoy's *The Death of Ivan Ilych*, for instance, has imagined a patient's unwilling slide toward death. He or she is also able to entertain the possibility that the horrors of illness are not entirely physical; an apparently self-possessed, successful, now terminally ill patient may lack the support of family and friends, but in some sense may have before him the discovery of an authentic life. Years of practice may provide wisdom to equal this awareness, but, unrelieved and unassessed, those years may also callous the physician who has "seen it all."

To cultivate practical wisdom in the diagnosis and treatment of patients, physicians are taught to employ a narratively mediated casuistry. The residencies are long apprenticeships that at once foster the accumulation of a large number of cases and guide the new physician in their judicious use. Most physicians, especially those who enter primary-care practice, extend this case-based reasoning beyond diagnosis and treatment to judgment about the people who are ill and how best to provide effective care. To some degree, physicians are prepared for this by attention given during their education to "case management."[13] To increase this preparation, the Association of American Medical Colleges' 1984 report on the general professional education of the physician, *Physicians for the Twenty-First Century* (the GPEP Report) calls for, among other things, more experience in ambulatory care.[14] Such exposure gives students an opportunity to begin to acquire a fuller collection of cases, one that takes into account the vagaries—economic, social, psychological—of the human beings who are ill. For effective and satisfying practice, a collection of life histories is needed. Years of experience taking care of patients may foster the clinical wisdom necessary to handle difficult cases well, but meanwhile for physicians, as for all of us, the vicarious experience offered by narrative increases familiarity with the range of human character and the life outside office and hospital to which medicine aims to restore them.

In particular, literature is a source of knowledge about the operation of values through time. Just as medical narrative is the repository of much of the professional ethos learned by students and residents, so narrative generally— biography, fiction, history—shapes moral sensibility and models clinical distance. Several recent works by physicians make this point. Robert Coles's recent book, *The Call of Stories: Teaching and the Moral Imagination*, extends the argument of his 1979 *New England Journal of Medicine* essay, "Medical Ethics and Living a Life," to argue the centrality of narrative in moral education.[15] In the earlier essay, which addresses the moral life of physicians, he distinguishes ethical reflection from its more academic cousin, ethical analysis, and recommends reading novels about doctors (*Middlemarch*, *Arrowsmith*, *Wonderland*) for their account of the peril that lies in wait for the unsuspecting and, particularly, the idealistic practitioner. In *Stories of Sickness*, Howard Brody finds an even more practical role for literature in delineating the responses to illness that govern the lives of patients.[16] With an eye to developing an ethical analysis grounded in the story of the patient's life, he surveys (among other works) *Philoctetes*, *The Magic Mountain*, *The Metamorphosis*, and *Cancer Ward* to offer physicians what he sees as the best available knowledge of the experience of illness. Arthur Kleinman samples the life stories of actual patients in *The Illness Narratives*, a framed collection of mostly autobiographical accounts of the experience of chronic disease.[17] They are ordinary, sad, sometimes frustrating, often heroic, always revealing about the role of medicine and health-care practitioners in the lives of the ill. Kleinman argues persuasively that the patient's culture is one the physician must enter carefully, ethnographically, openly.

Narrative, in medicine and out, cultivates the power of observation. No one is able to observe carefully details that are not known to exist or have dropped from memory. Narrative's clinical usefulness is most obvious in constructing the history of the patient's present illness. Into what category does this illness fall, the physician asks, and does it differ in any way from the index case of its kind? What has the patient's life been? How has it contributed to this illness? How will it help or hinder recovery? The conviction that such information is important must be acquired. It is not standard equipment that comes with a medical education. Medical students in the late-twentieth century find in Francis Peabody's 1927 essay, "The Care of the Patient," a still painfully accurate description of information overload and the neglect of the patient. But it is the actual experiences of illness in which they begin to take part that, for them, will illustrate the diagnostic and therapeutic importance of life histories and make Peabody's essay most persuasive. Familiarity with the life stories in fiction and drama fill out the range of human possibility. Many medical students, for example, have not known well a vigorous, healthy old person; outside their families few have known someone alert but steadily failing. Stereotypes abound, and narrative about the lives of the elderly—D. L. Coburn's "The Gin Game," Alice Adams' *Second Chances*—can subject them to scrutiny. William L. Morgan, Jr., has given the pages about Aunt Leonie from Proust's *Swann's Way* to his residents in internal medicine. Near ninety, she takes to her bed and from that vantage rules the family and the neighborhood. When she naps, "three streets away, a tradesman who had to hammer nails into a packing-case would send first to Françoise to make sure that my aunt was not resting."[18] "Is Aunt Leonie ill?" the residents were asked. "What should be the goals of her physician?"

A narrative view of medicine does not neglect the biomedical sciences. Instead, it adds to that scientific view a privileged humility in the care of the patient through its recognition that the larger biological story in which each human being participates moves, with or without medical attention, from birth to death. We may organize our life stories, plot them, change their course, shorten them—but their direction and the fact of their end are givens. For most of us these days disease is a small part of life—and as we move into our seventh decade, a not unexpected part. The perspective offered by this larger pattern of life-narrative can ease the expectation (held as often by physician as patient) that death can be defeated. By locating us all on the common narrative trajectory from birth to death, narrative restores the physician to the proper place in the patient's story. This larger view may offer help to the physician in thinking about and communicating prognostics: the meaning of the illness for the patient may be set by the medical meaning—the diagnosis and the customary course of treatment—but it is not confined to it. Likewise, even in chronic or fatal illnesses, the meaning of the illness is not the meaning of the patient's life. Thus hope need not always be construed as the hope of cure,[19] and therapy may be tailored to the patient's life stage and wishes.[20] Of little use is the enthusiastic athleticism that expects of every patient a no-holds-barred struggle against death or a run for the record books that is often more valuable to clinical pioneers than to their patients. Good physicians offer their patients all that is appropriate, urge them to make use of technological advances that are promising in their case, soothe fears, alleviate pain, persuade. But they do not lose sight of the lives out of which patients' choices come and into which medical therapy must intrude.

Narrative also offers physicians a way of confronting the pain of a patient's illness or loss. A physician who does not ask an elderly patient about her family because the answer may be, "There's no one; my husband died this last year," might justify the reluctance by saying that such a question will "make her sad." But such a physician has either a maimed sense of what causes sadness or a partial, perhaps rigid, sense of life patterns. The real threat is surely that the physician will feel the pain of her sadness. This is at once egocentric and much too self-deprecating: the question does not cause the sadness and may in fact do much to ease its pain. The bereaved are not all sad in the same way. Some are sad for reasons that can be addressed, some feel guilt and anger, which the physician as an arbiter of normality can render less painful. Many are stronger and healthier for having their sadness acknowledged. Physicians who read more than the bodies of their patients and are acquainted with more life stories than the ones in which they are asked to intervene are better prepared to see an individual's life as a moral trajectory and have a firmer grasp of the challenges we face at every stage of life.[21] They see more clearly the mix of pain, pleasure, and loss in most people's lives and know what, if anything, suffering may be good for. An understanding of the human condition gained from literary narrative may make it easier to meet patients' suffering with an educated innocence, an openness of observation that in itself can be some comfort to the patient. "It is difficult/to get the news from poems," William Carlos Williams wrote, "yet men die miserably every day/ for lack/ of what is found there."[22]

By enabling us to envision the whole of life into which our lives somehow must fit, literature, like Achilles' shield, offers us a little shelter from inevitable pain. Although it was forged for his protection, neither the shield nor the representation of life's wholeness emblazoned on it can save the hero from the death he is bound for or restore his friend Patroclus to life. Achilles knows it: his mother has told him that he will not long survive Hector, and it is Hector he means to kill. The gods know it: Hephaistos acknowledges the hero's mortality even as he agrees to set to work on the shield. The shield has a problematic, even paradoxical usefulness. It alters Achilles's fate not one bit. He nevertheless cannot return to battle without it. It will cover him meanwhile, enabling him to fight again, protecting him while he goes on to meet his fate.

Like Achilles, we readers retain our vulnerability. There is no protection from death or from the loss of the people whom we love—even for those of us who become physicians. In a literal sense, the work of art, like Achilles's shield, may be entirely useless. It serves no practical purpose in the world, but it is difficult, sometimes impossible, to return to battle without it. It sets its bearer's action in a context of meaning: violence and death are all around, but although inescapable, they are not all. The work of art shelters us meanwhile from injury and pain. Physicians, whose profession is not a protection from human suffering but a deliberate exposure to it, stand in need of that shield. Literature enables medical students and seasoned clinicians alike to face the onslaught of experience, bearing a knowledge of life that is both painfully particular and clearsightedly whole. Reminded that the single individual has only a small part in the scheme of things, no matter how heroic, they may better equip themselves with a protecting, protected concern for those who seek their help. Literature's representation of life provides its readers a little space between themselves and the onslaught in which to see clearly both their own deeds and the lives of others.

Notes

1. Leo Tolstoy, "The Death of Ivan Ilych," *The Death of Ivan Ilych and Other Stories*, trans. Aylmer Maude (New York: Signet, 1960), p. 121.

2. Albert R. Jonsen and Stephen Toulmin, *The Abuses of Casuistry* (Berkeley: University of California Press, 1988), was preceded by Toulmin's "The Tyranny of Principles," *Hastings Center Report* 11 (1981), 30–39, and Jonsen's "Casuistry in Clinical Ethics," *Theoretical Medicine* 7 (1986), 65–74.

   See also Warren Thomas Reich, "Caring for Life in the First of It: Moral Paradigms for Perinatal and Neonatal Ethics," *Seminars in Perinatology* 11 (1987), 279–87, and Howard Brody, *Stories of Sickness* (New Haven: Yale University Press, 1988). All are on the faculty of a medical school; Brody is also a physician. In a new preface to his first book, *The Place of Reason in Ethics* [1950] (Chicago: University of Chicago Press, 1986), Stephen Toulmin describes the influences that drew him away from analytical philosophy's way of conducting moral philosophy toward a more historical, contextual proto-casuist method.

3. Leon Kass, *Toward a More Natural Science: Biology and Human Affairs* (New York: Free Press, 1985), p. 12.

4. The term "pathography" is Anne Hunsaker Hawkins's; see "Two Pathographies: A Study in Illness and Literature," *Journal of Medicine and Philosophy* 9 (1984), 231–52.

5. Brian Clark, *Whose Life Is It Anyway?* (Derbyshire, Eng.: Amber Lane Press, 1978).

6. Robert B. White and H. Tristram Engelhardt, Jr., "A Demand to Die," *Hastings Center Report* 5 (1975), 9–10.

7. Larry Kramer, *The Normal Heart* (New York: New American Library, 1985).

8. William M. Hoffman, *As Is* (New York: Vintage, 1985).

9. Eric J. Cassell, "The Nature of Suffering and the Goals of Medicine," *New England Journal of Medicine* 306 (1982), 639–45.

10. Terry Mizrahi, *Getting Rid of Patients: Contradictions in the Socialization of Physicians* (New Brunswick, N.J.: Rutgers University Press, 1986).

11. John Berger, *A Fortunate Man*, with photographs by Jean Mohr (New York: Holt, 1967), p. 103.

12. William Carlos Williams, "The Autobiography" (1951), in *The William Carlos Williams Reader*, ed. M. L. Rosenthal (New York: New Directions Press, 1966), p. 307.

13. A. C. Dornhurst argues for a pragmatic approach to medical education in "Information Overload: Why Medical Education Needs a Shake-up," *Lancet* 2 [8245] (1981), 513–14.

14. Association of American Medical Colleges Project on the General Professional Education of the Physician, "Physicians for the Twenty-First Century," *Journal of Medical Education* 59 (1984), no. 11, part 2.

15. Robert Coles, *The Call of Stories: Teaching and the Moral Imagination* (Boston: Houghton Mifflin, 1989); "Medical Ethics and Living a Life," *New England Journal of Medicine* 301 (1979), 444–46.

16. Howard Brody, *Stories of Sickness* (New Haven: Yale University Press, 1988).

17. Arthur Kleinman, *The Illness Narratives: Suffering, Healing, and the Human Condition* (New York: Basic Books, 1988).

18. Marcel Proust, *Remembrance of Things Past: Swann's Way*, trans. C. K. Scott Moncrief and Terence Kilmartin (New York: Vintage, 1982).
19. Howard Brody, "Hope," *Journal of the American Medical Association* 246 (1981), 1411–12.
20. See Daniel Callahan, *Setting Limits: Medical Goals in an Aging Society* (New York: Simon and Schuster, 1987), and Brody, "The Physician-Patient Relationship as a Narrative," in *Stories of Sickness*, pp. 171–81.
21. David Burrell and Stanley Hauerwas, "From System to Story: An Alternative Pattern for Rationality in Ethics," *Knowledge, Value and Belief*, vol. 2: *The Foundations of Ethics and Its Relationship to Science*, ed. H. Tristram Engelhardt, Jr., and Daniel Callahan (Hastings-on-Hudson, N.Y.: The Hastings Center, 1977), pp. 111–52.
22. William Carlos Williams, "Asphodel, That Greeny Flower," in *Reader*, pp. 73–74.

Narrative and Medicine

Rita Charon

Rita Charon is a general internist and literary scholar at Columbia University, New York, New York, where she practices internal medicine and directs the Program in Narrative Medicine. Dr. Charon's research explores doctor–patient communication and the use of "narrative competence" to improve the quality of medical care. She is the author of *Narrative Medicine: Honoring the Stories of Illness* (New York: Oxford University Press, 2006) and the editor (with Martha Montello) of *Stories Matter: The Role of Narrative in Medical Ethics* (New York: Routledge, 2002). The article that follows develops the idea of "narrative competence" and recommends that health professionals acquire this skill.

A 36-year-old Dominican man with a chief symptom of back pain comes to see me for the first time. As his new internist, I tell him, I have to learn as much as I can about his health. Could he tell me whatever he thinks I should know about his situation? And then I do my best not to say a word, not to write in his chart, but to absorb all that he emits about his life and his health. I listen not only for the content of his narrative, but for its form—its temporal course, its images, its associated subplots, its silences, where he chooses to begin in telling of himself, how he sequences symptoms with other life events. I pay attention to the narrative's performance—the patient's gestures, expressions, body positions, tones of voice. After a few minutes, he stops talking and begins to weep. I ask him why he cries. He says, "No one has ever let me do this before."

More and more health care professionals and patients are recognizing the importance of the stories they tell one another of illness. As my colleagues and I in the Program in Narrative Medicine are discovering, not only is diagnosis encoded in the narratives patients tell of symptoms, but deep and therapeutically consequential understandings of the persons who bear symptoms are made possible in the course of hearing the narratives told of illness. Such fields as medical interviewing, primary care, literature and medicine, and relation-centered or patient-centered care have revolved around these "tellings"—whether the patient's private account in the office, the intern presenting on visit rounds, or the physician dictating a death summary after decades of now-ended care.[1] In turn,

---

From Rita Charon, "Narrative and Medicine," *New England Journal of Medicine* 2006;350 (9): 862–864. Reprinted with permission of the Massachusetts Medical Society. Copyright © 2006 Massachusetts Medical Society.

doctors have learned about therapeutic listening from practitioners of oral history, trauma studies, autobiography, and psychoanalysis. Only in the telling is the suffering made evident. Without the telling, not only treatment but suffering, too, might be fragmented. We have only to read the masterly accounts of illness written by patients—Jean-Dominique Bauby's *The Diving Bell and the Butterfly*, Christina Middlebrook's *Seeing the Crab*, John Hull's *Touching the Rock*—to grasp the power of the telling of illness and to accept our obligation to learn how to receive these stories.

Like their patients, some doctors have learned that it helps to represent, in words, what they go through in practicing medicine. More and more doctors write about their practices not in scientific reports but in narrative records of meaningful human interactions. In books and essays for medical and lay audiences or in private notes for themselves, doctors describe the emotional and personal aspects of their care of particular patients. Some authors report that such writing helps them to comprehend both their patients' ordeals and their own lives with the sick.[2] By rendering whole that which they observe and undergo, doctor-writers can reveal transcendent truths, exposed in the course of illness, about ordinary human life.

In the effort to help doctors understand what they and their patients experience in the presence of illness, medical educators have been paying increasing attention to narrative competence, defined as the set of skills required to recognize, absorb, interpret, and be moved by the stories one hears or reads. This competence requires a combination of textual skills (identifying a story's structure, adopting its multiple perspectives, recognizing metaphors and allusions), creative skills (imagining many interpretations, building curiosity, inventing multiple endings), and affective skills (tolerating uncertainty as a story unfolds, entering the story's mood). Together, these capacities endow a reader or listener with the wherewithal to get the news from stories and to begin to understand their meanings.

When a doctor practices medicine with narrative competence, he or she can quickly and accurately hear and interpret what a patient tries to say. The doctor who has narrative competence uses the time of a clinical interaction efficiently, wringing all possible medical knowledge from what a patient conveys about the experience of illness and how he or she conveys it. Not only the story of an illness, but the illness itself unfolds as a narrative. A disease has a characteristic time course, a complex mixture of causality and contingency, singular differences from and generic sameness to related diseases, a textual tradition within which it can be understood, and even a metaphorical system that reveals it (consider, for example, the complex metaphorical meanings of the word "immunity"). Narrative competence gives the doctor not only the means to understand the patient, but fresh means to understand the disease itself.

To enter a story is to make room for its teller, and the doctor with narrative skills habitually confirms the patient's worth in the process of attending seriously to what he or she tells. Such a doctor will demonstrate concern for a patient while concentrating on what the patient says and, as a result, can achieve the genuine intersubjective contact required for an effective therapeutic alliance. Narrative competence includes an awareness of the ethical complexity of the relationship between teller and listener, a relationship marked by duty toward privileged knowledge and gratitude for being heard.

What was once considered a civilizing veneer for the gentleman physician—reading literature, studying humanities, writing in literary ways about practice—is now being recognized as central to medical training for empathy and reflection.[3] Capacities that medicine now sometimes lacks—attunement to patients' individuality, sensitivity to emotional or cultural dimensions of care, ethical commitment to patients despite fragmentation and subspecialization, acknowledgment and then prevention of error—may be provided through a rigorous development of narrative skills. Perhaps strengthening the narrative competence of doctors might help them to achieve such elusive goals as humanism and professionalism by providing them with graduated skills in adopting patients' points of view, imagining what they endure, deducing what they need, and reflecting on what the physicians themselves undergo in caring for patients.[4]

Narrative skills that are important to medical practice have been identified, and methods of teaching them have been developed. Programs are emerging in "narrative medicine" or "narrative based medicine" to teach specific aspects of narrative competence. Such training efforts encourage health care professionals and students to write about their patients in nontechnical language, helping them to uncover and understand their implicit feelings toward and knowledge of their patients. These programs provide rigorous training in reading literary texts to supply health professionals with the equipment to interpret and make sense of the stories of others. Clinicians who receive such training often encourage patients to write—or, as with the young man in my office, to tell in uninterrupted narrative flow about their illnesses, demonstrating the therapeutic benefit for patients of such narration.

The mechanisms and processes by which narrative training benefits health professionals are currently under active investigation in federally funded research projects. Research on the outcomes of narrative writing in medical schools is under way in a number of settings, and the consequences of narrative training for multidisciplinary health care teams are just now being tracked. Investigators are examining the outcomes of therapeutic narrative practices in the ongoing care of the chronically ill.[5] The potential costs of these new narrative practices, including the increased time it may take to meet with new patients and challenges to patient confidentiality, are being recognized and addressed.

Many previous efforts have been made, from Francis Peabody on, to listen to patients and to care about what happens to them. My patient with back pain—actually a composite of several new patients I saw recently—revealed to me the connections among his symptoms, his illiteracy, his failures as a breadwinner, his familial losses, and his life in an alien culture. Armed with such knowledge, available to me most expediently through his personal narrative, I confirmed the gravity of all he told me and shared my optimism that things could improve for him. Together, we made a plan to investigate his educational needs and to evaluate his musculoskeletal symptoms. The more medicine understands the complexities of illness, the better clinicians can formulate their roles with respect to patients, both in technical dimensions and in dimensions of meaning. Narrative studies, many physicians are beginning to believe, can provide the "basic science" of a story-based medicine that can honor the patients who endure illness and nourish the physicians who care for them.

From the Program in Narrative Medicine, College of Physicians and Surgeons of Columbia University, New York.

Notes

1. Hunter K. *Doctors' stories: the narrative structure of medical knowledge*. Princeton, N.J.: Princeton University Press, 1991.
2. Verghese A. The physician as storyteller. *Ann Intern Med* 2001;135:1012–7.
3. Halpern J. *From detached concern to empathy: humanizing medical practice*. New York: Oxford University Press, 2001.
4. Greenhalgh T, Hurwitz B. *Narrative based medicine: dialogue and discourse in clinical practice*. London: BMJ Books, 1998.
5. Pennebaker JW. Telling stories: the health benefits of narrative. *Lit Med* 2000;19:3–18.

Nice Story, But So What?
Narrative and Justification in Ethics

John Arras

John Arras is Porterfield Professor of Biomedical Ethics and Director of the Bioethics Minor Program at the University of Virginia, Department of Philosophy, Charlottesville, Virginia. He is also a Fellow at the Hastings Center and former clinical bioethics faculty at Albert Einstein College of Medicine/Montefiore Medical Center. Professor Arras is coeditor (with Ezekiel J. Emanuel, Robert A. Crouch, Jonathan D. Moreno, and Christine Grady) of *Ethical and Regulatory Aspects of Clinical Research: Readings and Commentary* (Johns Hopkins University Press, 2003) and (with Bonnie Steinbock and Alex Jack London) of *Ethical Issues in Modern Medicine*, 6th edition (McGraw-Hill, 2003). He is the editor of *Bringing the Hospital Home: Ethical and Social Implications of High-Tech Home Care* (Johns Hopkins University Press, 1995). This essay explores the relationship between narrative ethics and ethical justification. Arras supports adding narrative analyses as a necessary corrective to principle- and theory-based methods of analysis.

Everywhere one looks in the academy these days, theory is out and stories are in. . . . Many scholars in the social sciences and humanities seem particularly eager to jettison the last vestiges of the Enlightenment ideals of objectivity, rationality, truth, and universality as these pertain both to matters epistemological and axiological. The consensus seems to be that, just as our ability to know is profoundly circumscribed by the contingencies of time, place, and our own psychological makeup (all knowledge is thus "local"), so our values are said to reach no farther than the bounds of our community or nation. Furthermore, it is maintained that any attempt to extend the boundaries of either our knowledge or our values is not just wrong and ill-fated (because, given our finitude, it cannot be accomplished), but also dangerous because it will inevitably amount to an imposition of our ways of knowing and valuing upon others. Thus, the belief in objectivity and universality that once drove the so-called "Enlightenment project"—a belief that such great but diverse thinkers as Voltaire, Rousseau, Kant, Locke, and Marx once viewed as profoundly liberatory—is now the object of a profound suspicion. Behind the search for universality must lie the will to dominate, to bend others to our ways of thinking and valuing. Objectivity and universality, once thought to be the key to our common deliverance from the narrowness and stupidity of local custom, have come to be seen as the seeds of tyranny.

In the place of the Western mind's traditional quest for the objective and universal laws undergirding nature, history, and morals, we now find the flourishing of narrative, storytelling, anecdote, and autobiography. Here too the argument is both epistemological and moral. All knowing is necessarily bound up with a narrative tradition of one kind or other; and all valuing grows out of and expresses the stories that constitute us as members of a particular family, community, or nation. . . .

This flourishing of narrative has brought about what literary critic David Simpson has conceived as a major shift in the "balance of trade" among academic departments of universities.[1] Literature has emerged as the major exporter of methods and themes to other departments once dominated by more objectivist and scientific tendencies. Indeed, the traditionally sharp boundaries between such academic subjects as history, anthropology, literature, and philosophy have recently yielded to make way for the triumph of a "literary culture" that now appears to dominate the academy. . . .

The field of bioethics is beginning to take its own narrative turn. Long dominated by the aspirations to objectivity and universality as embodied in its dominant "principlist" paradigm, bioethics is now witnessing an explosion of interest in narrative and storytelling as alternative ways of structuring and evaluating the experiences of patients, physicians, and other health care professionals. To be sure, the universalist mantra of autonomy, beneficence, and justice still holds sway in many quarters, and its principal defenders have proved to be quite adept at ingesting or co-opting much recent criticism without giving up on their central claims regarding the pivotal role of principles in the moral life.[2] Still, one wonders whether the current plethora of conferences, journal issues, and articles devoted to narrative bioethics presages not merely an important *shift* away from "principle-driven" ethics, a movement that has been proceeding apace for some time now under the auspices of casuistry and feminism, but also the imminent triumph of the literary sensibility in a field that has traditionally wished to appear as a source of "hard knowledge." . . .

. . . I shall attempt in the present essay a modest typology of narrative ethics. There are several different conceptions of "narrative ethics," and each carries significantly different implications for the question of moral justification. As we shall see, some conceptions are relatively modest and unthreatening to the claims of principles and theory, while the more robust versions of narrative ethics threaten to replace the regnant paradigm.

. . . We will first consider narrative as a supplement to (or ingredient of) principle-driven approaches to ethics. From this angle, narrative is seen as an indispensable and ubiquitous feature of the moral landscape. Here narrative not only allows us to delineate moral problems in a concrete fashion, but also plays an important role in the formulation of moral principles and the depiction of character. Then we will briefly inspect the view, powerfully articulated by Alasdair MacIntyre and Stanley Hauerwas, that narrative functions principally as the very ground of all moral justification. Narrative functions here not merely as a supplement or handmaiden to principles and theory, but also as the exclusive basis of ethical rationality itself. Finally, we shall canvass the place of narrative within a distinctly "postmodern" ethical stance, where narrative and the authenticity of the narrator appear to play the role of substitutes for ethical justification. As we shall see, each formulation of narrative ethics poses a progressively greater challenge to currently dominant ways of thinking about the role of stories in moral justification.

Narrative as Supplement to an Ethic of Principles

The most benign and least controversial version of "narrative ethics" asserts that an ethic of principles and theory cannot stand alone, that it must be supplemented by an understanding of the narrative structure of human action in order to achieve a more fully rounded and complete ethic. This assertion itself rests upon three distinct observations about the relationship between narrative and ethics: (1) that narrative elements are deeply embedded in all forms of moral reasoning; (2) that our responses to stories are the ground out of which principles and theories grow; and (3) that narrative is the only medium in which a concern for character and virtue can be intelligibly discussed.

A. The Pervasiveness of Narrative in Ethical Reasoning

. . . [S]ome advocates of narrative ethics insist that stories and moral theorizing are mutually interpenetrating and interdependent. They point out that moral narratives often embody a kind of argument (a "moral"), while much ethical argument is pervaded by narrative elements; and they claim that a keener awareness of these narrative elements embedded in all moral reasoning will permit a more reflective and penetrating mode of moral analysis.

Rita Charon, a physician and literary scholar, highlights a number of ways in which a heightened literary consciousness can augment our reasoning skills in a field like clinical bioethics.[3] For the practicing physician, Charon notes, closer attention to the narrative elements in the situation—and in particular to the patient's own story—would permit the recognition of ethical issues that often go unnoticed. What's really troubling a particular patient—for example, the likely impact of scheduled surgery on her ability to maintain her roles as worker, wife, and mother—will often not find its way into the dominant form of medical narrative, the medical chart. Although the chart and other forms of medical discourse, such as the truncated language of clinical rounds, pretend to have achieved a high level of universality and scientific objectivity, they often screen out the very meanings that the disease or illness has for the patient.[4] In the absence of an understanding of the existential implications of the patient's condition and the meanings of various treatment alternatives, the physician is likely not even to recognize moral tensions or problems latent in the medical encounter.

Charon also usefully points out that the various skills and sensitivities of the literary critic are indispensable in coming to terms adequately with the whole gamut of medical narratives, including not only the chart but also all the stories that caregivers, patients, family members, and authors tell about their experiences surrounding a particular "case." In particular, she notes, closer attention to the way in which medical narratives are presented—including, for example, the way in which the various elements are framed, the content is selected, and the author's point of view is established—can help us read more deeply and critically. Quoting with approval the German literary theorist Walter Benjamin ("The traces of the storyteller cling to the story the way the handprints of the potter cling to the clay vessel"), Charon argues that sensitivity to such questions as authorship and point of view constitute, along with several other important skills, a kind of "narrative competence" that is a prerequisite to doing good ethics.

For Rita Charon, then, "narrative ethics" essentially means a mode of moral analysis that is attentive to and critically reflective about the narrative elements of our experience. It is important to note, however, that Charon's plea for a narrative ethics is not meant as a fundamental challenge to an ethic driven by principles and theories. On the contrary, she explicitly wishes to leave intact the basic structure of principle-driven ethics.[5] On this view, narrative competence is recommended as a supplement, as a way to improve our use of the existing methods of moral analysis by gearing their deployment to the rich particularity of patients' lives. Principles retain their normative force; narrative sensitivity just makes them work better. "Narrative ethics" on this gloss is thus not a newer, better kind of ethics; it simply allows us to apply principles with greater sensitivity and precision.

B. Narrative as Ground and Object of Ethical Principles

A different conception of the relationship between narrative and moral justification, but a conception still faithful to the depiction of narrative ethics as a supplement to principles, might be sought in the notion of reflective equilibrium. This approach to moral methodology was first articulated in the early work of John Rawls,[6] and has since been the subject of much amplification and commentary at the hands of other ethical theorists.[7] I and many others have written elsewhere about this conception of moral justification within the context of bioethics,[8] so I will content myself here with a very brief sketch merely sufficient to make my point about the connection between narrative ethics and reflective equilibrium.

As Rawls and his followers depict it, reflective equilibrium offers an alternative picture of moral justification to the sort of "top-down" account favored by moral "deductivists." Deductivists view the process of moral justification as involving a unidirectional movement from preexisting theories and principles to their "application" at the level of the case. To justify an action or policy on this account is simply to bring it under the relevant theory, principle, or moral rule. According to the partisans of reflective equilibrium, this unidirectional picture distorts or totally ignores the pivotal role of intuitive, case-based judgments of right and wrong. To be sure, the sort of judgments they have in mind are not to be confused with just any responses to cases, no matter how prejudiced, ill-considered, or subject to coercion they might be. Rather, they are referring to those intuitive responses in which we have the most confidence, like those embedded in the conclusions that slavery or the killing of innocent children are wrong. Rawls referred to this class of intuitive responses as our "considered judgments." It is precisely these judgments, it is claimed, that give concrete meaning, definition, and scope to moral principles and that provide critical leverage in refining their articulations.

The partisans of reflective equilibrium claim, in effect, that principles and cases have a dialectical or reciprocal relationship. The principles provide normative guidance, while the cases provide considered judgments. The considered judgments, in turn, help shape the principles that then provide more precise guidance for more complex or difficult cases. Principles and cases thus coexist in creative tension or "reflective equilibrium." Ethical justification is then sought not in any kind of correspondence between our ethical judgments and some sort of transcendent realm of ethical norms or kingdom of ends, but rather in the overall meshing or coherence achieved among our intuitions about cases, our rules, principles, moral theories, and nonmoral theories about society, personhood, and so on.

Now the reason for bringing up this business of reflective equilibrium in the context of the present essay is that the cases about which we have these considered judgments are themselves narratives. They tell stories about what's happening in and around people's bodies and about their social relationships, stories that prominently feature some sort of moral dilemma or conflict. So, rather than viewing stories as being essentially remote from the realms of principle and theory—or, in the "man of science's" words, as "savage, primitive, underdeveloped, backward"[9] and so on—the advocates of this coherentist approach to moral justification would have us view narrative and stories as intimately bound up with the most sophisticated renderings of principle- and theory-driven moral reasoning. For no matter how far we progress toward the ethereal realms of principle and theory, we ought never to lose sight of the fact that all of our abstract norms are in fact distillations (and, yes, refinements) of our most fundamental intuitive responses to stories about human behavior. Our moral vocabulary and the very contours of our moral universe are shaped by the stories that we hear at our parents' knees. Principles and theories do not emerge full-blown from some empyrean realm of moral truth; rather, they always bear the marks of their history, of their coming-to-be through the crucible of stories and cases.

Thus, the defenders of a coherentist theory of moral justification, a theory aptly captured in the metaphor of reflective equilibrium, would claim, like Rita Charon, that narrative and moral theory are not alternatives, but are rather in-

separable elements in a perpetual to-and-fro movement from stories to principles and back again. According to both Charon and these moral coherentists,
"narrative ethics" is not a new way of doing ethics, but is rather a recognition
and full appreciation of the debt that principle- and theory-driven modes of discourse owe to stories. Here too, then, narrative ethics works to supplement,
rather than supplant, a principled approach to ethics.

C. Narrative and the Depiction of Character[10]

While some partisans of narrative ethics advance very strong and controversial
claims,[11] I think that all would agree that an appropriately complete story or history is a prerequisite to any responsible moral analysis. Before we attempt to
judge, we must understand, and the best way to achieve the requisite understanding is to tell a nuanced story.

Thus, when we debate the issue of assisted suicide, for example, we should
do so not as some sort of abstract, asocial, and timeless proposition, but rather
in the context of a full-bodied case. Dr. Timothy Quill's well-known case study
of Diane, a patient requesting assisted suicide, provides an excellent illustration
of this narrative approach.[12] Instead of focusing on the derivation and specification of principles, Dr. Quill gives us a rich picture of the "players" and their
characters. There was first and foremost his patient, Diane, a courageous but
fearful cancer patient seeking control of her dying process, a woman who had already overcome a previous cancer threat and her own debilitating alcoholism;
and there was Dr. Quill himself, who emerges as a competent and clearly compassionate physician torn between loyalties to his patient and the ethics of his
profession, a man courageous enough to "take small risks for people he cares
about." He explores the roles that the players occupy: a doctor trained to preserve life rather than end it, a patient who is also a wife, mother, and respected
friend. He tells us about their prior and ongoing relationship, how he had witnessed and rejoiced over Diane's past triumphs over adversity and anguished
with her over the current threat. He describes his own doubts and hopes for
Diane's future and the future of their ongoing relationship. He wonders whether
prescribing a lethal dose might restore her spirits and give her more emotional
comfort in her final struggle. And he alludes to the institutional and social context, albeit in my opinion not sufficiently,[13] with references to the current state
of the law.

Although a reconstructed principlist might object at this point that all the
above matters can and should be folded into a principlistic analysis as components of "the case," it remains true, I think, that the partisans of moral theory
and principlism have not given many of these issues their due. This is especially
true of Quill's concern to sketch the moral character of his players, the nature of
their past and future relationships, and the fine details of their institutional and
social context. As Bernard Williams has argued, most of the received moral theories operate with impoverished or empty conceptions of the individual.[14] In
order to bring the moral individual into clearer focus, he claims, we must attend

to his or her differential particularity, to the desires, needs, and "ground proj-ects" that coalesce into the character of the person. But if we are concerned with the depiction, understanding, and assessment of character, we can do so only by telling and retelling stories.[15]

It is important to note, however, that a salutary concern for the role of char-acter in ethics need not precipitate a wholesale rejection of principles and the-ory. Although some commentators have contended that an appropriate concern for character and its narrative environment should lead us to reject principle-based ethics,[16] one could just as well view reflection on character as a necessary supplement or extension of an ethic of principles. So understood, narrative ethics emerges once again as an adjunct to standard, principle-based ways of doing ethics.

Historical Narrative and Ethical Justification

The second major conception of narrative ethics I want to consider is a good deal less accommodating to principle-based ethics and poses a greater challenge to principlism's conceptions of moral justification.[17] This view, perhaps best rep-resented by Alasdair MacIntyre and Stanley Hauerwas, constitutes a frontal as-sault on the so-called Enlightenment project of establishing a rational basis for ethics beyond the constraints of traditions and culture. According to such crit-ics, reason unmoored to a historical community with its own specific canons of rationality is incapable of providing an adequate basis for morality. Reason and rationality, they claim, are always characteristic of a certain historical tradition, whether it be that of Ancient Greece, medieval Paris, or eighteenth-century Ed-inburgh. Our capacity to view things as reasonable, valuable, noble, appropriate, interesting, and so on is developed within the context of a certain narrative tra-dition that subtly shapes all of our knowing and valuing. Thus reason and ra-tionality will take on as many forms as there are basic historical traditions; there is no one model of rationality that might be used as a critical vantage point from which to pass judgment on the vast panoply of what Wittgenstein called "forms of life." The Enlightenment project of making ethics "scientific," objective, and rational by stripping it of all subjective elements borne by narrative is, they con-clude, a philosophical dead end.

In place of the Enlightenment's deracinated conception of reason, the champions of historical narrative would found ethics on stories and tradition. To be sure, they acknowledge that not just any story will qualify as a ground for our ethical life. Rather, they have in mind what one might call "foundational stories" such as the tradition of Greek or Norse epic poetry, the Bible and traditions of biblical commentary (such as the Talmud and Mishnah), or Confucianism. No matter how much one may strive for a universal and objective picture of things, they claim, at some point one simply has to have faith in a story.[18] The reason-ing has to end somewhere, and it ends where it began, with a narrative account of who we are as a people and how we got to be this way. Importantly, even the Enlightenment-inspired projects that attempt to rise above the particularities and vagaries of tradition and culture often betray a nascent awareness of the importance of narrative by portraying themselves as the inheritors of a distinct philosophic tradition, for example, of liberalism, utilitarianism, or social contractarianism.[19]

On this rival view, ethical justification is a matter of squaring one's actions with a social role (or roles) that is, in turn, justified by a fundamental narrative. Far from being justified before some court of abstract reason, our actions are ultimately sanctioned by appeal to the norms, traditions, and social roles of a particular social group. Obversely, according to MacIntyre and Hauerwas, to lack such a distinctive story is to lack a rationale for one's actions, character, and life.[20] For example, a doctor contemplating Timothy Quill's narrative might well object to the latter's embrace of physician-assisted suicide on the ground that throughout history, beginning with the Hippocratic Oath, physicians have defined themselves exclusively as healers, rather than as healers who might on occasion also kill their patients. When confronted with the proposition that our laws against physician-assisted suicide ought to be changed, such a doctor might well respond, not by invoking this or that principle or philosophical theory, but rather by recalling the physician's role in our society, which is, in turn, explicated and justified by an account of the Hippocratic historical tradition.[21]

This aspect of narrative ethics, understood in this stronger sense, generates an ethic that is highly concrete and effectively action-guiding in a manner unavailable to such standard Enlightenment theories as utilitarianism and Kantianism. Because the latter develop their criteria of right and wrong in a realm beyond the particularities of any particular time and place, they provide significant critical leverage; but they do so at the price of an abstractness and remoteness that often render them incapable of definitively guiding action in specific circumstances. The winds of utilitarianism notoriously blow in all sorts of different directions,[22] often simultaneously justifying contradictory positions on important matters of individual morality and public policy, as does Kant's categorical imperative, which seems to function better (at best) as a necessary condition of morality, telling us what we cannot do, rather than as a sufficient condition, telling us what we must do in specific circumstances. For the partisans of this more robust version of narrative ethics, one's story effectively provides the rationale for one's action. ("We are doctors. We don't kill!" "We help the needy, just as Christ bade us to do in the story of the Good Samaritan.")

The difficulties inherent in this particular narrativist project are predictable and serious. While the concreteness of the fundamental narrative indisputably paves the way for a truly practical ethic, it also sets the limits for any given story and thereby serves, in spite of itself, as a vehicle of transcendence beyond the merely local to other stories telling of other times, places, and ways of knowing and valuing. In the first place, foundational stories not only tell us who we are; they also tell us who we are not. In telling us the story of "our people" with our own particular exemplars of good and evil, for example, such stories also tell us about other peoples against whom we define ourselves. We usually do not define ourselves tout court; rather, we define ourselves against neighboring families, tribes, cities, states, and nations. Thus, the Israelites defined themselves against the gentiles, Protestants defined themselves against Catholics, Southern whites against African-Americans, and, in the neighborhood where I grew up, the Irish defined themselves against everyone else. At the heart of our own self-conception, then, lies a conception of the Other.[23]

Now, ordinarily this Other figures in our own self-conception not as a subject with his or her own story to tell, but rather as an objectified element in our own story. Thus, for contemporary Palestinians, the only relevant story is the history of their oppression at the hands of the Jewish state; conversely, for contemporary Israelis the relevant foundational story is the history of Palestinian aggression and terrorism. The subjects of these historical narratives are thus locked in a perpetual struggle, not only over land, but also over the meaning of their common history. This kind of struggle for narrative supremacy can obviously go on for a long time; sometimes (as in the Balkans) it can last for centuries. Once the realization sinks in, however, that the Other is not about to simply go away, the road to moral and political progress will usually involve an attempt on the part of warring traditions to hear and attempt to understand the story of the other party. But once one actually sits down to listen to the other's story, one opens oneself not simply to the possibilities for acquiring sympathy and tolerance, but also to the possibility for radical self-transformation. It could well turn out, once I have heard your story, that I judge it to be a better story than the one I was taught as a child.

In this way, an awareness of other stories leads to an awareness of the limits of our own. Obviously, we must begin with our own story, which we learn at our parents' knees and which conditions our entire outlook, but contact with the wider world of other stories usually leads us to question our own story and the various social roles to which it gives rise. Thus, a physician trained in the Hippocratic tradition might be exposed to her patients' stories of suffering, which themselves point to a wider political story of individual freedom struggling to free itself from traditional constraints imposed by the heavy hands of religion, custom, and professional codes of ethics. Such a physician might then experience a genuine moral conflict. In addition to her initial repugnance for physician-assisted suicide ("We're doctors. We don't kill.")—a repugnance founded upon her social role dictated by the story of Western medicine—she may now be attracted by other social roles (for example, that of patient advocate) generated by other stories (for example, that of the tradition of political liberalism). This physician then must confront the difficult business of choosing between social roles with their corresponding foundational stories. Whatever she decides, once the complexity of modern societies is acknowledged, a narrative ethic in this stronger sense no longer seems to offer a ready-made action-guiding solution. Just as the moral theorist must attempt to sort out, say, the respective attractions of various competing prima facie obligations in a complex situation of moral choice, so the proponent of narrative ethics not only must ask which story should control her actions in a given situation but eventually must confront the ultimate question of what makes any story morally compelling and worthy of our allegiance. How, in other words, are we to know that the story with which we begin is a "good story" or a better story than the available alternatives?

One way to solve this deep and vexing problem is to set out criteria for the evaluation of stories. Burrell and Hauerwas, for example, contend that "[t]he test of each story is the sort of person it shapes."[24] They elaborate on this answer by positing four additional desiderata that any good story, they assert, will have to display:

1. power to release us from destructive alternatives;
2. ways of seeing through current distortions;
3. room to keep us from having to resort to violence;
4. a sense for the tragic: how meaning transcends power.[25]

While one could quibble with this list of criteria by questioning either the appropriateness of each item or the comprehensiveness of the entire set, the more basic problem for narrative ethics involves the very idea of resorting to a set of abstract criteria for resolving conflicts among plausible stories. For if we are truly able to pick and choose among competing stories by deploying a set of criteria, then it would appear that the criteria themselves, and not the narratives, are fundamental to the critical function of ethics. Although the above list does appear to be rather idiosyncratic, we could easily translate some of its criteria into the traditional language of principles and theory. Thus, criteria one and three above could be recast into the language of "nonmaleficence" (that is, do no harm). The second criterion (bearing on release from distortions) could quite plausibly be read as a restatement of Marx's strictures against "false conscious-ness," a critical position owing more to a theory of social reality and its ideolog-ical distortions than to any narrative.[26] By supplementing Burrell and Hauerwas's list with, say, a principle of "beneficence," with respect for individuality or au-tonomy, or perhaps with an ideal of "human flourishing'" we could compile a set of criteria that might look something like W. D. Ross's list of prima facie du-ties,[27] which could then be applied to the various fundamental stories competing for our allegiance. Narrative ethics must on this account have recourse to an in-dependent set of abstract criteria bearing either on the rightness of actions or on the kinds of characters that our stories ought to foster. But the problem with this approach is obviously that it forfeits the supremacy of narrative over abstract principle, thereby returning us to the more benign conception of narrative ethics as a supplement to (or dialectically incorporated ingredient of) principles and theory.

Another way of sorting out the rival claims of competing stories, one more consonant with the whole idea of a robust narrative ethics, is to claim that some narratives do a better job of solving the problems that have claimed the atten-tion of other narratives. As developed by Alasdair MacIntyre, this claim boils down to the notion that the only corrective for a bad, inadequate, or incoherent narrative is a better narrative, not some set of abstract principles. MacIntyre de-velops this suggestion through his conception of "epistemological crises"[28] in

which the members of a narrative tradition come to see that tradition as ultimately unable to resolve its problems or inner tensions. MacIntyre views the
fundamental narratives as engaged in a quest to discern "the good life for man."
At a certain stage in its development, a narrative tradition may experience an
epistemological crisis or breakdown in which its resources no longer prove adequate to the task at hand. At this point, the members of such a tradition might
look to other narratives as resources for solving the very problems that had
proved so intractable within their inherited story. Adherents of the original narrative may find that the new tradition shows them not only a new story with new
social roles to supplant the old ones, but also a way out of their former epistemological impasse.

Importantly, MacIntyre contends that when an outside narrative assumes
this role, epistemological and moral progress has taken place. We have not
merely witnessed the abandonment of one story and accompanying social roles
for another story and other roles. If that were all that has happened, then we
could speak only of the temporal succession of one story by another and narrative ethics would have to remain silent on the fundamental question of which
story might be better than another, thereby settling for a disquieting relativism.
Rather, MacIntyre wants to claim that we have moved from a relatively narrow
and (by now) dysfunctional narrative to a more encompassing and more adequate story that effectively solves the problems of the first tradition. When this
happens, we have, in effect, moved from the particular to the (more) universal
without abandoning our commitment to narrative as the driving force behind
ethics. In other words, narrative ethics can remain critical without ultimately
abandoning narrative in the fashion of Burrell and Hauerwas.

In order to maintain this position, MacIntyre must insist that the adherents
of the faltering story must be able to see the succeeding story as holding the key
to the resolution of their former problems. They must, moreover, be able to see
the new story as constituting an advance over the old story in terms that would
be comprehensible to the adherents of the old story.[29] Without this sort of linkage, we would be back to a mere succession of stories instead of the hoped for
moral progress from the particular to the more encompassing view. While this
is not the place for a full-blown critique of MacIntyre's position, it should be
noted that his view on narrative and justification is controversial and problematic. In particular, . . . it is hard to understand how the substitution of a new
foundational narrative for a faltering story could leave intact all those old modes
of thought and evaluation that are supposed to evaluate the new story in terms
of the old.[30]

Narrative and Postmodern Ethics

So far we have canvassed two distinct approaches to the question of narrative
ethics and its relation to ethical justification. . . . As I shall try to show in this
final section, the postmodern storyteller has come to see narrative not as a sub-
strate, but rather as a substitute for the entire enterprise of moral justification.

A. What Is a Postmodern Ethic?

As I (dimly) understand it, "postmodernism" can be understood from one angle as a wholesale retreat not only from traditional theories—such as Marxism, Freudian psychoanalysis, or utilitarianism—but also from attempts at achieving some sort of grand coherence in our epistemological, ethical, and social views. In the place of theory and overarching coherence, the postmodernist asserts the virtues of the *petit récit* or "little narrative." . . . [T]he postmodernist. . . . seeks a kind of legitimation through the telling and retelling of stories.[31]

Richard Rorty's endorsement of an "ironist culture" provides an illuminating example of this eclipse of explanation and justification by narrative.[32] While Rorty concedes that on the most mundane level, within a particular narrative or historical tradition (for example, the common law), we can still make use of the notion of justification, he argues that at the more global level, where rival narratives, vocabularies, and traditions clash, we cannot speak meaningfully of justifying any one of these rival views by anchoring it in the bedrock of a true theory of history, human nature, or the natural world. When confronted with a sustained narrative that now shows signs of budding incoherence or newly perceived insensitivity to the sufferings of others—for example, a society (such as ours) that has traditionally and systematically degraded women—Rorty's "liberal ironist" must resort, not to logical argument, but rather to a kind of poetic redescription that allows us to see the world in new ways. Instead of presenting one's interlocutor with a logical argument that cannot be denied on pain of self-contradiction, the feminist must work with other like-minded people to forge a new vocabulary, a new set of meanings, and encourage others to begin to describe the world in similar ways.[33] For Rorty, then, the poet, not the traditional philosopher, is the vanguard of the human species.[34]

The ultimate goal of Rorty's culture of liberal ironism is not the replacing of falsity and distortion with truth (about "Man," "human nature," "History," "Reality"), but rather the mere continuation of the "conversation." Whereas both explanation and justification seek and require closure at some point[,] . . . Rorty's notion of conversation desires only its own continuation in a limitless quest for novelty. It refuses to seek a final resting place in some moral, social, or scientific bedrock that will put an end to disputation and conversation once and for all. One important ethical maxim that Rorty would have us derive from his notion of conversation—a kind of postmodern categorical imperative, if you will—is thus that we should always strive to keep "moral space" open for more dialogue.[35]

Another way to depict the implications of postmodernism for ethics is to describe it as an "ethics of voice."[36] In contrast to the standard brands of Enlightenment ethics that highlight either the content (for example, utilitarianism) or form (for example, Kantianism) of what is said, postmodernist ethics seems to be primarily concerned with who gets to tell the story. More specifically, the postmodern categorical imperative seems to come down to an insistence that everyone gets to tell his or her own story. Thus, Arthur Frank, a self-described postmodernist, sets out in his remarkable book, *The Wounded Storyteller*,[37] to rescue the first-person illness narratives of his fellow cancer sufferers from the "colonialism" of modernist (that is, scientific) medicine. According to Frank, those who suffer should be allowed and encouraged to speak for themselves, to find their own voice, rather than submit to the reductionistic and objectifying categories of modern medicine. Instead of the professionals' "case studies," narratives that objectify the experience and sufferings of people grappling with illness, Frank advocates the "case story" in which the ill are allowed to discover for themselves what it means to be a good person by telling and then reflecting on their own story.[38]

At this point, one might very well be moved to exclaim, "Nice story, but so what?" What is the connection, in other words, between all this storytelling and what might quaintly be called "the truth"? According to Frank, the stories that convey the subjective quest of the ill person "are their own truth," and he confesses to being unsure "what a 'false' personal account would be."[39] While prepared to grant that some personal narratives might be "evasive," Frank considers this evasiveness to be their truth. The more I reconstruct (distort?) the details of my own story, the more I manifest the truth of my desire to have experienced a different narrative course in my life (p. 22). Against an ethic of principles and rules, Frank claims that narrative ethics offers the ill person the freedom or "permission" to allow his story to lead in a variety of different directions in order to facilitate the process of self-discovery through the trial of illness (p. 160). And lest the reader begin to wonder about the potentially solipsistic consequences of such a view of truth in narrative, Frank concedes in the end that narratives are ultimately based upon an appeal to something more than our desires: "What is testified to remains the really real," he writes, "and in the end what counts are duties towards it" (p. 138). Still, although the act of providing "testimony" forges a connection for Frank to the "really real," this particular kind of postmodern testimony makes no pretense of grasping the whole or presenting a full panorama connecting my testimony to that of others. We are left with the ill person's *petit récit*.

As I mentioned at the beginning of this essay, this postmodernist privileging of the "little story" has both an epistemological and ethical dimension. We ought to favor such narratives, first, because we can't do any better. It is an epistemological error to believe that we can transcend the local, anchoring our science and ethics on the bedrock of the objectively and universally real. But we also have an ethical motivation to prefer the "little story" in the tendency of larger or more "totalizing" narratives, such as Marxism or Frank's portrayal of modern medicine, to silence, coerce, or, at the extreme, physically annihilate those who do not conform to their norms and expectations. As Lyotard mordantly observes, "The nineteenth and twentieth centuries have given us as much terror as we can take. We have paid a high enough price for the nostalgia of the whole and the one."[40]

B. Some Problems and Reservations

While I do not consider myself sufficiently well versed as yet in the literature of postmodernism to hazard a global assessment of this movement and its implications for ethics, the brief sketch I have presented above should provide us with a rich agenda for further elaboration, reflection, and critique. I will therefore limit myself to the expression of some initial doubts and worries regarding the promise of a postmodern ethic founded on *petits récits*.

1. The Threat of Subjectivism.   As developed by Arthur Frank, postmodern ethics risks sacrificing ethics at the altar of personal self-development. Not entirely satisfied with Rita Charon's portrayal of narrative ethics as a necessary adjunct to an ethic of principles, Frank argues that beyond the delimited sphere of "patienthood," in which the suffering individual is subjected to the norms and projects of health care providers, narrative ethics achieves autonomy and completeness in its own sphere, which is the sphere of "personal becoming" (p. 158). As noted above, this ethic cannot provide us with guidelines or principles; instead it provides each suffering individual with the moral space and "permission" to develop his or her story in ways that seem appropriate to her or his own life. While Frank says many important and interesting things on this theme, the overall effect of his argument seems to privilege the search for individual coherence over one quite central function of ethics, traditionally conceived, which is the passing of judgment on actions, policies, and character traits.

For example, Frank confesses to being unsure what a "false" personal account might be. I must admit to having a lot less trouble on this score. Although each and every one of us no doubt shades the truth or even intentionally distorts crucial facts in the stories we tell about our own lives, one need think only of the life story or personal testimony of Ronald Reagan to find a staggering example of duplicity and self-deception. As recounted and amply demonstrated in Gary Wills's fine biography,[41] Reagan was chronically and systematically incapable of telling fact from fiction about any of the defining events in his own life. Whether the issue concerned his boyhood days in Illinois, his wartime "service" in Hollywood, his behavior as president of the Screen Actors' Guild during the McCarthy era, or (I might add) during the Iran-Contra affair, Reagan seemed congenitally incapable of telling a story about his own life that was even remotely related to what had actually happened. In each and every case, the story told had more to do with what Reagan wished were true than it did to people and events in what might be referred to as the real world. But this, of course, should come as no great surprise. Although Reagan was perhaps more doggedly systematic in his penchant for self-deception and buffing his personal record than most people, all of us tell stories that deviate in greater or lesser measure from what really happened. Come to think of it, psychiatrists would probably be out of a job if all of us were more truthful, self-aware, and trustworthy in the stories we tell about ourselves.[42] If correspondence to what actually happened— making due allowance, of course, for the necessity and vagaries of interpretation—is an indispensable measure of the verisimilitude of stories, then it would seem that we have no more reason to place unquestioning trust in these "little narratives" than in some of the theorists' metanarratives.[43]

... We encounter here a broader and more fundamental problem with narrative as a vehicle for ethics. We have already seen, through Rita Charon's work, just how important narrative is to ethics as traditionally conceived. Narrative provides us with a rich tapestry of fact, situation, and character on which our moral judgments operate. Without this rich depiction of people, their situations, their motives, and so on, the moral critic cannot adequately understand the moral issue she confronts, and any moral judgments she brings to bear on a situation will consequently lack credibility. To paraphrase Kant, ethics without narrative is empty. But if all we do is strive to comprehend, if we are exclusively concerned with discerning coherence within a person's narrative, then we have no moral space left over for moral judgment. And this becomes a problem as soon as we realize that some internally coherent stories may yet be morally repugnant and fit objects for moral disapproval. To round out the allusion to Kant, we might say here that ethics without judgment is not ethics. Pace Frank, stories may well have their own (internal) truth, but that is not the only truth with which we must be concerned if we mean also to do ethics. The partisans of narrative ethics must therefore begin to think harder about the implications of "bad coherence" for their enterprises.[44]

2. Localism and Social Criticism.

A related problem for Frank, and more generally for postmodernism, is the temptation to fetishize "little narratives" at the expense of broader social understanding and critique. While there is unquestionably an important place for such narratives, ... it is also no doubt true that an overemphasis on the little story can render us purblind to larger social patterns and events that must also be grasped and understood if we are to achieve a fully rounded and adequate picture of our social world. It is an enduring temptation for Frank and the postmodernists, in their single-minded embrace of creativity, empathy, and compassion, uncritically to buy into the essentially romantic myth of the isolated individual or group, and thereby to ignore larger patterns and relationships that a more critical and socially attuned approach might recognize. At the very least, someone like Frank should be concerned not just with individual stories, but also with larger sets of relationships or recurring patterns that might cast new light on these stories and suggest common strategies for social improvement.

One might also entertain doubts in this connection about the extent of the postmodern critique of transcending the local. It is certainly true, as Lyotard points out, that a "totalizing" mentality has often led to oppression of dissenting minorities, but it is equally true that the rationalist tendencies of the Enlightenment tradition have also had a profoundly liberatory effect in many instances. ...

Consider the case of feminism. In contrast to Rorty's assessment,[45] many thoughtful feminists see theirs as an essentially modernist movement opposed to the arbitrary authority of men over women, as an assault on every social norm or institution resting on the ideology of male superiority.[46] Whereas Lyotard views a dogged attachment to the local and the *petit récit* as a liberation from the enslavement of "master narratives," these feminists see the Enlightenment ideals of freedom and equality as liberatory from the enslavement of women manifested in just about every local culture known hitherto. While communitarians like MacIntyre uncritically accept the social roles handed down by tradition and foundational narratives, feminists are bound, in the words of Sabina Lovibond, sooner or later to call the parish boundaries into question.[47] For them, the total reconstruction of society along rational lines—assuming the rationality of gender equality—is not so much a shopworn and discarded Enlightenment ideal as an indispensable blueprint for fundamental and desperately needed social reform. Crucial to this agenda is the critical idea of false consciousness, that is, the ability of dominant social classes to impose their own values and ideals on all other groups so that the latter are often impaired in their ability to discern their own true best interests. For many contemporary feminists, the gradual but systematic transcendence of pervasive local rationales for male domination constitutes the first order of business in social theory. To be sure, these feminists seek out and honor the individual experiences of individual women; that indeed is a large part of what "consciousness-raising" is all about. But it is also about linking the common experiences of individual women into a cohesive and global social critique and accompanying program for large-scale social action. For such activists and theorists, the postmodern attachment to the local represents a fundamental threat to feminism as a critical theory of society. . . .

Success?

In this essay I have canvassed three distinct approaches to the relationship of narrative ethics to ethical justification. I have come to the provisional conclusion that the first approach, which conceived of narrative as an essential element in any and all ethical analyses, constitutes a powerful and necessary corrective to the narrowness and abstractness of some widespread versions of principle- and theory-based ethics. The second approach, staked out by Burrell, Hauerwas, and MacIntyre, is initially plausible, but risks either falling back into a more principled version of ethics (compare Burrell and Hauerwas's search for abstract criteria) or sinking into a relativistic slough of incommensurable fundamental narratives (MacIntyre). Finally, I have argued that self-consciously postmodern approaches to ethics risk mistaking the authenticity of the narrator for ethical truth, and often ignore the larger social picture. While it should be obvious by now that narrative and narrative methods of inquiry are pervasive and indispensable for ethical analysis, it remains less clear whether any of the more fundamental assaults on principle- or theory-based ethics can be successful. Narrative is thus indisputably a crucial element of all ethical analysis, and we would all do better to be more self-conscious about the literary nature of ethical understanding and assessment. It remains to be seen, however, whether narrative will ever be in a position to supplant an ethic also undergirded by principles and theory.

Notes

1. David Simpson, *The Academic Postmodern and the Rule of Literature: A Report on Half Knowledge* (Chicago: University of Chicago Press, 1995).
2. Indeed, Beauchamp and Childress have made so many concessions to casuists, feminists, communitarians, and other critics in the fourth edition of their monumental and justly praised *Principles of Biomedical Ethics* (New York: Oxford University Press, 1994) that this observer has been moved to describe them as the "Borg of bioethics." (In the late lamented TV series, "Star Trek: The Next Generation," the Borg was an enormously powerful and unrelenting entity that subsisted by attacking and then assimilating other beings and whole civilizations into its neural network. The Borg's mantra was, "Resistance is futile. You will be assimilated!")
3. Rita Charon, "Narrative Contributions to Medical Ethics: Recognition, Formulation, Interpretation, and Validation in the Practice of the Ethicist," in E. R. DuBose, R. Hamel, and L. J. O'Connell, eds., *A Matter of Principles? Ferment in U.S. Bioethics* (Valley Forge, PA: Trinity Press International, 1994), pp. 260–283.
4. For a well-developed autobiographical account of how such a truncated medical vision can adversely affect the care (and lives) of patients, see Oliver Sacks, *A Leg to Stand On* (New York: Summit Books, 1984).
5. "The principlist methods of ethical inquiry remain as the structure for clarifying and adjudicating conflicts among patients, health providers, and family members at the juncture of a quandary. The principles upon which bioethics decisions have been based . . . continue to guide ethical action within health care" (p. 277).
6. John Rawls, A *Theory of Justice* (Cambridge, MA: Harvard University Press, 1971), pp. 48–51.
7. Norman Daniels, "Wide Reflective Equilibrium and Theory Acceptance in Ethics," *Journal of Philosophy*, vol. 76 (1979):256; "Wide Reflective Equilibrium in Practice," in L.W. Sumner and J. Boyle, eds., *Philosophical Perspectives on Bioethics* (Toronto: University of Toronto Press, 1996), pp. 96–114. See also Dwight Furrow, *Against Theory* (New York: Routledge, 1995), ch. 1.
8. My account here draws on my previous article, "Principles and Particularity: The Roles of Cases in Bioethics," *Indiana Law Journal*, vol. 69 (1994):992ff.
9. Jean-Francois Lyotard, *The Postmodern Condition: A Report on Knowledge* (Minneapolis: University of Minnesota Press, 1994), at p. 27.
10. This section is drawn from my article, "Principles and Particularity: The Roles of Cases in Bioethics," pp. 983–1014.

11. Nussbaum, for example, argues that narrative is the only proper medium for some philosophical issues. See "Introduction: Form and Content, Philosophy and Literature," in *Love's Knowledge: Essays on Philosophy and Literature* (New York: Cambridge University Press, 1990), pp. 3–53.

12. Timothy Quill, "Death and Dignity: A Case of Individualized Decision Making," *New England Journal of Medicine*, vol. 324 (1991):691ff.

13. Indeed, in my opinion, Quill's major failing is to have inadequately considered the implications of introducing the practice of assisted suicide within the context of a society that fails to provide adequate health care, including pain relief and treatment for depression, to millions of potential candidates.

14. Bernard Williams, "Persons, Character, Morality," in *Moral Luck* (New York: Cambridge University Press, 1981), pp. 1–19.

15. For a more fully developed statement of the fit between narrative and the depiction of character, see Tobin Siebers, *Morals and Stories* (New York: Columbia University Press, 1992), p. 15.

16. David Burrell and Stanley Hauerwas, "From System to Story: An Alternative Pattern for Rationality in Ethics," in H.T. Engelhardt, Jr., and Daniel Callahan, eds., *Knowledge, Value and Belief* (Hastings-on-Hudson, NY: The Hastings Center, 1997).

17. I say "conceptions" here to underscore the fact that principlism embraces both correspondence and coherentist approaches to moral justification.

18. Samuel Fleischacker, *The Ethics of Culture* (Ithaca: Cornell University Press, 1994).

19. See John Rawls, *Political Liberalism* (New York: Columbia University Press, 1993), pp. xxi–xxix.

20. Alasdaire MacIntyre, *After Virtue* (Notre Dame: Notre Dame University Press, 1981), p. 201: "[M]an is in his actions and practice, as well as in his fictions, essentially a story-telling animal. . . . I can only answer the question 'What am I to do?' if I can answer the prior question 'Of what story or stories do I find myself a part?'"

21. Willard Gaylin et al., "Why Doctors Must Not Kill," *Journal of the American Medical Association*, vol. 259 (1988):2139–2140.

22. George Sher, "Justifying Reverse Discrimination in Employment," *Philosophy & Public Affairs*, vol. 4, no. 2 (1975):159.

23. This dialectical aspect of self-definition has received its most memorable expression in Nietzsche's *On the Genealogy of Morals*, trans. W. Kaufmann (New York: Vintage, 1969). See especially the "First Essay: Good and Evil, Good and Bad," pp. 24–56. See also Hegel's *Phenomenology of Mind*, trans. William Wallace and A. V. Miller (Oxford: Clarendon Press, 1971).

24. "From System to Story," p. 136.

25. *Ibid.*, at p. 137.

26. Karl Marx, "Preface to a Critique of Political Economy," in D. McLellan, ed., *Karl Marx: Selected Writings* (Oxford: Oxford University Press, 1977), pp. 388–391.

27. W. D. Ross, *The Right and the Good* (Oxford: Clarendon Press, 1930).
28. Alasdair MacIntyre, *Whose Justice? Which Rationality?* (Notre Dame: University of Notre Dame Press, 1988), p. 362. MacIntyre also addresses this theme in "The Relationship of Philosophy to Its Past," in Richard Rorty, Jerome B. Schneewind, and Quentin Skinner, eds., *Philosophy in History* (New York: Cambridge University Press, 1984), p. 44.
29. MacIntyre plays out this theme in the context of the philosophy of science in the following way: "[T]his solution can now be formulated as a criterion by means of which the rational superiority of one large-scale body of theory to another can be judged. One large scale of theory—say, Newtonian mechanics—may be judged decisively superior to another—say, the mechanics of medieval impetus theory, if and only if the former body of theory enables us to give an adequate and by the best standards we have true explanation of why the latter body of theory both enjoyed the successes and victories that it did and suffered the defeats and frustrations that it did, where success and failure, victory and defeat are defined in terms of the standards for success and failure, victory and defeat provided by what I earlier called the internal problematic of the latter body of theory. . . . It is success and failure, progress and sterility in terms both of the problems and the goals that were or could have been identified by the adherents of the rationally inferior theory." See "The Relationship of Philosophy to Its Past," p. 43.
30. For a fuller development of this criticism, see Furrow, *Against Theory*, pp. 49–59.
31. Simpson, *The Academic Postmodern*, p. 62. In this connection Lyotard remarks, "narrative knowledge does not give priority to the question of its own legitimation and . . . certifies itself in the pragmatics of its own transmission without having recourse to argumentation and proof" (p. 27).
32. Richard Rorty, *Contingency, Irony, and Solidarity* (New York: Cambridge University Press, 1989); see also Rorty, *The Consequences of Pragmatism* (Minneapolis: University of Minnesota Press, 1982).
33. Richard Rorty, "Feminism and Pragmatism," in Grethe B. Peterson, ed., *The Tanner Lectures on Human Values*, vol. 13 (Salt Lake City: University of Utah Press, 1992), pp. 3–22.
34. Rorty, *The Consequences of Pragmatism*, p. 150; *Contingency, Irony, and Solidarity*, p. 20.
35. Margaret Urban Walker, "Keeping Moral Spaces Open," *Hastings Center Report* (March/April, 1993).
36. Arthur Frank, *The Wounded Storyteller* (Chicago: University of Chicago Press, 1995), p. xiii. Later on, Frank writes, "The idea of telling one's own story as a responsibility to the commonsense world reflects what I understand as the core morality of the postmodern" (p. 17).
37. *Ibid.*
38. Postmodernism is characterized by Simpson as exhibiting a nostalgia for the preprofessional. See Simpson, *The Academic Postmodern*, p. 47.
39. Frank, *The Wounded Storyteller*, p. 22.

40. Lyotard, *The Postmodern Condition*, p. 81. In this connection Simpson notes that even Hegel, the great Satan of postmodernism, saw a need to allocate some role in his totalizing system to the individual, to "the little guy." According to Simpson, the distinctive claim of postmodernism is that "[i]n these times we are all little guys" (*The Academic Postmodern*, p. 60).

41. Gary Wills, *Reagan's America: Innocents at Home* (Garden City, NY: Doubleday, 1987).

42. For an amusing novelistic portrayal of this fallible human tendency, see Irvin D. Yalom, *Lying on the Couch* (New York: Basic Books, 1996).

43. Simpson is particularly discerning on this point: "There is nothing whatever in our participation in little narratives, our own or those of a few natural hearts or professional colleagues or fellow sufferers, that guarantees an avoidance of the blind spots or even of critical errors. Telling one's own story, or the story of one's imagined group or subculture, with an implicit or explicit reliance on the dubious category of 'experience,' has in itself no more or less authority than the grandest of grand narratives" (*The Academic Postmodern*, p. 30).

44. In Frank's defense, one might recall his apparent acknowledgement of a referent for narrative ethics beyond the subjectivity of the individual storyteller: "What is testified to remains the really real, and in the end what counts are duties towards it" (p. 138). The problem with this, however, is that "what is testified to" remains precisely the pain and suffering of the individual storyteller, presumably as interpreted by that storyteller. So what began as a possible link to some tangible external check on the truth of stories ends up being one more manifestation of the subjectivistic nature of Frank's approach.

45. Rorty, "Feminism and Pragmatism."

46. Sabina Lovibond, "Feminism and Postmodernism," *New Left Review*, no. 178 (November/December 1989):5–28; Seyla Benhabib, "Feminism and the Question of Postmodernism," in *Situating the Self: Gender, Community and Postmodernism in Contemporary Ethics* (New York: Routledge, 1992), pp. 203–241.

47. Lovibond, "Feminism and Postmodernism," p. 22.

THE CHALLENGE OF USING METHODS IN CLINICAL SETTINGS

ETHICS COMMITTEES

Hospital Ethics Committees:
Is There a Role?

Robert M. Veatch

Robert M. Veatch is a philosopher by training. He is Professor of Medical Ethics and the former Director of the Kennedy Center for Ethics at Georgetown University. He has published extensively on topics in clinical ethics and the ethics of medical practitioners. In this article he considers four functions of ethics committees. He rules out an active decision-making role for ethics committees but endorses other functions such as developing and recommending ethics-related policies, advising and helping clarify ethically ambiguous cases, and confirming prognoses. In the course of assessing appropriate ethics committee functions, he illuminates concerns that ethics committees need to consider further.

O n March 31, 1976, the New Jersey Supreme Court announced its decision in the case of Karen Quinlan. In addition to the personal impact of that decision on those involved in the case, the court's opinion had a significant impact on the broader public. The proposal to establish what Chief Justice Richard J. Hughes called a hospital "Ethics Committee" gave a major impetus to a new

Robert M. Veatch. "Hospital Ethics Committees: Is There A Role?" Abridged from *The Hasting Center Report*, Volume 7, 1977, pp. 22–25. Reprinted by permission.

trend. Before the decision, there had been a few attempts to establish committees at the local hospital level to make, review, or advise in decisions regarding the care of the terminally ill, and since then the idea has been given substantial attention.

It is, therefore, an appropriate moment to ask a number of questions. . . .

. . . should hospital ethics committees function primarily to review technical, medical facts, such as prognoses, or should they review the ethical and other value issues involved in the actual treatment-stopping decision once the prognosis has been determined?

Possible Tasks for Hospital Ethics Committees

Four general types of hospital committees can be identified: committees to review the ethical and other values involved in individual patient care decisions, committees to make larger ethical and policy decisions, committees for counseling, and prognosis committees.

1.   Committees to review ethical and other values in individual patient care decisions   The committee originally proposed by Karen Teel and similar committees that have been proposed at other institutions are designed to review the appropriateness of decisions pertaining to the care of individual patients, in particular, determining when it is appropriate to stop treatment. Such a committee would take into account the patient's condition, but would move well beyond that to make decisions about whether treatment is appropriate, reasonable, or "ordinary."

According to many ethical traditions in medicine, a useless or gravely burdensome treatment is expendable. The tradition of Catholic moral theology defines such treatments as "extraordinary." Clearly, however, decisions about which treatments are expendable because they are unreasonable involve questions of ethics and other values. To say that a treatment is useless is to say that it will serve no appropriate or fitting purpose. It is an open question, for instance, whether a treatment which would sustain an individual's life in coma is useful or not. According to some ethical views that emphasize the duty to prolong life without asking questions about the quality of that life, such intervention may well be deemed useful. However, other, and in my view more plausible, ethical views emphasize that biological life *per se* should not be preserved unless other capacities and qualities are also present.

Deciding what counts as an expendable treatment or under what circumstances treatment should be discontinued once a prognosis has been determined is clearly a question of ethical and other value judgment[s]. If there is to be a committee at all to decide these questions, it should be a very broad one representing a range of ethical and other values. Some might argue that the committee should be made up of the ethically wisest people in the community, although selecting them would raise problems. However, individual patients who are competent have the right to refuse any medical treatment that is proposed for their own good. In the case of the incompetent patient, there is a recognized range of privacy and integrity for which the courts recognize familial discretion. Thus the question arises: why should there be any committee at all if the sole purpose is to review the wisdom of the decision? The conclusion seems inescapable that hospital committees have no appropriate role in actually making treatment-stopping decisions when the issue is one of whether the care is reasonable or unreasonable. They may still have counseling and other roles, to be considered below.

2. Committees to make larger ethical and policy decisions   Hospital committees may serve a second function, however. Questions arise that clearly involve ethical and other values, yet which in principle cannot be resolved by referring to the individual patient for the patient's own decision or, in the case of the incompetent, to the guardian and family. Institutional review boards for the protection of human subjects in biomedical research are an example of committees established to deal with this kind of question. Such committees must decide what information must be disclosed for a consent to be adequately informed for the research. In principle, one cannot ask the individual whether a particular risk should be disclosed without in the process disclosing that risk.

Another basic question faced by such institutional review boards is whether a hospital or other research facility should permit research even if the patient consents. Some hospitals might decide that research is sufficiently dangerous or useless that as a matter of ethical principle the research cannot be tolerated even with adequate consent.

A similar policy question is allocation of scarce hospital resources. In the 1960s when hemodialysis machines were scarce, committees were established at some hospitals to decide which patients would receive dialysis treatment. Those judgments involved medical criteria, but ethical judgments were central. Deciding which of two patients should receive a kidney machine when there is only one machine available cannot be left to the individual patient. Dialysis machines are no longer scarce, but there is a similar situation in neonatal intensive care units. Similarly a hospital might have a committee to make policy decisions about whether to build a new intensive care unit or to remodel the emergency room. Deciding whether obstetrics and abortion facilities or an alcoholism treatment unit should be established are other examples.

It seems reasonable that these policy-making committees should be broadly based, representing a wide range of ethical and sociological positions within the community or those responsible for the hospital. Different perspectives are important, since the committees are serving as agents for the broader community.

3. Counseling committees   A third kind of committee might be called a counseling committee. It could be established to deal with specific terminally ill patients, but for the purpose of counseling and support rather than actual decision making. The hospital might have available an ongoing committee made up of a psychiatrist, a psychologist, a social worker, chaplain, and others with moral counseling skills. This committee might meet to discuss ongoing problems of the care of the terminally ill, but also make itself available for counseling when necessary.

Several existing committees, such as the Optimum Care Committee at the Massachusetts General Hospital, exist primarily to provide counsel (*New England Journal of Medicine*, 295 [August 12, 1976], 362–64). Many of the existing committees, however, see themselves as providing counsel not to the patient, but to the physician. Clearly, if one holds that the patient (or his agent in cases when he is incompetent) is the primary decision maker, it would be appropriate for the counseling committee to provide services for the patient instead of or in addition to the physician. For some patients such a committee might not be necessary, since moral counseling is likely to come from outside the hospital—from one's clergyman, family, or friends. Such a counseling committee would not be composed of representatives of a cross-section of ethical and sociological positions, but of individuals who have appropriate counseling skills.

Other functions related to counseling might also be appropriate. Although the committee should not have a role in making the actual decision to stop treatment, occasionally there may be cases where the decision made by a parent or other guardian is so questionable that the physician, nurse, or other hospital personnel are convinced that it should be reviewed. The morally and legally appropriate course is to bring the matter to court. If the court finds that the parental judgment is so unreasonable that it cannot be tolerated, it will appoint a new guardian for the purposes of authorizing the treatment. In such cases, however, the hospital staff member may want some guidance before deciding to initiate the court review. A hospital committee, especially one that was broadly representative of the community's moral sensitivities, could provide a sounding board for the health care professional who had such doubts. The committee could even initiate the court review proceedings itself. In such cases, however, the court, not the committee, would finally override the guardian's judgment.

4. Prognosis committees   The emphasis in the *Quinlan* opinion that the so-called "Ethics Committee" was in fact to confirm a prognosis has led to suggestions that such committees should really be called "Prognosis Committees." In January 1977 in New Jersey the Attorney General, the Health Commissioner, the head of the State Licensing Board, and several medical professional organizations jointly endorsed guidelines for prognosis committees called "Guidelines for Health Care Facilities to Implement Procedures Concerning the Care of Comatose Non-Cognitive Patients." The committee's purpose is to confirm the prognosis that no reasonable possibility exists of the patient's return to a cognitive, sapient state.

The guidelines propose that the committee include physicians trained in general surgery, medicine, neurosurgery or neurology, anesthesiology, pediatrics (if so indicated), and two additional physicians from outside the hospital staff. Such a committee should be made up of those with the relevant medical and other scientific skills for establishing the prognosis. The New Jersey guidelines provide for no lay committee members; but if lay people without such medical skills were on such a committee, their only function would be to ask questions and provide a modest public presence.

The New Jersey guidelines are recommendations that have no official weight. In fact, one of the main thrusts of the *Quinlan* opinion was to recognize that the consensus of the medical profession need not be binding on guardians or the state in approving guardian treatment refusals. By analogy one might question whether the consensus of the medical professional societies and other health-related professionals would be binding in recommendations about the establishment of a prognosis committee.

Second, the guidelines state that "the attending physician, guided by the committee's decision with the concurrence of the family, may then proceed with the appropriate course of action and, if indicated, shall personally withdraw life-support systems." In this recommendation the New Jersey guidelines clearly go beyond the *Quinlan* opinion. It is my understanding that this provision was added at the urging of nurses and others who might be ordered by physicians to actually stop the life-support apparatus.

Furthermore, the guidelines say that the attending physician *may* then proceed rather than that he *shall* proceed, leaving open the question of what would happen should such a physician decide against stopping the life-support apparatus. Presumably the physician would normally be the one who stops an ongoing treatment if the patient or the patient's agent so decides. Especially when a court has already reviewed the specific case or cases of the same type, however, the physician should not have the discretion in deciding whether to follow such instructions. He may, of course, feel morally obliged to withdraw from the case and normally would be permitted to do so provided a suitable replacement can be found to provide professional medical support. To say that the physician *may* then proceed, however, implies that he may also continue treatment which legally, since consent is lacking, has the quality of an assault and morally the quality of violating the patient's, the agent's, or the family's autonomy. It also forecloses the possibility that someone else might be the more appropriate person for the task. While nurses or other hospital personnel should never be forced to participate in the treatment stopping, there may be cases where they or others who are not directly connected with the hospital would be the more appropriate ones. In some circumstances the patient himself, the family, a clergyman, or someone else in a special relationship with the patient may be more appropriate.

In spite of these reservations, the New Jersey guidelines are generally sound and the most concrete help available for a hospital in the process of establishing a committee. The commitment to prognosis review is clearly stated. The committee to review prognosis should be used whenever there is a difficult technical situation or concern that the individual physician's judgment needs review. As long as the committee does not mistakenly generalize its responsibilities and move into the area of approving or disapproving of the treatment decision arrived at by the patient or the patient's agent, the guidelines should be helpful.

There is one final problem, however. Deciding the prognosis of a patient may not be completely a technical question. We have become increasingly aware of a blurring between facts and values. When one is attempting to make a judgment of prognosis involving such vague terms as "reasonable hope," and "cognitive, sapient state," questions of value may impinge upon even the determination of prognosis. That may be one reason why a committee was established in the first place: to avoid depending exclusively on a single physician's evaluation of the prognosis. Those who advocate a prognosis committee should be aware of this difficulty. If a case were to be made for lay membership on a prognosis committee, this would provide one of the grounds.

These are just a few issues raised by proposals for hospital ethics committees to deal with decisions pertaining to the care of the terminally ill. Some of the proposals for hospital committees seem to me to be dangerous and misguided, for example, the use of a committee to approve or disapprove a treatment-stopping decision made by a patient or made by an agent for an incompetent patient. Other committee models make more sense. The committee made up of lay people to establish hospital policy or make resource allocation decisions, the counseling committee made up of those with counseling skills, and the prognosis committee made up of those with appropriate technical skills, are all reasonable ideas and could serve important functions. Hospital ethics committees are a new development, and it is still unclear which types will gain support and how they will evolve. The issues they pose are significant ones, and their resolution will merit further study.

The Inner Workings of an Ethics Committee:
Latest Battle Over Jehovah's Witnesses

Ruth Macklin

Ruth Macklin received her Ph.D. in Philosophy from Case Western Reserve University in 1968. She is currently Division Head and Professor in the Division of Philosophy and History of Medicine, Department of Epidemiology and Population Health at the Albert Einstein College of Medicine. She is also the Dr. Shoshanah Trachtenberg Frackman Faculty Scholar in Biomedical Ethics at Albert Einstein College of Medicine and serves as an adviser to the World Health Organization and the Joint United Nations Programme on HIV/AIDS. In this article she describes the evolution of a policy concerning consent for blood transfusion in Jehovah's Witnesses. The Ethics Committee's deliberations and discussions with non–committee members resulted in several revisions to the section on pregnancy. The section ultimately became an explication of the competing principles at stake; it serves as a reminder that ethical analysis does not yield consensus about the best course of action.

Is there anything new in the ongoing saga of Jehovah's Witness patients who seek to conform to the dictates of their religion, which prohibits transfusion of whole blood or blood products (including autologous transfusions, removal and replacement of the patient's own blood). . . .

. . . That issue is the right of a pregnant Jehovah's Witness to refuse a blood transfusion, resulting in the likelihood of her death and that of the fetus. . . .

Policy Regarding Consent for Blood Transfusion in the Jehovah's Witness

Under New York State Public Health Law (Section 2805-d) an informed consent discussion should be conducted by the responsible physician with any patient about to undergo treatment. The patient should be told all the information that a reasonable person would consider material to the decision to accept treatment. Information must be clearly and understandably presented in language the patient can reasonably be expected to understand. Information may not be withheld because of concern that disclosure could cause the patient to refuse treatment. The elements of the informed consent discussion should include:

Ruth Macklin. "The Inner Workings of an Ethics Committee: Latest Battle over Jehovah's Witnesses." Abridged from *Hastings Center Report*, Volume 18, 1988, pp. 15–20. Reprinted by permission.

a) The nature of the patient's illness
b) The nature and purposes of the proposed treatment, including:
   1. Risks or consequences
   2. Benefits
   3. Alternatives, including no treatment, and the risks of alternatives
c) The opportunity to question the proposed treatment

The Jehovah's Witness and Blood Transfusion Consent

A. Adult

The risks, benefits, and alternatives, if any, to a proposed blood transfusion must be explained, as described above, to an adult Jehovah's Witness. Any adult patient who is not incapacitated has the right to refuse treatment no matter how detrimental such a refusal may be to his health.

Special Circumstances

1. Capacity

If there are reasonable grounds to doubt the capacity of the patient to understand the risks, or benefits of and alternatives to transfusion, psychiatric evaluation must be obtained. If the evaluation confirms the patient's capacity to understand, transfusion will not be administered.

If the patient is found to lack the capacity to make the decision to refuse transfusion, transfusion will be withheld only if there is clear and convincing evidence of the patient's wish to reject treatment, such as the following. . .

a) If there is a document recently executed by the patient which directs unequivocally that transfusion should be withheld under all circumstances, or

b) If, prior to intervening incapacity during the hospitalization, the current chart documents the patient's unequivocal and consistent refusal to accept transfusion under any circumstances, or

c) If, in a patient who presents incapacitated, there is documentation in a prior hospital record within one year of an unequivocal and consistent refusal to accept transfusion under any circumstances.

Under any other circumstances, transfusion will be given.

2. Voluntariness

Any Jehovah's Witness who refuses transfusion will be offered an opportunity to discuss this refusal with a physician not directly involved in providing care, in order to ensure that the patient's decision is freely and voluntarily made. This physician should ordinarily be a psychiatrist unless the patient prefers otherwise.

3. Emergencies

In the event a Jehovah's Witness presents to the hospital with an immediate need for life-saving blood transfusion, transfusion will be given unless condition a, b, or c of Section 1 (Capacity) obtains.

In the event a life-threatening emergency requiring transfusion arises in the course of hospitalization, the same standard (Section 1, Capacity) will apply.

It is inappropriate to wait until a foreseeable emergency need for transfusion arises in order to avoid an informed consent discussion with a Jehovah's Witness. There is a positive obligation reasonably to anticipate the development of such an emergency need for transfusion.

B. Children

Parents have the right to consent to care for their dependent children; they do not have a coequal right to refuse care for their dependent children (Family Court Act Section 233). Parents do not have the right to deny minor children transfusions that are deemed medically necessary. In the event that a parent withholds consent for transfusion, the hospital administration must be contacted immediately and asked to seek a court order for transfusion. In the event that medical judgment holds any delay to be immediately life-threatening to the child or would produce irreversible harm, transfusion should be given.

For purposes of this policy a child is defined as anyone below the age of eighteen years. Emancipated minors will be treated as adults. (N.Y. State Public Health Law 2504 states: A minor parent or married minor may give consent. According to case law, a minor in the military may give consent; and a financially independent minor, living alone, may give consent.)

C. Pregnancy

In the case of a pregnant Jehovah's Witness's refusal of transfusion, policies relating to adult patients will apply before the third trimester. If the pregnancy has entered the third trimester, the State's interest in life requires the decision to be referred to the courts for adjudication. If in such a case medical judgment holds any delay to constitute an immediate threat to the pregnancy, transfusion should be given. Competent adult patients have the right to refuse medical treatment. This right extends to pregnant women. However, some members of society assert that the fetus has "interests" or "rights" that compete with the rights of the mother to control her own body. In general, the rights of the mother are clearly acknowledged to take precedence in early pregnancy. As gestation advances, it becomes increasingly difficult for some members of society to ignore the "interests of the fetus." Because of the dilemma that arises out of these opposing interests, physicians have an obligation to disclose from the outset if, under specified circumstances, they would be unable to honor a patient's wishes.

Because society and the law have not resolved the conflict between fetal and maternal interests, the policy cannot establish clear guidelines for action where a clinician's interpretation of the interests of the fetus are in conflict with the wishes of the mother. The clinician and patient, with ethical consultation, must seek to resolve such conflicts within the context of the doctor-patient relationship and resort to other means for conflict resolution, including hospital administration, when necessary. Every physician has the right and the obligation to try to turn the care of such a patient over to another caregiver if the patient's wishes are incompatible with the physician's professional and ethical values. This course of action is ethically superior to the coercion of an unwilling patient.

Behind Closed Doors:
Promises and Pitfalls of Ethics Committees

Bernard Lo

Bernard Lo is a Professor of General Internal Medicine at the University of California-San Francisco Medical School. He received fellowship training in ethics at the University of California-San Francisco while in the Robert Wood Johnson Clinical Scholars Program. He has contributed to the clinical ethics literature with articles invoking different methodologies, including ethical analysis, anthropological observation, survey research, and quantitative health services research. In the following article he primarily discusses the threats to optimal ethics committee function. Examples include imprecise goals, restricted access (who can request a consultation, who can attend a meeting, who can review the results of the discussion), and "groupthink." At the conclusion the author specifies several useful criteria for evaluating both the process by which ethics committees review cases and the results of their deliberations.

Hospital ethics committees have been hailed as providing a promising way to resolve ethical dilemmas in patient care. Although ethics committees may have various tasks, such as confirming prognoses, educating care givers, or developing hospital policies, their most innovative role is making recommendations in individual cases.[1-5] This role has been supported by the President's Commission for the Study of Ethical Problems in Medicine and Biomedical and Behavioral Research, the American Medical Association, and the American Hospital Association. Strictly speaking, such recommendations are not binding, but they undoubtedly carry great weight, especially if they are cogently justified.[6] It is predicted that most ethics committees will make recommendations in particular cases[3] and that the courts will respect them.[3]

Ethics committees may offer an attractive alternative to the courts.[3-5] The judicial system may be too slow for clinical decisions.[7,8] Moreover, the adversarial judicial process may polarize physicians, patients, and families,[9] whereas ethics committees may reconcile divergent views. The 1986 New York State

Bernard Lo. "Behind Closed Doors: Promises and Pitfalls of Ethics Committees." Reprinted from *The New England Journal of Medicine*, Volume 317, 1987, pp. 46–50. Reprinted by permission of *The New England Journal of Medicine*. Copyright © 1987, Massachusetts Medical Society.

Supported in part by a grant (1 P50 MH42459-01) from the National Institute of Mental Health and a grant from the Commonwealth Foundation.

Task Force on Life and the Law encouraged resolving patient care dilemmas at the hospital level, rather than turning to the courts, and suggested that ethics committees might mediate such disagreements.[10]

Although I support ethics committees, several questions trouble me. First, are these committees ethical? The goals and procedures of some committees may conflict with established ethical principles. Second, is agreement by committees always desirable? Group dynamics may lead to flawed information, reasoning, or recommendations. Third, are these committees effective? Like other medical innovations, they need to be rigorously evaluated.

Goals and Procedures of Ethics Committees

The very name suggests that ethics committees base their recommendations on ethical principles and rational deliberation, rather than on mere custom, political power, or self-interest. A consensus on medical decision making has emerged in the medical literature, court decisions, and reports of the President's Commission.[1,7,8,11,12] According to this consensus, competent patients should give informed consent or refusal to the recommendations of physicians. Care givers need not accede to patient requests for treatments, however, if there are no medical indications. In cases in which patients are incompetent, decisions should be based on their previously expressed preferences or, if such preferences are unclear or unknown, on their best interests. The goals of some ethics committees, however, may conflict with these ethical guidelines. Goals vary substantially among committees.[1-3,13-15] Some do not have explicit goals. One committee has said, "We have never formally stated in writing the exact purpose or purposes of our committee but have decided to proceed in an informal manner. . . . We felt that to formalize our objectives might be counterproductive to the work of our committee."[14] But as ethics committees mature, and especially as they wish to serve as alternatives to the courts, they need to define their goals more clearly. Some so-called ethics committees have as goals confirming prognoses, providing emotional support for care givers, or reducing legal liability for physicians or hospitals.[1-3,13-15] One hospital administrator has even suggested that the ethics committee be used as a public relations "tool" for justifying unpopular decisions to discontinue unprofitable services.[16] Although committees on quality assurance, staff support, risk management, or public relations are important, there is little reason for patients, their surrogates, or the public to accept their recommendations about patient care.

After clarifying goals, committees can establish procedures. Ethics committees must decide who can refer cases or attend meetings. Many committees limit participation by patients and families. According to a 1982 survey, only 25 percent of ethics committees that reviewed cases allowed patients to bring cases to the committee. Only 19 percent of committees allowed patients to attend meetings, whereas 44 percent allowed family members to do so.[17] Limiting access to committee proceedings may seem desirable. It may be sound political strategy to overcome initial resistance to the ethics committee within the hospital. For example, attending physicians may fear that their authority will be undermined if patients, families, or nurses can ask the committee to review cases. Restricting access may also facilitate frank discussions by care givers and committee members about sensitive topics. In addition, discussions with other health professionals may help physicians to clarify their thinking before they talk to patients or families.

Restricted discussions, however, may not be accepted by patients, families, and society. Patients or surrogates who disagree with physicians are unlikely to regard the committee as impartial if they may not convene the committee or present their views directly, whereas physicians may do so. Disagreements that reach ethics committees usually involve important personal issues—even questions of life and death. In such vital decisions, patients and their proxies are not likely to accept recommendations by a committee whose members they have not met or that seems to meet behind closed doors. The composition of ethics committees may not reassure patients that their wishes and interests are represented. Typically, most members of ethics committees are physicians, who may assess the importance of medical problems or the risks and benefits of treatment differently from patients.[18,19] Patients or surrogates who disagree with the committee's recommendations may say that the composition of the committee was biased against them.

Some committees meet with patients or family members who take the initiative and request meetings. But people who need the most help in expressing their preferences or interests may be the least likely to request a meeting. They may be cognitively impaired or unable to navigate the medical system, or there may be cultural, language, or educational barriers. Hence, it is desirable for the committee to take steps to inform patients, as well as care givers, of its work. Such information is particularly important if the committee can review a case without the consent of the parties. Mandatory review has been recommended, for example, when withholding life-sustaining treatment from neonates or from incompetent adults without surrogates is being considered.[20] A pamphlet about the committee might be distributed when patients are admitted. Patients or surrogates who are concerned that committee discussions or recommendations may invade their privacy can then express those concerns in advance. Before the committee discusses a case, it should inform patients or surrogates and invite them to participate in the deliberations.

Most ethics committees also restrict the access of nurses. The 1982 survey found that only 31 percent of committees allowed nurses to present cases, and only 50 percent allowed nurses to attend meetings.[17] But it may be advisable to increase the access of nurses. Nurses have close contact with patients and families and may take the role of patient advocates.[21] They may raise previously overlooked issues, contribute new information, or express the questions and viewpoints of patients and families. Disagreements by nurses with physicians' orders often indicate a need to reconsider decisions.[22]

Because ethics committees are touted as an alternative to the courts, it may be useful to compare their safeguards with those in legal procedures.[23] The legal system notifies parties of the proceedings, allows them to give evidence, and ensures representation for patients. If the patient is incompetent, the court may appoint a guardian ad litem to represent the interests of the patient or to argue for continuing treatment. Moreover, parties are notified of the decision and the reasons for it, so that the decision can be reviewed or appealed. Ethics committees that make recommendations may not need safeguards that are as elaborate as those in a legal system that makes binding decisions. But for ethics committees to be accepted as a quicker and less acrimonious alternative to the courts, they must be perceived to be as fair as the courts.

In order for ethics committees to assist in decision making, their recommendations and the reasons for them must be known by all parties. In addition to communicating with the patient or surrogate and the attending physician, a representative of the committee might write a note in the medical record, so that nurses, consultants, and physicians understand the committee's recommendation and reasoning. Ethics committees, however, may seem reluctant to allow their recommendations to be reviewed. Some committees do not note their recommendations and reasoning in the medical record. In addition, articles about ethics committees discuss how to reduce the liability of individual committee members by keeping records from being "discoverable"—that is, from being subpoenaed in civil suits.[20,24] Such apparent secrecy may evoke the suspicion that the committee is more concerned with protecting physicians, the hospital, or itself than with helping patients.

Pitfalls of Committee Discussions

Pressures on ethics committees to reach agreement may lead to recommendations that are ethically questionable. Agreement or even consensus does not confer infallibility. For example, in the 1960s, hospital committees selected patients with chronic renal failure for treatment with life-prolonging dialysis machines, which were limited in number. When it was disclosed that criteria of social worth were implicitly applied, these committee decisions were criticized as being unfair and discriminatory.[25]

In some circumstances, committees may impair rather than improve decision making. Political scientists and psychologists have shown that committees may inadvertently pressure members to reach consensus, avoid controversial issues, underestimate risks and objections, or fail to consider alternatives or to search for additional information.[26,27] In other words, committees may not serve their intended function of considering diverse viewpoints and arguments. Such undesirable qualities of committee discussions, which have been called "groupthink," may lead to grave errors in judgment.

Ethics committees may fall victim to groupthink. First, these committees may reach consensus too easily, by not adequately considering patients' preferences. Despite the ideal of informed consent, patients are often not involved in decisions about their care.[28–30] Second, committees may accept secondhand information uncritically. Physicians appreciate that medical consultants should take new histories, examine patients, and review x-ray films and scans.[31,32] Similarly, an ethics committee should scrutinize information about the medical situation and the patient's preferences. Conclusions and inferences, rather than primary data, may be presented. For instance, patients may be described as "terminal" or "hopelessly ill," or it may be reported that an incompetent patient would not want "heroic care." Since such phrases are ambiguous and potentially misleading, committees should require and, if necessary, seek out more specific information. Third, ethics committees may overlook imaginative means of resolving disagreements. Disputes over patient care are not always caused by conflicts of ethical principles or obligations. They may also result from misunderstandings, stress, or lack of attention to the details of care.[22] Despite stalemates over conflicting ethical principles or duties, agreements on particular recommendations for patient care may be possible.[33]

Ethics committees should appreciate that they work under conditions that predispose them to groupthink. A rapid recommendation may be needed despite uncertain information and conflicting values and interests. Such clinical urgency may press the committee to reach agreement. The committee may feel attacked by various groups: attending physicians who fear that their power is being usurped, nurses who think that they are given unreasonable orders, administrators who wish to control costs, or risk managers who want to avoid legal difficulties. If committee chairpeople are forceful leaders who control discussions, they may unintentionally discourage frank debate and disagreement. Tendencies toward groupthink may be reinforced if access to the committee is limited.

Ethics committees that recognize the dangers of groupthink can take steps to avoid them. First, committees can guard against premature agreement. The chairperson may explicitly ask that doubts and objections be expressed or may appoint members to make the case against the majority. Second, committees can scrutinize any secondhand information they receive. To understand the patient's preferences, the committee might talk with the patient or proxy directly, invite the patient or surrogate to participate in some discussions, or assign a committee member to act as a patient advocate. Third, the committee can look for innovative ways to settle disputes. Improved communication may resolve disagreements. Families, nurses, or house staff may accept the attending physician's decisions after they hear the reasons for it and have an opportunity to ask questions. Alternatively, a compromise may be negotiated.[34] For example, a patient who threatens to sign out of a cardiac care unit may agree to further treatment if he or she is given more control over the timing of the administration of medications and nursing care, and if one physician and one nurse take responsibility for answering his or her questions.

Evaluating Ethics Committees

Ultimately, the question of whether ethics committees are useful is an empirical one. Before consulting ethics committees can be considered to be a standard decision-making procedure rather than a promising innovation, they need to be evaluated. Because enthusiastic anecdotes about innovations may not be confirmed in controlled trials, pleas have been made to evaluate new technological procedures, such as angioplasty, before they are accepted and put into wide use.[35] Institutional innovations should also be evaluated, even if they seem to be obviously beneficial. For instance, hospices were expected to provide more humane and less expensive care for patients with terminal illnesses. Controlled studies, however, suggest that hospice care may not differ substantially from current conventional care and may be more expensive.[36-38]

As in any evaluation, deciding on clinically meaningful outcomes and designing unbiased studies require thought and planning. I suggest several criteria for evaluating both the process by which ethics committees review cases and the results of their deliberations. First, patients and their surrogates should have access to the ethics committees. Specifically, they should be able to ask the committees to review their cases and to meet with the committees if they desire. Second, recommendations by the committee and the reasons for them should be available to the parties in each case. Generally, a note in the medical record would be required. Third, recommendations by ethics committees and actual decisions by attending physicians should be consistent with ethical and legal guidelines. The gold standard should be the widespread ethical consensus that has emerged on many issues.[39] Evaluations might focus on whether ethics committees reduce discrepancies between this consensus and actual decisions by physicians. For instance, studies indicate that care givers often fail to discuss management options with patients or the surrogates of incompetent patients.[28-30] Ethics committees should recommend such discussions when appropriate. If their recommendations have an effect on care givers, fewer decisions will be made without such discussions with patients or their surrogates. Committees should also increase informed refusals of care by patients. Moreover, committees should decrease decisions based on ambiguous or uncorroborated secondhand information about the indications for treatment or about patient preferences. Fourth, parties in disagreements should be satisfied with the process of review and with the recommendations of the ethics committee. Although the degree of satisfaction of care givers with ethics consultations has been studied,[40] it is also important to determine the reactions of patients or their surrogates. Finally, ethics committees that make recommendations should have their own internal systems of review, to ensure that the suggested criteria are met.

In summary, the promise that ethics committees will resolve dilemmas about patient care and avoid legal disputes needs to be examined critically. If recommendations by ethics committees are to be accepted by patients, families, society, and the courts, the wishes and interests of patients must be represented and ethical guidelines must be followed. Committees can take active steps to reduce the risk of groupthink. Empirical studies may indicate what kinds of committees improve decisions relating to patient care and in which clinical circumstances.

References

1. President's Commission for the Study of Ethical Problems in Medicine and Biomedical and Behavioral Research. *Deciding to forego life-sustaining treatment: a report on the ethical, medical, and legal issues in treatment decisions.* Washington, D.C.: Government Printing Office, 1983.
2. Cranford RE, Doudera AE, eds. *Institutional ethics committees and health care decision making.* Ann Arbor, Mich.: Health Administration Press, 1984.
3. Bayley SC, Cranford RE. Ethics committees: what we have learned. In: Friedman E, ed. *Making choices: ethics issues for health care professionals.* Chicago: American Hospital Publishing, 1986:193–9.
4. Lynn J. Roles and functions of institutional ethics committees: the President's Commission's view. In: Cranford RE, Doudera AE, eds. *Institutional ethics committees and health care decision making.* Ann Arbor, Mich.: Health Administration Press, 1984:22–30.
5. Committee on Ethics and Medical-Legal Affairs. Institutional ethics committee's [*sic*]: roles, responsibilities, and benefits for physicians. *Minn Med* 1985; 68:607–12.
6. Siegler M. Ethics committees: decisions by bureaucracy. *Hastings Cent Rep* 1986; 16(3):22–4.
7. Lo B, Dornbrand L. The case of Claire Conroy: Will administrative review safeguard incompetent patients? *Ann Intern Med* 1986; 104:869–73.
8. Lo B. The Bartling case: protecting patients from harm while respecting their wishes. *J Am Geriatr Soc* 1986; 34:44–8.
9. Burt RA. *Taking care of strangers: the rule of law in doctor-patient relations.* New York: Free Press, 1979.
10. New York State Task Force on Life and the Law. Do not resuscitate orders: the proposed legislation and report of the New York State Task Force on Life and the Law, April 1986.
11. Wanzer SH, Adelstein SJ, Cranford RE, et al. The physician's responsibility towards hopelessly ill patients. *N Engl J Med* 1984; 310:955–9.
12. Lo B, Jonsen AR. Clinical decisions to limit treatment. *Ann Intern Med* 1980; 93:764–8.
13. Levine C. Questions and (some very tentative) answers about hospital ethics committees. *Hastings Cent Rep* 1984; 14(3):9–12.
14. Kushner T, Gibson JM. Institutional ethics committees speak for themselves. In: Cranford RE, Doudera AE, eds. *Institutional ethics committees and health care decision making.* Ann Arbor, Mich.: Health Administration Press, 1984:96–105.
15. Fost N, Cranford RE. Hospital ethics committees: administrative aspects. *JAMA* 1985; 253:2687–92.
16. Summers JW. Closing unprofitable services: ethical issues and management responses. *Hosp Health Serv Adm* 1985; 30:8–28.
17. Youngner SJ, Jackson DL, Coulton C, Juknialis BW, Smith E. A national survey of hospital ethics committees. *Crit Care Med* 1983; 11:902–5.

18. Friedin RB, Goldman L, Cecil RR. Patient–physician concordance in problem identification in the primary care setting. *Ann Intern Med* 1980; 93: 490–3.
19. McNeil BJ, Weichselbaum R, Pauker SG. Fallacy of the five-year survival in lung cancer. *N Engl J Med* 1978; 299:1397–401.
20. Winslow GR. From loyalty to advocacy: a new metaphor for nursing. *Hastings Cent Rep* 1984; 14(3):32–40.
21. Robertson JA. Ethics committees in hospitals: alternative structures and responsibilities. *Conn Med* 1984;48:441–4.
22. Lo B. The death of Clarence Herbert: withdrawing care is not murder. *Ann Intern Med* 1984; 101:248–51.
23. Baron C. The case for the courts. *J Am Geriatr Soc* 1984; 32:734–8.
24. Cranford RE, Hester FA, Ashley BZ. Institutional ethics committees: issues of confidentiality and immunity. *Law Med Health Care* 1985; 13:52–60.
25. Fox RC, Swazey JP, eds. *The courage to fail: a social view of organ transplants and dialysis*. Chicago: University of Chicago Press, 1974:240–79.
26. Janis IL, Mann L. *Decision-making: a psychological analysis of conflict, choice, and commitment*. New York: Free Press, 1977.
27. George A. Towards a more soundly based foreign policy. In: *Commission on the Organization of the Government for the Conduct of Foreign Policy*, appendix B. Washington, D.C., Government Printing Office, 1975.
28. Lidz CW, Meisel A, Osterweis M, Holden JL, Marx JH, Munetz MR. Barriers to informed consent. *Ann Intern Med* 1983; 99:539–43.
29. Bedell SE, Pelle D, Maher PL, Cleary P. Do-not-resuscitate orders for critically ill patients in the hospital: How are they used and what is their impact? *JAMA* 1986; 256:233–7.
30. Goldman L, Lee T, Rudd P. Ten commandments for effective clinicians. *Arch Intern Med* 1983; 143:1753–5.
31. Lo B, Saika G, Strull W, Thomas E, Showstack J. 'Do not resuscitate' decisions: a prospective study at three teaching hospitals. *Arch Intern Med* 1985; 145:1115–7.
32. Tumulty PA. *The effective clinician: his methods and approach to diagnosis and care*. Philadelphia: W.B. Saunders, 1973:45–8.
33. Beauchamp TL, Childress J. *Principles of biomedical ethics*. 2nd ed. New York: Oxford University Press, 1983.
34. Steinbrook R, Lo B. The case of Elizabeth Bouvia: Starvation, suicide, or problem patient? *Arch Intern Med* 1986; 146:161–4.
35. Mock MB, Reeder GS, Schaff HV, et al. Percutaneous transluminal coronary angioplasty versus coronary artery bypass: Isn't it time for a randomized trial? *N Engl J Med* 1985; 312:916–9.
36. Kane RL, Wales J, Bernstein L, Leibowitz A, Kaplan S. A randomised controlled trial of hospice care. *Lancet* 1984; 1:890–4.
37. Kane RL, Bernstein L, Wales J, Rothenberg R. Hospice effectiveness in controlling pain. *JAMA* 1985; 253:2683–6.

38. Birnbaum HG, Kidder D. What does hospice cost? *Am J Public Health* 1984; 74:689–97.
39. Jonsen AR. A concord in medical ethics. *Ann Intern Med* 1983; 99:261–4.
40. Perkins HS, Saathoff BS. How do ethics consultations benefit clinicians? *Clin Res* 1986; 34:831A. abstract.

ETHICS CONSULTATION

Ethics Consultation:
Skills, Roles, and Training

John La Puma and David L. Schiedermayer

John La Puma and Dr. Schiedermayer are general internists. In the following article, these authors delineate the myriad of skills and roles for an ethics consultant. They emphasize that ethics consultants should be accountable for the process and outcome of their work. They also assert that clinical judgment, based on experience with patients and the natural history of diseases, is a prerequisite for effective ethics consulting. Thus they indirectly imply that the role of ethics consulting is primarily for clinician-ethicists. However, they recognize that nonclinicians could acquire the skills of an ethics consultant with years of clinical experience. This article was interpreted by some as suggesting that ethics consultants should be physician-ethicists.

What Legitimates Ethics Consultation?

Moral authority for ethics consultation arises from several sources. The primary justification for ethics consultation derives from the mandate to protect and foster shared decision making in the clinical setting.[1] Physicians should share health care decisions with well-informed patients who can understand their diagnoses, prognoses, and the various alternatives of proposed treatment and of nontreatment, and who can make decisions.[2] When an ethical problem arises, ethics consultation should be used to assure that issues are clarified so that decision making can be shared.

Ethics consultants' demonstrated ability to help resolve ethical dilemmas in patient care legitimates the use of ethics consultation. Ethics consultants have practical expertise in the clinical arena and are increasingly recognized as members of the health care team. The physician's need for analysis and advice in individual cases, the institution's need for counsel in patient-related policy issues, and the patient's need for an advocate further legitimate the use of ethics consultation. Courts and presidential commissions have recommended that clinicians seek appropriate assistance in making moral decisions.[3,4] The American

John La Puma and David L. Schiedermayer. "Ethics Consultation: Skills, Roles, and Training." Abridged from *Annals of Internal Medicine*, Volume 114, 1991, pp. 155–160. Reprinted by permission.

College of Physicians and the American Medical Association have recognized that protecting and enhancing shared doctor-patient decision making is an ethical responsibility.[5,6] Physicians' concerns about liability and payers' concerns about the costs of care have fueled the search for special expertise.

Ethics consultants should be accountable for the process and outcome of their work. Having an institutional locus of accountability is reasonable, although the specific lines of authority and reporting relationship will differ according to each institution's structure and mission.[7] Vision and commitment are necessary to support the consultant in synergistic ventures with health care professionals specializing in other clinical areas.[8] A consultant may wish to report to or be sponsored by the medical staff executive committee, the department chairperson, the chief executive officer, the dean, or the board of trustees. Accountability keeps the consultant honest and humble and permits the consultant to work effectively within an institution. Finally, in the clinical model of ethics consultation, the consultant is accountable to his or her patients and their physicians.

Consultants should inform institutional ethics committees of relevant clinical activities. Ethics committees can use the consultant's knowledge of individual cases to reflect on larger trends and, when needed, suggest institutional policy; in addition, the committee may be able to provide the consultant with a multidisciplinary critique of his or her work.

Ethics Consultants and Ethics Committees

Ethics consultants are professionals with specialized training and experience that equip them to identify, analyze, and help resolve moral problems that arise in the care of individual patients. Consultants have the specific task of collecting disparate, but essential, aspects of a patient's medical course and personal history. The professional in charge of gathering the relevant data, identifying opposing arguments and values, and restoring a central ethical focus to a case makes the consultant's role "ethical" in nature.[9] Assisting physicians in developing structured, coherent, and humane strategies for identifying, analyzing, and resolving ethical dilemmas is the clinical ethics consultant's special responsibility.[9]

Ethics consultants may choose to work with ethics committees (Table 1). The consultant is often the chairperson or co-chairperson of the committee and may go to the bedside, do the consultation, and report back to the committee at its regularly scheduled meeting. The consultant-chairperson may form a consulting subcommittee of several members, or the entire committee may meet to consider cases, either at the bedside or in a committee room. . . .

The Ethics Consultant's Clinical Skills

The consultant should be able to identify and analyze moral problems in a patient's care; use reasonable clinical ethical judgment in solving these problems;

TABLE 1   Institutional Credibility, Sponsorship, and Relationships for
          Ethics Consultants

Institutional credibility
   Clinical, practical, and ethical expertise
   Fellowship training in medical ethics
   Demonstrated patient advocacy
   Legal and professional acceptance
   Mastery of medical ethical information and patient-related policy issues

Institutional sponsorship
   Medical staff executive committee
   Department chairperson
   Chief executive officer
   Dean
   Board of trustees

Institutional relationships
   Chairperson of ethics committee
   Chairperson of consulting subcommittee
   Consultant on hospital policy
   Liaison with hospital legal office
   Educator of and advisor to various hospital committees, such as ethics, quality
       assurance, utilization review, and institutional review

communicate effectively with health care professionals, patients, and families; negotiate and facilitate negotiations; and teach medical students, house staff, and attending physicians how to identify, analyze, and resolve similar problems in similar cases (Table 2).

The ability to analyze and separate the ethical questions in a complex case is among the most important of the ethics consultant's skills.[10] Data gathering usually begins with an interview with and examination of the patient, followed by a review of the medical record and hospital course and interviews with physicians, nurses, family members, and others of importance to the patient. Through consultation, ethical issues are often identified and clarified: In one series, the consultant identified a mean of 3.0 issues per case and was "very important" or "somewhat important" in clarifying ethical issues in 94% of cases.[11] Considerable change in case management has been reported in 18 of 44 cases at county and Veterans Affairs hospitals,[12] 20 of 51 cases at university hospitals,[11] and 53 of 104 cases at community hospitals.

Clinical judgment, based on both long experience with many patients and familiarity with the natural histories of many diseases,[13,14] is difficult to acquire. Skill in clinical judgment underlies effective consultation, enabling the consultant to make the medical distinctions that are technically and morally relevant in each case. The consultant considers the care of a particular patient in a particular circumstance with a particular illness, as particularity is the hallmark of good medical practice.

TABLE 2   Skills and Roles for Ethics Consultants

Fundamental skills
   Identify and analyze clinical ethical problems
   Use and model reasonable clinical judgment
   Communicate with and educate team, patient, and family
   Negotiate and facilitate negotiations
   Teach and assist in problem resolution

Appropriate roles
   Professional colleague
   Patient advocate
   Case manager
   Negotiator
   Educator

Excellent interpersonal and communication skills are necessary for ethics consultants. Consultants can teach and model effective communication (listening, reflecting, encouraging discussion) and appropriate attitudes (respect, compassion, and courteousness). Ethics consultants use both verbal and nonverbal communication as diagnostic and therapeutic tools.[15]

The ethics consultant must be especially competent in helping to resolve interpersonal conflicts in patient care. Emotionally charged situations may be identified as "ethical dilemmas," but are more usually the result of miscommunication.[16] The consultant must be able to negotiate—at the bedside, in hospital conference rooms, and with administrators and third-party payers. The consultant's expertise includes the ability to facilitate understanding, emphasize common interests instead of opposing positions, and remain tactful while suggesting a course of action. The consultant must consider the interests of patients, doctors, nurses, and administrators, because the clinical setting is a place of compromise. The consultant's ability to resolve cases in conflict hinges largely on mediation skills.[17]

Finally, the ethics consultant teaches medical students, housestaff, and attending physicians how to identify, analyze, and resolve ethical problems in similar cases.[18,19] Case process and case synthesis are inextricably integrated in ethics consultation: Both illustrate how ethical issues change over time. In addition, the consultant's written report may provide a detailed case analysis. Appended references of didactic and practical value allow requesting physicians to consider several views as they construct their own frameworks for decision making.

The Ethics Consultant's Roles

The consultant's roles may properly include those of professional colleague, educator, negotiator, advocate, and case manager (Table 2). The ethics consultant is a professional colleague. Rudd describes a professional colleague as "someone with whom to share the case's complexity and from whom discernible help will emerge."[20] The consultant's clinical judgment and ability to analyze ethical issues in individual cases identify the consultant as a professional colleague. The consultant should tailor the information, perspective, critique, or reassurance that he or she provides to help the requesting physician.[21] As Goldman and colleagues[22] note, the effective consultant communicates directly and nonthreateningly with the requesting physician.

Teaching ethical decision making to physicians is a central goal of ethics consultation.[23] The ethics consultant recognizes the requesting physician's ability and experience in analyzing and managing ethical dilemmas and provides effective, individualized instruction. The consultant then emphasizes principles that may apply to similar future cases.

The role of negotiator requires effective interpersonal and communication skills. The consultant can try to be a consensus-builder, but reasonable persons may disagree about the decisions made in a particular case.[24] The consultant acts as a rational, clear-headed participant who seeks to help disagreeing parties come to morally permissible conclusions. More often than not, disagreeing parties can agree on a practical solution, although their reasons for agreeing will be different.[25] The role of negotiator may properly include using persuasion, because ethics consultants have a professional obligation to effect morally permissible outcomes.

When a patient's situation mandates it, the consultant must be a patient advocate. The ethics consultant's primary duty is to the patient, but he or she also has duties to the requesting physician to be timely, clear, and specific.[18] Dual loyalty can be risky for the consultant, especially if he or she opposes the wishes or actions of family members, legal proxies, or physicians. When a patient's interests seem threatened by planned treatment, financial constraints, legal proceedings, or an unreliable proxy, the consultant's obligation may extend to confronting the family or physician, appealing economic constraints, and pursuing legal appeals.[26-28] Such actions may be difficult and time-consuming, but when harm to a patient seems imminent, consultants should try to prevent it.

The ethics consultant will seldom be required to manage a patient's case, even when a patient, family, or physician requests it. The attending physician should retain decision-making responsibility and authority, using the consultant's ongoing involvement as needed.[29] Ethics consultants should be prepared to help manage difficult cases when a patient's medical interests are threatened or when a patient, family, or professional colleague requires the consultant's skills in case management.

Ethics consultants can anticipate some pressure to assume other roles in the clinical setting. These roles properly belong to others, however, and should be referred to persons with the needed expertise. Ethics consultants may be asked to act as a case conscience (this role belongs to all physicians managing the case);

case counsel (this role belongs to the legal office or the patient's attorney); case quality reviewer (this role belongs to hospital quality assurance); case psychoanalyst (this role belongs to a psychiatrist or psychologist); or case clergy (this role belongs to the hospital chaplain).

Difficulties for Ethics Consultants

Several general objections to ethics consultants have been raised.[30] Whether "objective" advice can be given by ethics "experts" and, if so, how this expertise is acquired are debated.[31,32] The long-term effects of ethics consultation in the hospital are unknown.[33] Trained in moral philosophy, not in decision making, philosopher-ethicists may lack clinical judgment. They may be aloof, unavailable, or uncomfortable in the clinical setting. Alternatively, a physician-ethicist may focus on problem solving and neglect important social, philosophical, or theologic aspects of a case.

A second objection is financial: Ethics consultants presently generate little or no revenue. Although consultants' revenue-generating potential may increase with use of the resource-based relative value scale (a weighting scale that increases compensation for cognitive work), whether ethics consultants ought to be paid as well as at what rate and by whom are unresolved questions of practical, political, and moral import.[34] As an institution-based service, like radiology or anesthesiology, consultants require costly malpractice coverage. If cost-savings criteria are used to evaluate ethics consultation, morality may become a charade for cost-cutting, to the patient's disadvantage.

Third, ethics consultants' risk for legal liability is unknown. We have previously suggested a standard of care for ethics consultants.[9] To our knowledge, however, an ethics consultant has not yet been sued. In 1986, charges were brought against an ethics committee in southern California; the suit was dismissed in 1990, but reportedly has dissuaded the committee from reconvening (Ross JW. Personal communication).

Fourth, questions remain about intrusion into the doctor-patient relationship. Who has the authority to request a consult? For whom does the consultant work? These questions are controversial. In our view, physicians may ask ethics consultants to speak with families, third-party payers, or patients; patients may speak with consultants directly.

The consultant should be able to answer requests from many quarters, but the primary physician engages and dismisses the consultant. In a clinical model of ethics consultation, the consultant works for both the physician and the patient. If a team member wants an ethics consultation, suggesting it first to the primary physician may promote an open dialogue and help to resolve the problem. If the suggestion is not taken, the team member can appeal the refusal to his or her supervisor. Uninvited consultants should not intercede in cases: Ethics consultants should not be moral policemen.

Finally, whether ethics consultants must be physicians or may also be non-physicians is controversial. Nonphysicians may have the years of clinical experience necessary for the development of clinical judgment; if this is not the case, the clinical expertise of a physician colleague is required. More important than a medical degree is a consultant's ability to acquire and use the necessary skills and fulfill the appropriate roles of the ethics consultant. A professional who wishes to do ethics consultation should be trained in those skills and roles.

Training and Certification in Ethics Consultation

Training program curricula should provide the necessary skills for consultation practice. Ethics consultants need substantial patient care and hospital experience,[35] instruction in case law and legal processes,[36] practice in casuistic moral reasoning and ethical decision making,[37] and knowledge of medical humanism and humanistic behavior.[38-40] The experience of consulting with a skilled, well-trained mentor, reading carefully about the patient's medical and ethical presentation, and following the patient's case to its conclusion, constitutes a practical, established process of medical learning.

Who should train as an ethics consultant? Ideal candidates are clinicians who are expert in their own medical discipline and who have or wish to gain the skills and play the roles of the consultant.[41,42] Such candidates include physicians who are completing a primary care residency or who are the ethics committee chairperson or co-chairpersons.

To acquire the clinical skills of an ethics consultant, nonphysicians require several years of clinical experience and routine participation with medical teams in clinics, hospital rooms, and special care units. Training in different medical settings provides the necessary foundation for understanding the diversity and details of many medical illnesses and for developing clinical judgment. The complexity of the doctor-patient relationship; the individuality of patients' and families' preferences, goals, and interests; and the exigencies of hospitals, health care professionals, and third-party payers are best appreciated when observed firsthand.

To become ethics consultants, most nonphysicians and many physicians would require training in medical humanism, clinical psychology, medical sociology, and health law. Essential topics in medical humanism include integration of the qualities of integrity, respect, and compassion with bedside behavior; in clinical psychology, differentiation between organic and functional illnesses, recognition of differing doctor-patient relationships in different medical specialties, and determination of patient decision-making capacity; in medical sociology, comprehension of the special language, interrelationships, and hierarchies of hospital medicine, nursing, and medical social work; and, in health law, case and statutory law relevant to life-sustaining treatment, advance directives, and surrogate decision making.

Physicians who wish to become ethics consultants require training in moral reasoning and ethical decision making. Training must provide opportunities to reflect on and critique clinical ethical dilemmas, discover and discuss multidisciplinary perspectives, and learn and apply techniques of facilitation and negotiation. Continuing to hold primary clinical responsibilities during training is a direct, vital way of appreciating ethical dilemmas in patient care.

Whether clinical ethicists can or must have certification in a new medical field is controversial. Certification requires a defined body of useful clinical knowledge and an evaluation process that determines whether the candidate has mastered the knowledge and possesses a specified level of clinical competency. The American Board of Internal Medicine criteria for a new discipline include a significant scientific base and clearcut relation to internal medicine or its subspecialties; a recognition of the discipline in the medical, academic, and scientific communities; the potential for a significant number of practitioners in a well-defined practice; a requirement for formal training with prescribed standards; and improved patient care.[43] Ethics consultants, particularly those who are physicians, have begun to meet several of these criteria (for example, an identifiable base of scientific knowledge and improved clinical practice). Practical, political, and professional questions remain, however, about the incorporation of nonphysician ethics consultants (currently, the majority of ethics consultants) into a field of expertise in medicine.

Conclusion

The ethics consultant's role will continue to evolve. We favor a clinical model of ethics consultation, the process and outcome of which require continued study. Empiric data and critical review are necessary to evaluate the utility and limitations of consultation. An important question is whether patients, families, and physicians find ethics consultation to be beneficial. The issue of specialty certification in ethics consultation also requires further consideration and debate.

The consultant teaches the analytic, interpersonal, and communication skills that physicians need to solve ethical problems. The consultant assists in the decision-making process as a negotiator or advocate when the physician, the patient, or the family requires such assistance. Finally, the consultant is a clinical colleague with specialized training and experience who is available for consultation. Consultants who are competent in clinical ethics and who can use their skills and knowledge to assist patients and physicians at the bedside should be trained and available to assist patients, families, and physicians.

References

1. President's Commission for the Study of Ethical Problems in Medicine and Biomedical and Behavioral Research. *Making Health Care Decisions: The Ethical and Legal Implications of Informed Consent in the Patient-Practitioner Relationship*. Washington, DC. U.S. Government Printing Office; 1982.

2. Jonsen AR, Siegler M, Winslade WJ. *Clinical Ethics: A Practical Approach to Ethical Decisions in Clinical Medicine.* 2d ed. New York: Macmillan; 1986.
3. The National Commission for the Protection of Human Subjects of Biomedical and Behavioral Research. *The Belmont Report: Ethical Principles and Guidelines for the Protection of Human Subjects of Research.* Washington, DC: U.S. Government Printing Office; 1978:DHEW pub no (OS) 78-0012, 78-0013, 78-0014.
4. *President's Commission for the Study of Ethical Problems in Medicine and Biomedical and Behavioral Research.* Washington, DC: U.S. Government Printing Office; 1983.
5. Council on Ethical and Judicial Affairs of the American Medical Association. *Current Opinions.* Chicago: American Medical Association; 1989.
6. Ethics Committee, American College of Physicians. American College of Physicians ethics manual Part I. History: the patient; other physicians. *Ann Intern Med* 1989;111:245–52.
7. La Puma J. Clinical ethics, mission and vision: practical wisdom in health care. *Hospital and Health Services Admin.* 1990;35:321–6.
8. La Puma J. Researching for-profit research, the obligations of hospital ethicists. *Clin Res.* 1989;37:569–73.
9. La Puma J, Toulmin SE. Ethics consultants and ethics committees. *Arch Intern Med.* 1989;149:1109–12.
10. Rothenberg LS. Clinical ethicists and hospital ethics consultants: the nature of the "clinical" role In: Fletcher JC, Quist N, Jonsen AR, eds. *Ethics Consultation in Health Care.* Ann Arbor, Michigan: Health Administration Press; 1989:19–35.
11. La Puma J, Stocking CB, Silverstein MD, DiMartini A, Siegler M. An ethics consultation service in a teaching hospital: utilization and evaluation. *JAMA.* 1988;260:808–11.
12. Perkins HS, Saathoff BS. Impact of medical ethics consultations on physicians: an exploratory study. *Am J Med.* 1988;85:761–5.
13. Jonsen AR. Do no harm. *Ann Intern Med.* 1978;88:827–32.
14. Tumulty PA. What is a clinician and what does he do? *N Engl J Med.* 1970;283:20–4.
15. Cassell EJ. *Talking With Patients.* v. 1 and 2. Cambridge, Massachusetts: MIT Press; 1985.
16. Waitzkin H. Doctor-patient communication, clinical implications of social scientific research. *JAMA.* 1984;252:2441–6.
17. Drane JF. Hiring a hospital ethicist. In: Fletcher JC, Quist N, Jonsen AR. *Ethics Consultation in Health Care.* Ann Arbor, Michigan: Health Administration Press; 1989:117–33.
18. Self DJ, Lynn-Loftus GT. A model for teaching ethics in a family practice residency. *J Fam Pract.* 1983;16:355–9.
19. Barnard D. Residency ethics teaching: a critique of current trends. *Arch Intern Med.* 1988;148:1836–8.

20. Rudd P. Problems in consultation medicine: the generalist's reply. *J Gen Intern Med*. 1988;3:592–5.
21. Merli GJ, Weitz HW. The medical consultant. *Med Clin North Am*. 1987;71:353–5.
22. Goldman L, Lee T, Rudd P. Ten commandments for effective consultations. *Arch Intern Med*. 1983;143:1753–5.
23. Culver CM, Clouser KD, Gert B, et al. Basic curricular goals in medical ethics. *N Engl J Med*. 1985;312:253–6.
24. Moreno J. What means this consensus? Ethics committees and philosophic tradition. *The Journal of Clinical Ethics*. 1990;1:38–43.
25. Toulmin SE. The tyranny of principles. *Hastings Cent Rep*. 1981;11: 31–9.
26. La Puma J, Schiedermayer DL, Toulmin SE, Miles SH, McAtee J. The standard of care: a case report and ethical analysis. *Ann Intern Med*. 1988; 108:121–4.
27. La Puma J, Cassel CK, Humphrey H. Ethics, economics, and endocarditis: the physician's role in resource allocation. *Arch Intern Med*. 1988;148: 1809–11.
28. Schiedermayer DL, La Puma J, Miles SH. Ethics consultations masking economic dilemmas in patient care. *Arch Intern Med*. 1989;149:1303–5.
29. La Puma J, Schiedermayer DL. Outpatient clinical ethics. *J Gen Intern Med*. 1989;4:413–9.
30. Nielsen K. On being skeptical about applied ethics. In: Ackerman TF, Graber GC, Reynolds CH, eds. *Clinical Medical Ethics: Exploration and Assessment*. New York: University Press of America; 1987;95–116.
31. Phillips DF. Physicians, journalists, ethicists, explore their adversarial, interdependent relationship. *JAMA*. 1988;260:751–7.
32. Perkins H. Teaching medical ethics during residency. *Academic Medicine*. 1989;64:262–6.
33. Siegler M, Singer PA. Clinical ethics consultation: Godsend or "God squad." *Am J Med*. 1988;85:759–60.
34. Purtilo RB. Ethics consultation in the hospital. *N Engl J Med*. 1984;311: 983–6.
35. Siegler M. Cautionary advice for humanists. *Hastings Cent Rep*. 1981;11: 19–20.
36. Burt R. *Taking Care of Strangers: the Rule of Law in Doctor-Patient Relations*. New York: Free Press; 1979.
37. Jonsen AR, Toulmin S. *The Abuse of Casuistry: A History of Moral Reasoning*. Berkeley: University of California Press; 1988.
38. American Board of Internal Medicine Subcommittee on Evaluation of Humanistic Qualities of the Internist. Evaluation of humanistic qualities in the internist. *Ann Intern Med*. 1983;99:720–4.
39. Arnold RM, Povar G, Howell J. The humanities, humanistic behavior and the humane physician: a cautionary note. *Ann Intern Med*. 1987;106:313–8.

40. The American Board of Internal Medical Subcommittee on Humanistic Qualities. *A Guide to the Awareness and Evaluation of Humanistic Qualities in the Internist*. Portland: American Board of Internal Medicine; 1990.

41. Siegler M, Pellegrino EJ, Singer PA. Clinical medical ethics. *J Clin Eth*. 1990;1:5–9.

42. Perkins H. Clinical ethics fellowships. *SGIM Newsletter*. 1989;12:6.

43. American Board of Internal Medicine. *Criteria for the Definition of New Medical Areas*. Portland: American Board of Internal Medicine; 1984.

Facilitating Medical Ethics Case Review:

What Ethics Committees Can Learn from Mediation and Facilitation Techniques

Mary Beth West and Joan McIver Gibson

Mary Beth West is an attorney and Joan Gibson is a retired Senior Bioethicist at Institute for Ethics, University of New Mexico. They received a grant from the National Institute for Dispute Resolution in Washington, D.C. to assess how ethics committees could benefit from facilitation and mediation techniques. In the accompanying article they discuss how several methods of alternative dispute resolution (shuttle facilitation, mediation, and large-group facilitation) can facilitate ethics case consultations as well as other internal committee processes.

Case Consultation Models and Dispute Resolution Processes

Parallels can be drawn between case consultation models and three facilitated processes—shuttle facilitation, mediation, and large-group facilitation. This section explores those parallels.

Case Consultation and Shuttle Facilitation

In some healthcare institutions, medical ethics issues may be addressed without getting the interested parties together in a meeting. For example, a social worker, physician, chaplain, or another [ethics] committee member may meet with parties individually and resolve the issue before it gets to the committee. Alternatively, the [entire] committee may meet with the physician, while a committee member may talk individually with the family, patient, surrogate, or other party. In another format, the committee may designate one or more committee members as "liaisons" to meet with each party individually as a prelude to a meeting in which the committee develops advice outside the presence of the parties.

These models bear some similarities to shuttle facilitation—the process in which a facilitator works toward resolution through separate meetings with the

---

Mary Beth West and Joan McIver Gibson. "Facilitating Medical Ethics Case Review: What Ethics Committees Can Learn from Mediation and Facilitation Techniques." Abridged from *Cambridge Quarterly Healthcare Ethics*, Volume 1, 1992, pp. 64–74. Reprinted with the permission of Cambridge University Press.

parties. A shuttle facilitator meets individually with each party to explore under-lying interests and needs. Options for resolution and eventual consensus are also developed as the facilitator shuttles between or among the parties. One major distinction between shuttle facilitation and the types of ethics committee con-sultation models described above, however, is that ethics committees do not nec-essarily base their strategies on the underlying assumption that they are neutral facilitators. Nor do they attempt to balance the power between or among the participants. For example, committees may meet with one party without meet-ing with the other, or may invite one party to the committee meeting, while meeting with the other only outside the framework of the committee. To the ex-tent that a committee sees its function as helping parties work toward a resolu-tion of issues, the lack of balance and neutrality, or the parties' perception of lack of balance and neutrality, may make it more difficult for the committee to fulfill that function.

Case Consultation and Mediation

Where committees bring together parties with a subset of two or three commit-tee members, case consultation may resemble a mediation or small-group facil-itation. Whether this type of consultation in fact exhibits the characteristics of mediation or facilitation, however, depends on how the committee members perceive their roles in the case consultation. To the extent that the committee members are without independent interests and views concerning the issues, and to the extent that they see their roles as helping the parties work toward resolu-tion, their actions may resemble those of a neutral mediator or facilitator. Be-cause committees often view their roles as educational, however, one or more committee members in case review may have independent views or expertise to offer. Where committee members are "interested" participants, it may be diffi-cult for those members to act as facilitators. Even "interested" participants can act as facilitators, however, where they perceive their roles as neutral and use their expertise to help the parties explore interests and solutions rather than to advocate particular ideas or resolutions.

The role of facilitator in small-group case consultation is to create an at-mosphere of trust in which the parties are able to express and explore their un-derlying interests and needs, identify solutions, and work toward consensus. Some committees appear to view their roles as encompassing some but not all of these elements. For example, some committees see themselves as a resource to help parties explore ideas, but not to help them work toward consensus on a particular solution or plan. The functions of such committees may resemble mediation, without the steps of agenda setting and reaching resolution, and follow-up.

The extent to which small-group case consultation resembles mediation or facilitated settlement may also depend on the nature of the parties. If one party is a medical professional and the other is a patient or family member, the patient or family member may view himself or herself as an "outsider" and may view the committee and the medical professional as essentially one entity. In that case, the situation may appear as "4 or 5 on 1" rather than as a more balanced mediation or facilitation involving facilitators and two or more parties. Because the committee members are not perceived as neutral, it becomes significantly more difficult for the committee to fulfill a facilitation role. A committee that wishes to fulfill the role of facilitator must consider ways to minimize power imbalances among parties.

Case Consultation and Large-Group Facilitation

Where committees bring together parties and a number of committee members who may have positions or views to offer concerning the matter at issue, the process more closely resembles large-group mediation or facilitation. In a case involving differences between a physician and a patient's family concerning withdrawal of life support, for example, the physician, individual members of the patient's family, and various committee members (such as the chaplain, an ethicist, and a neurologist) may each have positions or interests in the resolution. While the chaplain, ethicist, and other medical personnel are not "parties" in the true sense, one or more of them may have views he or she feels should be recognized and taken into account in the discussion. An ethics committee meeting that brings together all these players faces some of the same challenges as a large-group facilitation of a public policy issue.

Facilitators in large-group meetings face significant challenges. They must create an atmosphere of trust in which the parties can explore their underlying interests and needs, in which the parties and other participants in the group are heard and understood, and in which parties are able to explore options for resolution within the framework of the applicable interests and needs. Ethics committees face similar challenges but with important differences. Large-group facilitations are often led by one or two neutral facilitators, whose sole responsibilities are to nurture the atmosphere of trust and to move the participants through the steps necessary to reach consensus. In the ethics committee setting, on the other hand, the person running the meeting may be one of the participants with a position or interest rather than a designated neutral facilitator. Even if the person running the meeting does not have an interest in the outcome, he or she may not have been trained in facilitation techniques. This adds to the difficulty in moving the process through the steps necessary to reach the committee's goal.

Recommendations for Ethics Committees in Case Consultation

Understanding the relationship among committee role, source(s) of power, and process is a prerequisite for successful case consultation. Even committees with defined roles have given little thought to how process assists case consultation.

In doing so, committees should focus on three primary stages of consultation. These are intake, the consultation itself, and follow-up.

Intake

Intake is critical. This stage offers the opportunity to diagnose the issue and determine the most effective process or method for its resolution. In addition, many cases are resolved at this stage. Hospital chaplains, social workers, physicians, and others who are members of ethics committees reported that they are able to resolve issues at this stage, often using shuttle facilitation techniques without involving the committee.

Little attention has been given to intake as an integral and constitutive function of ethics committee process. Although some committees have designated a person to call meetings, others could not describe a "normal" method by which issues arrive before and are presented to the committee. Even in committees using a designated process to convene meetings, the person exercising that function often fails to consider the best process for handling specific types of cases.

The intake stage can play an important diagnostic role. In the dispute resolution context, courts throughout the country[1] are setting up multidoor programs designed to offer a variety of opportunities for resolution, depending on the nature of the dispute and the parties. A careful intake process staffed by a person trained in techniques for diagnosis and resolution of problems could help committees identify interested parties, identify the interests and nature of the problem, and set up a case consultation structure and process designed most effectively to address those issues requiring committee attention. Careful intake also offers the potential of resolving some issues without the need for committee involvement.

We recommend that each ethics committee designate one member as its intake specialist. That member should in most cases be a hospital employee or someone who is readily available in the institution on short notice. The intake specialist should receive training in techniques for diagnosis and resolution of problems and in dispute resolution processes. Although formal training in diagnosis of bioethics conflicts/issues has not yet been developed, training in general facilitation and mediation techniques would be helpful. The process used by such intake specialists would likely involve the following stages, which closely resemble those designed by the trainers of the American Bar Association for referrals in multidoor courthouse programs: identification of interested parties; introduction, making each party comfortable, and establishing rapport; gathering information and maintaining an open, sensitive climate; problem clarification and summary; review of possible processes for addressing the issue; and selection of the option.[2]

Once the intake specialist has identified the interested parties, he or she can meet with them together or separately to determine the outlines of the issue and to determine which committee process would be most appropriate. In this preliminary process, some disputes and dilemmas will, no doubt, be resolved.

Case Consultation

Models for case consultation processes should vary, depending on the committee's perceived role in its institution, the nature of the issue, and the parties involved. Several critical elements are involved. These include, first, matters of form: the place the consultation is held, the number of people involved, the relationships of those people to the issue, whether both parties attend, and whether or not one or more members are present as neutral facilitators. Second are elements of process. Here the committee needs to determine how best to design the consultation to meet its goals. If the goal is to assist the parties in reaching consensus on a solution, for example, perhaps the process should incorporate the basic stages of mediation.

Committees should develop form and process guidelines that reflect their views of their roles in case consultation. This section outlines several models. Additional work is necessary to refine these models and to develop intake and consultation processes specifically designed for ethics committees. The ideal appears to involve design of several forms and processes that can be used flexibly by each committee, depending on the nature of the issue, the parties involved, and the goals for that consultation.

Issue resolution   For committees who see their roles as attempting to facilitate resolution, the process should be designed to most closely resemble a mediation or facilitated settlement. This model would call for one or two members of the committee trained as a neutral facilitator(s) to meet with the parties. Keeping in mind the persons necessary for resolution, the group should be as small as possible; the meeting could involve only the parties and the facilitator or could also involve a few other committee members. Alternatively, other committee members with views or interests could be brought in as "experts" to assist the parties at the appropriate time in the process.

The facilitator(s) should be flexible in designing and carrying out the process but should basically structure it along the lines of the mediation stages: introduction and explanation of the process, information gathering and issue identification, agenda setting and reaching resolution, and agreement. The facilitator(s) may meet with parties separately in caucus and should be willing to bring in others with views or expertise that will be helpful to the parties in considering the issues. The facilitator(s) should also be flexible in helping the parties design interim solutions or steps where final resolution is not attainable.

For example, assume that the family of a comatose patient takes the position that life-support treatment should be terminated immediately. The doctor, on the other hand, may insist that more information is needed before making that kind of decision. Delving beneath the outcome sought by the family, the participants might find such a position based on the anger, frustration, and pain that comes from dealing with what the family perceives as an uncaring, unresponsive institution rather than from a basic wish to let the patient die. The family's underlying interest may be in ending a situation that is unbearable for the family members. On the other hand, in advocating delay, the physician may be looking not so much for more information but rather for emotional and legal support. The doctor's underlying interest may be to avoid legal liability or to be faithful to a moral commitment to sustain a patient's life. In this situation, where the patient's family members feel that their concerns have not been heard and responded to by the physicians and hospital staff, the facilitator(s) might help the family and hospital design a method for future communication that must be implemented before the family is ready to consider the more basic treatment issues.

Issue exploration   Committees who see their roles as helping the parties explore issues, but not necessarily as assisting them in reaching resolution, may structure the process somewhat differently. However, it would still be helpful to have a trained member of the committee act as a neutral facilitator. The committee, however, may meet in a larger group and may meet with both parties together or with the parties separately. The committee should start with an introduction and explanation of the process and should then attempt to gather information and help the parties identify and explore the issues. Committee members and the facilitator(s) may also assist the participants in determining potential solutions but would not see their role as attempting to help the parties reach consensus on a solution.

Education   Finally, committees who see their primary roles as education of the parties may use yet a different process. Those committees may choose to meet in large session, including all the committee members (and possible outsiders) who have views to offer. Such committees may meet with the parties, either at the same time or separately, or may choose to meet without the parties. In the latter case, the committee's views or advice would be transmitted to the parties by one or more committee members. Involvement of a person trained as a facilitator would be helpful where the committee meets with one or more of the parties. Where the committee does not, facilitation training is less critical.

Where the committee sees its role as educational and simply offers its advice and views to one or both parties, the parties still must reach consensus to proceed. In some cases, provision of information alone may pave the way for resolution. In others, however, the parties may need further assistance. Such assistance could be provided by a trained intake specialist or by one of the committee members trained in communication and facilitation.

Follow-up

Few of the committees surveyed follow up on case consultations. Occasionally a report is made at the next full meeting of the committee. Only rarely do committees formally follow up with the parties to determine what action was taken or to offer further assistance. Chaplains or social workers may occasionally perform this function informally.

We recommend that the intake specialist formally follow up with participants after case consultations. The purpose of the follow-up would be twofold.
First, it would provide information to the committee concerning the outcome of
the case consultation. Second, it would make further assistance available to the
parties should they need such assistance to reach consensus or should additional
issues arise after the formal case consultation.

Conclusion

. . . Facilitation and mediation techniques *can* be helpful in case review consultations as well as in other internal committee processes. *How* those techniques
can best assist committees depends on an understanding of each committee's
role in its institution, the applicable source(s) of committee power, the types of
cases typically coming before the committee, and the committee's goals in consultation. Processes currently in use by ethics committees resemble certain
forms of facilitation, such as shuttle facilitation, mediation, and large-group facilitation. The ideal format might encompass a more flexible approach designed
to reflect and respond to the variety of issues, parties, and institutional roles that
a single committee may confront.

We recommend that committees review and analyze the processes they use,
their level of success with case consultations, and the cause and effect relationship between the two. As the initial step, committees should consider training
one or more of their members in facilitation and communication techniques.
Using these newly trained members, committees should then review and appropriately revise the structure and techniques used in case consultation, paying
special attention to the roles and activities of intake and follow-up. Based on our
preliminary research, it is clear that these stages play a key role in individual
consultations as well as in committees' overall relationships to their institutions.
It is also clear that when preliminary intake and postconsultation follow-up are
undertaken, it is only in a most abbreviated fashion. Finally, it will be important
for committees periodically to review and evaluate the success of revised procedures in responding to their needs and goals.

References

1. National Institute of Justice. Toward the multi-door courthouse—dispute
   resolution intake and referral. Washington, D.C.: NIJ Reports; 1986(Jul);
   SNI 198.
2. See above. National Institute of Justice, 1986.

Ethics Consultation in U.S. Hospitals:
A National Survey

Ellen Fox, Sarah Myers, and Robert A. Pearlman

Ellen Fox is Director of the National Center for Ethics in Health Care, Department of Veterans Affairs, in Washington, D.C. Sarah Myers works in the Division of Health Policy and Clinical Effectiveness at Cincinnati Children's Medical Center, Cincinnati, Ohio. Robert Pearlman is Professor of Medicine at the University of Washington and Chief of the Ethics Evaluation Service for the National Center for Ethics in Health Care. In this essay, the authors surveyed 600 general U.S. hospitals to find out about the prevalence, practitioners, and process of ethics consultation. They found that although ethics consultation services are widespread, most individuals providing this service have no formal supervised training in ethics consultation. Moreover, most ethics consultation services lack any formal process for evaluating their services. Finally, most individuals providing ethics consultations are white and non-Hispanic. This study points to the need for more rigorous training and evaluation of ethics consultation. It also highlights the need for greater diversity among ethics consultants.

Introduction

During its three-decade history (Kosnick 1974; Rosner 1985), ethics consultation has evolved into an organized and widely accepted health care service endorsed by respected commissions and professional groups (President's Commission 1983; American Hospital Association 1986; American Medical Association 1985; American Society for Bioethics and Humanities 1998). Despite the rapid growth of ethics consultation, little is known about this clinical service in terms of its actual prevalence or practices. Previous studies on ethics consultation have consisted mainly of descriptions of ethics consultation performed within a single institution (Dowdy et al. 1998; Orr et al. 1996; Waisel et al. 2000) or small group of institutions (Csikai et al. 1998; Godkin et al. 2005; Schneiderman et al. 2003; Touhy 2006). Although there have been several national studies of ethics committees (American Hospital Association 1997; Youngner 1983; Bernt 2006; McGee et al. 2001), they provide relatively little information on ethics consultation *per se*. The study reported here is the first to describe in detail how ethics consultation is practiced in general hospitals throughout the U.S.

The National Study on Ethics Consultation in U.S. Hospitals was launched with the aim of providing baseline data to facilitate future efforts at evaluation and quality improvement. We administered a detailed survey to key informants within a random, stratified sample of all U.S. general hospitals to enable us to make estimates about current practices nationwide. We addressed the following questions:

1. What is the prevalence of ethics consultation services (ECSs) in U.S. hospitals?
2. Who performs ethics consultation and what are the backgrounds and training of these individuals?
3. How do these services function?
4. Are a hospital's bed size, ownership, and teaching role related to the characteristics of its ECS?

Methods

Definition

Ethics consultation was defined as "a service provided by a committee, team, or individual to address the ethical issues involved in a specific, active clinical case" (Tulsky and Fox 1996, 112).

Survey Questionnaire Development

A survey questionnaire addressing our key research questions was developed based on the research literature on ethics consultation and input from a working group of 20 experts in the fields of bioethics, clinical ethics consultation, evaluation, and health services research (Fox and Tulsky 1996). In addition, pilot studies of the survey instrument and protocols for identifying the best informant were conducted on a sample of hospitals (n = 47). The questionnaire and protocol for data collection were refined based on feedback from the interviewer and survey respondents.

The final instrument contained 41 primary questions and 15 potential follow-up questions for a total of 56 items. Representative questions include:

"How many individuals have performed ethics consultations for the institution in the past year?" (Respondents were asked to specify a number.)
"What ethics-related training was received by the individual(s) who perform consultation?" (Respondents were asked to indicate a number for each of several choices such as "Completed a fellowship or graduate degree program in bioethics . . .")
"In practice, how is information gathered in an ethics consultation?" (Respondents were asked to estimate the frequency for each of several choices, including "One-on-one discussions with members of the clinical staff." Response categories were "always," "usually," "about half the time," "infrequently," and "never.")

"Is there a formal process for evaluating the ECS that involves the collection and analysis of data on consultations performed?" (Respondents who answered yes were asked to describe the process used for the evaluation in an open-ended format.)

Sample

The population studied consisted of all general hospitals participating in the American Hospital Association's 1998 annual survey of hospitals (n = 5,072). Demographic data describing each hospital, including academic affiliation, bed size, and ownership category, were obtained from this database for use during sample selection and data analysis. Hospital characteristics are shown in Table 1.

The final random sample included 600 hospitals, or 12% of all general hospitals in the American Hospital Association's database (Table 1). Since we predicted that larger hospitals would perform more ethics consultations, we developed a weighting scheme to over-sample these hospitals. To determine the number of hospitals sampled in each bed size category, we multiplied the frac-

TABLE 1   Description of U.S. General and Sample Hospitals*

	U.S. General Hospitals (n = 5,072)	Sampled Hospitals (n = 600)	Participating Hospitals (n = 519)
	% (N)	% (N)	% (N)
Bed Size			
1–99	45 (2,264)	12 (75)	12 (62)
100–199	25 (1,270)	20 (119)	19 (100)
200–299	13 (678)	19 (113)	18 (94)
300–399	8 (385)	16 (93)	17 (86)
400–499	4 (191)	10 (60)	10 (53)
≥500	6 (284)	23 (140)	24 (124)
Ownership			
Local government	24 (1,205)	15 (92)	16 (82)
Church, not-for-profit	12 (604)	17 (102)	18 (91)
Other, not-for-profit	46 (2,339)	51 (305)	52 (269)
Private, for profit	13 (678)	12 (69)	10 (50)
Federal government	5 (247)	5 (32)	4 (27)
Academic Affiliation			
None	78 (3,950)	53 (320)	52 (270)
Medical school/residency	16 (789)	28 (168)	29 (148)
Membership in Council of Teaching Hospitals	7 (332)	19 (112)	19 (101)

*Percentages may exceed 100% due to rounding.

tion of total beds in that category (# of beds in category/# of beds in all U.S. hospitals) by 600 (# of hospitals desired for the sample).

Data Collection

The research protocol was approved by the institutional review boards at George Washington University and at the Washington, DC VA Medical Center. Interviewers used a pre-tested protocol to identify the "best informant" about ethics consultation (i.e., "the person most actively involved in ethics consultation") in each hospital in the sample. Once the interviewer secured the name of a "best informant," the interviewer attempted to reach this person by phone. If repeated attempts to contact this person were unsuccessful, or on request by potential participants, a written version of the survey was sent via fax or mail. Data collection took place between September 1999 and May 2000.

Once a best informant was reached, the interviewer confirmed that the individual was "the person most actively involved in ethics consultation" within the facility. Confirmed best informants were asked whether the hospital had an ECS according to our study definition. In hospitals with ECSs, consenting best informants completed either a telephone interview (n = 444, 86%) or a written survey (n = 75, 14%). When hospitals had more than one ECS, informants were asked to refer only to the ECS with the highest consultation volume when answering questions.

Data Analysis

We analyzed the survey data using SAS, version 6.12, and STATA. Where appropriate, the data were weighted using the degrees of freedom method in order to make inferences about the entire population of U.S. general hospitals. To determine the weight for each bed size category, we divided the fraction of total hospitals in that category (# of hospitals in category/# of hospitals in total sample) by the fraction of total hospital beds in that category (# of beds in that category/# of beds in total sample).

We estimated the prevalence of ethics consultation nationally by extrapolating from the proportion of participating hospitals that reported having an ECS. We generated descriptive statistics describing the characteristics of those ECSs that were identified. Chi-square statistics were employed to compare counts and frequencies across categories of hospitals. Finally, we employed analysis of variance and regression analysis to identify relationships between the characteristics of hospitals and the characteristics of their ECSs, as well as relationships between different ECS characteristics.

Results

Six hospitals (1%) were closed during data collection. Of the 594 remaining hospitals, 519 completed all or part of the survey (87.4%). Characteristics of participating hospitals are shown in Table 1. Participating hospitals were similar to nonparticipating hospitals in bed size and academic affiliation, but nonparticipating hospitals were more often private, for profit (23% versus 10%; $P < .01$).

The results presented throughout the remainder of this section are estimates for the entire population of U.S. general hospitals. These estimates were determined by weighting the sampling adjustments made prior to analysis.

Characteristics of "Best Informants"

The majority of "best informants" about ethics consultation (56%) were chairs of the ethics committee or ECS. Their official titles most often related to medical staff (23%), hospital administration (16%), nursing (9%), quality improvement/utilization review (10%) or chaplaincy (13%). Just 4% of "best informants" had official job titles that included the term "ethics" or a variation thereof, such as "bioethics" or "ethicist."

Prevalence of Ethics Consultation Services

Eighty-one percent of all U.S. general hospitals had ECSs (Table 2), and an additional 14% were in the process of developing them. The prevalence of ECSs varied significantly across hospital categories. The prevalence of ECSs was

TABLE 2   Estimated Prevalence of Ethics Consultation Services in
U.S. General Hospitals

Demographic Category	% ($n = 519$)
Bed Size[†]	
1–99	65
100–199	92
200–299	97
300–399	97
400–499	100
≥500	100
Ownership[†]	
Local government	57
Church, not-for-profit	94
Other, not-for-profit	86
Private, for profit	98
Federal government	100
Academic Affiliation	
None	77
Medical school/residency	94
Membership in Council of Teaching Hospitals	100

[†]$P < .001$, chi-square test

100% in hospitals with over 400 beds, in VA and other federal government hospitals, and in hospitals belonging to the Council of Teaching Hospitals (COTH). Six percent of hospitals had more than one ECS; of those, a large majority (80%) had two.

General Characteristics of Ethics Consultation Services

When respondents were asked which of three models of ethics consultation best described their ECS, most responded that consultations were generally performed by a small team of individuals (68%) as opposed to a full ethics committee (23%) or an individual consultant (9%). Although respondents were not asked to specify the frequency with which their ECS used each of these models, many volunteered that they used more than one model, or a combination of models, depending on the consultation. While 27% had been active for fewer than five years, 47% had been active for 5–10 years, and 26% had been active for more than 10. Hospitals with more than 300 beds were more likely to have had ECSs for more than 10 years than smaller hospitals (42% versus 22%, P ≥ .001). A majority of ECSs were primarily accountable either to the hospital administration (36%) or the medical staff (29%).

Table 3 summarizes how respondents characterized the importance of 10 potential goals of the ethics consultation within their institutions. The goals that were most often characterized as primary were "intervening to protect patient rights" (94%), "diffusing real or imagined conflicts" (77%), and "causing a change in patient care that improves quality" (75%). The top three primary goals did not differ according to ownership, but did differ according to academic affiliation: "protecting patient rights" was less likely to be a primary goal in hospitals belonging to COTH (83% versus 95%, P <.05).

Salary support was provided specifically for ethics consultation in only 16% of hospitals. Nonetheless, at 83% of hospitals, respondents were of the opinion

TABLE 3   Goals of Ethics Consultation (%)

	Primary Goal	Secondary Goal	Not an Explicit Goal
Intervening to protect patient rights	94	5	1
Resolving real or imagined conflicts	77	22	1
Changing patient care to improve quality	75	19	6
Increasing patient/family satisfaction	68	26	6
Educating staff about ethical issues	59	37	4
Preventing ethical problems in the future	59	36	5
Meeting a perceived need of staff	50	35	14
Providing moral support to staff	47	47	6
Suspending unwanted or wasteful treatments	41	40	19
Reducing the risk of legal liability	40	49	11

that the financial support devoted to ethics consultation was sufficient. The perception of adequate financial support for the ECS diminished in hospitals that had more consultations (e.g., larger COTH hospitals). Examples of explanations for why financial support was or was not sufficient are included in Figure 1.

The median number of ethics consultations performed by ECSs during the year prior to survey was three (range 0–300). Twenty-two percent of ECSs performed zero consultations in the previous year; 90% performed fewer than 25. The median number of consultations varied according to bed size: one (0–99 beds), 3 (100–199 beds), six (200–299 beds), 10 (300–399 beds), 12 (400–499 beds), and 15 (≥500 beds). The number of consultations performed by ECSs in federal government hospitals and in COTH hospitals exceeded the number for other hospitals by a factor of four or greater (median 12 versus 3, $P < .05$; median 18 versus 3, $P < .01$), even after controlling for bed size.

Characteristics of Active Ethics Consultation Services

The remaining results are for ECSs that had performed at least one ethics consultation in the year prior to the survey. ECSs that had never performed an ethics consultation or that had not performed an ethics consultation in more than a year were excluded from the analyses.

Workload of Ethics Consultation Services

Across ECSs, the average number of individuals who performed a given ethics consultation ranged from one to 18 (median 4). The total number of individuals who had performed ethics consultations in the past year ranged from one to 36 (median 8). The number of person-hours spent on the average consultation ranged from one to 120 (median 6). This number varied by hospital ownership category, with ECSs in federal government hospitals spending fewer person-hours per consultation than those in other hospitals (median of 4 versus 6; $P < .01$). Both the average number of individuals who performed a single consultation and the number of person-hours spent on an average consultation varied according to the primary model of consultation used. In ECSs using the full ethics committee model, an average of 9 individuals spent 15 person-hours per consultation, compared with 4 individuals and 7 person-hours for the small team model, and 2 individuals and 4 person-hours for the individual consultant model, respectively ($P < .01$). Even after controlling for the consultative model, the number of individuals involved in a single consultation varied according to academic affiliation: in COTH hospitals fewer individuals performed each consultation than in other hospitals surveyed (3 versus 4; $P < .01$).

The average time that elapsed between the initial consultation request and the completion of the case ranged from one to 672 hours (median 12). In 25% of ECSs, the average time that elapsed was under three hours; in 11%, it was over 24.

Sufficient (n = 307)	Not Sufficient (n = 119)
• Ethics is one of seven priorities within mission integration.	• Ethics is not valued enough for the institution to pay for it.
• A foundation…is committed to raising money. We have resources needed and then some.	• Like other ethics consultation services we have no budget, we just exist.
• All of the ethics committee members are salaried employees. It is part of the job, but salary is not earmarked for the committee.	• Everyone on ethics committee is a volunteer—they often work on weekends or after work, but they are not compensated.
• The administration is committed to it. They spent $1500 on training for each member.	• There is not a push administratively. They say "why go to training if you did 6 years ago?"
• It seems to work. There is a real interest when we call a meeting. Attendance is incredible, no matter when it is called.	• It takes lots of time. Things are getting tighter and tighter. Fortunately we have a lot of interested people.
• We have $2,000–$2,500 per year. We would like more, but we are a 164-bed hospital.	• We are a huge facility and there should be an official ethicist who is paid and full time.
• We have a budget to do education, provide speakers, in-services and to receive literature and membership in organizations.	• Would like money for conferences, educational materials, seminars relating to bioethics. We have none now.
• We're doing a good job, not spending money.	• We had a [volunteer] ethicist, but he left.
• We don't get any money. We don't need any.	• No support except when JCAHO's here.
• Often when a case comes up with an ethical dilemma, it can be handled appropriately by the risk manager.	• Lack of training reflects the lack of financial support.
• We are supported adequately by our salaries so we do not mind.	• Support is grossly insufficient. All work is on a volunteer basis.

Figure 1. Examples of Explanations for Why Financial Support for Ethics Consultation Was or Was Not Sufficient.

When extrapolated to all general hospitals in the United States, these data suggest that in a one-year period approximately 29,000 individuals devoted more than 314,000 hours to performing over 36,000 ethics consultations.

Individuals Performing Ethics Consultation

Averaged across ECSs, the individuals who performed ethics consultation during the year prior to survey were described as follows: 54% were female, 90% were white non-Hispanic, 4% black non-Hispanic, 3% Hispanic, and the rest were from other ethnic backgrounds. Individuals performing ethics consultation were almost all physicians (34%), nurses (31%), social workers (11%), chaplains (10%), or administrators (9%). Fewer than 4% were attorneys, other health care providers, lay persons, or "other" (e.g., philosophers, theologians). These percentages did not differ significantly by bed size, academic affiliation, or consultation model used, with the following exceptions: hospitals with fewer than 100 beds had a lower percentage of "other health care providers" performing ethics consultation than other hospitals (1%, P <.01); COTH-affiliated hospitals had a lower percentage of nurses (25%, P <.01) and administrators (5%, P <.05) than other hospitals; and hospitals with ECSs that performed more than 25 consultations in the previous year were more likely than other hospitals to have ethics consultants in the "other" category (9%, P <.05).

Another way of characterizing the individuals who perform ethics consultation is by the percentage of hospitals whose ECSs included individuals from various backgrounds. In this regard, 94% of hospitals had one or more physicians performing ethics consultation, 91% had nurses, 71% had social workers, 70% had chaplains, 61% had administrators, 32% had attorneys, 25% had other health care providers, and 23% had lay people. With respect to years of experience, 9% of ECSs had performed consultation for <1 year, 53% for 1–5 years, 27% for 5–10 years, 10% for >10 years, and 2% unknown.

As for ethics-specific training, 5% had completed a fellowship or graduate degree program in bioethics, 41% had learned to perform ethics consultation with formal, direct supervision by an experienced member of an ECS, and 45% had learned independently, without formal, direct supervision by an experienced member of an ECS. Individuals on ECSs in hospitals with no academic affiliation were significantly more likely to have learned without formal supervision than individuals in academically affiliated or COTH hospitals (54%, 39%, 22%, respectively; P <.001). Similarly, individuals on ECSs in local government or in private, for-profit hospitals were more likely to have learned without formal supervision than individuals in religiously affiliated, other non-profit, or Federal government hospitals (63% and 54% vs. 30%, 47%, and 40%, P <.001). And individuals on ECSs that generally used a full ethics committee model were more likely to have learned without formal supervision than individuals on ECSs that generally used a small team model or an individual consultant model (62% vs. 46% vs. 30%, P = .001).

Access to Ethics Consultation

Almost all ECSs (95%) indicated that anyone could request a consultation. Among those with restrictions on who could request consultations, requests were allowed most often from attending physicians (98%), family members (94%), patients (92%), and nurses (92%), and less often from house officers (73%) and medical students (51%). Sixty-two percent of ECSs accepted anonymous requests for consultation.

Few ECSs required permission prior to ethics consultation from either the patient or surrogate (24%) or from the attending physician (9%). On the other hand, most ECSs did require notification of the patient or surrogate (59%) and/or the attending physician (76%) prior to a consultation.

A minority of ECSs (40%) used a call schedule. Call schedule usage was significantly associated with consultation volume ($P <.001$), and with academic affiliation independent of consultation volume ($P <.05$).

How Ethics Consultation Services Are Publicized

The ECS was publicized in 95% of hospitals, most often through staff publications (75%) or through written materials distributed to all patients (70%). Publicizing the ECS, and the number of different publicizing methods used, were significantly related to consultation volume ($P <.01$).

How Ethics Consultation Services Gather Information

ECSs used a variety of methods to gather information about specific cases. Ninety-two percent of ECSs "usually" or "always" had one-on-one discussions with members of the clinical staff, whereas 78% "usually" or "always" had one-on-one discussions with patients or family members. And while 87% of ECSs "usually" or "always" gathered information through direct examination of the patient's medical record, only 54% "usually" or "always" gathered information through direct examination of the patient. Sixty-seven percent of ECSs "usually" or "always" gathered information through group meetings involving clinical staff, compared to meetings involving the patient (29%), or family members (48%).

The ethics consultation model used was related to the information-gathering process. "Always" having one-on-one discussions with the patient or family was more common among ECSs that generally used an individual consultant model (61%) or a small team model (45%) than it was among ECSs that generally used a full committee model (22%, $P <.01$). The same pattern was evident but weaker with "always" having one-on-one discussions with the clinical staff (individual consultant model 75%, small team model 69%, full committee model 52%; $P = .09$). The ethics consultation model used did not, however, relate to whether information was gathered by directly examining the medical record, seeing the patient, or having meetings involving members of the clinical staff, patient and/or family.

Actions Recommended by Ethics Consultation Services

The actions recommended as an end result of ethics consultations varied widely, both between and within ECSs (see Table 4). In 14% of the hospitals, the ECS recommended a single best course of action 100% of the time, while in 25% of hospitals, the ECS never did. In 16% of hospitals the ECS described a range of acceptable actions 100% of the time; in 22% they never did. In 65% of the hospitals the ECS made some form of recommendation 100% of the time; in 6% they never did. On average ECSs recommended a single best course of action 46% of the time, described a range of acceptable actions 41% of the time, and made no recommendation 13% of the time.

Approximately half of ECS (51%) never determined recommendations through voting, but 8% did so "always" and 20% did so "at least half the time." Although 87% of ECSs indicated that they used the same decision-making model for every case, respondents were rarely able to describe the model they used in specific terms. The model most commonly identified by name was the "4-box" or "Jonsen/Siegler/Winslade" model (Jonsen et al. 2002), which was mentioned by only 15 ECSs (3%).

Record Keeping and Reporting by Ethics Consultation Services

Averaged across all ECSs, consultations were documented in the medical record 72% of the time (in brief notes 43% of the time and in detailed analyses 29% of the time). With respect to internal record keeping, consultations were recorded 93% of the time (brief notes 24%, detailed analyses 59%, other 10%). Larger hospitals and COTH hospitals more frequently documented the consultation in both the medical record and internal files ($P < .01$).

The majority of hospitals (91%) always reported the results of an ethics consultation to the director of the ECS and to all members of the ECS (73%), while most (64%) never reported the results in writing to the patient and/or

TABLE 4   Actions Recommended in Ethics Consultation: Percentage of Ethics Consultation Services Reporting Various Frequencies of Three Different End Results

End Result	\multicolumn{7}{c}{Frequency with Which Each End Result Occurred}						
	0% of the time	1–20% of the time	21–40% of the time	41–60% of the time	61–80% of the time	81–99% of the time	100% of the time
Recommend a best course of action	25	10	11	15	19	5	14
Specify a range of acceptable actions	22	15	20	15	12	1	16
Make no specific recommendation	65	17	7	4	1	1	6

family. Approximately 26% of ECSs always reported the results of a case to an authority external to the ECS, such as hospital administration, medical staff governance, or an institutional board.

Follow-up by Ethics Consultation Services

While ECSs occasionally participated actively throughout the patient's stay (13%), most of the time they merely stayed informed about the case by periodically contacting the involved parties (41%), or did not remain involved at all unless another consultation was requested (43%).

Evaluation Performed by Ethics Consultation Services

Only 28% of ECSs reported that they had a "formal process for evaluating the ECS that involved the collection and analysis of data on consultations performed." In 13% of ECSs, this evaluation process consisted exclusively of an internal review and discussion of case records. Only 7% of ECSs surveyed participants in the consultation process, and only 4% were formally evaluated by anyone outside of the ECS. The type of evaluation process used was not significantly related to bed size, academic affiliation, ownership, or ethics consultation model.

Discussion

This study reveals several important findings. First, the prevalence of ethics consultation in general hospitals in the United States is quite high. Fully 95% of hospitals either had an ECS or were in the process of developing one. This is remarkable, especially considering that almost half of U.S. hospitals have fewer than 100 beds. The vast majority of hospitals, large and small, offer ethics consultation, including all hospitals with more than 400 beds, all federal government hospitals, and all hospitals belonging to the Council of Teaching Hospitals. Thus while little is known about the quality or effectiveness of ethics consultation, the service has become a routine part of U.S. health care.

Another important finding is that considerable time and resources are invested in ethics consultation in the United States. Specifically, the data suggest that, within the more than 6,000 general hospitals in this country, approximately 29,000 individuals devote more than 314,000 hours to ethics consultation each year. And these numbers underestimate the total amount of time and resources devoted to ethics consultation in this country, since they do not include consultations performed in specialty hospitals, nursing homes, home care agencies, or other health care settings. Moreover, some general hospitals have multiple ECSs (such as one for adult medicine, one for pediatrics, and one for psychiatry), but these data describe only the most active ECS within each hospital.

The study also suggests that the most commonly used model of ethics consultation is the small team model, with a median of four individuals performing each consultation. Interestingly, the total number of person-hours spent on the average consultation varied dramatically according to the primary model of consultation used—from four in ECSs that primarily used the individual consultant model, to seven in those that primarily used the small team model, to 15 in those that primarily used the full ethics committee model. While it appears from these data that the individual consultant model is by far the most efficient, further study is required to determine which model is most effective under what circumstances.

It appears from this study that a very high proportion of the individuals who perform ethics consultation in U.S. hospitals are clinicians, especially doctors and nurses, compared to a very small proportion of nonclinicians such as philosophers, theologians, and attorneys. Some may find this surprising, since nonclinicians tend to be much more highly represented among ethics consultants who serve on consensus panels, publish articles in the scholarly literature, and/or are members of professional organizations such as the American Society for Bioethics and Humanities. In addition, previous studies of ethics consultants have mostly relied on convenience samples (e.g., attendees at an academic meeting [Fox and Stocking, 1993] or personal contacts of the investigators [Self et al., 1993]), which may have contributed to a skewed perception of ethics consultants. By highlighting the difference between individuals who perform ethics consultation in this country and the subset of those individuals who are most active in the academic bioethics community, our findings underscore the need to learn more about this broader population of ethics consultants and how their perspectives and experiences may differ from those who are most prominent in speaking and writing about ethics consultation practices.

With respect to education and training, fewer than half of the individuals who perform ethics consultation had learned to perform ethics consultation with formal, direct supervision by an experienced member of an ECS, and only one in 20 had completed a fellowship or graduate degree program in bioethics. While the American Society for Bioethics and Humanities has suggested that certain skills, knowledge, and character traits are required to perform ethics consultation competently (ASBH 1998), and many assume that formal ethics education is needed to develop these competencies, it should be noted that the effects of ethics education on the moral reasoning of health care professionals have not been well established (Bardon 2004). It may be that on-the-job experience is much more important than formal training. If this is the case, then the fact that many ECSs have a very low level of consultation activity may be cause for concern. Is there sufficient activity to develop and maintain the competencies required for the ethics consultation? Future studies are needed both to assess competencies and to determine how best to develop and retain them.

In certain respects, ethics consultation practices were quite consistent across all consultation services. For example, the vast majority of ECSs had as a primary goal "intervening to protect patient rights," would accept consultation requests from "anyone," and made efforts to publicize their service. Consultation services routinely gathered information from the patient's medical record and from members of the clinical staff. And almost all kept records of their ethics consultations in internal files.

In other respects, however, ethics consultation practices varied considerably among ECSs. For example, about half of ECSs made a habit of seeing the patient, while the remainder did so infrequently or not at all. In some cases, variation may be cause for concern, at least to the extent that ECSs are not following widely recognized standards (Wolf 1991). For example, while under most circumstances professional guidelines consider patient or surrogate notification mandatory (American Medical Association 2005; American Society for Bioethics and Humanities 1998), notification was required by only slightly more than half of ECSs. While some experts hold that ethics consultation should never involve voting (Fletcher and Hoffman 1994; Moss and Glover 2006), almost half of ECSs used this method at least some of the time. And while the need for evaluation of ethics consultation activities is widely acknowledged (American Academy of Pediatrics 2001; American Society for Bioethics and Humanities 1998; Fletcher and Hoffman 1994; Fox, E., and J. A. Tulsky 1996), almost three quarters of ECSs had no such a process in place.

Ethics consultation practices varied not only among ECSs, but also within ECSs—that is, from one consult to the next. Some of this variation is expected, because each consultation is unique. On the other hand, many ECSs did not have explicit policies or procedures and were unable to describe the methods they used in specific terms. This is troubling, since almost all guidelines and consensus statements agree that ECSs should have clear standards and follow them consistently.

Finally, and perhaps most fundamentally, there was wide variation in terms of the end result of the ethics consultation process. While most ECSs always recommended specific actions, others did not make recommendations at all. Of those that made recommendations, some routinely offered a range of possible options, whereas others routinely recommended a single best course of action. This raises the possibility that beyond procedural differences, ECSs differ more fundamentally in their understandings of the appropriate role of ethics consultation in clinical decision making. This possibility requires further study.

This study has several important limitations. The data are several years old and may be outdated. Informants reported, largely from memory, activity during the previous year, which introduces a potential recall bias. In addition, the survey contained mostly multiple choice questions with a finite number of response categories and therefore may have failed to capture the full range of possible responses.

Despite these limitations, this study is the first to provide a representative snapshot of national ethics consultation practices. It also raises several concerns that suggest a need for additional research. While the prevalence of ECSs in U.S. hospitals is quite high, there appear to be wide variations in practice, a lack of formal training, and few mechanisms for quality control.

To ensure the quality and consistency of ethics consultation practices, we believe there need to be clear standards for ethics consultation practice, educational resources to assist ECSs in implementing those standards, and tools to evaluate whether the standards are being met. Efforts to fill these needs are already under way by the Clinical Ethics Task Force of the American Society for Bioethics and Humanities (www.asbh.org) and by the National Center for Ethics in Health Care of the Veterans Health Administration (http://www.va.gov/IntegratedEthics).

References

American Academy of Pediatrics, Committee on Bioethics. 2001. Policy statement: Institutional ethics committees. *Pediatrics* 107: 205–209.

American Hospital Association. 1986. Guidelines for hospital ethics committees. In *Handbook for hospital ethics committees*, ed., J. W. Ross. Chicago: American Hospital Association.

American Hospital Association. AHA annual survey database, Fiscal Year 1998.

American Medical Association, Council on Ethical and Judicial Affairs. 1985. Guidelines for ethics committees in health care institutions. *JAMA* 235: 2698–9.

American Medical Association, Council on Ethical and Judicial Affairs. 2005. *Code of medical ethics: Current opinions with annotations, 2004–2005*, Section 9.115. Chicago: American Medical Association.

American Society for Bioethics and Humanities, Task Force on Standards for Bioethics Consultation. 1998. *Core competencies for health care ethics consultation: The report of the American Society for Bioethics and Humanities*. Glenview, IL: American Society for Bioethics and Humanities.

Bardon, A. 2004. Ethics education and value prioritization among members of U.S. hospital ethics committees. *Kennedy Institute of Ethics Journal* 14(4): 395–406.

Bernt, F., P. Clark, J. Starrs, and P. Talone. 2006. Ethics committees in Catholic hospitals. A new study assesses their role, impact, and future in CHA-member hospitals. *Health Progress* 87(2): 18–25.

Brennan, T. A. 1988. Ethics committees and decisions to limit care: The experience at Massachusetts General Hospital. *JAMA* 260: 803–7.

Csikai, E. L. 1998. The status of hospital ethics committees in Pennsylvania. *Cambridge Quarterly of Healthcare Ethics* 7(1): 104–7.

Dowdy, M. D., C. Robertson, and J. A. Bander. 1998. A study of proactive ethics consultation for critically and terminally ill patients with extended lengths of stay. *Critical Care Medicine* 26(2): 252–9. Erratum in *Critical Care Medicine* 26(11): 1923.

Fletcher, J. C., and D. E. Hoffman. 1994. Ethics committees: Time to experiment with standards. *Annals of Internal Medicine* 120: 335–8.

Fox, E., and Stocking, C. 1993. Ethics consultants' recommendations for life-prolonging treatment of patients in a persistent vegetative state. *JAMA*. 270:2578–82.

Fox, E., and J. A. Tulsky. 1996. Evaluation research and the future of ethics consultation. *The Journal of Clinical Ethics* 7: 146–9.

Godkin, M. D., K. Faith, R. E. Upshur, S. K. MacRae, C. S. Tracy, and the PEECE Group. 2005. Project examining effectiveness in clinical ethics (PEECE): Phase 1—Descriptive analysis of nine clinical ethics services. *Journal of Medical Ethics* 31(9): 505–12.

Jonsen, A. R., Siegler, M., and Winslade, W. J. 2002. *Clinical ethics: A practical approach to ethical decisions in clinical medicine (5th ed.).* New York: McGraw Hill.

Kelly, S. E., P. A. Marshall, L. M. Sanders, T. A. Raffin, and B. A. Koenig. 1997. Understanding the practice of ethics consultation: Results of an ethnographic multi-site study. *The Journal of Clinical Ethics* 8(2): 136–49.

Kosnik, A. R. 1974. Developing a health facility medical-moral committee. *Hospital Progress* 55: 40–4.

Moss, A. H. and Glover, J. J. West Virginia Network of Ethics Committees: Starting an ethics committee, questions to consider and first steps. http://www.hsc.wvu.edu/chel/wvnec/index.htm. Last accessed July 31, 2006.

Orr, R. D., K. R. Morton, D. M. DeLeon, and J. C. Fals. 1996. Evaluation of an ethics consultation service: Patient and family perspective. *The American Journal of Medicine* 101(2): 135–41.

President's Commission for the Study of Ethical Problems in Medicine and Biomedical and Behavioral Research. 1983. *Deciding to forego life-sustaining treatment: a report on the ethical, medical, and legal issues in treatment decisions.* Washington: U.S. Government Printing Office.

Rosner, F. 1985. Hospital medical ethics committees: A review of their development. *JAMA* 253: 2693–7.

Schneiderman, L. J., T. Gilmer, H. D. Teetzel, D. O. Dugan, J. Blustein, R. Cranford, K. B. Briggs, G. I. Komatsu, P. Goodman-Crews, F. Cohn, and E. W. Young. 2003. Effect of ethics consultations on nonbeneficial life-sustaining treatments in the intensive care setting: A randomized controlled trial. *JAMA* 290(9): 1166–72.

Self, D. J., Skeel, J. D, and Jecker N. S. 1993. A comparison of the moral reasoning of physicians and clinical medical ethicists. *Academic Medicine.* 68: 852–55.

Tulsky, J. A., and E. Fox. 1996. Evaluating ethics consultation: Framing the questions. *Journal of Clinical Ethics* 7: 109–15.

Tuohey, J. 2006. Ethics consultation in Portland. Providence Health's Oregon region has created an ethics consult service. *Health Progress* 87(2): 36–41.

Waisel, D. B., S. E. Vanscoy, L. H. Tice, K. L. Bulger, J. O. Schmelz, and P. J. Perucca. 2000. Activities of an ethics consultation service in a Tertiary Military Medical Center. *Military Medicine* 165(7): 528–32.

Wolf, S. M. 1991. Ethics committees and due process: Nesting rights in a community of caring. *Maryland Law Review* 50: 798–858.

Youngner, S. J. 1983. A national survey of hospital ethics committees. *Critical Care Medicine* 11(11): 902–5.

What *Is* Medical Ethics Consultation?

Giles R. Scofield

Giles Scofield, JD, MA, is a clinical ethicist at the Centre for Clinical Ethics in Toronto. In the selection below, Scofield argues that the scope and limits of ethics consultation is not well defined. Because ethics consultation has no clear boundaries, it is impossible to know when boundaries have been crossed, and whether such crossings are justified. Scofield also notes that, unlike other consultation services in the health care setting, ethics consultation lacks a credentialing process of any kind. Whereas ethics consultants typically are regarded as professionals, they lack all of the formal attributes traditionally associated with a profession, such as a code of ethics, standards of accreditation, and conditions of licensure. Scofield concludes that the field of ethics consultation is without any means of ensuring that its practitioners have expertise or perform in a competent manner.

[The essay begins by noting that bioethics lacks a formal professional status. Ethics consultants, in particular, are outliers because they work in the clinical setting, where they are involved in the care of patients. Unlike physicians, nurses, social workers, and others in their midst, the ethics consultant's job requires no specific credentials or licensure, and no specific training or education.]

. . . Although medical ethics consultation emerged as a self-styled "nascent profession" in the 1980s, it has yet to attain "formal" professional status. It lacks *all* of the formal attributes traditionally associated with a profession: a code of ethics, standards of accreditation, and conditions of licensure. There is not even a statute that prohibits unqualified individuals from engaging in the unauthorized practice of medical ethics consultation. Thus, although everybody knows *that* and *who* they are, nobody knows quite *what* they are, or *what* they do, much less what business they have being who they are and doing what they do.

This is a curious state of affairs, for several reasons. In the early 1990s, John C. Fletcher, an early and persistent advocate of medical ethics consultation, agonized that some kind of "ethics disaster" would befall the field unless it abandoned its laissez-faire approach to the practice of ethics consultation and did something—though never clear what—both about how medical ethics consultants do their work and how they are trained and educated. Moreover, because of the substantial and seemingly desirable benefits that come with an occupation becoming a "profession," it is odd that the field thinks and talks about itself and

acts as if it already were a profession, but that it cannot and will not profession-
alize itself formally. Finally, because ethics consultants work with medical professionals and in health care institutions, where credentialing and accreditation and
the other attributes commonly associated with professionalism are integral components of insuring that everything gets done in the right way and by the right
people, one would expect medical ethics consultants to seek—and for others to
insist that they obtain—formal, professional status.

The more one looks into this curious state of affairs, the more curious it
becomes. At about the same time that Fletcher, along with some others, agonized
about what might happen if the field did not get its house in order, the Society for
Health and Human Values and the Society for Bioethics Consultation [SHHV-
SBC] created a task force whose mission was to look into, think about, and report
back on these issues. After due deliberation, that same task force decided *against*
professionalization, i.e., against certifying those who provide ethics consultation,
against accrediting the programs that train them, and against adopting a code of
ethics to govern them. Instead, the task force decided to adopt precatory, as
opposed to mandatory, standards: the so-called "core competencies."[1] Ostensibly,
this did not mean that Fletcher's concerns were misplaced, but only that they
were—for some reason or other—ill-timed. Now the alarm has been sounded
once again, in an article and a series of accompanying commentaries that recently
appeared in the *American Journal of Bioethics*. Although some reason exists to
believe that the field really means business this time—as well as some reason to
believe that the field may be crying wolf once again—there is even more reason
to believe both that the field may be a wolf dressed in sheep's clothing, *and* that
it is trying to pull the wool over everyone's eyes, including its own. To understand
how and why this is so, we need to examine the matter more closely. [The essay
next takes up a related point. The scope and limits of the activity of ethics consultation are vague and poorly defined. There is no clear consensus among
bioethicists about what is and is not the role of the clinical ethics consultant.]

What's My Line?

"Professions are separated by a clear *boundary* from other occupations."[2]

. . . As everyone knows, defining one's "scope of practice," "turf," jurisdiction,
and cognitive and practical "domain" is how professions define themselves in relation to one another. . . . Turf is also how prospective clients know who and who
not to turn to for advice or assistance. And yet, when it comes to knowing what
medical ethics consultants do or what medical ethics consultation is, the field is no
closer to knowing what that is or what they do than it was and they were a decade
ago. . . . [O]ne might have expected the task force to answer the question and
thereby settle the matter in a definitive, authoritative manner, it did not; on the
contrary, it only added to the confusion. In its official report, the task force said:

> Health care ethics consultation is a service provided by an individual or group to help
> patients, families, surrogates, health care providers, or other involved parties address
> uncertainty or conflict regarding value-laden issues that emerge in health care.[3]

. . . If this is what ethics consultation is and what ethics consultants do, then it is difficult, if not impossible to know what it is *not*, or with what ethics consultants cannot, do not, should not, and will not concern themselves. Small wonder that there seems to be no limit to what they can and will involve themselves in . . . The truth, it seems, does not exist where medical ethics consultation is concerned, but only different and differing interpretations as to what it means for someone to be a medical ethics consultant, as well as what it is that one needs to know and to be able to do in order to become, to be, and to remain one. Consistent with this state of affairs, a recently published book on the subject describes the current state of the field in the following manner:

> What should be the role of clinical ethicists? There is a clear consensus they can and should play an important part in policy development and staff education. There is not, however, similar agreement regarding how case consultations should be handled. Should ethicists merely analyze problems and outline relevant value dimensions? Should they become patient advocates? Physician or hospital advocates? Should they give prescriptive recommendations as to best ethical choices?[4]

Even a full decade after the task force completed its report, no one can or does know what they do or what they know; thus, a criticism made of consultants generally applies with equal force to medical ethics consultants as well.

> [It] is a tale of mystery and imagination. Nobody seems to know quite what it is, let alone whether it delivers value for money. The consultants do their best to maintain the mystique, pleading client confidentiality and hiding behind such terms as 'value propositions' and 'service offerings' . . . It is hard to avoid the conclusion that, along with . . . advice, the industry dispenses a little witchcraft as well.[5]

All of which leaves us where we began, with a curious state of affairs, one made all the more curious by the very nature of the field that underlies and gives rise to this practice. According to medical ethicists themselves, "[A]mbiguity and vagueness are the enemies of analytic philosophy and consequently of applied ethics." If ambiguity and vagueness are the "enemies of bioethics," then would not medical ethicists have to be . . . idiots, if not quite simply mad, to allow, if not encourage and cultivate, the impression that there is something vague, ambiguous, and even mysterious about what they know and what they do? And yet, as the preceding discussion shows, that is precisely what they do, have done, and seem incapable of not doing. If they are neither idiots nor mad, then there must be some reason for, some method to, their "madness."

This is why it makes eminent sense to heed Paul Starr's advice, which is that we should not simply take the field's claims at face value, but that we should instead be skeptical of, question, and even challenge both the claims and the denials that the field's practitioners make, both in terms of what they say about themselves and in terms of what they insinuate about the rest of us. It is precisely because the field has little reason to be and every reason not to be appropriately self-critical that others have more reasons to be and fewer reasons not to be skeptical, if not suspicious, of the field. Because establishing a field's professional jurisdiction—its turf—is a, if not the, *sine qua non* of establishing its professional identity, is the field's persistent inability to set definite limits or boundaries—as to both its cognitive domain and scope of practice—simply the result of its being a "new profession" going through a painful and protracted adolescence? Or is there more to this—and something other to this—than what meets the eye? Is the field's awkward progress towards professionalization just "the way it is" insofar as the "evolution" of a new profession is concerned? Or might "intelligent design" have something to do with it?

Boundaries? What Boundaries?

. . . To answer these and other questions, one finds oneself having to discern if there might be something more and something other to what medical ethicists mean by what they say—an undertaking which is easier said than done, given the manner in which they tend to express themselves. Consider, for example, how they talk about what is supposed to be the preferred approach to medical ethics consultation, i.e., the so-called ethics facilitation model, which the task force described in the following manner:

> We believe that an ethics facilitation approach is most appropriate for health care ethics consultation in contemporary society. The ethics facilitation approach is informed by the context in which ethics consultation is provided and involves two core features: identifying and analyzing the nature of value uncertainty and facilitating the building of consensus. By 'consensus,' we mean agreement by all involved parties, whether that agreement concerns the substantively morally optimal solution, or, more typically, who should be allowed to make the decision.[6]

. . . As is the case with every other description of what it is that medical ethics consultation is, this description answers the question only by begging further questions. What does it mean, exactly and precisely, for ethics facilitation to be "informed by the context in which it is provided"? What is a "substantively morally optimal solution"? Who says so, and on what basis? . . .

Because relatively little insight into the matter will likely be gained by trying to understand what this text means at this point in the discussion, I want to bracket this section and postpone analyzing it for now. Instead, I first want to focus attention on what is an arguably more interesting, revealing, and potentially more penetrable portion of this text. The section in question concerns itself with how the task force explained and tried to justify its decision to embrace the so-called ethics facilitation model, which it did, among other things, in the course of rejecting the so-called "authoritarian model" and the so-called "pure facilitation model" in favor of its own so-called "ethics facilitation model," all while providing reasons for doing so.

. . . [T]he reasons the task force gave were as follows:

> We believe that an ethics facilitation approach is most appropriate for health care ethics consultation in contemporary society. . . . The ethics facilitation approach recognizes the societal boundaries for morally acceptable solutions. In contrast to the

authoritarian approach, ethics facilitation emphasizes an inclusive-consensus-building process. It respects individuals' rights to live by their values by not displacing moral decision-making authority or representing the personal views of the consultant only. In contrast to the pure facilitation approach, ethics facilitation recognizes that societal values, law, and institutional policy have implications for a morally acceptable consensus. The ethics facilitation approach is consistent with both the pluralistic context within which ethics consultation is done and the rights of individuals to live by their own values, recognizing that there are definite boundaries within which decisions must be made and helping to ensure that these boundaries are not transgressed.[7]

This passage is interesting for several reasons. One is that . . . the task force says this about both the authoritarian and the pure facilitation approaches: "We are not claiming that anyone actually does ethics consultation in either of these two ways."[8] Instead, the task force was just describing and discussing these approaches for "illustrative purposes."[9]

That being the case, whether one chooses to take the task force at its word or not, one is left wondering, instead of knowing, what to make of what it says about the "authoritarian" and the "pure facilitation" approaches. Do they exist in reality? Or are they figments of the task force's imagination, conjured up and invoked for the sake of making an argument . . . ? If neither the authoritarian nor the pure approach exists in reality, can and does the ethics facilitation approach exist in reality? If so, in what kind of and in whose reality does it exist? Can and does anyone actually do ethics consultation in *this* way? Or does the description of the ethics facilitation approach also serve what can only be regarded as an illustrative purpose? If so, then what exactly does it illustrate?

On the other hand, if the task force was intended to describe a phenomenon that manifests itself as a continuum, ranging from a too-heavy-handed to a too-light-handed approach to ethics facilitation, with the ethics facilitation model as just-right-handed, then other problems might also arise, one of which is that it defeats the purpose of describing a continuum, because it treats the facilitation model *as if* it were not a part of this continuum. Instead, it stands apart from the continuum, as if it were and could only be part of the solution, and were not and could not be any part of the problem. Moreover, if one follows the task force's thinking, one must conclude that the ethics facilitation model is not and cannot be pure. This begs the general question of what its "impurities" consist of, and the specific concern of whether they consist in part of some form or forms of "authoritarianism." No matter how one reads the task force's description, it proves to be a problematic text, as evidenced by how it has been read and interpreted by others.

What also makes this passage interesting is the matter-of-fact, relatively un-self-conscious manner in which the task force mentions—in passing and without further elaboration—that "definite boundaries" exist and that "these boundaries are not [to be] transgressed." *What* boundaries? What is a boundary? What is a boundary transgression? Where do *these* words and this language come from? And what do they mean? Although these are good questions, the report itself does not address, much less answer them . . .

[An important point the essay makes is that bioethicists have a responsibility, which they fail to meet, to establish clear boundaries for ethics consultation. Without clear boundaries, it becomes impossible to hold ethicists accountable when boundaries are crossed. More generally, it becomes much more difficult to say what is and is not ethically acceptable behavior.]

What about boundary transgressions? Probe just a little bit further into how words such as boundaries and boundary transgressions are used, and one will discover *both* that there exists a considerable, but manageable literature on boundaries, boundary crossings, and boundary violations *and* that this is a topic that medical ethics consultants themselves occasionally write about, presumably because they know and profess to know something, if not a good deal, about it. There is, it seems, a good deal more to the task force's offhand remark than what meets the eye.

According to this literature, "[A] boundary may be parsimoniously defined as the 'edge' of appropriate behavior," meaning appropriate *professional* behavior.[10] As it turns out, knowing what a clinician's professional role is and is not supposed to be plays a significant role in determining what and where the "edge of appropriate behavior is," as well as why it is where it is. Understanding one's role is to understand one's limits, and vice versa. Accordingly, one's "scope of practice" plays a significant role in determining what it means to be a certain kind of professional, that there are limits to what one may do as a professional, to what it means to behave in a professionally appropriate manner, what those limits are, and why those limits are what and where they are. Simply put, a professional knows—and is supposed to know—his or her "limitations."

Once one knows what boundaries are, as well as something about why boundaries are, one can learn what distinguishes a boundary crossing from a boundary violation. A boundary violation "represents a *harmful* crossing, a transgression, of a boundary," and a "boundary crossing," in contrast, occurs when someone crosses the line, but does so in a manner that does not hurt, and may instead be helpful to a patient or client. Whereas the former is inexcusable and unjustifiable, the latter ordinarily may be excusable and justifiable. Although the concepts of boundary, boundary crossings, and boundary violations can be and often are somewhat fuzzy, they are and must be, if only for practical reasons, discernible, knowable, and enforceable. Discerning the one from the other is not a no-brainer, but neither is it rocket science, as the following passage illustrates.

We have suggested that boundary transgressions can be divided into two broad categories. The first is a "boundary crossing," a benign variant where the ultimate effect of the deviation from the usual . . . behavior may be to advance the therapy in a constructive way that does not harm the patient. . . .Some of these [boundary crossings] may be fully appropriate human responses to unusual events that might involve physical contact. A patient stumbled as she was leaving the office and fell to the floor. The therapist helped the patient up and made sure that she was all right. A patient entered her therapist's office and announced that she had just received news that her son had died. The patient reached out to embrace the therapist, and the therapist accepted the embrace as the patient sobbed. A failure to respond in a human way in such situations would very likely devastate the patient and might even lead to a premature interruption of the therapy. . . . The second broad category is the "boundary violation," when the transgression is clearly harmful to or exploitative of the patient. In contrast to the boundary crossing, it is usually not productively discussed by the therapist and patient and may be part of a repetitive practice. The harm may range from wasting time and therapeutic opportunity to inflicting severe trauma. Examples [of boundary violations] include the therapist who hugs the patient at the end of each session and tells him or her "I love you," or the therapist who asks the patient to pick up his or her dry cleaning, the repeated disclosure of the therapist's own personal problems in a way that burdens the patient, and, of course, overt sexual contacts.[11]

When it comes to boundary *crossings*, the rule may be said to be, "[W]hen in doubt, be human." When it comes to boundary *violations*, in contrast, the rule seems to be, "[W]hen in doubt, don't." When it comes to discerning the one from the other, the rule is the same: be professional, be a professional, use your head, i.e., think about what your role is and about how you can and should, and can but should not, fulfill that role . . .

Although discussions of boundary theory, boundary crossings, and boundary violations are commonly and prominently associated with mental health professionals—psychiatrists, psychologists, and social workers—these matters concern and are of concern to any and all of the so-called "helping professions," which include, but are not limited to physicians, hospital chaplains, and genetic counselors, to mention a few notable and relevant examples. Moreover, while such discussions tend to focus on such issues as gifts, dual roles, time management, billing practices, and inappropriate physical contact, e.g., sex, it is obviously also possible—and inappropriate—to overstep one's *metaphysical boundaries.*

That metaphysical boundaries exist and that it would be wrong for an ethics consultant to cross that line is clearly consistent with and implicit in the task force's decision to reject the so-called authoritarian model, on the grounds that it enables, allows, and encourages medical ethicists to intrude *too much* into the patient's, the family's, or an organization's "space." That such boundaries exist, and that they should not be violated, is confirmed, for example, by the respect owed to an individual's religious beliefs, by the belief that there is or should be some such thing as a "dignitary" tort, and by the manner in which we expect genetic counseling to be conducted, i.e., with due respect for the moral (metaphysical) autonomy of the person.

In sum, respecting a person's metaphysical integrity is tied to respecting a person's physical integrity. Consistent with this outlook, the task force rightly and explicitly said, elsewhere in the report, "Individuals . . . do not give up the right to live by their own moral values when they become patients or take up the practice of health care. These rights set *boundaries* that must be respected in ethics consultation." Thus, metaphysical boundaries do indeed exist.

Although a belief in the existence of metaphysical boundaries is clearly consistent with the task force's decision to prefer the ethics facilitation to the authoritarian approach, it is less obviously consistent with its decision to prefer the ethics facilitation to the so called pure facilitation model. After all, what could or would be wrong with an ethics consultant *not* violating an individual's metaphysical integrity? Overstepping someone's metaphysical boundary is self-evidently problematic in a way that not doing so is not, especially given the single example that task force offers for the purposes of "illustrating"—and then rejecting—the so-called pure facilitation approach. In that example, *others*, e.g., an incompetent patient's family, are simply allowed to ignore and override a patient's prior wishes, while an ethics consultant, who presumably can and does know what the incompetent patient would want, simply allows this to happen . . .

So What?

. . . Turn to the field of medical ethics consultation, and what one discovers is *not* that discussions about boundary issues are absent, but that medical ethicists publicly discuss the problems that boundary issues pose to *other* professionals, and try as best they can to bury those issues insofar as they concern them, what they do, or how they do it. If one examines the minutes of the SHHV-SBC Task Force, one will learn that the task force members were quite aware of boundary issues. Not only did they discuss the ways in which boundaries distinguish one profession from another, they got into an interesting discussion concerning what the minimum character traits for health care ethics case consultation should be. Interestingly enough, they preferred talking about character traits instead of virtues, because they believed "the latter is especially *theory bound*." . . .

The same group that focused on these issues was told that it would be "good for the character group work to lead to the development of a code of ethics for [health care ethics consultants]." At the next meeting, the following discussion occurred:

> [A]fter considering whether the character material should be cast [either] as minimally required or as aspirational, the TF [task force] thought that adopting either position exclusively would be a mistake. It was agreed that certain character traits should be cultivated both in training and throughout one's practice of hcec [health care ethics consultation]. Third, though there was disagreement about the development of a code of ethics for those conducting hcec, there was considerable agreement that developing one now would be premature.[12]

[In conclusion, the essay notes that keeping the boundaries of ethics consultation vague benefits ethics consultants themselves. Likewise, neglecting to set clear standards for education, training, and licensure, serves the interests of bioethicists, not the interests of patients, families, or other health professionals who rely on their expertise.]

. . . Had the field actually done what it had said it was going to—*had* to—do a decade ago, then it would have found itself having to police *itself* for the sake of others. Now, it can instead police *others* for the sake of itself, and simply spin this otherwise inexplicable turn of events as a good faith on its part to protect the public from harm. It would have harmed and not advanced the field's interests to police itself a decade ago, and it will advance, but not harm the field's interest to police others now, which explains both why the field can do what it did then and do what it is doing now, and why the field can still believe—and expect others to believe—that it is acting rationally, reasonably, and consistently, *not* irrationally, unreasonably, and inconsistently.

. . . Power explains why it needed to protect its own inept then, and why it says that it, by which it also means all the rest of us, needs to be protected from the so-called "inept" now. Back then, the scandal was about *us;* now the scandal is about *them.* That is the difference. What remains constant is what the field was then and is now trying to do, which was and is to "save the life of ethics," i.e., the livelihood of medical ethicists as consultants. . . .

Selected References

1. American Society for Bioethics and the Humanities, *Core Competencies for Health Care Ethics Consultation: The Report of the American Society for Bioethics and Humanities* (Glenview, IL: American Society for Bioethics Consultation, 1998) [hereinafter cited as ASBH]; see also, SHHV-SBC, *Discussion Draft of the Task Force on Standards for Bioethics Consultation* (undated document); see also, L. F. Post, J. Blustein, and N. N. Dubler, *Handbook for Healthcare Ethics Committees* (Baltimore: Johns Hopkins University Press, 2007).
2. M. Aulisio, "Minutes" (final version), SHHV-SBC Task Force on Standards for Bioethics Consultation, Meeting One, May 24–26, 1996 (July 18, 1996): at 4. (Emphasis added.)
3. See SHHV-SBC, *supra* note 1, at 3.
4. C. Meyers, *A Practical Guide to Clinical Ethics Consulting: Expertise, Ethos, and Power* (Lanham, MD: Rowman & Littlefield Publishers, Inc., 2007): at 11.
5. "Trimming the Fat," *The Economist* (Special Survey) 3, March 22, 1997.
6. M. P. Aulision, R. M. Arnold, S. J. Younger, "Health Care Ethics Consultation: Nature, Goals, and Competencies," *Annals of Internal Medicine* 133, no. 1 (2000): 59–69, at 61.
7. *Id.* (Aulisio, Arnold, and Youngner), at 61.
8. ASBH, *supra* note 18, at 5, n. 9.
9. See ASBH, *supra* note 18, at 5; Aulisio, Arnold and Youngner, *supra* note 6, at 60.
10. T. G. Gutheil and G. O. Gabbard, "Misuses and Misunderstandings of Boundary Theory in Clinical and Regulatory Settings," *American Journal of Psychiatry* 155, no. 3 (1998): 409–414, at 410; see also, T. G. Gutheil and G. O. Gabbard, "The Concept of Boundaries in Clinical Practice: Theoretical and Risk Management Dimensions," *American Journal of Psychiatry* 150, no. 2 (1993): 189–190.
11. See Gutheil and Gabbard ("Misuses and Misunderstandings"), *supra* note 10, at 410; see also, T. G. Gutheil, "Boundary Issues and Personality Disorders," *Journal of Psychiatric Practice* 11, no. 2 (2005): 88–96.
12. M. Aulisio, "Minutes of Meeting Four," SHHV-SBC Task Force on Standards for Bioethics Consultation, May 30–31, 1997, at 5.

QUESTIONING THE METHODS OF BIOETHICS

ॐ

A Defense of Ethical Relativism

Ruth Benedict

Ruth Benedict (1887–1948) was an American anthropologist and author of *Patterns of Culture* (1935), a comparative study of diverse cultures. In the selection below, Benedict puts forth the view that what is and is not behaviorally normal is culturally determined. She defends this view by pointing to the cultural practices of different peoples. Appealing to the same argument, she advances the additional claim that moral distinctions, such as right and wrong, must also be culturally determined. Benedict thus holds that the statement "it is morally right" means the same thing as the statement "it is habitual."

Modern social anthropology has become more and more a study of the varieties and common elements of cultural environment and the consequences of these in human behavior. For such a study of diverse social orders, primitive peoples fortunately provide a laboratory not yet entirely vitiated by the spread of a standardized worldwide civilization. Dyaks and Hopis, Fijians and Yakuts are significant for psychological and sociological study because only among these simpler peoples has there been sufficient isolation to give opportunity for the development of localized social forms. In the higher cultures the standardization of custom and belief over a couple of continents has given a false sense of the inevitability of the particular forms that have gained currency, and we need to turn to a wider survey in order to check the conclusions we hastily

From Ruth Benedict, "Anthropology and the Abnormal," *Journal of General Psychology* 10 (1934): 59–82. Reprinted by permission of Heldref Publications.

base upon this near-universality of familiar customs. Most of the simpler cultures did not gain the wide currency of the one which, out of our experience, we identify with human nature, but this was for various historical reasons, and certainly not for any that gives us as its carriers a monopoly of social good or of social sanity. Modern civilization, from this point of view, becomes not a necessary pinnacle of human achievement but one entry in a long series of possible adjustments.

These adjustments, whether they are in mannerisms like the ways of showing anger, or joy, or grief in any society, or in major human drives like those of sex, prove to be far more variable than experience in any one culture would suggest. In certain fields, such as that of religion or of formal marriage arrangements, these wide limits of variability are well known and can be fairly described. In others it is not yet possible to give a generalized account, but that does not absolve us of the task of indicating the significance of the work that has been done and of the problems that have arisen.

One of these problems relates to the customary modern normal-abnormal categories and our conclusions regarding them. In how far are such categories culturally determined, or in how far can we with assurance regard them as absolute? In how far can we regard inability to function socially as diagnostic of abnormality, or in how far is it necessary to regard this as a function of the culture?

As a matter of fact, one of the most striking facts that emerges from a study of widely varying cultures is the ease with which our abnormals function in other cultures. It does not matter what kind of "abnormality" we choose for illustration, those which indicate extreme instability, or those which are more in the nature of character traits like sadism or delusions of grandeur or of persecution, there are well-described cultures in which these abnormals function at ease and with honor, and apparently without danger or difficulty to the society.

The most notorious of these is trance and catalepsy. Even a very mild mystic is aberrant in our culture. But most peoples have regarded even extreme psychic manifestations not only as normal and desirable, but even as characteristic of highly valued and gifted individuals. This was true even in our own cultural background in that period when Catholicism made the ecstatic experience the mark of sainthood. It is hard for us, born and brought up in a culture that makes no use of the experience, to realize how important a role it may play and how many individuals are capable of it, once it has been given an honorable place in any society. . . .

Cataleptic and trance phenomena are, of course, only one illustration of the fact that those whom we regard as abnormals may function adequately in other cultures. Many of our culturally discarded traits are selected for elaboration in different societies. Homosexuality is an excellent example, for in this case our attention is not constantly diverted, as in the consideration of trance, to the interruption of routine activity which it implies. Homosexuality poses the problem very simply. A tendency toward this trait in our culture exposes an individual to all the conflicts to which all aberrants are always exposed, and we tend to iden-

tify the consequences of this conflict with homosexuality. But these consequences are obviously local and cultural. Homosexuals in many societies are not
incompetent, but they may be such if the culture asks adjustments of them that
would strain any man's vitality. Wherever homosexuality has been given an honorable place in any society, those to whom it is congenial have filled adequately
the honorable roles society assigns to them. Plato's *Republic* is, of course, the
most convincing statement of such a reading of homosexuality. It is presented as
one of the major means to the good life, and it was generally so regarded in
Greece at that time.

The cultural attitude toward homosexuals has not always been on such a
high ethical plane, but it has been very varied. Among many American Indian
tribes there exists the institution of the berdache, as the French called them.
These men-women were men who at puberty or thereafter took the dress and
the occupations of women. Sometimes they married other men and lived with
them. Sometimes they were men with no inversion, persons of weak sexual en-
dowment who chose this role to avoid the jeers of the women. The berdaches
were never regarded as of first-rate supernatural power, as similar men-women
were in Siberia, but rather as leaders in women's occupations, good healers in
certain diseases, or, among certain tribes, as the genial organizers of social affairs. In any case, they were socially placed. They were not left exposed to the
conflicts that visit the deviant who is excluded from participation in the recognized pattern of his society.

The most spectacular illustrations of the extent to which normality may be
culturally defined are those cultures where an abnormality of our culture is the
cornerstone of their social structure. It is not possible to do justice to these possibilities in a short discussion. A recent study of an island of northwest Melane-
sia by Fortune describes a society built upon traits which we regard as beyond
the border of paranoia. In this tribe the exogamic groups look upon each other
as prime manipulators of black magic, so that one marries always into an enemy
group which remains for life one's deadly and unappeasable foes. They look
upon a good garden crop as a confession of theft, for everyone is engaged in
making magic to induce into his garden the productiveness of his neighbors';
therefore no secrecy in the island is so rigidly insisted upon as the secrecy of a
man's harvesting of his yams. Their polite phrase at the acceptance of a gift is,
"And if you now poison me, how shall I repay you this present?" Their preoccupation with poisoning is constant; no woman ever leaves her cooking pot for
a moment untended. Even the great affinal economic exchanges that are characteristic of this Melanesian culture area are quite altered in Dobu since they are
incompatible with this fear and distrust that pervades the culture. They go farther and people the whole world outside their own quarters with such malignant
spirits that all-night feasts and ceremonials simply do not occur here. They have
even rigorous religiously enforced customs that forbid the sharing of seed even
in one family group. Anyone else's food is deadly poison to you, so that commu-
nality of stores is out of the question. For some months before harvest the whole
society is on the verge of starvation, but if one falls to the temptation and eats

up one's seed yams, one is an outcast and a beachcomber for life. There is no coming back. It involves, as a matter of course, divorce and the breaking of all social ties.

Now in this society where no one may work with another and no one may share with another, Fortune describes the individual who was regarded by all his fellows as crazy. He was not one of those who periodically ran amok and, beside himself and frothing at the mouth, fell with a knife upon anyone he could reach. Such behavior they did not regard as putting anyone outside the pale. They did not even put the individuals who were known to be liable to these attacks under any kind of control. They merely fled when they saw the attack coming on and kept out of the way. "He would be all right tomorrow." But there was one man of sunny, kindly disposition who liked work and liked to be helpful. The compulsion was too strong for him to repress it in favor of the opposite tendencies of his culture. Men and women never spoke of him without laughing; he was silly and simple and definitely crazy. Nevertheless, to the ethnologist used to a culture that has, in Christianity, made his type the model of all virtue, he seemed a pleasant fellow. . . .

. . . Among the Kwakiutl it did not matter whether a relative had died in bed of disease or by the hand of an enemy; in either case death was an affront to be wiped out by the death of another person. The fact that one had been caused to mourn was proof that one had been put upon. A chief's sister and her daughter had gone up to Victoria, and either because they drank bad whiskey or because their boat capsized, they never came back. The chief called together his warriors, "Now I ask you, tribes, who shall wail? Shall I do it or shall another?" The spokesman answered, of course, "Not you, Chief. Let some other of the tribes." Immediately they set up the war pole to announce their intention of wiping out the injury, and gathered a war party. They set out, and found seven men and two children asleep and killed them. "Then they felt good when they arrived at Sebaa in the evening."

The point which is of interest to us is that in our society those who on that occasion would feel good when they arrived at Sebaa that evening would be the definitely abnormal. There would be some, even in our society, but it is not a recognized and approved mood under the circumstances. On the Northwest Coast those are favored and fortunate to whom that mood under those circumstances is congenial, and those to whom it is repugnant are unlucky. This latter minority can register in their own culture only by doing violence to their congenial responses and acquiring others that are difficult for them. The person, for instance, who, like a Plains Indian whose wife has been taken from him, is too proud to fight, can deal with the Northwest Coast civilization only by ignoring its strongest bents. If he cannot achieve it, he is the deviant in that culture, their instance of abnormality.

This head-hunting that takes place on the Northwest Coast after a death is no matter of blood revenge or of organized vengeance. There is no effort to tie up the subsequent killing with any responsibility on the part of the victim for the death of the person who is being mourned. A chief whose son has died goes visiting wherever his fancy dictates, and he says to his host, "My prince has died today, and you go with him." Then he kills him. In this, according to their interpretation, he acts nobly because he has not been downed. He has thrust back in return. The whole procedure is meaningless without the fundamental paranoid reading of bereavement. Death, like all the other untoward accidents of existence, confounds man's pride and can only be handled in the category of insults.

Behavior honored upon the Northwest Coast is one which is recognized as abnormal in our civilization, and yet it is sufficiently close to the attitudes of our own culture to be intelligible to us and to have a definite vocabulary with which we may discuss it. The megalomaniac paranoid trend is a definite danger in our society. It is encouraged by some of our major preoccupations, and it confronts us with a choice of two possible attitudes. One is to brand it as abnormal and reprehensible, and is the attitude we have chosen in our civilization. The other is to make it an essential attribute of ideal man, and this is the solution in the culture of the Northwest Coast.

These illustrations, which it has been possible to indicate only in the briefest manner, force upon us the fact that normality is culturally defined. An adult shaped to the drives and standards of either of these cultures, if he were transported into our civilization, would fall into our categories of abnormality. He would be faced with the psychic dilemmas of the socially unavailable. In his own culture, however, he is the pillar of society, the end result of socially inculcated mores, and the problem of personal instability in his case simply does not arise.

No one civilization can possibly utilize in its mores the whole potential range of human behavior. Just as there are great numbers of possible phonetic articulations, and the possibility of language depends on a selection and standardization of a few of these in order that speech communication may be possible at all, so the possibility of organized behavior of every sort, from the fashions of local dress and houses to the dicta of a people's ethics and religion, depends upon a similar selection among the possible behavior traits. In the field of recognized economic obligations or sex tabus this selection is as nonrational and subconscious a process as it is in the field of phonetics. It is a process which goes on in the group for long periods of time and is historically conditioned by innumerable accidents of isolation or of contact of peoples. In any comprehensive study of psychology, the selection that different cultures have made in the course of history within the great circumference of potential behavior is of great significance.

Every society, beginning with some slight inclination in one direction or another, carries its preference farther and farther, integrating itself more and more completely upon its chosen basis, and discarding those types of behavior that are uncongenial. Most of those organizations of personality that seem to us most incontrovertibly abnormal have been used by different civilizations in the very foundations of their institutional life. Conversely, the most valued traits of our normal individuals have been looked on in differently organized cultures as aberrant. Normality, in short, within a very wide range, is culturally defined. It is primarily a term for the socially elaborated segment of human behavior in any culture; and abnormality, a term for the segment that that particular civilization does not use. The very eyes with which we see the problem are conditioned by the long traditional habits of our own society.

It is a point that has been made more often in relation to ethics than in relation to psychiatry. We do not any longer make the mistake of deriving the morality of our locality and decade directly from the inevitable constitution of human nature. We do not elevate it to the dignity of a first principle. We recognize that morality differs in every society, and is a convenient term for socially approved habits. Mankind has always preferred to say, "It is morally good," rather than "It is habitual," and the fact of this preference is matter enough for a critical science of ethics. But historically the two phrases are synonymous.

The concept of the normal is properly a variant of the concept of the good. It is that which society has approved. A normal action is one which falls well within the limits of expected behavior for a particular society. Its variability among different peoples is essentially a function of the variability of the behavior patterns that different societies have created for themselves, and can never be wholly divorced from a consideration of culturally institutionalized types of behavior.

Each culture is a more or less elaborate working-out of the potentialities of the segment it has chosen. In so far as a civilization is well integrated and consistent within itself, it will tend to carry farther and farther, according to its nature, its initial impulse toward a particular type of action, and from the point of view of any other culture those elaborations will include more and more extreme and aberrant traits.

Each of these traits, in proportion as it reinforces the chosen behavior patterns of that culture, is for that culture normal. Those individuals to whom it is congenial either congenitally, or as the result of childhood sets, are accorded prestige in that culture, and are not visited with the social contempt or disapproval which their traits would call down upon them in a society that was differently organized. On the other hand, those individuals whose characteristics are not congenial to the selected type of human behavior in that community are the deviants, no matter how valued their personality traits may be in a contrasted civilization.

The Dobuan who is not easily susceptible to fear of treachery, who enjoys work and likes to be helpful, is their neurotic and regarded as silly. On the Northwest Coast the person who finds it difficult to read life in terms of an insult contest will be the person upon whom fall all the difficulties of the culturally unprovided for. The person who does not find it easy to humiliate a neighbor, nor to see humiliation in his own experience, who is genial and loving, may, of course, find some unstandardized way of achieving satisfactions in his society, but not in the major patterned responses that his culture requires of him. If he is born to play an important role in a family with many hereditary privileges, he can succeed only by doing violence to his whole personality. If he does not succeed, he has betrayed his culture; that is, he is abnormal.

I have spoken of individuals as having sets toward certain types of behavior, and of these sets as running sometimes counter to the types of behavior which are institutionalized in the culture to which they belong. From all that we know of contrasting cultures it seems clear that differences of temperament occur in every society. The matter has never been made the subject of investigation, but from the available material it would appear that these temperament types are very likely of universal recurrence. That is, there is an ascertainable range of human behavior that is found wherever a sufficiently large series of individuals is observed. But the proportion in which behavior types stand to one another in different societies is not universal. The vast majority of individuals in any group are shaped to the fashion of that culture. In other words, most individuals are plastic to the moulding force of the society into which they are born. In a society that values trance, as in India, they will have supernormal experience. In a society that institutionalizes homosexuality, they will be homosexual. In a society that sets the gathering of possessions as the chief human objective, they will amass property. The deviants, whatever the type of behavior the culture has institutionalized, will remain few in number, and there seems no more difficulty in moulding the vast malleable majority to the "normality" of what we consider an aberrant trait, such as delusions of reference, than to the normality of such accepted behavior patterns as acquisitiveness. The small proportion of the number of the deviants in any culture is not a function of the sure instinct with which the society has built itself upon the fundamental sanities, but of the universal fact that, happily, the majority of mankind quite readily take any shape that is presented to them. . . .

The Challenge of Cultural Relativism

James Rachels

James Rachels (1941–2003) was University Professor at the University of Alabama at Birmingham, Department of Philosophy, from 1977 until his death in 2003. He chaired the Department of Philosophy from 1977 to 1979 and served as Dean of the School of Humanities from 1978 to 1983. His books include *Can Ethics Provide Answers?: And Other Essays in Moral Philosophy* (Rowman and Littlefield, 1997), *Created from Animals: The Moral Implications of Darwinism* (Oxford University Press, 1990), and *The End of Life: Euthanasia and Morality* (Oxford University Press, 1986). His work focused on issues in ethical theory, such as ethical egoism and the rule against killing, and also on practical moral problems, such as euthanasia and animal rights.

> Morality differs in every society, and is a convenient term for socially approved habits.
>
> Ruth Benedict, *Patterns of Culture* (1934)

How Different Cultures Have Different Moral Codes

Darius, a king of ancient Persia, was intrigued by the variety of cultures he encountered in his travels. He had found, for example, that the Callatians (a tribe of Indians) customarily ate the bodies of their dead fathers. The Greeks, of course, did not do that—the Greeks practiced cremation and regarded the funeral pyre as the natural and fitting way to dispose of the dead. Darius thought that a sophisticated understanding of the world must include an appreciation of such differences between cultures. One day, to teach this lesson, he summoned some Greeks who happened to be at his court and asked what it would take for them to eat the bodies of their dead fathers. They were shocked, as Darius knew they would be, and replied that no amount of money could persuade them to do such a thing. Then Darius called in some Callatians and, while the Greeks listened, asked them what it would take for them to burn their dead fathers' bodies. The Callatians were horrified and told Darius to not even mention such an awful thing.

This story, recounted by Herodotus in his History, illustrates a recurring theme in the literature of social science: Different cultures have different moral

From James Rachels, *The Elements of Moral Philosophy*, 4th edition (New York: McGraw-Hill, 2003). Reprinted with permission of McGraw-Hill. Copyright © 2003 McGraw-Hill Companies.

codes. What is thought right within one group may be utterly abhorrent to the members of another group, and vice versa. Should we eat the bodies of the dead or burn them? If you were a Greek, one answer would seem obviously correct; but if you were a Callatian, the opposite would seem equally certain.

It is easy to give additional examples of the same kind. Consider the Eskimos. The Eskimos are the indigenous peoples of Alaska, northern Canada, Greenland, and northeastern Siberia. (Today, none of these groups call themselves "Eskimos"; however, I'll use that term because it's the only one that refers to this scattered Arctic population.) Traditionally, Eskimos have lived in small settlements, separated by great distances. Prior to the 20th century, the outside world knew little about them. Then explorers began to bring back strange tales.

Eskimo customs turned out to be very different from our own. The men often had more than one wife, and they would share their wives with guests, lending them out for the night as a sign of hospitality. Moreover, within a community, a dominant male might demand—and get—regular sexual access to other men's wives. The women, however, were free to break these arrangements simply by leaving their husbands and taking up with new partners—free, that is, so long as their former husbands chose not to make trouble. All in all, the Eskimo custom was a volatile practice that bore little resemblance to what we call marriage.

But it was not only their marriages and sexual practices that were different. The Eskimos also seemed to have less regard for human life. Infanticide, for example, was common. Knud Rasmussen, one of the most famous early explorers, reported that he met one woman who had borne 20 children but had killed 10 of them at birth. Female babies, he found, were especially liable to be killed, and this was permitted simply at the parents' discretion, with no social stigma attached to it. Old people as well, when they became too feeble to contribute to the family, were left out in the snow to die. So, in Eskimo society, there seemed to be remarkably little respect for life.

To most Americans, these were disturbing revelations. Our own way of living seems so natural and right that for many of us it is hard to conceive of others living so differently. And when we do hear of such things, we tend immediately to categorize the other peoples as "backward" or "primitive." But to anthropologists, there was nothing particularly surprising about the Eskimos. Since the time of Herodotus, enlightened observers have accepted the idea that conceptions of right and wrong differ from culture to culture. If we assume that our ethical ideas will be shared by all peoples at all times, we are merely being naive.

Cultural Relativism

To many thinkers, this observation—"Different cultures have different moral codes"—has seemed to be the key to understanding morality. The idea of universal truth in ethics, they say, is a myth. The customs of different societies are all that exist. These customs cannot be said to be "correct" or "incorrect," for that implies that we have an independent standard of right and wrong by which they may be judged. But no such independent standard exists; every standard is culture-bound. The sociologist William Graham Sumner, writing in 1906, put it like this:

> The "right" way is the way which the ancestors used and which has been handed down. The tradition is its own warrant. It is not held subject to verification by experience. The notion of right is in the folkways. It is not outside of them, of independent origin, and brought to test them. In the folkways, whatever is, is right. This is because they are traditional, and therefore contain in themselves the authority of the ancestral ghosts. When we come to the folkways we are at the end of our analysis.

This line of thought, more than any other, has persuaded people to be skeptical about ethics. Cultural Relativism, as it has been called, challenges our belief in the objectivity and universality of moral truth. It says, in effect, that there is no such thing as universal truth in ethics; there are only the various cultural codes, and nothing more. Moreover, our own code has no special status; it is merely one among many. As we shall see, this basic idea is really a compound of several different thoughts.

It is important to distinguish the various elements of Cultural Relativism because, on analysis, some parts turn out to be correct, while others are mistaken. The following claims have all been made by cultural relativists:

1. Different societies have different moral codes.
2. The moral code of a society determines what is right within that society; that is, if the moral code of a society says that a certain action is right, then that action is right, at least within that society.
3. There is no objective standard that can be used to judge one society's code as better than another's. In other words, there is no "universal truth" in ethics; there are no moral truths that hold for all people at all times.
4. The moral code of our own society has no special status; it is but one among many.
5. It is mere arrogance for us to judge the conduct of other peoples. We should adopt an attitude of tolerance toward the practices of other cultures.

These five propositions may seem to go together, but they are independent of one another, in the sense that some of them might be false even if others are true. Indeed, two of them appear to be inconsistent with each other. The second proposition says that right and wrong are determined by the norms of a society; the fifth proposition says that we should always be tolerant of other cultures. But what if the norms of a society favor intolerance? For example, when the Nazi army marched into Poland on September 1, 1939, thus beginning World War II, this was an intolerant action of the first order. But what if it was in line with Nazi ideals? A cultural relativist, it seems, cannot criticize the Nazis for being intolerant, if all they're doing is following their own moral code.

Given that cultural relativists take pride in their tolerance, it would be ironic if their theory actually supported people in intolerant cultures being intolerant of other societies. However, it need not have that implication. Properly understood, Cultural Relativism holds that the norms of a culture reign supreme *within the bounds of the culture itself.* Thus, once the German soldiers entered Poland, they became bound by the norms of Polish society—norms that obviously excluded the mass slaughter of innocent Poles. "When in Rome," the old saying goes, "do as the Romans do." Cultural relativists agree.

The Cultural Differences Argument

Cultural Relativism is a theory about the nature of morality. At first blush, it seems quite plausible. However, like all such theories, it may be subjected to rational analysis; and when we analyze Cultural Relativism, we find that it is not as plausible as it initially appears to be.

The first thing to notice is that at the heart of Cultural Relativism is a certain form *of argument.* Cultural relativists argue from facts about the differences between cultural outlooks to a conclusion about the status of morality. Thus, we are invited to accept this reasoning:

(1) The Greeks believed it was wrong to eat the dead, whereas the Callatians believed it was right to eat the dead.
(2) Therefore, eating the dead is neither objectively right nor objectively wrong. It is merely a matter of opinion that varies from culture to culture.

Or, alternatively:

(1) The Eskimos see nothing wrong with infanticide, whereas Americans believe infanticide is immoral.
(2) Therefore, infanticide is neither objectively right nor objectively wrong. It is merely a matter of opinion, which varies from culture to culture.

Clearly, these arguments are variations of one fundamental idea. They are both special cases of a more general argument, which says:

(1) Different cultures have different moral codes.
(2) Therefore, there is no objective "truth" in morality. Right and wrong are only matters of opinion, and opinions vary from culture to culture.

We may call this the Cultural Differences Argument. To many people, it is persuasive. But from a logical point of view, is it sound?

It is not sound. The problem is that the conclusion does not follow from the premise—that is, even if the premise is true, the conclusion still might be false. The premise concerns what people *believe*—in some societies, people believe one thing; in other societies, people believe something else. The conclusion, however, concerns what *really is the case.* The trouble is that this sort of conclusion does not follow logically from this sort of premise.

Consider again the example of the Greeks and Callatians. The Greeks believed it was wrong to eat the dead; the Callatians believed it was right. Does it follow, from *the mere fact that they disagreed*, that there is no objective truth in the matter? No, it does not follow; it could be that the practice was objectively right (or wrong) and that one of them was simply mistaken.

To make the point clearer, consider a different matter. In some societies, people believe the earth is flat. In other societies, such as our own, people believe the earth is spherical. Does it follow, from the mere fact that people disagree, that there is no "objective truth" in geography? Of course not; we would never draw such a conclusion because we realize that, in their beliefs about the world, the members of some societies might simply be wrong. There is no reason to think that if the world is round everyone must know it. Similarly, there is no reason to think that if there is moral truth everyone must know it. The fundamental mistake in the Cultural Differences Argument is that it attempts to derive a substantive conclusion about a subject from the mere fact that people disagree about it.

This is a simple point of logic, and it is important not to misunderstand it. We are not saying that the conclusion of the argument is false. That is still an open question. The logical point is merely that the conclusion does not follow from the premise. This is important, because in order to determine whether the conclusion is true, we need arguments in its support. Cultural Relativism proposes this argument, but it is fallacious. So it proves nothing.

The Consequences of Taking Cultural Relativism Seriously

Even if the Cultural Differences Argument is unsound, Cultural Relativism might still be true. What would it be like if it were true? In the passage quoted above, William Graham Sumner states the essence of Cultural Relativism. He says that there is no measure of right and wrong other than the standards of one's society: "The notion of right is in the folkways. It is not outside of them, of independent origin, and brought to test them. In the folkways, whatever is, is right." Suppose we took this seriously. What would be some of the consequences?

1. We could no longer say that the customs of other societies are morally inferior to our own. This, of course, is one of the main points stressed by Cultural Relativism. We would have to stop condemning other societies merely because they are "different." So long as we concentrate on certain examples, such as the funerary practices of the Greeks and Callatians, this may seem to be a sophisticated, enlightened attitude.

However, we would also be barred from criticizing other, less benign practices. For example, the Chinese government has a long tradition of repressing political dissent within its own borders. At any given time, thousands of political prisoners in China are doing hard labor, and in the Tiananmen Square episode of 1989, Chinese troops slaughtered hundreds, if not thousands, of peaceful protesters. Cultural Relativism would preclude us from saying that the Chinese government's policies of oppression are wrong. (We could not even say that a society that respects free speech is *better* than Chinese society, for that would imply a universal standard of comparison.) The failure to condemn *these* practices does not seem enlightened; on the contrary, political oppression seems wrong wherever it occurs. Nevertheless, if we accept Cultural Relativism, we have to regard such social practices as immune from criticism.

2. *We could decide whether our actions are right or wrong just by consulting the standards of our society.* Cultural Relativism suggests a simple test for determining what is right and what is wrong: All we need to do is ask whether the action is in line with the code of the society in question. Suppose a resident of India wonders whether her country's caste system—a system of rigid social hierarchy—is morally correct. All she has to do is ask whether this system conforms to her society's moral code. If it does, there is nothing to worry about, at least from a moral point of view.

This implication of Cultural Relativism is disturbing because few of us think that our society's code is perfect—we can think of all sorts of ways in which it might be improved. Yet Cultural Relativism not only forbids us from criticizing the codes of *other* societies; it also stops us from criticizing our own. After all, if right and wrong are relative to culture, this must be true for our own culture just as much as for other cultures.

3. *The idea of moral progress is called into doubt.* We think that at least some social changes are for the better. Throughout most of Western history, the place of women in society was narrowly circumscribed. Women could not own property; they could not vote or hold political office; and they were under the almost absolute control of their husbands. Recently, much of this has changed, and most people think of it as progress.

But if Cultural Relativism is correct, can we legitimately view this as progress? Progress means replacing a way of doing things with a better way. But by what standard do we judge the new ways as better? If the old ways were in accordance with the social standards of their time, then Cultural Relativism would say it is a mistake to judge them by the standards of another era. Eighteenth-century society was a different society from the one we have now. To say that we have made progress implies a judgment that present-day society is better, just the sort of transcultural judgment that, according to Cultural Relativism, is unacceptable.

Our ideas about social reform will also have to be reconsidered. Reformers such as Martin Luther King, Jr., have sought to change their societies for the better. Within the constraints imposed by Cultural Relativism, there is only one way this might be done. If a society is not living up to its own ideals, the reformer may be regarded as acting for the best in promoting those ideals. After all, those ideals are the standard by which we judge his or her proposals. But no one may challenge the ideals themselves, for they are by definition correct. According to Cultural Relativism, then, the idea of social reform makes sense only in this limited way.

These three consequences of Cultural Relativism have led many thinkers to reject it as implausible on its face. It does make sense, they say, to condemn some practices, such as slavery and anti-Semitism, wherever they occur. It makes sense to think that our own society has made some moral progress, while admitting that it is still imperfect and in need of reform. Because Cultural Relativism implies that these judgments make no sense, the argument goes, it cannot be right.

Why There Is Less Disagreement Than It Seems

The original impetus for Cultural Relativism comes from the observation that cultures differ dramatically in their views of right and wrong. But how much do they actually differ? It is true that there are differences. However, it is easy to overestimate the extent of those differences. Often, when we examine what seems to be a dramatic difference, we find that the cultures do not differ nearly as much as it appears.

Consider a culture in which people believe it is wrong to eat cows. This may even be a poor culture, in which there is not enough food; still, the cows are not to be touched. Such a society would appear to have values very different from our own. But does it? We have not yet asked why these people will not eat cows. Suppose it is because they believe that after death the souls of humans inhabit the bodies of animals, especially cows, so that a cow may be someone's grandmother. Shall we say that their values are different from ours? No; the difference lies elsewhere. The difference is in our belief systems, not in our values. We agree that we shouldn't eat Grandma; we simply disagree about whether Grandma could be a cow.

The point is that many factors work together to produce the customs of a society. The society's values are only one of them. Other matters, such as the religious and factual beliefs held by its members, and the physical circumstances in which they live, are also important. We cannot conclude, then, merely because customs differ, that there is disagreement about values. The difference in customs may be attributable to some other aspect of social life. Thus, there may be less disagreement about values than there appears to be.

Consider again the Eskimos, who killed perfectly healthy infants, especially girls. We do not approve of such things; in our society, a parent who kills a baby will be locked up. Thus, there appears to be a great difference in the values of our two cultures. But suppose we ask why the Eskimos did this. The explanation is not that they had no affection for their children or lacked respect for human life. An Eskimo family would always protect its babies if conditions permitted. But the Eskimos lived in a harsh environment, where food was in short supply. A fundamental postulate of Eskimo thought was this: "Life is hard, and the margin of safety small." A family may want to nourish its babies but be unable to do so.

As in many traditional societies, Eskimo mothers would nurse their infants over a much longer period than mothers in our culture—for four years, and perhaps even longer. So, even in the best of times, there were limits to the number of infants one mother could sustain. Moreover, the Eskimos were nomadic; unable to farm in the harsh northern climate, they had to move about in search of food. Infants had to be carried, and a mother could carry only one baby in her parka as she traveled and went about her outdoor work.

Infant girls were more readily disposed of for two reasons. First, in Eskimo society, the males were the primary food providers—they were the hunters—and it is obviously important to maintain a sufficient number of food providers. But there was an important second reason as well. Because the hunters suffered a high casualty rate, the adult men who died prematurely far outnumbered the women who died early. If male and female infants had survived in equal numbers, the female adult population would have greatly outnumbered the male adult population. Examining the available statistics, one writer concluded that "were it not for female infanticide . . . there would be approximately one-and-a-half times as many females in the average Eskimo local group as there are food-producing males."

So among the Eskimos, infanticide did not signal a fundamentally different attitude toward children. Instead, it arose from the recognition that drastic measures are sometimes needed to ensure the family's survival. Even then, however, killing the baby was not the first option considered. Adoption was common; childless couples were especially happy to take a more fertile couple's "surplus." Killing was the last resort. I emphasize this in order to show that the raw data of anthropologists can be misleading; it can make the differences in values between cultures appear greater than they are. The Eskimos' values were not all that different from our own. It is only that life forced choices upon them that we do not have to make.

How All Cultures Have Some Values in Common

It should not be surprising that, despite appearances, the Eskimos were protective of their children. How could it be otherwise? How could a group survive that did not value its young? It is easy to see that, in fact, all cultural groups must protect their infants. Babies are helpless and cannot survive if they are not given extensive care for a period of years. Therefore, if a group did not care for its young, the young would not survive, and the older members of the group would not be replaced. After a while, the group would die out. This means that any cultural group that continues to exist must care for its young. Infants who are not cared for must be the exception rather than the rule.

Similar reasoning shows that other values must be more or less universal. Imagine what it would be like for a society to place no value on truth telling. When one person spoke to another, there would be no presumption that she was telling the truth, for she could just as easily be lying. Within that society, there would be no reason to pay attention to what anyone says. (I ask you what time it is, and you say, "Four o'clock." But there is no presumption that you are speaking truly; you could just as easily have said midnight. So I have no reason to pay attention to your answer. In fact, there was no point asking you in the first place.) Communication would be extremely difficult, if not impossible. And because complex societies cannot exist without communication among their members, society would become impossible. It follows that in any complex society there must be a presumption in favor of truthfulness. There may, of course, be exceptions to this rule; that is, there may be situations in which it is thought to be permissible to lie. Nevertheless, these will be exceptions to a rule that is in force in the society.

Here is one more example of the same type. Could a society exist in which there was no prohibition on murder? What would this be like? Suppose people were free to kill one another at will, and no one thought there was anything wrong with it. In such a "society," no one could feel safe. Everyone would have to be constantly on guard, and to survive they would have to avoid other people as much as possible. This would inevitably result in individuals trying to become as self-sufficient as possible; after all, associating with others would be dangerous. Society on any large scale would collapse. Of course, people might band together in smaller groups with others whom they could trust not to harm them. But notice what this means: They would be forming smaller societies that did acknowledge a rule against murder. The prohibition of murder, then, is a necessary feature of all societies.

There is a general theoretical point here, namely, that *there are some moral rules that all societies must have in common, because those rules are necessary for society to exist.* The rules against lying and murder are two examples. And, in fact, we do find these rules in force in all viable cultures. Cultures may differ in what they regard as legitimate exceptions to the rules, but this disagreement exists against a broad background of agreement. Therefore, it is a mistake to overestimate the amount of difference between cultures. Not every moral rule can vary from society to society.

Judging a Cultural Practice to Be Undesirable

In 1996, a 17-year-old named Fauziya Kassindja arrived at Newark International Airport in New Jersey and asked for asylum. She had fled her native country of Togo, a small West African nation, to escape what people there call "excision." Excision is a permanently disfiguring procedure that is sometimes called "female circumcision," although it bears little resemblance to the Jewish practice. More commonly, at least in Western media, it is referred to as "female genital mutilation."

According to the World Health Organization, the practice is widespread in 26 African nations, and 2 million girls each year are painfully excised. In some instances, excision is part of an elaborate tribal ritual, performed in small villages, and girls look forward to it because it signals their acceptance into the adult world. In other instances, the practice is carried out in cities on young women who desperately resist.

Fauziya Kassindja was the youngest of five daughters in a devoutly Muslim family. Her father, who owned a successful trucking business, was opposed to excision, and he was able to defy the tradition because of his wealth. His first four daughters were married without being mutilated. But when Fauziya was 16, he suddenly died. She then came under the authority of her aunt, who arranged a marriage for her and prepared to have her excised. Fauziya was terrified, and her mother and oldest sister helped her escape.

In America, Fauziya was imprisoned for nearly 18 months while the authorities decided what to do with her. During this time, she was subjected to humiliating strip searches, denied medical treatment for her asthma, and generally treated like a dangerous criminal. Finally, she was granted asylum, but not before she became the center of a controversy about how we should regard the cultural practices of other peoples. A series of articles in *The New York Times* encouraged the idea that excision is a barbaric practice that should be condemned. Other observers were reluctant to be so judgmental. Live and let live, they said; after all, our culture probably seems just as strange to them.

Suppose we are inclined to say that excision is bad. Would we merely be imposing the standards of our own culture? If Cultural Relativism is correct, that is all we can do, for there is no culture-neutral moral standard to which we may appeal. But is that true?

Is There a Culture-Neutral Standard of Right and Wrong?

There is much that can be said against excision. Excision is painful and results in the permanent loss of sexual pleasure. Its short-term effects can include hemorrhage, tetanus, and septicemia. Sometimes the woman dies. Long-term effects can include chronic infection, scars that hinder walking, and continuing pain.

Why, then, has it become a widespread social practice? It is not easy to say. The practice has no obvious social benefits. Unlike Eskimo infanticide, it is not necessary for the group's survival. Nor is it a matter of religion. Excision is practiced by groups from various religions, including Islam and Christianity, neither of which commends it.

Nevertheless, a number of reasons are given in its defense. Women who are incapable of sexual pleasure are less likely to be promiscuous; thus, there will be fewer unwanted pregnancies in unmarried women. Moreover, wives for whom sex is only a duty are less likely to be unfaithful to their husbands; and because they are not thinking about sex, they will be more attentive to the needs of their husbands and children. Husbands, for their part, are said to enjoy sex more with wives who have been excised. (The women's own lack of enjoyment is said to be unimportant.) Men will not want unexcised women, as they are unclean and immature. Above all, excision has been done since antiquity, and we may not change the ancient ways.

It would be easy, and perhaps a bit arrogant, to ridicule these arguments. But notice an important feature of this whole line of reasoning: It attempts to justify excision by showing that excision is beneficial—men, women, and their families are said to be better off when women are excised. Thus, we might approach this reasoning, and excision itself, by asking whether this is true: Is excision, on the whole, helpful or harmful?

In fact, this is a standard that might reasonably be used in thinking about any social practice: Does the practice promote or *hinder* the welfare of the people whose lives are affected by it? And, as a corollary, we may ask if there is an alternative set of social arrangements that would do a better job of promoting their welfare. If there is, we may conclude that the existing practice is deficient.

But this looks like just the sort of independent moral standard that Cultural Relativism says cannot exist. It is a single standard that may be brought to bear in judging the practices of any culture, at any time, including our own. Of course, people will not usually see this principle as being "brought in from the outside" to judge them, because all viable cultures value human happiness.

Why, Despite All This, Thoughtful People May Be Reluctant to Criticize Other Cultures

Although they are personally horrified by excision, many thoughtful people are reluctant to say it is wrong, for at least three reasons. First, there is an understandable nervousness about interfering in the social customs of other peoples. Europeans and their cultural descendants in America have a shabby history of destroying native cultures in the name of Christianity and enlightenment. Recoiling from this shameful record, some people refuse to criticize other cultures, especially cultures that resemble those that have been wronged in the past. There is a difference, however, between (a) judging a cultural practice to be deficient and (b) thinking that we should announce that fact, conduct a campaign, apply diplomatic pressure, and send in the troops. The first is just a matter of trying to see the world clearly, from a moral point of view. The second is something else entirely. Sometimes it may be right to "do something about it," but often it will not be.

Second, people may feel, rightly enough, that they should be tolerant of other cultures. Tolerance is, no doubt, a virtue—a tolerant person is willing to live in peaceful cooperation with those who see things differently. But there is nothing in the nature of tolerance that requires us to say that all beliefs, all religions, and all social practices are equally admirable. On the contrary, if we did not think that some were better than others, there would be nothing for us to tolerate.

Finally, people may be reluctant to judge because they do not want to express contempt for the society being criticized. But again, this is misguided: To condemn a particular practice is not to say that the culture on the whole is contemptible or that it is generally inferior to any other culture, including one's own. It could have many admirable features. In fact, we should expect this to be true of most human societies—they are mixtures of good and bad practices. Excision happens to be one of the bad ones.

What Can Be Learned from Cultural Relativism

So far, in discussing Cultural Relativism, I have dwelled on its mistakes. I have said that it rests on an unsound argument, that it has implausible consequences, and that the extent of moral disagreement is far less than it suggests. This all adds up to a rather thorough repudiation of the theory. Nevertheless, it is still an appealing idea, and you may have the feeling that all this is a little unfair. The theory must have something going for it—why else has it been so influential? In fact, I think there is something right about Cultural Relativism, and there are two lessons we should learn from the theory, even if we ultimately reject it.

First, Cultural Relativism warns us, quite rightly, about the danger of assuming that all our preferences are based on some absolute rational standard. They are not. Many (but not all) of our practices are merely peculiar to our society, and it is easy to lose sight of that fact. In reminding us of it, the theory does us a service.

Funerary practices are one example. The Callatians, according to Herodotus, were "men who eat their fathers"—a shocking idea, to us at least. But eating the flesh of the dead could be understood as a sign of respect. It could be taken as a symbolic act that says: We wish this person's spirit to dwell within us. Perhaps this was the understanding of the Callatians. On this way of thinking, burying the dead could be seen as an act of rejection, and burning the corpse as positively scornful. If these ideas seem hard to imagine, then we may need to have our imaginations stretched. Of course, we may feel a visceral repugnance at the idea of eating human flesh. But what of it? This repugnance may be, as the relativists say, only a reflection of what is customary in our particular society.

There are many other matters that we tend to think of in terms of objective right and wrong that are really nothing more than social conventions. During the 2004 Super Bowl halftime show, Justin Timberlake ripped off part of Janet Jackson's costume, thus exposing one of her breasts to the audience. CBS quickly cut to an aerial view of the stadium, but as far as the country was concerned, the damage had been done. Half a million viewers complained, and the federal government fined CBS $550,000. In America, a publicly exposed breast is scandalous. In other cultures, however, such displays are common. Objectively speaking, the display of a woman's breast is neither right nor wrong. Cultural Relativism begins with the valuable insight that many of our practices are like this—that is, they are only cultural products. Then it goes wrong by inferring that, because some practices are like this, all must be.

The second lesson has to do with keeping an open mind. In the course of growing up, each of us has acquired some strong feelings: We have learned to think of some types of conduct as acceptable, and we have learned to reject others. Occasionally, we may find those feelings challenged. For example, we may have been taught that homosexuality is immoral, and we may feel quite uncomfortable around gay people and see them as alien and perverted. But then someone suggests that this may be a mere prejudice; that there is nothing evil about homosexuality; that gay people are just people, like anyone else, who happen, through no choice of their own, to be attracted to members of the same sex. Because we feel so strongly about the matter, we may find it hard to take this line of reasoning seriously. Even after we listen to the arguments, we may still have the unshakable feeling that homosexuals must be unsavory.

Cultural Relativism provides an antidote for this kind of dogmatism. When he tells the story of the Greeks and Callatians, Herodotus adds:

> For if anyone, no matter who, were given the opportunity of choosing from amongst all the nations of the world the set of beliefs which he thought best, he would inevitably, after careful consideration of their relative merits, choose that of his own country. Everyone without exception believes his own native customs, and the religion he was brought up in, to be the best.

Realizing this can help us broaden our minds. We can see that our feelings are not necessarily perceptions of the truth—they may be nothing more than the result of cultural conditioning. Thus, when we hear it suggested that some element of our social code is *not* really the best, and we find ourselves instinctively resisting the suggestion, we might stop and remember this. Then we will be more open to discovering the truth, whatever it might be.

We can understand the appeal of Cultural Relativism, then, even though the theory has serious shortcomings. It is an attractive theory because it is based on a genuine insight: that many of the practices and attitudes we think so natural are really only cultural products. Moreover, keeping this thought firmly in view is important if we want to avoid arrogance and keep an open mind. These are important points, not to be taken lightly. But we can accept them without accepting the whole theory.

Back to the Five Claims

Let us now return to the five tenets of Cultural Relativism that were listed earlier. How did they fare in our discussion?

1. Different societies have different moral codes.

This is certainly true, although there are some values that all cultures share, such as the value of truth telling, the importance of caring for the young, and the prohibition on murder. Also, when customs differ, the underlying reason will often have more to do with the factual beliefs of the cultures than with their values.

2. The moral code of a society determines what is right within that society; that is, if the moral code of a society says that a certain action is right, then that action is right, at least within that society.

Here we must bear in mind the difference between what a society *believes* about morals and what is *really true*. The moral code of a society has a lot to do with what people in that society believe to be right. However, that code, and those people, can be in error. Earlier, we considered the example of excision—a barbaric practice endorsed by many societies. Consider two more examples. In 2005, a young woman from Australia was convicted of trying to smuggle nine pounds of marijuana into Indonesia. For that crime, she was sentenced to 20 years in prison. Under Indonesian law, she might have received death. In 2002, an unwed mother in Nigeria was sentenced to be stoned to death for having had sex out of wedlock. It is unclear whether Nigerian values, on the whole, approve of this verdict, since it was later overturned. But when the verdict was read, everyone in the courtroom shouted their approval.

Cultural Relativism holds, in effect, that societies are morally infallible—in other words, that the morals of a culture can never be wrong. But when we see that societies can and do endorse grave injustices, we see that societies, like their members, can be in need of moral improvement.

3. There is no objective standard that can be used to judge one society's code as better than another's. In other words, there is no "universal truth" in ethics; there are no moral truths that hold for all people at all times.

It is difficult to think of ethical principles that hold for all people at all times. However, if we are to criticize the practice of slavery, or stoning, or genital mutilation, and if we do not retreat from the idea that such practices are really and truly wrong, we must appeal to principles that are not tethered to one society's peculiar outlook. Earlier I suggested one such principle: that it always matters whether a practice promotes or hinders the welfare of the people affected by it.

4. The moral code of our own society has no special status; it is but one among many.

It is true that the moral code of our society has no special status. After all, our society has no heavenly halo around its borders; our values do not have any special standing just because we happen to believe them. However, to say that the moral code of one's own society "is merely one among many" seems to deny the possibility that one moral code might be better or worse than some others. Whether the moral code of one's own society "is merely one among many" is, in fact, an open question. That code might be one of the best; it might be one of the worst.

5. It is mere arrogance for us to judge the conduct of other peoples. We should adopt an attitude of tolerance toward the practices of other cultures.

There is much truth in this, but the point is overstated. We are often arrogant when we criticize other cultures, and tolerance is generally a good thing. However, we shouldn't be tolerant of everything. Human societies have done terrible things, and it is a mark of progress when we can say that those things are in the past.

New Malaise:
Bioethics and Human Rights in the Global Era

Paul Farmer and Nicole Gastineau Campos

Paul Farmer is a medical anthropologist and physician. He is currently the Presley Professor of Medical Anthropology in the Department of Social Medicine at Harvard Medical School. He is also the founding director of Partners In Health, an international charity organization that provides health care services, research, and advocacy for those who are ill and living in poverty. Nicole Gastineau Campos was a research assistant at Partners In Health and a graduate student at the Harvard School of Public Health when this essay was written. In this essay the authors call attention to the fact that bioethics is an activity of industrialized nations and, as such, has concerned itself primarily with problems affecting the affluent, such as the problem of stopping excessive or futile treatments. They argue that too little attention is paid to global health inequities, such as the distribution of health care between rich and poor nations and the ethics of transnational research. They conclude that bioethics needs a new focus on human rights and the poor.

. . . How Far Has Bioethics Really Come?

From 1932 to 1972, the U.S. Public Health Service conducted the infamous Tuskegee Syphilis Study in Alabama. The researchers recorded the natural history of syphilis in an attempt to learn more about the disease by following six hundred men, of whom about four hundred had syphilis, throughout their lifetimes. All were African American, many were sharecroppers, and most lived in poverty. Despite the discovery of a cure for the disease in 1947—to this day, syphilis is treated with penicillin—subjects were never offered that very inexpensive drug, even though they had joined the study assuming that they would be treated. Nor were they informed of the study's real purpose.[1]

The Tuskegee experiment ended in 1972—amid public outrage—when the *Atlanta Constitution* and the *New York Times* ran front-page stories on the study. In a critical reassessment, historian Allan Brandt notes, "The entire study had been predicated on nontreatment. Provision of effective medication would have violated the rationale of the experiment—to study the natural course of the disease until death."[2] It took the U.S. government decades to acknowledge its wrongdoing; President Clinton's public apology came in 1997. Today, references

Excerpted from *The Journal of Law, Medicine, and Ethics* 32, Summer 2004: 243–251. Reproduced by permission of the American Society of Law, Medicine, and Ethics. All rights reserved.

to "Tuskegee" are shameful reminders of a not so distant past, when race and low socioeconomic status of human subjects made a host of ethical violations— concerning everything from informed consent to lack of treatment with standardized regimens—acceptable in the quest for scientific truth. Surely, institutional review boards would never allow such an unethical study to take place in our times.

Yet on March 29, 2000, Reuters reported the following story:

> A study of more than 15,000 people in Uganda that has raised ethical questions about AIDS research in poor countries concluded that the risk of spreading AIDS through heterosexual sex rose and fell with the amount of virus in the blood.
>
> The study, in Thursday's issue of the *New England Journal of Medicine*, also confirmed earlier research suggesting that circumcision guarded against the spread of H.I.V., the virus that causes AIDS.
>
> The research was controversial, not because of its conclusions, but because of its methodology. Unlike studies of H.I.V. in developed countries, the volunteers in the Uganda study were not offered treatment, nor did doctors inform the healthy spouse of an infected person that his or her partner harbored the virus.
>
> Instead, the team led by Dr. Thomas Quinn of the National Institute of Allergy and Infectious Diseases, tested the volunteers and tracked the spread of their illness.[3]

The randomized controlled trial conducted in Uganda between November 1994 and October 1998 examined the relationship between serum viral load, concurrent sexually transmitted diseases, and other known and putative HIV risk factors (for example, male circumcision and several socio-demographic and behavioral factors). The research team screened 15,127 individuals in a rural district, of whom 415 were identified as HIV-positive with an initially HIV-negative partner. The researchers then tracked these serodiscordant couples for thirty months, following the viral load of the infected partner and the rate of seroconversion among the previously uninfected partners. The study concludes that "viral load is the chief predictor of the risk of heterosexual transmission of HIV-1." Such a finding "raises the possibility that reductions in viral load brought about by the use of antiretroviral drugs could potentially reduce the rate of transmission." Quinn and colleagues called for *more research* "to develop and evaluate cost-effective methods, such as effective and inexpensive antiretroviral therapy or vaccines, for reducing viral load in HIV-infected persons."[4]

The Ugandan study has occasioned a good deal of comment. Some of it appeared in the same issue of the *New England Journal of Medicine:* "Tragically," noted a researcher from another U.S. university, "results such as these could be obtained only in places with a very high incidence and prevalence of the virus and few practical or affordable means of preventing transmission. . . . The challenge now is to use these results *to develop prevention strategies* that can benefit everyone, especially those who participated in the research."[5]

Develop *prevention* strategies. This sounds eminently reasonable at first blush. But were more research and the development of prevention strategies the only real challenges emerging from this and other studies? Prevention strategies had already failed those infected during the course of the Ugandan study; prevention strategies were hardly the "challenge" at hand for "those who participated in the research." Around the time the Ugandan study was published, our group, Partners In Health—a charity that provides medical services to the poor of rural Haiti, Lima, Chiapas, Siberia, and Roxbury, Massachusetts—held a conference in rural Haiti. The conference was attended mostly by women living in poverty, several of them also living with HIV. The women at this meeting raised a very different set of challenges. As one asked, "What about those of us who already have HIV? Are we merely to wait for death?" Another participant said simply, "Treatment is important for sick people."

. . . It is not our intention here to focus overmuch on this particular study. Indeed, as Quinn and colleagues were quick to point out,". . .[f]our institutional review boards in the United States and Uganda had approved the study and a data safety and monitoring board from the National Institutes of Health, composed of U.S. and Ugandan representatives, monitored their work. At no time was it recommended that the researchers provide antiretroviral treatment to the participants."[6] What we are suggesting, however, is that ethical codes and review boards are not always helpful. . . . They set their sights on compliance with the law, not pragmatic ethical engagement with the interests of the study subjects. By their actions, they may endorse an unacknowledged consensus that in fact all humans are not created equal and that this inequality accounts for both differential distribution of disease and differential standards of care, making it acceptable to perform research in some parts of the world that would be unethical in the United States.

In an editorial that appeared in the *Lancet* in 1997, in response to the AIDS clinical trials being conducted in developing countries—in this case, studies involving what many argued were unethical placebo controls in AZT trials attempting to develop a cheaper drug regimen to prevent mother-child transmission of HIV—the editorial's author blasted ethicists for ignoring inequalities: "the ethics industry can be oddly parochial, paying little attention to events in the developing world that would demand vigorous pronouncements had they happened on its own doorstep."[7] The *Lancet* editors went so far as to suggest that this lack of attention is deliberate rather than careless: "Did ethicists know nothing about these trials . . . or was the fate of impoverished Africans thought not worthy of ethical consideration?"[8]

Indeed, as far as the poor and marginalized are concerned, bioethics has failed on numerous occasions, be it from lack of influence or failure to consider social justice. In this paper, we call for bioethics to adopt a new focus. Given that we are living in a global world order, global health equity, more than ever before, must be a goal of any serious ethical charter. We believe bioethics should be guided by equity. This process can begin only when full social complexity is restored to the ethical problems we discuss here. We also address some specific bioethical issues concerning transnational research and human rights. We discuss the overlay of social justice onto bioethics. Ultimately, we aim to show that bioethics must deal with difficult questions regarding social and economic rights in order to retain its authority to judge medical and technological developments. With a new focus on human rights, bioethics can engage other disciplines to ensure that social and economic rights for the poor are realized.

Embedding Bioethics in a Social Context

Despite claims of objectivity, science and medicine are both fundamentally social in nature. The research problems and questions we choose to pursue, the methods we follow, and the practical applications that result from scientific discovery are all determined by society's priorities and interests. *On Being a Scientist*, a joint publication of the National Academy of Science, the National Academy of Engineering, and the Institute of Medicine, put it this way:

> Scientific knowledge emerges from a process that is intensely human, a process indelibly shaped by human virtues, values and limitations and by social contexts . . . Science is a social enterprise . . . and takes place within a broad social and historical context, which gives substance, direction, and ultimately meaning to the work of individual scientists . . . Individual knowledge only becomes general knowledge after discussion, collaboration, peer review, judgment and incorporation into already accepted knowledge—an ongoing process of review and revision that minimizes individual subjectivity.[9]

As implied here, the choices we as a society make regarding scientific research also reflect our ethical priorities—our "virtues, values and limitations." Surely it is a problem when, according to Benatar, Daar, and Singer, "The interests of powerful nations, those who fund research and perhaps even the interests of many researchers often outweigh the interests of research subjects or society as a whole." They go on to note that "the value placed on acquiring new knowledge exceeds that placed on how best to apply existing knowledge."[10]

Why is it so important to embed bioethics in a social context? Benatar, Daar and Singer suggest that "The emphasis on military research and the neglect of diseases that afflict billions of people living in abject misery reflects a value system that marginalizes and devalues with impunity the lives of more than half of the world's population."[11] Michael Selgelid worries that "a situation analogous to the 10/90 divide in medical research [a phenomenon by which the diseases accounting for 90 percent of the global burden of disease receive less than 10 percent of research funds] holds true for research in bioethics. A quick flip through most bioethics texts and journals . . . reveals attention on abortion, euthanasia, assisted reproduction, genetics, and doctor-patient relationships." Selgelid notes that "Greatly lacking, in comparison, is discussion of ethical issues involving infectious disease and the (related) health care situation in the developing world."[12] Just as science does not always focus on the most serious and widespread problems, neither, it seems, does ethics.

In fact, conventional bioethics has difficulty addressing broad issues of inequity for at least four reasons. First, ethics draws strength from experience-distant disciplines such as philosophy, lending ethics debates a curious, at times almost silly, tenor, as Larry Churchill has noted:

> Bioethical disputes—as measured by the debates in journals and conferences in the United States—often seem to be remote from the values of ordinary people and largely irrelevant to the decisions they encounter in health care. In this sense, philosophical theorizing might be considered harmless entertainment, which if taken too seriously would look ridiculous, as several Monty Python skits have successfully demonstrated.[13]

Owing perhaps to the emphasis on fine distinctions and thought experiments, there have been few attempts to ground bioethics in political economy, history, anthropology, sociology, and the other contextualizing disciplines (although each of these would have no doubt lent its own native silliness).

Second, bioethics has been to a large extent a phenomenon of industrialized nations. This has facilitated the process of erasing discussions about the poor, since most of them live elsewhere. Thus, the great majority of the world's ethical dilemmas—and, in our opinion, the most serious ones—are not discussed at all by the very discipline claiming expertise in such matters.

The third reason that medical ethics and bioethics have been mum on vast inequalities is that experts have dominated public discourse on these matters, drowning out the voices of those who have far more direct experience of the problems bioethics addresses. To again cite Churchill:

> Ethics, understood as the capacity to think critically about moral values and direct our actions in terms of such values, is a generic human capacity. Except for sociopaths, it is common to all of us, and skill in ethics does not lend itself easily to encapsulation in theoretical categories, core competencies, or a professional specialty.[14]

A fourth reason behind the silence of bioethics on such issues is in part an unavoidable but geographically and economically circumscribed one. In the hospital we are asked to address the "quandary ethics" of individual patients. In the affluent countries where medical and bioethics have blossomed, these have often been elderly patients for whom further care may be deemed futile, even though the machine of "care" grinds on, leaving family and providers feeling a bit ground up themselves; other "high tech" tertiary care-driven issues abound. These are not, to say the least, the urgent medical ethics questions in resource-poor settings.

We are not suggesting that the ethical focus on the individual patient should change; what should change, rather, is that millions are still denied the chance to become patients and to have bioethics' "individual focus" trained on them. Beyond the administrative borders erected around catchment areas or states or nations, millions die—not from too much care or inappropriate care but rather from no care at all. One gets the sense, in attending ethics rounds and reading the now-copious ethics literature, that these have-nots are an embarrassment to the ethicists, for the problems of poverty and racism and a lack of national health insurance figure far too rarely in a literature dominated by seemingly endless discussions of brain death, organ transplantation, xenotransplantation, cloning, and care at the end of life. When the end of life comes early—from death in childbirth, say, or from tuberculosis or infantile diarrhea—the scandal is immeasurably greater, but these tragedies meet with far too little discussion in the medical and bioethics literature.

In an era of increased communication, this selective attention is easily recognized for the absurdity it is. Many among the world's poor already seem to have noticed that ethicists are capable of endlessly rehashing the perils of too much care, while each year millions die what the patients who visit our clinic in Haiti call "stupid deaths." Inattention may be expedient, certainly, but ignored populations are less easy to silence now. Communications are different—and by and large better—in the global era; oppressed populations have the means to talk back. The sheer burden of unnecessary suffering and premature death, and the current trend toward even further entrenchment of social inequality, give the oppressed more and more urgent reasons to address us.

Thus even without recourse to ethical reasoning we find the world revealed to us as it really is: a place in which the majority of serious bioethics issues go unremarked by experts in this field.

Transnational Epidemics, Transnational Research, and Human Rights in the Global Era

As humans, we are all vulnerable to sickness; but some groups are far more vulnerable than others, as every serious epidemiological study has shown. For the poor in affluent countries, it is possible to document the impact of inferior health care services. For example, in a large study examining the relation between poverty and medical treatment of acute myocardial infarction in the United States, Rathore and colleagues found that black, female, and poor patients received inferior care during hospitalization and were not consistently offered even inexpensive therapies upon discharge.[15] Given the close association between race and class in the United States, it is also relevant to mention the growing body of evidence revealing significant racial and ethnic disparities in health outcomes, in type and quality of health care that African Americans and members of other minorities receive—even after certain access-related factors such as education, income, and insurance status are controlled.[16]

In many resource-poor countries, it is often possible to document a complete absence of modern medical care.[17] The 2002 Human Development Indicators illuminate the problem.[18] Take Chad, for example, which has only three physicians for every 100,000 people; the same is true for Burkina Faso and Eritrea (data collected from 1990 to 1999). Seventeen other African countries that reported figures (Angola, Benin, Cameroon, Central African Republic, Côte d'Ivoire, The Democratic Republic of Congo, The Gambia, Ghana, Lesotho, Mali, Niger, Senegal, Sierra Leone, Sudan, Tanzania, Togo, and Zambia), one

Asian country (Nepal), and Haiti have a physician-to-population ratio that is worse than 10:100,000, whereas the United States has 279 physicians for every 100,000 people. These ratios do not entirely vary as a function of a country's gross national product. Relatively poor Cuba, for example, has more physicians per capita population than any country in the world except Italy—and Cuba is also judged to have the most equitably financed health care system in Latin America.

A few decades ago, the impact of today's injustice would have been significant, but not invariably a matter of life and death. That is, people lived and died, many of them unjustly, but even the well-to-do lived in fear of microbes that could kill them, as Nancy Tomes reminds us in her book about infectious diseases at the turn of the nineteenth century.[19] And although the nonpoor always did better than the poor, pneumococcal pneumonia and tuberculosis came with a high case-fatality rate, regardless of social station.

The situation is very different now, in large part because medicine is realizing the benefits of being the "youngest science," as Lewis Thomas has written.[20] Using the basic sciences to develop new therapies and the scientific method to evaluate their efficacy reminds us that the fight over equal access to ineffective or marginally effective interventions is certainly no longer worth wasting time on. But biomedicine can at last offer the sick truly revolutionary new therapies. Antibiotics and vaccines can, for the fortunate few, virtually erase the risk of mortality from polio, tetanus, measles, pneumonia, staphylococcal and other bacterial infections, diarrheal disease, malaria, and tuberculosis. Even HIV disease has been rendered, for those with access to therapy, a readily treatable disease. In a word: we can do it.

The increased mobility of both people and information reminds us that epidemics of disease are transnational.[21] The recent SARS outbreaks add the reminder that capital—like information, misinformation, and fear—moves even more quickly than do airborne viruses. SARS emerged at a line of demarcation between a low-income area with abruptly privatizing healthcare (China) and a high-income area with reasonably good public health and hospitals (Hong Kong). The infection was considered a crisis issue because it affected areas of high economic traffic, whereas AIDS and tuberculosis are seen to threaten populations that are not "connected" or rich enough to matter. There are questions of scale, both in terms of burden of disease and of responses to it. Everywhere, in the papers and on the internet, were images of commuters in Asia wearing surgical masks and empty airplanes and marketplaces. These images speak volumes, for in parts of the world where tuberculosis and AIDS and malaria reap their grim harvest of 6 million lives a year, there are not many trains, airplanes, or even masks. Many of these 6 million deaths would be preventable—if only the fruits of research were available to those who need them most. Bioethics needs to concern itself with the fact that research resources were moved from AIDS in order to focus on SARS.[22] Bioethicists need to ask hard questions about why victims of HIV and tuberculosis who live in developing countries often face certain death when affluent countries possess effective therapies for their ailments that could he made available on terms that meet reasonable tests of economics. In short, since we can, we should.

Research universities and development agencies now also have global reach, and, just as epidemics are transnational, so too, increasingly, is research. But although pathogens readily cross borders, the fruits of research are often delayed. For example, it seems to be easy enough to use First World diagnostics—in the Ugandan study discussed above, sophisticated assays of viral load were available—but First World therapies are often deemed infeasible, too difficult, or "cost-ineffective." In other words, though we can, and should, do right, we choose not to. This is the core ethical problem.

What does bioethics have to say about such transnational research? The short answer: not enough, at least so far. This in spite of the demands expressed in the international Code of Medical Ethics, first drafted in Geneva in 1949 in the wake of the Nuremberg trials, that physicians not only place the well-being of research subjects above the supposed benefits to science and society but also that they declare, "I will not permit considerations of religion, nationality, race, party politics or social standing to intervene between my duty and my patient."[23] But is it not precisely "social standing" and "nationality," not to mention "race," that place Ugandans at risk for becoming AIDS research subjects *and* for receiving substandard medical care? By substandard, we mean lower than the care that the researchers would expect for themselves in the event that they were to contract HIV.

A 1997 article by Peter Lurie and Sidney Wolfe[24] triggered what have become increasingly vocal attacks on AIDS clinical trials being conducted in developing countries. The particular studies Lurie and Wolfe were concerned with used placebo controls in AZT trials conducted in developing countries, despite the scientific knowledge that a better treatment standard existed at the time. The study protocol was thus argued to be in violation of several documents on research ethics and treatment of human subjects, including the Declaration of Helsinki and the CIOMS *International Ethical Guidelines for Biomedical Research Involving Human Subjects*. Those who defend the use of a placebo group in these trials, including many researchers and a number of bioethicists, cite "local standards of care," suggesting that the placebo controls were no worse off because antiretrovirals were not available in their environment under normal circumstances. Udo Schüklenk cast this argument in a different light:

> In the real world there is no such thing as a fixed local standard of care. Rather, the local standard of care in, for example, India, is a standard of care determined by the prices set by Western pharmaceutical multinationals. The only reason why the [AZT placebo] trials took place at all is the pricing schedule set by the manufacturer of the drug. Glaxo-Wellcome therefore, more than anything else, determines what is described by Bioethicists and clinical researchers as the "local standard of care."[25]

The devastation wrought by HIV in sub-Saharan Africa—AIDS is now far and away the leading cause of adult death across the continent and has already orphaned fourteen million children there[26]—has brought the local standard-of-care argument to the forefront of medical and public debate in the past few years.

On one side of the debate, ethicist David Resnik defends the local-standard-of-care argument by differentiating it from Tuskegee: "In the Tuskegee study, subjects were denied an effective treatment for syphilis, i.e. penicillin, even after it became widely *available* to the general population in the U.S. An effective treatment for preventing the perinatal transmission of HIV is available to the general population in developed nations but is *not available* to the general population in developing countries." Resnik concludes that "Denying treatment when treatment is not available to the general population (but may be available in some populations) . . . is morally acceptable."[27] His further reasoning suggests that

> there should be some reasonable limits on beneficence in research or in therapy. If there were no limits on the demands of beneficence, then it would follow that physicians and physician/researchers would be required to provide all medical care that could benefit patients or research subjects . . . A more philosophical way of stating this point is that there is a difference between doing what is ethically required and doing what is ethically supererogatory.[28]

Most disturbing, however, is his acknowledgment of different standards for the advantaged and disadvantaged: "Since different nations have different social, economic and political conditions, the demands of beneficence may vary from one nation to another."[29] In spite of his acceptance of different standards for research in the North and the South, Resnik remarks that "The potential connection between ethics and economics is very disturbing: clinical trials cost less where ethical standards are 'lower.'"[30]

Those who defend different ethical standards for rich and poor may be content with managing inequalities, no matter who is responsible for creating and maintaining them. And in Resnik's astute comment that trials cost less "where ethical standards are 'lower'" is the crux of the problem. In response to Resnik, Zulueta asserts that "the abrogation of the duty of care, particularly towards vulnerable individuals and their future offspring, may set dangerous precedents and erode the barriers to exploitation."[31] Ethics debates regarding the standard of care offered in clinical trials are likely to arise anew as the result of the HIV vaccine trials currently taking place. Will subjects be offered treatment if they acquire HIV for whatever reason, or will researchers be able to hide behind the claim that they do not have the "clinical capacity to manage antiretroviral treatment?"[32] If an effective vaccine is found, will those in the developing world be able to receive it quickly and at minimal or no expense? Or will we perpetuate the outcome gap between patients in rich countries and non-patients in poor ones? Let us not forget that the success of the new treatments is directly predicated on clinical trials, increasingly carried out in "low-cost" environments.

It is no exaggeration to say that the majority of international biomedical research has inequality as its foundation. As Marcia Angell has argued:

Research in the Third World looks relatively attractive as it becomes better funded and regulations at home become more restrictive. Despite the existence of codes requiring that human subjects receive at least the same protection abroad as at home, they are still honored partly in the breach. The fact remains that many studies are done in the Third World that simply could not be done in the countries sponsoring the work. Clinical trials have become a big business, with many of the same imperatives. To survive, it is necessary to get the work done as quickly as possible, with a minimum of obstacles. When these considerations prevail, it seems as if we have not come very far from Tuskegee after all.[33]

If we are to behave ethically abroad as well as at home, we must maintain a new focus . . . we need a human rights approach that concerns itself *especially* with the poor. . . .

A New Bioethical Focus on Equity and Human Rights

How can bioethics begin to address these most difficult and important questions concerning human rights? Past failures do not preclude future success. Benatar argues that

> In defense of the Human Rights approach, as the single most powerful means of promoting human well-being, it can be argued that failure to achieve human rights more widely is not the result of an inadequate concept of human rights, but rather that the full potential of the human rights approach has not been achieved because of simplistic or insincere use of the term, and a lack of commitment by powerful nations to what a more wholesome concept of human rights means and implies for them as well as for others.[34]

By advocating sincerely for human rights, bioethicists can encourage those in other disciplines to apply principles of equity and social justice. To that end, we pose questions that we hope will elicit thought and commentary from those who wish to refocus the field of bioethics.

If access to health care is considered a human right, who is considered "human enough" to have that right?

Looking back over the concept of human rights, we can see that social inequalities have always been used to deny some people status as fully human. When the French promoted the "rights of the citizen," they certainly did not—and do not, for that matter—confer citizenship lightly. Thus human rights were, from the beginning, quite distinct from the rights of all humans *qua human*. And so it has continued, with the poor, with women, and with black people, for those of low caste, for people with disabilities, for children, or "aliens" from other nations, etc., denied the full complement of human rights. . .

Can bioethics help us develop a broader view of who gets sick and why? Of who has access to care and why?

. . . [W]e are suggesting a new mindfulness—as a moral and analytic stance—of the strikingly deterministic pathways to both sickness and care.

Medical ethicists and physicians might reply that this is an exercise best left to epidemiologists and to those who study health care systems. But we should not only pass this task on to other parties. First, we cannot always trust others to respect the rights of individual patients. Second, practitioners of many disciplines related to medicine have proven incapable of understanding the biosocial complexity that defines unequal health outcomes and health and human rights. Although disciplinary specialization has yielded great insights, we have tried to emphasize the cost of desocializing the concept of rights and the costs of desocializing ethics. Whether we consider Russian prisoners with drug-resistant tuberculosis, Haitian women who contract HIV when they go to the city to seek work, or Bostonians with AIDS living on the streets, the story is the same: a failure to understand social process leads to analytic failures, with significant implications for policy and practice.

How does the struggle for social and economic rights relate to, for example, a "Patients' Bill of Rights"?

Most charters of patients' rights ignore the sick *nonpatients* who never get into the exclusive club of those who actually receive modern health care. Patients rights are conceptualized on the pattern of consumer rights, not human rights. As stated above, the "bottom billion" who have no access at all to modern medicine, are neglected by bioethicists. Thus, in discussions of ethics, global health equity has become the elephant in the room that nobody seems to see.

But whether we are talking about uninsured U.S. citizens, African research subjects, or prisoners with drug-resistant tuberculosis, charters proclaiming health care as a right take on their full power only when we add a clause about the intention to include everybody. This ideal could be dismissed as pie-in-the-sky, but it seems better to set goals high than to sink to a pragmatism that leads inevitably to "ethical dilemmas" that apply to so few around the world.

What do the destitute sick have to say about medical ethics?

. . . Since the poor are those put at risk of sickness and then denied access to care, they are in many ways those most affected, though least represented, by bioethics codes. Within and across national boundaries, the destitute sick should be the primary judges of any code of medical ethics. Applying a "perfect" ethical code in one country alone is an impossibility in the global era. Again, we are led back to global health equity as a necessary component of any discussion of medical ethics and bioethics.

Human Rights in Bioethics—For Everybody

Perhaps the greatest challenge for bioethics is to *resocialize the way we see ethical dilemmas in medicine*. Restoring such problems to their full social complexity is our best vaccine against ignoring large segments of the human race.

... The true value of the human rights movement's central documents is revealed only when they serve to protect the rights of those who are most likely to have their rights violated. The proper beneficiaries of the Universal Declaration of Human Rights—however inexpedient this point might be in our age of individualism and affluence and relativism—are the poor and disempowered. Since the burden of disease is borne by the poor and the marginalized, we are offered a chance to contemplate the lot of most of humanity and to ask, simply enough, if by "everybody" we truly mean everybody. In the United States, the context of many ethical discussions is that we must do our best to manage our vast prosperity. But if Haiti and Uganda and Russia and Thailand are also part of our world, let us include them in the benefits, not just the fallouts and placebos, of our scientific and economic power. . . .

References

1. A. M. Brandt, "Racism and Research: The Case of the Tuskegee Syphilis Study," *Hastings Center Report 8*, no. 6 (1978): 21–29; D. Rothman, "Were Tuskegee and Willowbrook 'Studies in Nature'?" *Hastings Center Report 12*, no. 2 (1984): 5–7.
2. See Brandt, supra note 1, at 27.
3. "Criticized Research Quantifies the Risk of AIDS Infection," *New York Times*, March 30, 2000, at A16.
4. T. C. Quinn, M. J. Wawer, N. Sewankambo, et al. "Viral Load and Heterosexual Transmission of Human Immunodeficiency Virus Type 1," *New England Journal of Medicine* 342 (2000): 921–929, at 921, 927, 928.
5. M. S. Cohen, "Preventing Sexual Transmission of HIV—New Ideas from Sub-Saharan Africa," *New England Journal of Medicine* 342 (2000): 970–972, at 972 (emphasis added).
6. *See* Quinn, Wawer, Sewankambo et al., supra note 4.
7. Anonymous, "The Ethics Industry," *The Lancet* 350 (1997): 897.
8. *Id.*
9. National Academy of Science, National Academy of Engineering, Institute of Medicine, *On Being a Scientist* 2nd edition (Washington D.C.: National Academy Press, 1994).
10. S. R. Benatar, A. S. Daar, and P. A. Singer, "Global Health Ethics: The Rationale for Mutual Caring," *International Affairs* 79 (2003): 107–138, at 110.
11. *Id.*
12. M. Selgelid, "Ethics and Infectious Disease," *Bioethics* 19 (2005): 272–289.
13. L. R. Churchill, "Are We Professionals? A Critical Look at the Social Role of Bioethicists," *Daedalus* 128 (1999): 253–274, at 255.
14. *Id.*, at 259.
15. S. S. Rathore, A. K. Berger, K. P. Weinfurt, et al., "Race, Sex, Poverty, and the Medical Treatment of Acute Myocardial Infarction in the Elderly," *Circulation: Journal of the American Heart Association* 102 (2000): 642–648.

16. W. M. Byrd and L. A. Clayton, "An American Health Dilemma," Vol. 2, *Race Medicine, and Health Care in the United States, 1900–2000* (New York: Routledge, 2002); K. Fiscella, P. Franks, M. R. Gold, et al., "Inequality in Quality: Addressing Socioeconomic, Racial, and Ethnic Disparities in Health Care," *Journal of the American Medical Association* 283 (2000): 2579–2584; H. P. Freeman and R. Payne, "Racial Injustice in Health Care," *New England Journal of Medicine* 342 (1991): 1045–1047; E. C. Schneider, A. M. Zaslavsky, and A. M. Epstein, "Racial Disparities in the Quality of Care for Enrollees in Medicare Managed Care," *Journal of the American Medical Association* 287 (2002): 1288–1294; Institute of Medicine, *Care Without Coverage: Too Little, Too Late* (Washington, D.C.: National Academy Press, 2002); Institute of Medicine, *Unequal Treatment: Confronting Racial and Ethnic Disparities in Health Care* (Washington, D.C.: National Academy Press, 2002).

17. World Bank, *World Development Indicators*, Table 2.15 (Washington, D.C.: International Bank, 2001), at 98–100.

18. United Nations Development Programme, *Human Development Indicators* (New York, Oxford: Oxford University Press, 2002).

19. N. Tomes, *The Gospel of Germs: Men, Women, and the Microbe in American Life* (Cambridge, Mass.: Harvard University Press, 1998).

20. L. Thomas, *The Youngest Science. Notes of a Medicine-Watcher* (New York: Viking Press, 1983).

21. P. Farmer, "Social Inequalities and Emerging Infectious Diseases," *Emerging Infectious Diseases* 2 (1996): 259–269; P. Farmer, *Infections and Inequalities: The Modern Plagues* (Berkeley: University of California Press, 1999); P. Farmer, J. Bayona, M. Becerra, et al., "The Dilemma of MDRTB in the Global Era," *International Journal of Tuberculosis and Lung Disease* 2 (1998): 869–876; L. Garrett, *Betrayal of Trust: The Collapse of Global Public Health* (New York: Hyperion, 2000).

22. A. Regaldo and M. Schoofs, "AIDS Researcher HO, Others Set to Begin to Tackle SARS," *Wall Street Journal*, April 24, 2003, at A3.

23. World Medical Association, *World Medical Association International Code of Medical Ethics* amended by the Twenty-Second World Medical Assembly, Sydney, Australia, August 1968; and the Thirty-Fifth World Medical Assembly, Venice, Italy, October 1983; available at <http://www.wma.net/e/policy/b3.htm> (last visited May 11, 2004). On the impact of Nuremberg on Medical Ethics, sec R. J. Lifton, *The Nazi Doctors: Medical Killing and the Psychology of Genocide* (New York: Basic Books, 1986).

24. P. Lurie and S. M. Wolfe, "Unethical Trials of Interventions to Reduce Perinatal Transmission of the Human Immunodeficiency Virus in Developing Countries," *New England Journal of Medicine* 337 (1997): 853–856.

25. U. Schüklenk, "Protecting the Vulnerable: Testing Times for Clinical Research Ethics," *Social Science and Medicine* 51 (2000): 969–977, at 973.

26. Joint United Nations Programme on HIV/AIDS (UNAIDS), *Report on the Global HIV/AIDS Epidemic 2002* (Geneva: UNAIDS, 2002), at 8.

27. D. Resnik, "The Ethics of HIV Research in Developing Nations," *Bioethics* 12 (1998): 286–306, at 302.
28. *Id.*, at 303.
29. *Id.*, at 303.
30. *Id.*, at 306.
31. P. Zulueta, "Randomised Placebo-Controlled Trials and HIV-Infected Pregnant Women in Developing Countries. Ethical Imperialism or Unethical Exploitation?" *Bioethics* 15 (2001): 289–311, at 302.
32. R. H. Gray, T. C. Quinn, D. Serwadda et al., "The Ethics of Research in Developing Countries," *New England Journal of Medicine* 343 (2000): 361–362, at 361.
33. M. Angell, "The Ethics of Clinical Research in the Third World," *New England Journal of Medicine* 337 (1997): 847–849, at 849.
34. S. Benatar, "Human Rights in the Biotechnology Era," *BMC International Health and Human Rights* 2 (2002): 1–9, at 5, available at <http://www.biomedcentral.com/1472-698X/2/3>.

PART III

The Practice of Bioethics

Introduction to the Practice of Bioethics

Robert A. Pearlman

D uring the latter half of the last century, health care professionals and society faced ethical questions about the responsible use of new technologies such as in vitro fertilization, genetic testing, and the withholding and withdrawing of medical treatment that provided marginal or no meaningful benefit. New technologies for the beginning of life offered great hope to women previously unable to bear children and to parents wanting to minimize the risk of passing severe hereditary diseases on to their offspring. However, these technologies were often implemented before adequate testing or were offered without full informed consent. At the end of life, new technologies often offered marginal benefits accompanied by significant burdens, forcing patients and health care providers to struggle with ethical choices. Three major responses characterized society's efforts to apply ethical theory and principles to these new challenges presenting in the health care setting. These responses were the formation of ethics committees, the emergence of ethics consultants, and the formulation of ethics-related policies and guidelines.

The Historical Development of Ethics Advisory Committees

During the early 1960s the development of hemodialysis permitted treatment of persons with chronic kidney failure. To Belding Scribner, the technology's principal inventor, several ethical concerns arose relating to patient selection, termination of treatment by the physician or the patient, and death with dignity.[1] The Seattle Artificial Kidney Center's Admission and Policy Committee was formed to develop guidelines for screening and selecting dialysis candidates, anticipated to be more numerous than available dialysis machines. Membership on the committee was intended to reflect broad representation from the community. Despite these intentions, the majority of the members reflected white, upper-middle-class demographics. The criteria that emerged in the early, challenging, and exhausting deliberations about selection of dialysis candidates reportedly included factors such as sex, marital status, education, income, occupation, and past performance and future potential. These "social worth" criteria were criticized as prejudicial after publication and review of the workings of the committee.[2] One criticism stated, "The Pacific Northwest is no place for a Henry David Thoreau with bad kidneys."[3] This early committee experience highlighted two important lessons. First, diversity in committee membership is a safeguard to avoid the introduction of bias against classes of people. Second, public review of policies and procedures may help identify unexpected problems.

During the 1970s, ethics committees with a very different function were formed, owing partly to the decision of the New Jersey Supreme Court in the case of Karen Ann Quinlan. In *Quinlan*, the court stated that if the hospital ethics committee agreed "that there is no reasonable possibility of Karen's ever emerging from her present comatose condition to a cognitive, sapient state," the request of her parents, guardians, and attending physicians to remove life-sustaining treatment could be acted upon without fear of civil or criminal liability (see "In the Matter of Karen Quinlan," Part I, Section 1). Thus, early ethics committees functioned as prognosis committees, helping to ensure that patients in whom life-sustaining treatment was withheld or withdrawn qualified by virtue of being in a persistent vegetative state.

By the 1980s, the role of ethics committees had expanded. Court cases pertaining to decision making for severely brain-damaged patients (including newborns) led to federal regulations to prevent discrimination on the basis of handicap and a recommendation by the President's Commission for the Study of Ethical Problems in Medicine and Biomedical and Behavioral Research.[4-7] In *Deciding to Forego Life-Sustaining Treatment*, the Commission suggested institutional review such as the ethics committee as a mechanism to protect the interests of patients who lack decision-making capacity and to ensure their well-being and self-determination. This recommendation was grounded in the premise that the possibility of errors in the process of decision making for mentally incapacitated patients could be reduced with judicious use of a review process. The so-called Baby Doe regulations also strongly encouraged hospitals caring for newborn infants to establish infant care review committees to develop policies, monitor adherence through retrospective record review, and review cases on an emergency basis when withholding or withdrawing treatment was being considered.

Subsequent debate and policy initiatives extended the competent patient's right to accept or reject recommended treatment to all circumstances of decisional incapacity.[8] For example, family members of patients incapable of making decisions could speak for the previously competent patient and request the withholding or discontinuation of life-sustaining treatment, even if the patient was not in a persistent vegetative state.[9] This made the need for prognosis committees moot. The need for ethics committees, however, did not abate.[10]

Roles of Health Care Institution Ethics Committees

Ethics committees serve multiple purposes. In the broadest sense, they link societal values to medical practice. More specifically, they protect the rights and welfare of patients by promoting shared decision making. In 1983, the President's

Commission outlined four functions of ethics committees that remain prominent for many ethics committees today (President's Commission for the Study of Ethical Problems in Medicine and Biomedical Research, 1983):

- Review cases to confirm the responsible physician's diagnosis and prognosis of a patient's medical condition (which took on less importance as discussed above);
- Provide a forum for discussing broader social and ethics concerns raised by a particular case, and in so doing, teach professional staff how to identify, frame, and resolve ethics problems;
- Formulate policy and guidelines regarding such decisions; and
- Review decisions made by others about the treatment of specific patients, or make such decisions themselves.[7]

In responding to this charge, many ethics committees today are instrumental in developing educational programs for clinicians and patients (e.g., about patient rights and responsibilities), developing or reviewing ethics policies (e.g., decision making for patients who lack decision-making capacity, consent for HIV testing and counseling), and providing case consultations and retrospective case reviews. To accomplish these tasks, ethics committee members must develop and maintain sufficient expertise.

When ethics committees initially are formed, there may be debate about what type of training and education develops the necessary expertise and skills to conduct ethical analysis in the clinical setting. Committees may consider whether an ethicist is necessary to conduct consultations.[11] However, three lines of argument cast doubt on the need for and wisdom in recommending an ethics expert at every health care facility. First, physician educators worry about giving the message that ethical analysis and problem resolution are activities requiring outside expertise. As a result, physician trainees might not learn how to engage proficiently in these activities, and consequently not develop into morally responsible providers. Moreover, reliance on an expert might foster singular views about contentious issues and limit moral discourse. Second, the experience of committees demonstrates that although skills in ethical analysis and critical thinking are important, other skills are necessary. These other skills include adequate understanding of clinical medicine and decision making, as well as communication and interpersonal interaction (see La Puma and Schiedermayer, Part II, Section 2). Adequate knowledge of health care law, public policy, and cultural and religious traditions are also recommended as desirable attributes.[12,13] Third, few individuals have all of the requisite knowledge and skills to be an effective ethics consultant and, as a result, the approach to conducting ethics consultation usually involves more than one person. To ensure adequate knowledge, skills, and sensitivity to differences among people, the typical ethics committee and ethics consultation service are comprised of health care professionals from many disciplines, people with expertise in law and religious traditions, and representatives from the community that health care institutions serve.

In the last several years the National Center for Ethics in Health Care (Veterans Health Administration [VHA]) has developed and implemented the Integrated Ethics model.[14] The major differences in this three-pronged model of an ethics program, when compared to traditional ethics committees, are outlined below in the description of three core functions:

- Ethics consultation continues to be a principal component of an ethics program by responding to ethics questions in health care, but the focus is more expansive by addressing ethics questions throughout the organization—from the clinical case at the bedside to the decision or non-clinical policy that is developed in the boardroom;
- Preventive ethics, the second core function, aims to promote ethics quality by addressing systems-level factors that affect ethics practices through a quality improvement approach; and
- Ethical leadership, the third core function, aims to promote a positive ethics environment and culture.

This expanded and integrated model of an ethics program requires a structure that identifies and addresses ethical issues in the clinical and non-clinical settings and breaks down traditional "silos," provides a standardized and systematic approach to promoting ethical practices throughout health care organizations, promotes accountability, and uses assessment tools and data to foster quality improvement. The roles and responsibilities of participants in this model are available at the Integrated Ethics web site.[14]

Ethics Education

Ethics committee members (and especially those who function as ethics consultants) require sufficient education to educate other professionals about the ethical issues in health care, to influence behavior and improve health care ethics practices. Education also serves to prevent or minimize threats to optimal ethics committee functioning (see Lo, Part II, Section 2). Despite its advantages, education can be difficult to implement because the time and energy required to educate members of an ethics committee frequently takes longer and is greater than anticipated, for several reasons. First, the diversity of its members' backgrounds often translates into appreciable differences in baseline knowledge about ethics and clinical medicine; understandings of concepts invoked in ethical analysis; and cultural values. Second, committee members often are volunteers with an interest in ethics, but without protected time to participate in ethics committee activities. Third, some committee members are designated by their administrative superiors to participate, and they have both limited time available and the additional barrier of limited interest in ethics. The time and energy of educating an ethics committee is tantamount to teaching or having a class design a course for itself over a 1- to 3-year period. This challenge is more daunting when one considers turnover in committee membership and the need for education and training to be repeated periodically for new members.

Other challenges in educating ethics committee members include making resource materials accessible, addressing topics of interest, developing tools for rigorous analysis, promoting consistency across different cases, and identifying morally justifiable reasons for inconsistent recommendations. In addition, ethics committee members can benefit from education about barriers to effective inter-professional communication, respectful communication between professionals and patients and their families, the art of listening, and negotiation. Those who are responsible for ensuring the quality of ethics consultations need to identify and mitigate consultants' proficiency gaps.

An early educational task for most ethics committees is exploring the nature of ethics and ethical analysis. Thus, many ethics committee members learn key principles, such as beneficence, nonmaleficence, autonomy and justice, as well as ethical theories and methods of moral inquiry. It is equally important for committee members to become familiar with a systematic approach to case analysis. One common approach is the "four-box" method of identifying relevant case information.[15] With this approach, information is collected and organized into categories of medical indications, patient (or surrogate) preferences, quality of life considerations, and contextual features. Although this approach does not articulate a specific strategy for normative ethical analysis, it forces consistent consideration of certain topics in clinical case deliberations.

Another approach to ethical case analysis that some ethics committee members learn is the application of subjective expected utility theory from cognitive psychology.[16] This approach relies heavily on the diversity, integrity, and honesty of ethics committee members. The first step in this approach is to have the ethics committee clearly articulate the ethical question (e.g., "Should she treat?" or "Should she not treat?"). This an often difficult and worthwhile task that forces agreement about which question needs to be answered first. On each side of this question, supportive arguments, possibly reflecting principles, outcomes, values, and legal concerns, are identified. The next task in the analysis is for the committee members to argue and debate the merits of each, and ultimately to weigh these factors. By an inductive process, a morally acceptable course of action is identified. At this point in the process, recommendations can be reviewed for potential legal liability or institutional barriers.

VHA has developed a practical and systematic, step-by-step approach for performing an ethics consultation. It is called CASES, which stands for the five main steps (each main step has several secondary steps):

- **C**larify the consultation request (e.g., obtain preliminary information from the requestor, establish realistic expectations from the consultation process, formulate the ethics question);
- **A**ssemble the relevant information (e.g., identify the appropriate sources of information, gather the information systematically from each source, summarize the information and the ethics question);
- **S**ynthesize the information (e.g., engage in ethical analysis, identify the ethically appropriate decision maker, facilitate moral deliberation about ethically justifiable options);

- Explain the synthesis (e.g., communicate the synthesis to key participants, document the consultation in the health record and consultation service records);
- Support the consultation process (e.g., follow up with the participants, evaluate the consultation, identify underlying systems issues).[14]

After educating itself, the committee usually engages in or organizes educational activities for other members of the health care institution. The principal objective for most committees is to improve the quality of care for patients by augmenting the ability of clinicians to recognize, understand, and help resolve common and challenging ethics concerns in clinical practice. Ethics education also attempts to help health care professionals examine their own personal and professional moral commitments; equip themselves with sufficient philosophical, social, and legal knowledge to aid in clinical reasoning; and develop interactional skills to facilitate effective listening and communication with patients, families, and other professionals. Less frequently the goals of educational activities are to influence the organizational culture and to promote concepts of personal responsibility, professional integrity, comfort with uncertainty, an appreciation for the difference between authority and power, open-mindedness, and a nonintimidating milieu.

Educational activities often take the form of case-based ethics rounds, lectures, panel discussions, debates, role modeling, and development of pamphlets. Obviously, the format for educational activities and products varies depending on the target audience. Sometimes the audience is the clinical professional staff, and at other times it is patients and their family members. Recent guidelines from The Joint Commission promote mechanisms to educate patients about their rights to have advance directives and to know about relevant institutional and state policies.[17] Unfortunately, many of the educational materials that have been developed lack motivational appeal and primarily focus on the logistics of completing advance directives.

The development of ethics education in the clinical setting poses numerous challenges. First, the culture of learning in most medical centers is case-based or patient-centered. Thus, educational activities often need to be tailored to specific cases or the clinical interests and needs of the clinicians. This culture often inhibits discussion of ethical theory or the humanities, and yet such discussion may help provide a structure for thinking about new challenges. Second, effective teaching often occurs at the bedside. This requires having either consultants or clinicians with ethics expertise available during routine clinical teaching sessions. Third, many clinician-trainees consider ethics, humanistic medicine, and interactional skills less important than other aspects of clinical knowledge. Thus, attitudinal biases frequently interfere with clinician receptivity. Partial responses to this barrier include competent educators, education evaluations leading to refinement in teaching, and repeated role modeling and support by clinician mentors.

Case Consultation

Ethics case consultation usually occurs in one of three modalities: by an ethics committee, by a small subgroup of the committee working as an ethics consultation service, and by individual ethics consultants. The subgroup is the most common arrangement. Usually the members of the ethics consultation service respond to consult requests in a timely manner, and at a later time review and discuss their activities with the other members of the ethics committee (e.g., during the next scheduled meeting).

Clarification of the goals is an important priority before the initiation of consultations. In many health care institutions the goals are formulated to identify morally sound solutions to ethical problems that arise in the medical context and to help resolve ethical dilemmas. Some ethics committees and, by extension, ethics consultation services assume a more patient-centered role, such that the "interests of all parties, especially those of the incapacitated person, are adequately represented, and that the decision reached lies within the range of permissible alternatives."[7] Despite efforts to define clearly the goals of consultative work, many questions remain. Of paramount importance is whether cases are brought to review on a voluntary basis or whether reviews occur on all cases of a certain type (e.g., withholding artificial hydration and nutrition).

With the sharing of experience around the country, most U.S. hospitals have adopted the time-honored tradition of voluntary consultations as the most appropriate format and least antithetical to the medical culture (see Fox, Myers, and Pearlman, Part II, Section 2). More recently, many ethics policies specify the opportunity for patients and family members to request ethics consultations. Patient knowledge of this resource, however, remains elusive. Even fewer committees or consultation services as a matter of practice explicitly invite patients or their family members to participate in case discussions. Lack of patient involvement, however, procedurally undermines a common committee objective, the protection of patients.

The practical aspects of providing ethics consultations raise interesting questions: whether consultative recommendations should be advisory (optional) or mandatory, and whether the results of a consultation should be documented in the medical record or written and recorded elsewhere. In most institutions these questions have been resolved by having ethics consultation model clinical consultation services[18] (see also Fox, Myers, and Pearlman, Part II, Section 2). Thus, consultations usually are advisory and are written in the medical record.

It has been argued, however, that the one *special* task of ethics committees is to exercise authority to postpone medical decisions that the committee or the ethics consultation service counsels against, or to initiate judicial review of such decisions.[19] A minor challenge is the additional question of what to do about advisory recommendations when consensus is lacking after case deliberation. When a range of ethically acceptable decisions exists, recommendations delineating one course of action may appear to foreclose other options. To prevent this, consultation notes may present conflicting opinions and their rationale when several options emerge as ethically defensible. Moreover, as consultative advice is advisory and potentially at variance with professionals' understanding of power and authority, negotiation and facilitation techniques and skills are useful for ethics consultants (see West and Gibson, Part II, Section 2).

The ethical analysis that occurs in case consultations usually reflects both inductive and deductive reasoning. Although members may be well versed in distinct principles of biomedical ethics or casuistic analysis, the nature of committee membership (involving multiple disciplines) and the orientation of discussion (being case oriented) often foster analyses of the ethical problems that combine approaches. In these discussions common principles such as respect for persons and autonomy, beneficence and nonmaleficence, and fairness, are often considered. So too are the particularities of the case, experience with similar situations, the test of generalizability, the test of publicity (How would this read in tomorrow's newspaper?), legal constraints, and professional standards. When deliberations involve honest and explicit communication with critical questioning of assumptions, the end result is a rigorous process that represents the eclectic nature of contemporary ethical reasoning.

Besides the abilities to assess and analyze ethical issues and to negotiate and facilitate differences of opinion, what other skills are required to provide ethics consultation? One skill is the ability to listen and communicate effectively with patients, families, and other professionals. In addition, adequate knowledge of health care policies, institutions, systems, and law; relevant codes of ethics; and cultural and religious traditions are considered essential, according to the core competencies for ethics consultants described by the American Society for Bioethics and the Humanities.[13]

Hospital Policies

Health care institutions frequently involve ethics committees or programs to help develop policies pertaining to informed consent, use of new technologies, withholding and withdrawing life-sustaining treatment, medical record confidentiality, determination of decision-making incapacity, HIV testing and counseling, care of the dying, and advance care planning. Most committees quickly learn that subcommittee delegation fosters efficient progress. Many committees also learn that reviewing and critiquing other institutions' policies help prevent recreating the wheel.

Unfortunately, unforeseen challenges often await the implementation of policies. For example, even after "do-not-resuscitate" policies were implemented in hospitals, it became apparent that physicians frequently did not initiate communication about this until after the patient became incapacitated, and thus the conversation involved family members.[20] Similarly, hospitals responded to the Patient Self Determination Act's goal of educating patients about advance directives, but were forced to consider the work demands of hospital personnel. As a result, many hospital policies about advance directives became administrative tools to ensure procedural compliance with Joint Commission requirements. In another area, reproductive technologies, new developments forced repeated modifications in guidelines from the American Society of Reproductive Medicine and resultant hospital policies about the number of embryos that should be implanted with in vitro fertilization and the indications for use of egg donors. Empirical research identifying these and other challenges has enabled refinements in understanding of ethical problems and established the need for periodic review and modification of policies.

Proposed Ethics Standards From Non-Clinical Settings

Professional organizations, individual scholars, and state and federal initiatives also have contributed to the formulation of policies and guidelines relevant to the practice of bioethics. At the national level, the landmark *Roe v. Wade* Supreme Court decision in 1973 introduced a trimester framework for addressing the constitutionality of state abortion laws. The effect of this decision was to overturn 46 state laws governing abortion. More recently, the President's Council on Bioethics recommended to Congress several prohibitions pertaining to assisted reproduction, including prohibiting the creation ex vivo of any human embryo with the intent to transfer it to a woman's body to initiate a pregnancy.

In many circumstances the purposes of these initiatives have been to facilitate decision making for mentally incapacitated patients and to increase the likelihood that either previously expressed wishes were honored or, if prior wishes were unknown, that the best-interests standard for surrogate decision making and reasonableness prevails.[17,21-23] Policies, especially legislative ones, are blunt instruments of change, and therefore can lack sufficient specificity to provide guidance in situations with additional clinical complexity. Using policies about advance directives as an example, the Patient Self Determination Act (PSDA) was passed to promote awareness and respect for advance directives. The PSDA did not address the more important questions of how to talk to patients about their preferences for medical treatment in the face of decisional incapacity, and how to assess the level of a patient's decisional incapacity to know when to shift focus to an advance directive. The PSDA also did not address how to develop policies that are ethically and logistically feasible to facilitate decision making when a patient's wishes are not known, and how to adjudicate apparent conflicts between prior preferences and current best interests.

Recommendations from organizations or scholars occasionally present competing ideas that lead to better understanding of ethical issues. Unfortunately, different policies about the same issue in different settings may foster variability in practice based on the specifics of the accepted policy. Recently, this has been true for care near the end of life. Medical futility is a case in point. Some policies retain the language that physicians and nurses are not obligated to provide futile treatment.[24] Others identify value elements and thus require patient involvement to avoid bias.[25] Other policies with futility judgments regard the use of physiologically futile interventions as falling outside the standards of professional behavior.[26]

The Texas state legislature passed a statute that outlines an approach and criteria for handling clinical cases in which medical treatment is considered to be medically futile.[27] Although this statute does not apply outside Texas, as this and other policies are discussed and evaluated, the concept of medical futility becomes more refined in ethical deliberations. Another example of competing ideas advancing our understanding of ethical issues in care of the dying pertains to the role of quality of life in decision making. While many scholars have focused on the implications of physical, functional, and existential losses on quality of life, others have stressed the importance of maximizing existing capacities to enhance quality of life (see Pearlman et al. and Asch articles, Part III, Section 2).

Different policy proposals also can lead to frustration among patients and educators in bioethics. This exists with physician-assisted death. Competing policy recommendations from scholars and professional organizations about physician-assisted death have prevented consensus about decriminalizing or retaining the criminal status of this behavior. One group of scholars and clinicians has offered a policy that anchors its position to compassion and reduction of suffering, and then proceeds to offer suggestions to reduce the likelihood of abuse.[28] By contrast, other policies pertaining to physician-assisted death argue against physician involvement in this activity because of potential, future harms such as diminution of respect and trust in the profession or service organization, insidious pressure for older and disabled persons to avail themselves of assisted suicide, and inequitable access to this service. The relative inertia within the medical profession about changing the status of physician-assisted death has led to patient frustration and the resultant consumer initiatives to decriminalize physician-assisted death in the states of Washington and Oregon. Frustrations sometimes exist among educators in bioethics. An example is the American College of Physicians's policy about professional ethics. In one voice the College admonishes physicians to avoid conflicts of interest, but in another section discusses the physician's duty to render care *after* the relationship and its financial arrangements are secure.[22] This potential inconsistency undermines the educational value of such a policy when taken in its entirety.

Linking Ethics and Quality Assurance and Improvement

Like other activities in health care, ethics activities and products can be evaluated for their effectiveness.[29] With regard to ethics committee or program functions, empirical methods can be employed to characterize the scope and intensity of activities, assess effectiveness of educational programs, and identify which elements, approaches, or target audiences are associated with greater effectiveness. To date there is limited information about these relationships and their ramifications. Similar types of questions are relevant to evaluating policies. Do they ensure consistent, ethically acceptable behaviors, augment education, or result in the desired outcomes?

Ethics consultations are another activity that might benefit from a quality improvement approach. Currently, professional standards are lacking, quality control is questionable, and questions exist as to the value and effectiveness of ethics consultations. Some critics think that anyone can hang a shingle and call himself or herself an ethicist. In response to this gap in knowledge, the Agency for Health Care and Policy Research funded a conference in 1995 to promote a research agenda for quality assurance in ethics consultation.

If and when standards for ethics consultations develop, the standards should address the following potential concerns:

1. The consultation discusses a course of action that is at variance with an institution's policy and does not advise the consultee of the conflict with the institution's policy;
2. The consultation fails to consider a central element or component of the clinical case;
3. The consultation discusses a course of action that is against the law without informing the consultee of the potential liability;
4. The consultant breaches patient confidentiality;
5. The consultant fails to inform the consultee of new major insights or a change in advice obtained from the full committee's review; and
6. The consultant recommends a course of action that is discriminatory against a class of people by virtue of their age, gender, race, ethnicity, disability, or vulnerability.

These possible behaviors raise questions of irresponsibility. The challenge of quality assurance is to identify a way to reduce the likelihood of these events, promote quality service, and simultaneously not create a cumbersome monitoring system or meaningless activity that merely gives the appearance of promoting quality. In the absence of quality assurance efforts, the kinds of problems addressed by Bernard Lo (Part II, Section 2) can create obstacles to the functioning of ethics committees, programs, and consultants.

Recently two major health care organizations, the American Medical Association (AMA) and VHA, developed programs that promote ethics as a component of quality. In both of these programs ethics quality is assessed by comparing health care practices with performance measures. The AMA's Ethical Force program and VHA's Integrated Ethics initiative use assessment tools that focus on domains in ethics to identify opportunities for improving ethics quality. For example a facility might be interested in measuring how often health care providers ask patients whether health-related information can be shared with their family members before including them in discussions. And if a large gap between routine practice and best practice existed, the health care facility would utilize the methods of quality improvement to identify the systemic causes before initiating corrective actions. Thus, the linkage between ethics and quality improvement parallels the inseparable linkage between clinical and organizational ethics.

New and Future Challenges

Ethics committees assumed a central role in the emergence of clinical ethics. In the future, ethics programs will address clinical and organizational issues such as ethical issues related to fair allocation of resources, the cultural diversity of patients and the health care workforce, and availability of new genetic information. Policies and procedures also will be needed to respond to the challenges of a changing system of health care delivery, diversification of the patient population, and technological advances.

Managed Health Care

Pressures to control health care costs and provide access to people who are uninsured or underinsured, as well as the general sense that society is getting less than what it is paying for, is reshaping the delivery of health care in the United States. Managed care arrangements and constraining access to treatments that either provide marginal benefit or only benefit a few at great cost, are two new trends confronting ethics in health care.

Managed care will raise concerns as the old ethics of patient autonomy and fiduciary responsibility are mixed with organizational ethics, business ethics, and competing responsibilities. Physicians and nurses will seek guidance as they struggle over conflicts of interest and conscience. Without clear and explicit guidelines, health providers will experience conflicts about whether their primary responsibility should be the best interests of the patient, the well-being of society, or the success of the health plan. Informed consent policies and practices will be challenged by business interests not to divulge information about the availability of services in other health care plans. Patients likely will seek support from ethics committees or ethics programs to protect their newly identified "rights," identified in the 20th century, or to appeal policies that prevent their access to beneficial treatments.

In response to emerging ethical challenges that involve business and organizational elements, ethics committees and programs will have to clarify their missions and goals. The greatest challenge that will confront ethics committees may be identifying and managing their own conflicting roles and advocacy responsibilities. Will ethics committees be able to protect the interests of patients and simultaneously help articulate and nurture a positive health care ethics environment, including a high standard of business ethics? Questions like this are already being voiced. To address these types of questions, in some settings organizational ethics subcommittees have developed under the umbrella of an ethics committee; in other settings separate organizational ethics committees have been formed. It is hoped that ethics programs will serve as role models for health professionals by demonstrating how to protect patients and at the same time support institutional policies that have been developed with involvement of all stakeholders and that are fair, explicit, and accessible.

At the policy level, not only ethics committees and programs, but also federal and state groups, professional organizations, health care institutions, and scholars will attempt to develop policies that maximize health for society in general and are sensitive to fiscal constraints. Policies will attempt, for example, to distinguish basic from nonbasic health care, identify fair criteria and strategies for allocating resources, and develop fair mechanisms to develop and appeal policies. Policies that explicitly ration will be tested and challenged by politics, a culture that embraces rescue medicine, and the mainstream societal preference to avoid difficult choices. Simultaneously, policy development will need to safe-guard the integrity of the health care professions and patients' trust in their health care providers.

Cultural Diversity

Demographic changes in the composition of the country suggest a very different future American. In the near future White, Anglo-Saxon protestants will be the minority in a multicultural society. Increasing numbers of patients will speak lan-guages other than English and will have non-Western values. For example, the priority of individual autonomy is not embraced by all patients, and formalized informed consent practices may be offensive to others. See, for example, Carrese's and Rhode's discussion of Navajo patients' responses to informed consent prac-tices (Part III, Section 3). Physicians and nurses will seek guidance on how to be "ethical" (per their understanding of ethical principles and ethics-related policies) and culturally sensitive at the same time. Ethics consultants will have to help providers identify the underlying goals and values expressed in ethic policies and procedures, so that they can try to appeal to these fundamental goals in the con-text of another culture. At times it may seem like the integrity of the profession is challenged, and the role of the ethics consultant may ultimately be to help the health care providers understand the true meaning of integrity. The role of the ethics consultant will be first, to help identify strategies that support Western ideals while being sensitive and respectful of non-Western cultural beliefs; second, to help identify commonalities in values across cultures when strategies present cultural conflicts; and third, to identify the limits of acceptable practices that maintain the provider's professional integrity and personal moral beliefs.

Ethics consultants and committees will struggle with cultural diversity. Issues such as personal responsibility for health, respect for persons, the role of the community or the interpreter in decision making, and the very meaning of illness may become topics for ethics consultations and policy development. Cultural diversity and the resultant differences in meaning attributed to princi-ples and ideas will help refine ethical analysis, and ensure that it is not merely dominance of one cultural view over another.

Genetic Information

Advances in molecular biology and the genome project's goal of mapping DNA will force the health care profession to confront issues about the ethical use of medical information. The ability to screen genetically for risks of conditions has just started to challenge ethics committees and hospital policies. Similar to the heart transplant issue in the early 1980s, new technological advances will have to be evaluated in terms of their need, anticipated benefits, costs, and negative impact on other services. Future questions may focus on the value of knowing more about one's genetic makeup; problems of confidentiality when one family member's results implicate others; whether or not informed consent and genetic counseling accurately characterize the risks of learning genetic information; the relationship between probabilities for a population and the implications for an individual; and legitimate grounds for access to this knowledge in a health care system that is attempting to control health care costs. Empirical research will help identify the magnitude of anticipated and unanticipated consequences resulting from access to greater genetic knowledge.

Ethics committees and programs will be asked to help develop policies that delineate access to genetic testing and the handling of confidential information. Health care workers likely will seek help when they perceive that patients want confirmation of their anticipated future health. Health care professionals also will struggle with parents who want to know the genetic predisposition of their fetus to decide whether to carry a pregnancy to term. The challenge for those involved in the practice of bioethics will be balancing respect for autonomous desires with both societal pressure to control health care costs and committee members' personal views. An additional challenge will be for ethics programs to nurture a medical culture that prevents wholesale use of medical technology simply because it is available. It is possible for ethics committees and programs to try to create a framework for handling the challenges of these new medical advances. Ethics committees and programs, initiatives, and policies can be agents for responding to future ethical challenges.

References

1. Scribner, B. H. "Ethical problems of using artificial organs to sustain life." *Transactions of the American Society for Artificial Internal Organs* 10:209–212, 1964.
2. Alexander, S. "They decide who lives, who dies: medical miracle puts a moral burden on a small committee." *Life* 53:102ff, 1962.
3. Snaders, D., and Dukeminier, J. "Medical advance and legal lag: hemodialysis and kidney transplantation." *University of California Law Association Law Review* 15:357–413, 1968.
4. *Barber v. Superior Court*, 195 California Reporter. 484, 486 (California Appellate 1983).
5. *Infant Doe v. Baker*, No. 482 S 140 (Indiana Supreme Court, May 27, 1982).
6. "Nondiscrimination on the basis of handicap; procedures and guidelines relating to health care for handicapped infants; Final Rule," *49 Federal Register* 1622, 1623 (January 16, 1984) (codified at 45 CFR 84.55).
7. President's Commission for the Study of Ethical Problems in Medicine and Biomedical Research. *Deciding to Forego Life-Sustaining Treatment: Ethical, Medical and Legal Issues in Treatment Decisions*. Washington, D.C.: U.S. Government Printing Office, 1983.
8. Faden, R. R., Beauchamp, T. L., and King N. M. P. *A History and Theory of Informed Consent*. New York, NY: Oxford Press, 1986.

9. Buchanan, A. E., and Brock, D. W. *Deciding for Others: The Ethics of Surrogate Decisionmaking.* Cambridge, England: Cambridge University Press, 1989.

10. Cranford, R. E., and Doudera, A. E. "The emergence of institutional ethics committees." In: Cranford, R. E., and Doudera, A. E., eds. *Institutional Ethics Committees and Health Care Decision Making.* Ann Arbor, MI: Health Administration Press, 1984.

11. Singer, P. A., Pellegrino, E. D., and Siegler, M. "Ethics committees and consultants." *Journal of Clinical Ethics* 1:263–267, 1990.

12. Fletcher, J. C., and Hoffman, D. E. "Ethics committees: time to experiment with standards." *Annals of Internal Medicine* 120:335–338, 1994.

13. American Society for Bioethics and Humanities Task Force on Standards for Bioethics and Humanities. *Core Competencies for Health Care Ethics Consultation: The Report of the American Society for Bioethics and Humanities.* Glenview, IL: American Society for Bioethics and Humanities; 1998.

14. National Center for Ethics in Health Care. Veterans Health Administration. www.ethics.va.gov/IntegratedEthics.

15. Jonsen, A. R., Siegler, M., and Winslade, W. J. *Clinical Ethics,* 5th ed. New York, N.Y.: McGraw Hill, 2002.

16. Lusted, L. B. *Introduction to Medical Decision Making.* Springfield, IL: Charles C. Thomas, 1968.

17. The Joint Commission Accreditation Manual for Hospitals. RI.01.05.01. Oak Brook Terrace, IL: The Joint Commission, 2010.

18. Stadler, H. A., Morrissey, J. M., Williams-Rice, B., Tucker, J. E., Paige, J. A., McWilliams, J. E., and Kay, D. "HEC consortium survey: current perspectives of physicians and nurses." *Hospital Ethics Committee Forum* 6:269–281, 1994.

19. Capron, A. M. "Decision review: a problematic task." In: Cranford, R. E., Doudera, A. E., eds. *Institutional Ethics Committees and Health Care Decision Making.* Ann Arbor, MI: Health Administration Press, 1984.

20. Bedell, S. E. W., Pelle, D., Maher, P. L., and Cleary, P. D. "Do-not-resuscitate orders for critically ill patients in the hospital. How are they used and what is their impact?" *Journal of the American Medical Association* 256:233–237, 1986.

21. Braithwaite, S., and Thomasma, D. C. "New guidelines for foregoing life-sustaining treatment for incompetent patients: an anti-cruelty policy." *Annals of Internal Medicine* 104:711–725, 1986.

22. American College of Physicians Ethics Committee. "American College of Physicians' Ethics Manual," 3rd ed. *Annals of Internal Medicine* 117:947–960, 1992.

23. Omnibus Budget Reconciliation Act of 1990 (Western Supplement 1991). Pub. L. No. 101-508 4206, 4751. 1990.

24. Council on Ethical and Judicial Affairs, American Medical Association. "Guidelines for the appropriate use of do-not-resuscitate orders." *Journal of the American Medical Association* 265:1868–1871, 1991.

25. Curtis, R. C., Park, D. R., Krone, M. R., and Pearlman, R. A. "Use of the medical futility rationale in do not attempt resuscitation orders." *Journal of the American Medical Association* 273:124–128, 1995.

26. Tomlinson, T., and Brody, H. "Futility and the ethics of resuscitation." *Journal of the American Medical Association* 264:1276–1280, 1990.
27. Fine, R. L., and Mayo, T. W. "Resolution of futility by due process: early experience with the Texas Advance Directives Act." *Annals of Internal Medicine* 138:743–746, 2003.
28. Quill, T. E., Cassell, C. K., and Meier, D. E. "Care of the hopelessly ill: proposed clinical criteria for physician assisted suicide." *New England Journal of Medicine* 327: 1380–1384, 1992.
29. Hoffman, D. E. "Evaluating ethics committees: a view from the outside." Milbank Quarterly 71:677–701, 1993.

Ethical Topics at the Beginning of Life

Abortion

&

Roe v. Wade:
Majority Opinion and Dissent

United States Supreme Court

In the landmark 1973 U.S. Supreme Court decision, *Roe v. Wade*, excerpted below, the U.S. Supreme Court introduced a trimester framework for addressing the constitutionality of state abortion laws. It ruled that a state cannot interfere with a woman's right to choose abortion during the first trimester of pregnancy. However, during the second trimester a state's interest in protecting maternal health becomes compelling and thus certain restrictions are permissible. In the final trimester of pregnancy, the Court held that a state's interest in protecting potential human life becomes compelling, and states can prohibit abortion in the final twelve weeks of pregnancy. The effect of *Roe v. Wade* was to overturn 46 state laws governing abortion. The Court's ruling was controversial and remains so today.

U.S. Supreme Court

Roe et al. v. Wade, District Attorney of Dallas County Appeal from the United States District Court for the Northern District of Texas No. 70-18.

Argued December 13, 1971. Reargued October 11, 1972. Decided January 22, 1973

A pregnant single woman (Roe) brought a class action challenging the constitutionality of the Texas criminal abortion laws, which proscribe procuring or at-

Condensed and reproduced from 410 United States Reports 113. Decided January 22, 1973.

tempting an abortion except on medical advice for the purpose of saving the mother's life. . . .

MR. JUSTICE BLACKMUN delivered the opinion of the Court . . .

. . . It perhaps is not generally appreciated that the restrictive criminal abortion laws in effect in a majority of States today are of relatively recent vintage. Those laws, generally proscribing abortion or its attempt at any time during pregnancy except when necessary to preserve the pregnant woman's life, are not of ancient or even of common-law origin. Instead, they derive from statutory changes effected, for the most part, in the latter half of the 19th century. . . .

It is . . . apparent that at common law, at the time of the adoption of our Constitution, and throughout the major portion of the 19th century, abortion was viewed with less disfavor than under most American statutes currently in effect. Phrasing it another way, a woman enjoyed a substantially broader right to terminate a pregnancy than she does in most States today. At least with respect to the early stage of pregnancy, and very possibly without such a limitation, the opportunity . . . to make this choice was present in this country well into the 19th century. Even later, the law continued for some time to treat less punitively an abortion procured in early pregnancy. . . .

Three reasons have been advanced to explain historically the enactment of criminal abortion laws in the 19th century and to justify their continued existence. It has been argued occasionally that these laws were the product of a Victorian social concern to discourage illicit sexual conduct. Texas, however, does not advance this justification in the present case, and it appears that no court or commentator has taken the argument seriously.

A second reason is concerned with abortion as a medical procedure. When most criminal abortion laws were first enacted, the procedure was a hazardous one for the woman. This was particularly true prior to the . . . development of antisepsis. Antiseptic techniques, of course, were based on discoveries by Lister, Pasteur, and others first announced in 1867, but were not generally accepted and employed until about the turn of the century. Abortion mortality was high. Even after 1900, and perhaps until as late as the development of antibiotics in the 1940s, standard modern techniques such as dilation and curettage were not nearly so safe as they are today. Thus, it has been argued that a State's real concern in enacting a criminal abortion law was to protect the pregnant woman, that is, to restrain her from submitting to a procedure that placed her life in serious jeopardy.

Modern medical techniques have altered this situation. Appellants and various amici refer to medical data indicating that abortion in early pregnancy, that is, prior to the end of the first trimester, although not without its risk, is now relatively safe. Mortality rates for women undergoing early abortions, where the procedure is legal, appear to be as low as or lower than the rates for normal childbirth. Consequently, any interest of the State in protecting the woman from an inherently hazardous procedure, except when it would be equally dangerous for her to forego it, has largely disappeared. Of course, important state interests in the areas of health and medical standards do remain. . . . The State has a legitimate interest in seeing to it that abortion, like any other medical procedure, is performed under circumstances that insure maximum safety for the patient. This interest obviously extends at least to the performing physician and his staff, to the facilities involved, to the availability of after-care, and to adequate provision for any complication or emergency that might arise. The prevalence of high mortality rates at illegal "abortion mills" strengthens, rather than weakens, the State's interest in regulating the conditions under which abortions are performed. Moreover, the risk to the woman increases as her pregnancy continues. Thus, the State retains a definite interest in protecting the woman's own health and safety when an abortion is proposed at a late stage of pregnancy.

The third reason is the State's interest—some phrase it in terms of duty—in protecting prenatal life. Some of the argument for this justification rests on the theory that a new human life is present from the moment of conception. The State's interest and general obligation to protect life then extends, it is argued, to prenatal life. Only when the life of the pregnant mother herself is at stake, balanced against the life she carries within her, should the interest of the embryo or fetus not prevail. Logically, of course, a legitimate state interest in this area need not stand or fall on acceptance of the belief that life begins at conception or at some other point prior to live birth. In assessing the State's interest, recognition may be given to the less rigid claim that as long as at least potential life is involved, the State may assert interests beyond the protection of the pregnant woman alone. . . .

Parties challenging state abortion laws have sharply disputed in some courts the contention that a purpose of these laws, when enacted, was to protect prenatal life. Pointing to the absence of legislative history to support the contention, they claim that most state laws were designed solely to protect the woman. Because medical advances have lessened this concern, at least with respect to abortion in early pregnancy, they argue that with respect to such abortions the laws can no longer be justified by any state interest. There is some scholarly support for this view of original purpose. The few state courts called upon to interpret their laws in the late 19th and early 20th centuries did focus on the State's interest in protecting the woman's health rather than in preserving the embryo and fetus. . . .

The Constitution does not explicitly mention any right of privacy. In a line of decisions, however, . . . the Court has recognized that a right of personal privacy, or a guarantee of certain areas or zones of privacy, does exist under the Constitution. In varying contexts, the Court or individual Justices have, indeed, found at least the roots of that right in the First Amendment, . . . in the Fourth and Fifth Amendments, . . . in the penumbras of the Bill of Rights, . . . in the Ninth Amendment, . . . or in the concept of liberty guaranteed by the first section of the Fourteenth Amendment. . . . These decisions make it clear that only personal rights that can be deemed "fundamental" or "implicit in the concept of ordered liberty," . . . are included in this guarantee of personal privacy. They also make it clear that the right has some extension to activities relating to marriage, . . . procreation, . . . contraception, . . . family relationships, . . . and child rearing and education. . . .

This right of privacy, whether it be founded in the Fourteenth Amendment's concept of personal liberty and restrictions upon state action, as we feel it is, or, as the District Court determined, in the Ninth Amendment's reservation of rights to the people, is broad enough to encompass a woman's decision whether or not to terminate her pregnancy. The detriment that the State would impose upon the pregnant woman by denying this choice altogether is apparent. Specific and direct harm medically diagnosable even in early pregnancy may be involved. Maternity, or additional offspring, may force upon the woman a distressful life and future. Psychological harm may be imminent. Mental and physical health may be taxed by child care. There is also the distress, for all concerned, associated with the unwanted child, and there is the problem of bringing a child into a family already unable, psychologically and otherwise, to care for it. In other cases, as in this one, the additional difficulties and continuing stigma of unwed motherhood may be involved. All these are factors the woman and her responsible physician necessarily will consider in consultation.

On the basis of elements such as these, appellant and some amici argue that the woman's right is absolute and that she is entitled to terminate her pregnancy at whatever time, in whatever way, and for whatever reason she alone chooses. With this we do not agree. Appellant's arguments that Texas either has no valid interest at all in regulating the abortion decision, or no interest strong enough to support any limitation upon the woman's sole determination, are unpersuasive. The Court's decisions recognizing a right of privacy also acknowledge that some state regulation in areas protected by that right is appropriate. As noted above, a State may properly assert important interests in safeguarding health, in maintaining medical standards, and in protecting potential life. At some point in pregnancy, these respective interests become sufficiently compelling to sustain regulation of the factors that govern the abortion decision. The privacy right involved, therefore, cannot be said to be absolute. . . .

We, therefore, conclude that the right of personal privacy includes the abortion decision, but that this right is not unqualified and must be considered against important state interests in regulation.

We note that those federal and state courts that have recently considered abortion law challenges have reached the same conclusion. . . .

Although the results are divided, most of these courts have agreed that the right of privacy, however based, is broad enough to cover the abortion decision; that the right, nonetheless, is not absolute and is subject to some limitations; and that at some point the state interests as to protection of health, medical standards, and prenatal life, become dominant. We agree with this approach.

Where certain "fundamental rights" are involved, the Court has held that regulation limiting these rights may be justified only by a "compelling state interest,". . . and that legislative enactments must be narrowly drawn to express only the legitimate state interests at stake. . . .

The appellee and certain amici argue that the fetus is a "person" within the language and meaning of the Fourteenth Amendment. In support of this, they outline at length and in detail the well-known facts of fetal development. If this suggestion of personhood is established, the appellant's case, of course, collapses, . . . for the fetus' right to life would then be guaranteed specifically by the Amendment. The appellant conceded as much on reargument. On the other hand, the appellee conceded on reargument that no case could be cited that holds that a fetus is a person within the meaning of the Fourteenth Amendment. . . .

All this, together with our observation, supra, that throughout the major portion of the 19th century prevailing legal abortion practices were far freer than they are today, persuades us that the word "person," as used in the Fourteenth Amendment, does not include the unborn. . . .

As we have intimated above, it is reasonable and appropriate for a State to decide that at some point in time another interest, that of health of the mother or that of potential human life, becomes significantly involved. The woman's privacy is no longer sole and any right of privacy she possesses must be measured accordingly.

Texas urges that, apart from the Fourteenth Amendment, life begins at conception and is present throughout pregnancy, and that, therefore, the State has a compelling interest in protecting that life from and after conception. We need not resolve the difficult question of when life begins. When those trained in the respective disciplines of medicine, philosophy, and theology are unable to arrive at any consensus, the judiciary, at this point in the development of man's knowledge, is not in a position to speculate as to the answer. . . .

It should be sufficient to note briefly the wide divergence of thinking on this most sensitive and difficult question. There has always been strong support for the view that life does not begin until live birth. This was the belief of the Stoics. It appears to be the predominant, though not the unanimous, attitude of the Jewish faith. It may be taken to represent also the position of a large segment of the Protestant community, insofar as that can be ascertained; organized groups that have taken a formal position on the abortion issue have generally regarded abortion as a matter for the conscience of the individual and her family. As we have noted, the common law found greater significance in quickening. Physicians and their scientific colleagues have regarded that event with less interest and have tended to focus either upon conception, upon live birth, or upon the interim point at which the fetus becomes "viable," that is, potentially able to live outside the mother's womb, albeit with artificial aid. Viability is usually placed at about seven months (28 weeks) but may occur earlier, even at 24 weeks. . . .

In areas other than criminal abortion, the law has been reluctant to endorse any theory that life, as we recognize it, begins before live birth or to accord legal rights to the unborn except in narrowly defined situations and except when the rights are contingent upon live birth. . . . In short, the unborn have never been recognized in the law as persons in the whole sense. . . .

In view of all this, we do not agree that, by adopting one theory of life, Texas may override the rights of the pregnant woman that are at stake. We repeat, however, that the State does have an important and legitimate interest in preserving and protecting the health of the pregnant woman, whether she be a resident of the State or a nonresident who seeks medical consultation and treatment there, and that it has still another important and legitimate interest in protecting the potentiality of human life. These interests are separate and distinct. Each grows in substantiality as the woman approaches . . . term and, at a point during pregnancy, each becomes "compelling."

With respect to the State's important and legitimate interest in the health of the mother, the "compelling" point, in the light of present medical knowledge, is at approximately the end of the first trimester. This is so because of the now-established medical fact, . . . that until the end of the first trimester mortality in abortion may be less than mortality in normal childbirth. It follows that, from and after this point, a State may regulate the abortion procedure to the extent that the regulation reasonably relates to the preservation and protection of maternal health. Examples of permissible state regulation in this area are requirements as to the qualifications of the person who is to perform the abortion; as to the licensure of that person; as to the facility in which the procedure is to be performed, that is, whether it must be a hospital or may be a clinic or some other place of less-than-hospital status; as to the licensing of the facility; and the like.

This means, on the other hand, that, for the period of pregnancy prior to this "compelling" point, the attending physician, in consultation with his patient, is free to determine, without regulation by the State, that, in his medical judgment, the patient's pregnancy should be terminated. If that decision is reached, the judgment may be effectuated by an abortion free of interference by the State.

With respect to the State's important and legitimate interest in potential life, the "compelling" point is at viability. This is so because the fetus then presumably has the capability of meaningful life outside the mother's womb. State regulation protective of fetal life after viability thus has both logical and biological justifications. If the State is interested in protecting fetal life after viability, it may go so far as to proscribe abortion . . . during that period, except when it is necessary to preserve the life or health of the mother.

. . . To summarize and to repeat:

1. A state criminal abortion statute of the current Texas type, that excepts from criminality only a life-saving procedure on behalf of the mother, without regard to pregnancy stage and without recognition of the other interests involved, is violative of the Due Process Clause of the Fourteenth Amendment.

(a) For the stage prior to approximately the end of the first trimester, the abortion decision and its effectuation must be left to the medical judgment of the pregnant woman's attending physician.

(b) For the stage subsequent to approximately the end of the first trimester, the State, in promoting its interest in the health of the mother, may, if it chooses, regulate the abortion procedure in ways that are reasonably related to maternal health.

(c) For the stage subsequent to viability, the State in promoting its interest in the potentiality of human life . . . may, if it chooses, regulate, and even proscribe, abortion except where it is necessary, in appropriate medical judgment, for the preservation of the life or health of the mother.

2. The State may define the term "physician," . . . to mean only a physician currently licensed by the State, and may proscribe any abortion by a person who is not a physician as so defined.

    . . . The decision leaves the State free to place increasing restrictions on abortion as the period of pregnancy lengthens, so long as those restrictions are tailored to the recognized state interests. The decision vindicates the right of the physician to administer medical treatment according to his professional judgment up to the points where important . . . state interests provide compelling justifications for intervention. Up to those points, the abortion decision in all its aspects is inherently, and primarily, a medical decision, and basic responsibility for it must rest with the physician. If an individual practitioner abuses the privilege of exercising proper medical judgment, the usual remedies, judicial and intra-professional, are available. . . .

MR. JUSTICE REHNQUIST, dissenting.

. . . I would reach a conclusion opposite to that reached by the Court. I have difficulty in concluding, as the Court does, that the right of "privacy" is involved in this case. . . . A transaction resulting in an operation such as this is not "private" in the ordinary usage of that word. Nor is the "privacy" that the Court finds here even a distant relative of the freedom from searches and seizures protected by the Fourth Amendment to the Constitution, which the Court has referred to as embodying a right to privacy. . . .

    If the Court means by the term "privacy" no more than that the claim of a person to be free from unwanted state regulation of consensual transactions may be a form of "liberty" protected by the Fourteenth Amendment, there is no doubt that similar claims have been upheld in our earlier decisions on the basis of that liberty. . . . But that liberty is not guaranteed absolutely against deprivation, only against deprivation without due process of law. The test traditionally applied in the area of social and economic legislation is whether or not a law such as that challenged has a rational relation to a valid state objective. . . . But the Court's sweeping invalidation of any restrictions on abortion during the first trimester is impossible to justify under that standard, and the conscious weighing of competing factors that the Court's opinion apparently substitutes for the established test is far more appropriate to a legislative judgment than to a judicial one.

... The decision here to break pregnancy into three distinct terms and to outline the permissible restrictions the State may impose in each one, for example, partakes more of judicial legislation than it does of a determination of the intent of the drafters of the Fourteenth Amendment.

The fact that a majority of the States reflecting, after all, the majority sentiment in those States, have had restrictions on abortions for at least a century is a strong indication, it seems to me, that the asserted right to an abortion is not "so rooted in the traditions and conscience of our people as to be ranked as fundamental," ... Even today, when society's views on abortion are changing, the very existence of the debate is evidence that the "right" to an abortion is not so universally accepted as the appellant would have us believe. ...

A Defense of Abortion

Judith Jarvis Thomson

Judith Jarvis Thomson, Emeritus Professor in Philosophy at Massachusetts Institute of Technology, received her Ph.D. from Columbia University. She is a moral philosopher who is well known for her "What if . . . ?" thought experiments. She uses this technique to support the permissibility of abortion. By assuming that a fetus is a person and anchoring to contractual relationships, she argues in "A Defense of Abortion" that the fetus' right to life is really a right not to be killed unjustly.

. . . Opponents of abortion commonly spend most of their time establishing that the fetus is a person, and hardly any time explaining the step from there to the impermissibility of abortion. Perhaps they think the step too simple and obvious to require much comment. Or perhaps instead they are simply being economical in argument. Many of those who defend abortion rely on the premise that the fetus is not a person, but only a bit of tissue that will become a person at birth; and why pay out more arguments than you have to? Whatever the explanation, I suggest that the step they take is neither easy nor obvious, that it calls for closer examination than it is commonly given, and that when we do give it this closer examination we shall feel inclined to reject it.

I propose, then, that we grant that the fetus is a person from the moment of conception. How does the argument go from here? Something like this, I take it. Every person has a right to life. So the fetus has a right to life. No doubt the mother has a right to decide what shall happen in and to her body; everyone would grant that. But surely a person's right to life is stronger and more stringent than the mother's right to decide what happens in and to her body, and so outweighs it. So the fetus may not be killed; an abortion may not be performed.

It sounds plausible. But now let me ask you to imagine this. You wake up in the morning and find yourself back to back in bed with an unconscious violinist. A famous unconscious violinist. He has been found to have a fatal kidney ailment, and the Society of Music Lovers has canvassed all the available medical records and found that you alone have the right blood type to help. They have therefore kidnapped you, and last night the violinist's circulatory system was plugged into yours, so that your kidneys can be used to extract poisons from his

Excerpted from *Philosophy and Public Affairs*, 1 (1), Autumn 1971: 41–66. Reproduced with permission of Princeton University Press. Copyright © 1971 Princeton University Press.

blood as well as your own. The director of the hospital now tells you, "Look, we're sorry the Society of Music Lovers did this to you—we would never have permitted it if we had known. But still, they did it, and the violinist now is plugged into you. To unplug you would be to kill him. But never mind, it's only for nine months. By then he will have recovered from his ailment, and can safely be unplugged from you." Is it morally incumbent on you to accede to this situation? No doubt it would be very nice of you if you did, a great kindness. But do you have to accede to it? What if it were not nine months, but nine years? Or longer still? What if the director of the hospital says, "Tough luck, I agree, but you've now got to stay in bed, with the violinist plugged into you, for the rest of your life. Because remember this. All persons have a right to life, and violinists are persons. Granted you have a right to decide what happens in and to your body, but a person's right to life outweighs your right to decide what happens in and to your body. So you cannot ever be unplugged from him." I imagine you would regard this as outrageous, which suggests that something really is wrong with that plausible-sounding argument I mentioned a moment ago.

In this case, of course, you were kidnapped; you didn't volunteer for the operation that plugged the violinist into your kidneys. Can those who oppose abortion on the ground I mentioned make an exception for a pregnancy due to rape? Certainly. They can say that persons have a right to life only if they didn't come into existence because of rape; or they can say that all persons have a right to life, but that some have less of a right to life than others, in particular, that those who came into existence because of rape have less. But these statements have a rather unpleasant sound. Surely the question of whether you have a right to life at all, or how much of it you have, shouldn't turn on the question of whether or not you are the product of a rape. And in fact the people who oppose abortion on the ground I mentioned do not make this distinction, and hence do not make an exception in case of rape.

Nor do they make an exception for a case in which the mother has to spend the nine months of her pregnancy in bed. They would agree that would be a great pity, and hard on the mother; but all the same, all persons have a right to life, the fetus is a person, and so on. I suspect, in fact, that they would not make an exception for a case in which, miraculously enough, the pregnancy went on for nine years, or even the rest of the mother's life.

. . . Where the mother's life is not at stake, the argument . . . seems to have a much stronger pull. "Everyone has a right to life, so the unborn person has a right to life." And isn't the child's right to life weightier than anything other than the mother's own right to life, which she might put forward as ground for an abortion?

This argument treats the right to life as if it were unproblematic. It is not, and this seems to me to be precisely the source of the mistake. For we should now, at long last, ask what it comes to, to have a right to life. In some views having a right to life includes having a right to be given at least the bare minimum one needs for continued life. But suppose that what in fact is the bare minimum a man needs for continued life is something he has no right at all to be given? If

I am sick unto death, and the only thing that will save my life is the touch of Henry Fonda's cool hand on my fevered brow, then all the same, I have no right to be given the touch of Henry Fonda's cool hand on my fevered brow. It would be frightfully nice of him to fly in from the West Coast to provide it. It would be less nice, though no doubt well meant, if my friends flew out to the West Coast and carried Henry Fonda back with them. But I have no right at all against anybody that he should do this for me. Or again, to return to the story I told earlier, the fact that for continued life that violinist needs the continued use of your kidneys does not establish that he has a right to be given the continued use of your kidneys. He certainly has no right against you that you should give him continued use of your kidneys. For nobody has any right to use your kidneys unless you give him such a right; and nobody has the right against you that you shall give him this right—if you do allow him to go on using your kidneys, this is a kindness on your part, and not something he can claim from you as his due. Nor has he any right against anybody else that they should give him continued use of your kidneys. Certainly he had no right against the Society of Music Lovers that they should plug him into you in the first place. And if you now start to unplug yourself, having learned that you will otherwise have to spend nine years in bed with him, there is nobody in the world who must try to prevent you, in order to see to it that he is given something he has a right to be given.

Some people are rather stricter about the right to life. In their view, it does not include the right to be given anything, but amounts to, and only to, the right not to be killed by anybody. But here a related difficulty arises. If everybody is to refrain from killing that violinist, then everybody must refrain from doing a great many different sorts of things. Everybody must refrain from slitting his throat, everybody must refrain from shooting him—and everybody must refrain from unplugging you from him. But does he have a right against everybody that they shall refrain from unplugging you from him? To refrain from doing this is to allow him to continue to use your kidneys. It could be argued that he has a right against us that we should allow him to continue to use your kidneys. That is, while he had no right against us that we should give him the use of your kidneys, it might be argued that he anyway has a right against us that we shall not now intervene and deprive him of the use of your kidneys. I shall come back to third-party interventions later. But certainly the violinist has no right against you that you shall allow him to continue to use your kidneys. As I said, if you do allow him to use them, it is a kindness on your part, and not something you owe him.

The difficulty I point to here is not peculiar to the right to life. It reappears in connection with all the other natural rights; and it is something which an adequate account of rights must deal with. For present purposes it is enough just to draw attention to it. But I would stress that I am not arguing that people do not have a right to life, quite to the contrary, it seems to me that the primary control we must place on the acceptability of an account of rights is that it should turn out in that account to be a truth that all persons have a right to life.

I am arguing only that having a right to life does not guarantee having either a right to be given the use of or a right to be allowed continued use of another person's body—even if one needs it for life itself. So the right to life will not serve the opponents of abortion in the very simple and clear way in which they seem to have thought it would.

There is another way to bring out the difficulty. In the most ordinary sort of case, to deprive someone of what he has a right to is to treat him unjustly. Suppose a boy and his small brother are jointly given a box of chocolates for Christmas. If the older boy takes the box and refuses to give his brother any of the chocolates, he is unjust to him, for the brother has been given a right to half of them. But suppose that, having learned that otherwise it means nine years in bed with that violinist, you unplug yourself from him. You surely are not being unjust to him, for you gave him no right to use your kidneys, and no one else can have given him any such right. But we have to notice that in unplugging yourself, you are killing him; and violinists, like everybody else, have a right to life, and thus in the view we were considering just now, the right not to be killed. So here you do what he supposedly has a right you shall not do, but you do not act unjustly to him in doing it.

The emendation which may be made at this point is this: the right to life consists not in the right not to be killed, but rather in the right not to be killed unjustly. This runs a risk of circularity, but never mind: it would enable us to square the fact that the violinist has a right to life with the fact that you do not act unjustly toward him in unplugging yourself, thereby killing him. For if you do not kill him unjustly, you do not violate his right to life, and so it is no wonder you do him no injustice.

But if this emendation is accepted, the gap in the argument against abortion stares us plainly in the face: it is by no means enough to show that the fetus is a person, and to remind us that all persons have a right to life—we need to be shown also that killing the fetus violates its right to life, i.e., that abortion is unjust killing. And is it?

I suppose we may take it as a datum that in a case of pregnancy due to rape the mother has not given the unborn person a right to the use of her body for food and shelter. Indeed, in what pregnancy could it be supposed that the mother has given the unborn person such a right? It is not as if there were unborn persons drifting about the world, to whom a woman who wants a child says "I invite you in."

But it might be argued that there are other ways one can have acquired a right to the use of another person's body than by having been invited to use it by that person. Suppose a woman voluntarily indulges in intercourse, knowing of the chance it will issue in pregnancy, and then she does become pregnant; is she not in part responsible for the presence, in fact the very existence, of the unborn person inside her? No doubt she did not invite it in. But doesn't her partial responsibility for its being there itself give it a right to the use of her body? If so, then her aborting it would be more like the boy's taking away the chocolates, and less like your unplugging yourself from the violinist—doing so would be depriving it of what it does have a right to, and thus would be doing it an injustice.

And then, too, it might be asked whether or not she can kill it even to save her own life: If she voluntarily called it into existence, how can she now kill it, even in self-defense?

The first thing to be said about this is that it is something new. Opponents of abortion have been so concerned to make out the independence of the fetus, in order to establish that it has a right to life, just as its mother does, that they have tended to overlook the possible support they might gain from making out that the fetus is dependent on the mother, in order to establish that she has a special kind of responsibility for it, a responsibility that gives it rights against her which are not possessed by any independent person—such as an ailing violinist who is a stranger to her.

On the other hand, this argument would give the unborn person a right to its mother's body only if her pregnancy resulted from a voluntary act, undertaken in full knowledge of the chance a pregnancy might result from it. It would leave out entirely the unborn person whose existence is due to rape. Pending the availability of some further argument, then, we would be left with the conclusion that unborn persons whose existence is due to rape have no right to the use of their mothers' bodies, and thus that aborting them is not depriving them of anything they have a right to and hence is not unjust killing.

And we should also notice that it is not at all plain that this argument really does go even as far as it purports to. For there are cases and cases, and the details make a difference. If the room is stuffy, and I therefore open a window to air it, and a burglar climbs in, it would be absurd to say, "Ah, now he can stay, she's given him a right to the use of her house—for she is partially responsible for his presence there, having voluntarily done what enabled him to get in, in full knowledge that there are such things as burglars, and that burglars burgle." It would be still more absurd to say this if I had had bars installed outside my windows, precisely to prevent burglars from getting in, and a burglar got in only because of a defect in the bars. It remains equally absurd if we imagine it is not a burglar who climbs in, but an innocent person who blunders or falls in. Again, suppose it were like this: people-seeds drift about in the air like pollen, and if you open your windows, one may drift in and take root in your carpets or upholstery. You don't want children, so you fix up your windows with fine mesh screens, the very best you can buy. As can happen, however, and on very, very rare occasions does happen, one of the screens is defective; and a seed drifts in and takes root. Does the person-plant who now develops have a right to the use of your house? Surely not—despite the fact that you voluntarily opened your windows, you knowingly kept carpets and upholstered furniture, and you knew that screens were sometimes defective. Someone may argue that you are responsible for its rooting, that it does have a right to your house, because after all you could have lived out your life with bare floors and furniture, or with sealed windows and doors. But this won't do—for by the same token anyone can avoid a pregnancy due to rape by having a hysterectomy, or anyway by never leaving home without a (reliable!) army.

It seems to me that the argument we are looking at can establish at most that there are some cases in which the unborn person has a right to the use of its mother's body, and therefore some cases in which abortion is unjust killing. There is room for much discussion and argument as to precisely which, if any. But I think we should sidestep this issue and leave it open, for at any rate the argument certainly does not establish that all abortion is unjust killing.

There is room for yet another argument here, however. We surely must all grant that there may be cases in which it would be morally indecent to detach a person from your body at the cost of his life. Suppose you learn that what the violinist needs is not nine years of your life, but only one hour: all you need do to save his life is to spend one hour in that bed with him. Suppose also that letting him use your kidneys for that one hour would not affect your health in the slightest. Admittedly you were kidnapped. Admittedly you did not give anyone permission to plug him into you. Nevertheless it seems to me plain you ought to allow him to use your kidneys for that hour—it would be indecent to refuse.

Again, suppose pregnancy lasted only an hour, and constituted no threat to life or health. And suppose that a woman becomes pregnant as a result of rape. Admittedly she did not voluntarily do anything to bring about the existence of a child. Admittedly she did nothing at all which would give the unborn person a right to the use of her body. All the same it might well be said, as in the newly emended violinist story, that she ought to allow it to remain for that hour—that it would be indecent in her to refuse.

Now some people are inclined to use the term "right" in such a way that it follows from the fact that you ought to allow a person to use your body for the hour he needs, that he has a right to use your body for the hour he needs, even though he has not been given that right by any person or act. They may say that it follows also that if you refuse, you act unjustly toward him. This use of the term is perhaps so common that it cannot be called wrong; nevertheless it seems to me to be an unfortunate loosening of what we would do better to keep a tight rein on. Suppose that box of chocolates I mentioned earlier had not been given to both boys jointly, but was given only to the older boy. There he sits, stolidly eating his way through the box, his small brother watching enviously. Here we are likely to say "You ought not to be so mean. You ought to give your brother some of those chocolates." My own view is that it just does not follow from the truth of this that the brother has any right to any of the chocolates. If the boy refuses to give his brother any, he is greedy, stingy, callous—but not unjust. I suppose that the people I have in mind will say it does follow that the brother has a right to some of the chocolates, and thus that the boy does act unjustly if he refuses to give his brother any. But the effect of saying this is to obscure what we should keep distinct, namely the difference between the boy's refusal in this case and the boy's refusal in the earlier case, in which the box was given to both boys jointly, and in which the small brother thus had what was from any point of view clear title to half.

A further objection to so using the term "right" that from the fact that A ought to do a thing for B, it follows that B has a right against A that A do it for him, is that it is going to make the question of whether or not a man has a right to a thing turn on how easy it is to provide him with it; and this seems not merely unfortunate, but morally unacceptable. Take the case of Henry Fonda again. I said earlier that I had no right to the touch of his cool hand on my fevered brow, even though I needed it to save my life. I said it would be frightfully nice of him to fly in from the West Coast to provide me with it, but that I had no right against him that he should do so. But suppose he isn't on the West Coast. Suppose he has only to walk across the room, place a hand briefly on my brow—and lo, my life is saved. Then surely he ought to do it, it would be indecent to refuse. Is it to be said "Ah, well, it follows that in this case she has a right to the touch of his hand on her brow, and so it would be an injustice in him to refuse"? So that I have a right to it when it is easy for him to provide it, though no right when it's hard? It's rather a shocking idea that anyone's rights should fade away and disappear as it gets harder and harder to accord them to him.

So my own view is that even though you ought to let the violinist use your kidneys for the one hour he needs, we should not conclude that he has a right to do so—we should say that if you refuse, you are, like the boy who owns all the chocolates and will give none away, self-centered and callous, indecent in fact, but not unjust. And similarly, that even supposing a case in which a woman pregnant due to rape ought to allow the unborn person to use her body for the hour he needs, we should not conclude that he has a right to do so; we should conclude that she is self-centered, callous, indecent, but not unjust, if she refuses. The complaints are no less grave; they are just different. However, there is no need to insist on this point. If anyone does wish to deduce "he has a right" from "you ought," then all the same he must surely grant that there are cases in which it is not morally required of you that you allow that violinist to use your kidneys, and in which he does not have a right to use them, and in which you do not do him an injustice if you refuse. And so also for mother and unborn child. Except in such cases as the unborn person has a right to demand it—and we were leaving open the possibility that there may be such cases—nobody is morally required to make large sacrifices, of health, of all other interests and concerns, of all other duties and commitments, for nine years, or even for nine months, in order to keep another person alive.

We have in fact to distinguish between two kinds of Samaritan: the Good Samaritan and what we might call the Minimally Decent Samaritan. The story of the Good Samaritan, you will remember, goes like this

> A certain man went down from Jerusalem to Jericho, and fell among thieves, which stripped him of his raiment, and wounded him, and departed, leaving him half dead.
>
> And by chance there came down a certain priest that way; and when he saw him, he passed by on the other side.

> And likewise a Levite, when he was at the place, came and looked on him, and passed by on the other side.
>
> But a certain Samaritan, as he journeyed, came where he was; and when he saw him he had compassion on him.
>
> And went to him, and bound up his wounds, pouring in oil and wine, and set him on his own beast, and brought him to an inn, and took care of him.
>
> And on the morrow, when he departed, he took out two pence, and gave them to the host, and said unto him, "Take care of him; and whatsoever thou spendest more, when I come again, I will repay thee." (*Luke* 10:30–35)

The Good Samaritan went out of his way, at some cost to himself, to help one in need of it. We are not told what the options were, that is, whether or not the priest and the Levite could have helped by doing less than the Good Samaritan did, but assuming they could have, then the fact they did nothing at all shows they were not even Minimally Decent Samaritans, not because they were not Samaritans, but because they were not even minimally decent.

These things are a matter of degree, of course, but there is a difference, and it comes out perhaps most clearly in the story of Kitty Genovese, who, as you will remember, was murdered while thirty-eight people watched or listened, and did nothing at all to help her. A Good Samaritan would have rushed out to give direct assistance against the murderer. Or perhaps we had better allow that it would have been a Splendid Samaritan who did this, on the ground that it would have involved a risk of death for himself. But the thirty-eight not only did not do this, they did not even trouble to pick up a phone to call the police. Minimally Decent Samaritanism would call for doing at least that, and their not having done it was monstrous.

After telling the story of the Good Samaritan, Jesus said "Go, and do thou likewise." Perhaps he meant that we are morally required to act as the Good Samaritan did. Perhaps he was urging people to do more than is morally required of them. At all events it seems plain that it was not morally required of any of the thirty-eight that he rush out to give direct assistance at the risk of his own life, and that it is not morally required of anyone that he give long stretches of his life—nine years or nine months—to sustaining the life of a person who has no special right (we were leaving open the possibility of this) to demand it.

Indeed, with one rather striking class of exceptions, no one in any country in the world is legally required to do anywhere near as much as this for anyone else. The class of exceptions is obvious. My main concern here is not the state of the law in respect to abortion, but it is worth drawing attention to the fact that in no state in this country is any man compelled by law to be even a Minimally Decent Samaritan to any person; there is no law under which charges could be brought against the thirty-eight who stood by while Kitty Genovese died. By contrast, in most states in this country women are compelled by law to be not merely Minimally Decent Samaritans, but Good Samaritans to unborn persons inside them. This doesn't by itself settle anything one way or the other, because it may well be argued that there should be laws in this country—as there are in many European countries—compelling at least Minimally Decent Samaritanism. But it does show that there is a gross injustice in the existing state of the law. And it shows also that the groups currently working against liberalization of abortion laws, in fact working toward having it declared unconstitutional for a state to permit abortion, had better start working for the adoption of Good Samaritan laws generally, or earn the charge that they are acting in bad faith.

I should think, myself, that Minimally Decent Samaritan laws would be one thing, Good Samaritan laws quite another, and in fact highly improper. But we are not here concerned with the law. What we should ask is not whether anybody should be compelled by law to be a Good Samaritan, but whether we must accede to a situation in which somebody is being compelled—by nature, perhaps—to be a Good Samaritan. We have, in other words, to look now at third-party interventions. I have been arguing that no person is morally required to make large sacrifices to sustain the life of another who has no right to demand them, and this even where the sacrifices do not include life itself; we are not morally required to be Good Samaritans or anyway Very Good Samaritans to one another. But what if a man cannot extricate himself from such a situation? What if he appeals to us to extricate him? It seems to me plain that there are cases in which we can, cases in which a Good Samaritan would extricate him. There you are, you were kidnapped, and nine years in bed with that violinist is ahead of you. You have your own life to lead. You are sorry, but you simply cannot see giving up so much of your life to the sustaining of his. You cannot extricate yourself, and ask us to do so. I should have thought that in light of his having no right to the use of your body it was obvious that we do not have to accede to your being forced to give up so much. We can do what you ask. There is no injustice to the violinist in our doing so.

. . . My argument will be found unsatisfactory on two counts by many of those who want to regard abortion as morally permissible. First, while I do argue that abortion is not impermissible, I do not argue that it is always permissible. There may well be cases in which carrying the child to term requires only Minimally Decent Samaritanism of the mother, and this is a standard we must not fall below. I am inclined to think it a merit of my account precisely that it does not give a general yes or a general no. It allows for and supports our sense that, for example, a sick and desperately frightened fourteen-year-old schoolgirl, pregnant due to rape, may of course choose abortion, and that any law which rules this out is an insane law. And it also allows for and supports our sense that in other cases resort to abortion is even positively indecent. It would be indecent in the woman to request an abortion, and indecent in a doctor to perform it, if she is in her seventh month, and wants the abortion just to avoid the nuisance of postponing a trip abroad. . . .

Secondly, while I am arguing for the permissibility of abortion in some cases, I am not arguing for the right to secure the death of the unborn child. It is easy to confuse these two things in that up to a certain point in the life of the fetus it is not able to survive outside the mother's body; hence removing it from her body guarantees its death. But they are importantly different. I have argued that you are not morally required to spend nine months in bed, sustaining the life of that violinist; but to say this is by no means to say that if, when you un- plug yourself, there is a miracle and he survives, you then have a right to turn round and slit his throat. You may detach yourself even if this costs him his life; you have no right to be guaranteed his death, by some other means, if unplug- ging yourself does not kill him. . . .

Why Abortion Is Immoral

Don Marquis

Don Marquis received his Ph.D. in philosophy from the University of Indiana. He currently is a professor in the Department of Philosophy at the University of Kansas. In his article, "Why Abortion is Immoral," the author argues that killing is seriously wrong because it deprives another of a valuable "future-like-ours." He concludes that the wrongness of killing should be extended to fetal life.

. . . [W]e can start from the following unproblematic assumption concerning our own case: it is wrong to kill us. Why is it wrong? Some answers can be easily eliminated. It might be said that what makes killing us wrong is that a killing brutalizes the one who kills. But the brutalization consists of being inured to the performance of an act that is hideously immoral; hence, the brutalization does not explain the immorality. It might be said that what makes killing us wrong is the great loss others would experience due to our absence. Although such hubris is understandable, such an explanation does not account for the wrongness of killing hermits, or those whose lives are relatively independent and whose friends find it easy to make new friends.

A more obvious answer is better. What primarily makes killing wrong is neither its effect on the murderer nor its effect on the victim's friends and relatives, but its effect on the victim. The loss of one's life is one of the greatest losses one can suffer. The loss of one's life deprives one of all the experiences, activities, projects, and enjoyments that would otherwise have constituted one's future. Therefore, killing someone is wrong, primarily because the killing inflicts (one of) the greatest possible losses on the victim. . . . When I am killed, I am deprived both of what I now value which would have been part of my future personal life, but also what I would come to value. Therefore, when I die, I am deprived of all of the value of my future. Inflicting this loss on me is ultimately what makes killing me wrong. This being the case, it would seem that what makes killing any adult human being prima facie seriously wrong is the loss of his or her future.

. . . The claim that what makes killing wrong is the loss of the victim's future is directly supported by two considerations. In the first place, this theory

Condensed from *The Journal of Philosophy*, 86 (4), April 1989: 183–202. Reprinted with permission of The Journal of Philosophy, Inc. Copyright © 1989 Journal of Philosophy, Inc.

explains why we regard killing as one of the worst of crimes. Killing is especially wrong, because it deprives the victim of more than perhaps any other crime. In the second place, people with AIDS or cancer who know they are dying believe, of course, that dying is a very bad thing for them. They believe that the loss of a future to them that they would otherwise have experienced is what makes their premature death a very bad thing for them. A better theory of the wrongness of killing would require a different natural property associated with killing which better fits with the attitudes of the dying. What could it be?

The view that what makes killing wrong is the loss to the victim of the value of the victim's future gains additional support when some of its implications are examined. In the first place, it is incompatible with the view that it is wrong to kill only beings who are biologically human. It is possible that there exists a different species from another planet whose members have a future like ours. Since having a future like that is what makes killing someone wrong, this theory entails that it would be wrong to kill members of such a species. Hence, this theory is opposed to the claim that only life that is biologically human has great moral worth, a claim which many anti-abortionists have seemed to adopt. This opposition, which this theory has in common with personhood theories, seems to be a merit of the theory.

In the second place, the claim that the loss of one's future is the wrong-making feature of one's being killed entails the possibility that the futures of some actual nonhuman mammals on our own planet are sufficiently like ours that it is seriously wrong to kill them also. Whether some animals do have the same right to life as human beings depends on adding to the account of the wrongness of killing some additional account of just what it is about my future or the futures of other adult human beings which makes it wrong to kill us. No such additional account will be offered in this essay. Undoubtedly, the provision of such an account would be a very difficult matter. Undoubtedly, any such account would be quite controversial. Hence, it surely should not reflect badly on this sketch of an elementary theory of the wrongness of killing that it is indeterminate with respect to some very difficult issues regarding animal rights.

In the third place, the claim that the loss of one's future is the wrong-making feature of one's being killed does not entail, as sanctity of human life theories do, that active euthanasia is wrong. Persons who are severely and incurably ill, who face a future of pain and despair, and who wish to die will not have suffered a loss if they are killed. It is, strictly speaking, the value of a human's future which makes killing wrong in this theory. This being so, killing does not necessarily wrong some persons who are sick and dying. Of course, there may be other reasons for a prohibition of active euthanasia, but that is another matter. Sanctity-of-human-life theories seem to hold that active euthanasia is seriously wrong even in an individual case where there seems to be good reason for it independently of public policy considerations. This consequence is most implausible, and it is a plus for the claim that the loss of a future of value is what makes killing wrong that it does not share this consequence.

In the fourth place, the account of the wrongness of killing defended in this essay does straightforwardly entail that it is prima facie seriously wrong to kill children and infants, for we do presume that they have futures of value. Since we do believe that it is wrong to kill defenseless little babies, it is important that a theory of the wrongness of killing easily account for this. Personhood theories of the wrongness of killing, on the other hand, cannot straightforwardly account for the wrongness of killing infants and young children. Hence, such theories must add special ad hoc accounts of the wrongness of killing the young. The plausibility of such ad hoc theories seems to be a function of how desperately one wants such theories to work. The claim that the primary wrong-making feature of a killing is the loss to the victim of the value of its future accounts for the wrongness of killing young children and infants directly; it makes the wrongness of such acts as obvious as we actually think it is. This is a further merit of this theory. Accordingly, it seems that this value of a future-like-ours theory of the wrongness of killing shares strengths of both sanctity-of-life and personhood accounts while avoiding weaknesses of both. In addition, it meshes with a central intuition concerning what makes killing wrong.

The claim that the primary wrong-making feature of a killing is the loss to the victim of the value of its future has obvious consequences for the ethics of abortion. The future of a standard fetus includes a set of experiences, projects, activities, and such which are identical with the futures of adult human beings and are identical with the futures of young children. Since the reason that is sufficient to explain why it is wrong to kill human beings after the time of birth is a reason that also applies to fetuses, it follows that abortion is prima facie seriously morally wrong.

This argument does not rely on the invalid inference that, since it is wrong to kill persons, it is wrong to kill potential persons also. The category that is morally central to this analysis is the category of having a valuable future like ours; it is not the category of personhood. The argument to the conclusion that abortion is prima facie seriously morally wrong proceeded independently of the notion of person or potential person or any equivalent. Someone may wish to start with this analysis in terms of the value of a human future, conclude that abortion is, except perhaps in rare circumstances, seriously morally wrong, infer that fetuses have the right to life, and then call fetuses "persons" as a result of their having the right to life. Clearly, in this case, the category of person is being used to state the *conclusion* of the analysis rather than to generate the *argument* of the analysis.

. . . How complete an account of the wrongness of killing does the value of a future-like-ours account have to be in order that the wrongness of abortion is a consequence? This account does not have to be an account of the necessary conditions for the wrongness of killing. Some persons in nursing homes may lack valuable human futures, yet it may be wrong to kill them for other reasons. Furthermore, this account does not obviously have to be the sole reason killing is wrong where the victim did have a valuable future. This analysis claims only that, for any killing where the victim did have a valuable future like ours, having that future by itself is sufficient to create the strong presumption that the killing is seriously wrong.

One way to overturn the value of a future-like-ours argument would be to find some account of the wrongness of killing which is at least as intelligible and which has different implications for the ethics of abortion. Two rival accounts possess at least some degree of plausibility. One account is based on the obvious fact that people value the experience of living and wish for that valuable experience to continue. Therefore, it might be said, what makes killing wrong is the discontinuation of that experience for the victim. Let us call this the *discontinuation account*. Another rival account is based upon the obvious fact that people strongly desire to continue to live. This suggests that what makes killing us so wrong is that it interferes with the fulfillment of a strong and fundamental desire, the fulfillment of which is necessary for the fulfillment of any other desires we might have. Let us call this the *desire account*.[1]

. . . One problem with the desire account is that we do regard it as seriously wrong to kill persons who have little desire to live or who have no desire to live or, indeed, have a desire not to live. We believe it is seriously wrong to kill the unconscious, the sleeping, those who are tired of life, and those who are suicidal. The value-of-a-human future account renders standard morality intelligible in these cases; these cases appear to be incompatible with the desire account.

The desire account is subject to a deeper difficulty. We desire life, because we value the goods of this life. The goodness of life is not secondary to our desire for it. If this were not so, the pain of one's own premature death could be done away with merely by an appropriate alteration in the configuration of one's desires. This is absurd. Hence, it would seem that it is the loss of the goods of one's future, not the interference with the fulfillment of a strong desire to live, which accounts ultimately for the wrongness of killing.

It is worth noting that, if the desire account is modified so that it does not provide a necessary, but only a sufficient, condition for the wrongness of killing, the desire account is compatible with the value of a future-like-ours account. The combined accounts will yield an anti-abortion ethic. This suggests that one can retain what is intuitively plausible about the desire account without a challenge to the basic argument of this paper.

It is also worth noting that, if future desires have moral force in a modified desire account of the wrongness of killing, one can find support for an anti-abortion ethic even in the absence of a value of a future-like-ours account. If one decides that a morally relevant property, the possession of which is sufficient to make it wrong to kill some individual, is the desire at some future time to live—one might decide to justify one's refusal to kill suicidal teenagers on these grounds, for example—then, since typical fetuses will have the desire in the future to live, it is wrong to kill typical fetuses. Accordingly, it does not seem that a desire account of the wrongness of killing can provide a justification of a pro-choice ethic of abortion which is nearly as adequate as the value of a human-future justification of an anti-abortion ethic.

The discontinuation account looks more promising as an account of the wrongness of killing. It seems just as intelligible as the value of a future-like-ours account, but it does not justify an anti-abortion position. Obviously, if it is the continuation of one's activities, experiences, and projects, the loss of which makes killing wrong, then it is not wrong to kill fetuses for that reason, for fetuses do not have experiences, activities, and projects to be continued or discontinued. Accordingly, the discontinuation account does not have the anti-abortion consequences that the value of a future-like-ours account has. Yet, it seems as intelligible as the value of a future-like-ours account, for when we think of what would be wrong with our being killed, it does seem as if it is the discontinuation of what makes our lives worthwhile which makes killing us wrong.

Is the discontinuation account just as good an account as the value of a future-like-ours account? The discontinuation account will not be adequate at all, if it does not refer to the value of the experience that may be discontinued. One does not want the discontinuation account to make it wrong to kill a patient who begs for death and who is in severe pain that cannot be relieved short of killing. (I leave open the question of whether it is wrong for other reasons.) Accordingly, the discontinuation account must be more than a bare discontinuation account. It must make some reference to the positive value of the patient's experiences. But, by the same token, the value of a future-like-ours account cannot be a bare future account either. Just having a future surely does not itself rule out killing the above patient. This account must make some reference to the value of the patient's future experiences and projects also. Hence, both accounts involve the value of experiences, projects, and activities. So far we still have symmetry between the accounts.

The symmetry fades, however, when we focus on the time period of the value of the experiences, etc., which has moral consequences. Although both accounts leave open the possibility that the patient in our example may be killed, this possibility is left open only in virtue of the utterly bleak future for the patient. It makes no difference whether the patient's immediate past contains intolerable pain, or consists in being in a coma (which we can imagine is a situation of indifference), or consists in a life of value. If the patient's future is a future of value, we want our account to make it wrong to kill the patient. If the patient's future is intolerable, whatever his or her immediate past, we want our account to allow killing the patient. Obviously, then, it is the value of that patient's future which is doing the work in rendering the morality of killing the patient intelligible.

This being the case, it seems clear that whether one has immediate past experiences or not does no work in the explanation of what makes killing wrong. The addition the discontinuation account makes to the value of a human future account is otiose. Its addition to the value-of-a-future account plays no role at all in rendering intelligible the wrongness of killing. Therefore, it can be discarded with the discontinuation account of which it is a part.

The analysis [above] . . . suggests that alternative general accounts of the wrongness of killing are either inadequate or unsuccessful in getting around the anti-abortion consequences of the value of a future-like-ours argument. A different strategy for avoiding these anti-abortion consequences involves limiting the scope of the value of a future argument. More precisely, the strategy involves arguing that fetuses lack a property that is essential for the value-of a-future argument (or for any anti-abortion argument) to apply to them.

One move of this sort is based upon the claim that a necessary condition of one's future being valuable is that one values it. Value implies a valuer. Given this one might argue that, since fetuses cannot value their futures, their futures are not valuable to them. Hence, it does not seriously wrong them deliberately to end their lives.

This move fails, however, because of some ambiguities. Let us assume that something cannot be of value unless it is valued by someone. This does not entail that my life is of no value unless it is valued by me. I may think, in a period of despair, that my future is of no worth whatsoever, but I may be wrong because others rightly see value—even great value—in it. Furthermore, my future can be valuable to me even if I do not value it. This is the case when a young person attempts suicide, but is rescued and goes on to significant human achievements. Such young people's futures are ultimately valuable to them, even though such futures do not seem to be valuable to them at the moment of attempted suicide. A fetus's future can be valuable to it in the same way. Accordingly, this attempt to limit the anti-abortion argument fails.

Another similar attempt to reject the anti-abortion position is based on Tooley's claim that an entity cannot possess the right to life unless it has the capacity to desire its continued existence. It follows that, since fetuses lack the conceptual capacity to desire to continue to live, they lack the right to life. Accordingly, Tooley concludes that abortion cannot be seriously prima facie wrong.[2]

What could be the evidence for Tooley's basic claim? Tooley once argued that individuals have a prima facie right to what they desire and that the lack of the capacity to desire something undercuts the basis of one's right to it.[3] This argument plainly will not succeed in the context of the analysis of this essay, however, since the point here is to establish the fetus's right to life on other grounds. Tooley's argument assumes that the right to life cannot be established in general on some basis other than the desire for life. This position was considered and rejected in the preceding section of this paper.

One might attempt to defend Tooley's basic claim on the grounds that, because a fetus cannot apprehend continued life as a benefit, its continued life cannot be a benefit or cannot be something it has a right to or cannot be something that is in its interest. This might be defended in terms of the general proposition that, if an individual is literally incapable of caring about or taking an interest in some X, then one does not have a right to X or X is not a benefit or X is not something that is in one's interest.

Each member of this family of claims seems to be open to objections. . . . [O]ne may have a right to be treated with a certain medical procedure (because of a health insurance policy one has purchased), even though one cannot conceive of the nature of the procedure.[4] And, as Tooley himself has pointed out, persons who have been indoctrinated, or drugged, or rendered temporarily unconscious may be literally incapable of caring about or taking an interest in something that is in their interest or is something to which they have a right, or is something that benefits them. Hence, the Tooley claim that would restrict the scope of the value of a future-like-ours argument is undermined by counterexamples.[5]

Finally, Paul Bassen[6] has argued that, even though the prospects of an embryo might seem to be a basis for the wrongness of abortion, an embryo cannot be a victim and therefore cannot be wronged. An embryo cannot be a victim, he says, because it lacks sentience. His central argument for this seems to be that, even though plants and the permanently unconscious are alive, they clearly cannot be victims. What is the explanation of this? Bassen claims that the explanation is that their lives consist of mere metabolism and mere metabolism is not enough to ground victimizability. Mentation is required.

The problem with this attempt to establish the absence of victimizability is that both plants and the permanently unconscious clearly lack what Bassen calls "prospects" or what I have called "a future life like ours." Hence, it is surely open to one to argue that the real reason we believe plants and the permanently unconscious cannot be victims is that killing them cannot deprive them of a future life like ours; the real reason is not their absence of present meritation. Bassen recognizes that his view is subject to this difficulty, and he recognizes that the case of children seems to support this difficulty, for "much of what we do for children is based on prospects." He argues, however, that, in the case of children and in other such cases, "potentiality comes into play only where victimizability has been secured on other grounds."[7]

Bassen's defense of his view is patently question-begging, since what is adequate to secure victimizability is exactly what is at issue. His examples do not support his own view against the thesis of this essay. Of course, embryos can be victims: when their lives are deliberately terminated, they are deprived of their futures of value, their prospects. This makes them victims, for it directly wrongs them.

The seeming plausibility of Bassen's view stems from the fact that paradigmatic cases of imagining someone as a victim involve empathy, and empathy requires mentation of the victim. The victims of flood, famine, rape, or child abuse are all persons with whom we can empathize. That empathy seems to be part of seeing them as victims.

In spite of the strength of these examples, the attractive intuition that a situation in which there is victimization requires the possibility of empathy is subject to counterexamples. Consider a case that Bassen himself offers: "Posthumous obliteration of an author's work constitutes a misfortune for him only if he had wished his work to endure."[8] The conditions Bassen wishes to impose upon the possibility of being victimized here seem far too strong. Perhaps this author, due to his unrealistic standards of excellence and his low self-esteem, regarded his work as unworthy of survival, even though it possessed genuine literary merit. Destruction of such work would surely victimize its author. In such a case, empathy with the victim concerning the loss is clearly impossible.

Of course, Bassen does not make the possibility of empathy a necessary condition of victimizability; he requires only mentation. Hence, on Bassen's actual view, this author, as I have described him, can be a victim. The problem is that the basic intuition that renders Bassen's view plausible is missing in the author's case. In order to attempt to avoid counterexamples, Bassen has made his thesis too weak to be supported by the intuitions that suggested it.

Even so, the mentation requirement on victimizability is still subject to counterexamples. Suppose a severe accident renders me totally unconscious for a month, after which I recover. Surely killing me while I am unconscious victimizes me, even though I am incapable of mentation during that time. It follows that Bassen's thesis fails. Apparently, attempts to restrict the value of a future-like-ours argument so that fetuses do not fall within its scope do not succeed.

In this essay, it has been argued that the correct ethic of the wrongness of killing can be extended to fetal life and used to show that there is a strong presumption that any abortion is morally impermissible. If the ethic of killing adopted here entails, however, that contraception is also seriously immoral, then there would appear to be a difficulty with the analysis of this essay.

But this analysis does not entail that contraception is wrong. Of course, contraception prevents the actualization of a possible future of value. Hence, it follows from the claim that futures of value should be maximized that contraception is prima facie immoral. This obligation to maximize does not exist, however; furthermore, nothing in the ethics of killing in this paper entails that it does. The ethics of killing in this essay would entail that contraception is wrong only if something were denied a human future of value by contraception. Nothing at all is denied such a future by contraception, however.

Candidates for a subject of harm by contraception fall into four categories: (1) some sperm or other, (2) some ovum or other, (3) a sperm and an ovum separately, and (4) a sperm and an ovum together. Assigning the harm to some sperm is utterly arbitrary, for no reason can be given for making a sperm the subject of harm rather than an ovum. Assigning the harm to some ovum is utterly arbitrary, for no reason can be given for making an ovum the subject of harm rather than a sperm. One might attempt to avoid these problems by insisting that contraception deprives both the sperm and the ovum separately of a valuable future like ours. On this alternative, too many futures are lost. Contraception was supposed to be wrong, because it deprived us of one future of value, not two. One might attempt to avoid this problem by holding that contraception deprives the combination of sperm and ovum of a valuable future like ours. But here the definite article misleads. At the time of contraception, there are hundreds of millions of sperm, one (released) ovum and millions of possible combinations of all of these. There is no actual combination at all. Is the subject of the loss to be a merely possible combination? Which one? This alternative does not yield an actual subject of harm either. Accordingly, the immorality of contraception is not entailed by the loss of a future-like-ours argument simply because there is no nonarbitrarily identifiable subject of the loss in the case of contraception.

The purpose of this essay has been to set out an argument for the serious presumptive wrongness of abortion subject to the assumption that the moral permissibility of abortion stands or falls on the moral status of the fetus. Since a fetus possesses a property, the possession of which in adult human beings is sufficient to make killing an adult human being wrong, abortion is wrong. This way of dealing with the problem of abortion seems superior to other approaches to the ethics of abortion, because it rests on an ethics of killing which is close to self-evident, because the crucial morally relevant property clearly applies to fetuses, and because the argument avoids the usual equivocations on 'human life,' 'human being,' or 'person.' The argument rests neither on religious claims nor on Papal dogma. It is not subject to the objection of "speciesism." Its soundness is compatible with the moral permissibility of euthanasia and contraception. It deals with our intuitions concerning young children.

Finally, this analysis can be viewed as resolving a standard problem—indeed, *the* standard problem—concerning the ethics of abortion. Clearly, it is wrong to kill adult human beings. Clearly, it is not wrong to end the life of some arbitrarily chosen single human cell. Fetuses seem to be like arbitrarily chosen human cells in some respects and like adult humans in other respects. The problem of the ethics of abortion is the problem of determining the fetal property that settles this moral controversy. The thesis of this essay is that the problem of the ethics of abortion, so understood, is solvable.

References and Notes

1. Presumably a preference utilitarian would press such an objection. Tooley once suggested that his account has such a theoretical underpinning. See his "Abortion and Infanticide" *Philosophy and Public Affairs*, 11, 1 (1972): 37–65, at pp. 44–45.

2. Michael Tooley, "Abortion and Infanticide," *Philosophy and Public Affairs* 11, 1, 1972, 37–65, at pp. 46–47.

3. Michael Tooley, "Abortion and Infanticide," *Philosophy and Public Affairs* 11, 1, 1972, 37–65, at pp. 44–45.

4. John C. Stevens, "Must the Bearer of a Right Have the Concept of That to Which He Has a Right?" *Ethics*, xcv, 1 (1984):68–74.

5. Michael Tooley, "Abortion and Infanticide," *Philosophy and Public Affairs* 11, 1, 1972, 37–65, at pp. 47–49.

6. Paul Bassen, "Present Sakes and Future Prospects: The Status of Early Abortion," *Philosophy and Public Affairs*, xi, 4 (1982):322–326.

7. Paul Bassen, "Present Sakes and Future Prospects: The Status of Early Abortion," *Philosophy and Public Affairs*, xi, 4 (1982):322–326, at p. 333.

8. Paul Bassen, "Present Sakes and Future Prospects: The Status of Early Abortion," *Philosophy and Public Affairs*, xi, 4 (1982), p. 318.

ASSISTED REPRODUCTIVE TECHNOLOGIES

Assisted Reproduction

President's Council on Bioethics

The President's Council on Bioethics was created in 2001 to advise the President on bioethical issues that may emerge as a consequence of advances in biomedical science and technology. In "Reproduction and Responsibility: The Regulation of New Biotechnologies," the Council recommends a series of measures to advance public understanding of the challenges of new technologies affecting human reproduction. These measures fall into three categories: studies and collection of more data, oversight and self-regulation, and legislative action. In this chapter on assisted reproduction, the Council characterizes the techniques and practices of assisted reproduction and identifies ethical considerations, including patient vulnerability, the premature adoption of novel technologies into practice, and the effects of assisted reproduction on the child.

I. TECHNIQUES AND PRACTICES

Most methods of assisted reproduction involve five discrete phases: (1) collection and preparation of gametes; (2) fertilization; (3) transfer of an embryo or multiple embryos to a woman's uterus; (4) pregnancy; and (5) delivery and birth. We will discuss each phase separately. . . .

A. Collection and Preparation of Gametes

The precursors of human life are the gametes: sperm and ova. Parents seeking to conceive through assisted reproduction usually provide their own gametes. In the United States in the year 2001, 75.2 percent of the ART [assisted reproductive technology cycles] undertaken used never-frozen, nondonor ova or embryos and another 13.7 percent used frozen nondonor ova or embryos. Of the remaining 11.1 percent of cycles using donor embryos, the breakdown is as follows: 3.2 percent of the embryos were previously cryopreserved, and 8 percent were not. . . .

Excerpted from The President's Council on Bioethics, *Reproduction and Responsibility: The Regulation of New Biotechnologies* (Washington, D.C., March 2004), chapter 2, pp. 23–46. Many notes omitted.

Acquiring ova for use in artificial reproduction is significantly more oner-ous, painful, and risky than acquiring sperm (though its risks are still low in ab-solute terms). In the normal course of ovulation, one mature oocyte is produced per menstrual cycle. However, in assisted reproduction—to increase the proba-bility of success—many more ova are typically retrieved and fertilized. Thus, the ova source (who is usually also the gestational mother) undergoes a drug-induced process intended to stimulate her ovaries to produce many mature oocytes in a single cycle. This procedure, commonly referred to as "superovula-tion," requires the daily injection of a synthetic gonadotropin analog, accompa-nied by frequent monitoring using blood tests and ultrasound examinations. This treatment begins midway through the previous menstrual cycle and con-tinues until just before ova retrieval. The synthetic gonadotropin analogs give the clinician greater control over ovarian stimulation and prevent premature re-lease of the ova. . . .

When blood testing and ultrasound monitoring suggest that the ova are suf-ficiently mature, the clinician attempts to harvest them. This is typically achieved by ultrasound-guided transvaginal aspiration. In this procedure, a nee-dle guided by ultrasound is inserted through the vaginal wall and into the ma-ture ovarian follicles. An ovum is withdrawn (along with some fluid) from each follicle. This is an outpatient procedure. Risks and complications are low, but may include accidental puncture of nearby organs such as the bowel, ureter, bladder, or blood vessels, as well as the typical risks accompanying outpatient surgery (for example, risks related to administration of anesthesia, infection, etc.).

Once sperm and ova have been collected, they are cultured and treated to maximize the probability of success. . . .

B. Fertilization

Once the ova and sperm have been properly prepared, the clinician attempts to induce fertilization—the union of sperm and ovum culminating in the fusion of their separate pronuclei and the initiation of a new, integrated, self-directing or-ganism. It is common practice to attempt to fertilize all available ova. Fertiliza-tion can be achieved through a number of means including (1) "classical" IVF [in vitro fertilization], (2) gamete intrafallopian transfer (GIFT), (3) intracyto-plasmic sperm injection (ICSI). . . .

IVF is the most common method of artificial fertilization. In 2001, it was used by 99 percent of ART patients. As noted previously, both sperm and ovum are cultured to maximize the probability of fertilization. The ova are examined and rated for maturity in an effort to calculate the optimal time for fertilization. They are usually placed in a tissue culture medium and left undisturbed for two to twenty-four hours. . . . Once the gametes are adequately prepared, thousands of tiny droplets of sperm are placed in the culture medium containing a single ovum. After 24 hours, each of the oocytes is examined to determine whether fer-tilization has occurred.

GIFT was introduced in 1984 as an alternative to standard IVF. Today, attempts at fertilization via GIFT are rare. In 2001, they accounted for less than 1 percent of all attempts at fertilization used by ART patients. As the name suggests, fertilization using GIFT occurs within the woman's body. Ovarian stimulation and retrieval are performed in the same manner as in IVF. In a single procedure, ova are retrieved, combined with the sperm outside the body, and then transferred back into the fallopian tube where it is hoped that fertilization itself will occur. . . .

A new and increasingly popular technique for fertilization is intracytoplasmic sperm injection. As the name implies, with ICSI, ovum-sperm fusion is accomplished not by chance, but by injecting a single sperm directly into an oocyte. . . . A single sperm is selected and drawn into a thin pipette from which it is injected into the cytoplasm of the ovum cell.

ICSI is indicated in cases of severe male-factor infertility, in which male patients have either malformed sperm or an abnormally low sperm count. ICSI is also ideal for patients whose sperm would not otherwise penetrate the exterior of an oocyte. ICSI was used in 49.2 percent of all ART cycles in 2001. However, 42.2 percent of those ICSI cycles were undertaken by couples without male-factor infertility. The growing popularity of this technique most likely has to do with the wish to increase the control over, and success rates for, fertilization: ICSI, unlike standard IVF, guarantees the entrance of a single sperm directly into a single egg. . . .

[Following fertilization of harvested eggs, multiple embryos may be produced and available for transfer to the woman's uterus.] . . . [I]n many cases not all embryos are transferred. . . , [thus,] cryopreservation of embryos [for possible use in future cycles] has become an integral part of ART. . . . A recently reported study by the Society for Assisted Reproductive Technology and RAND estimates that 400,000 embryos are in cryostorage in the United States.

Most ART patients do not receive cryopreserved embryos. In 2001, only 14 percent of all ART cycles involved transfer of frozen embryos. The rate of live births for cycles using cryopreserved embryos is significantly lower than it is for never-frozen embryos (23.4 percent versus 33.4 percent). Experts estimate that only 65 percent of frozen embryos survive the thawing process. There are, however, incentives for couples to use cryopreserved embryos; doing so eliminates the cost and effort of further oocyte retrieval. This can decrease the cost of a future cycle by roughly $6,000. . . . Cryopreservation also reduces pressure to implant all embryos at once, thus reducing the risk of high-order multiple pregnancies.

C. Transfer

. . . Typically, the embryos are transferred on the second or third day after fertilization, at the four- to eight-cell stage. To maximize the probability of implantation, some clinicians cultivate embryos until the blastocyst stage (five days after fertilization) before transferring them to the uterus. . . .

Once the embryos have been selected and prepared, they are transferred into the uterus. The total number of embryos transferred per cycle varies, usually according to the age of the recipient. For women under 35, the average number of never-frozen embryos transplanted per transfer procedure was 2.8. For women 35 to 37, 38 to 40, and 41 to 42, the average numbers of never-frozen embryos transplanted per transfer procedure were, respectively, 3.1, 3.4, and 3.7. . . .

Typically embryos are transferred into the uterus using a catheter. The catheter is inserted through the woman's cervix and the embryos are injected into her uterus (along with some amount of the culture fluid). This procedure does not require anesthesia. Following injection, the patient must lie still for at least one hour. While the transfer procedure is regarded as simple, different practitioners tend to achieve different outcomes.

An alternative method of embryo transfer is zygote intrafallopian transfer (ZIFT). In ZIFT, the embryo is placed (via laparoscopy) directly into the fallopian tube, rather than into the uterus. In this way, it is similar to the transfer of gametes in GIFT. Some individuals opt for ZIFT on the theory that it enhances the likelihood of implantation, given that the embryo matures en route to the uterus, presumably as it would in natural conception and implantation. Additionally, many patients prefer ZIFT to GIFT because the process of fertilization and early development of the embryo may be monitored. However, ZIFT remains a rare choice, accounting for 0.8 percent of all ART cycles in 2001.

D. Pregnancy

Successful implantation of an embryo in the uterine lining marks the beginning of pregnancy. In 2001, 32.8 percent of the ART cycles undertaken resulted in clinical pregnancy. This number varied according to patient age. . . .

Multiple gestations are common among pregnancies facilitated by assisted reproductive technologies. The rate of multiple-fetus pregnancies from ART cycles using never-frozen, nondonor ova or embryos in 2001 was 36.7 percent. For the same time period, the multiple infant birth rate in the United States was 3 percent. The extraordinarily high rate of multiple pregnancies resulting from assisted reproduction is almost entirely attributable to the transfer of multiple embryos per cycle.

In an effort to reduce the risks of multiple pregnancy, practitioners sometimes employ a procedure termed "fetal reduction," the reduction in the number of fetuses in utero by selective abortion. Fetuses are selected for destruction based on size, position, and viability (in the clinician's judgment). The clinician, using ultrasound for guidance, inserts a needle through the mother's abdomen (transabdominal multifetal reduction) through the uterine wall. The clinician then administers a lethal injection to the heart of the selected fetus—typically potassium chloride. The dead fetus's body decomposes and is reabsorbed. To be effective, transabdominal multifetal reduction must be performed at ten to twelve weeks' gestation. In an alternative procedure, transvaginal multifetal reduction, a needle is inserted through the vagina. Transvaginal multifetal reduction must be performed between six and eight weeks gestation (eight weeks is recommended).

E. Delivery

In 2001, for never-frozen nondonor ova or embryos, the overall rate of live births per cycle was 27 percent (33.4 percent live births per transfer). Among these pregnancies, 82.2 percent resulted in live births. Of these resulting 21,813 live births, 35.8 percent were multiple infant births (32 percent twins and 3.8 percent triplets or more). . . .

F. Disposition of Unused Embryos

As mentioned above, in many cases of ART there are in vitro embryos that re-main untransferred following a successful cycle. There are five possible out-comes for such an embryo: (1) it may remain in cryostorage until transferred into the mother's uterus in a future ART cycle; (2) it may be donated to another person or couple seeking to initiate a pregnancy; (3) it may be donated for pur-poses of research; (4) it may remain in cryostorage indefinitely; or (5) it may be thawed and destroyed. . . .

II. ETHICAL CONSIDERATIONS

The development and practice of assisted reproductive technologies have yielded great goods. They have relieved the suffering of many who are afflicted with infertility, helping them to conceive biologically related children. Yet these activities also raise a variety of ethical issues. Some concern the well-being of the participants in assisted reproduction. . . .

The intersection of two key factors—patient vulnerability and novel (in some cases untested) technology—defines much of the arena of concern. First, assisted reproduction is generally practiced on patients who are experiencing great emotional strain. When it succeeds it can be a source of great joy—as it has been for tens of thousands of parents each year. But success is far from uni-versal, especially for older patients; and even when it happens, the process and the circumstances surrounding it can be difficult to bear. Those suffering from infertility often come to practitioners of assisted reproduction after prolonged periods of failure and dismay. This vulnerability may lead some individuals to take undue risks (such as to insist on transferring an unduly large number of em-bryos). The occasional irresponsible clinician may even pressure patients to take such risks, for the sake of improving his reportable success rates.

Second, some assisted reproductive technologies have been used in clinical practice without prior rigorous testing in primates or studies of long-term out-comes. IVF itself was performed on at least 1,200 women before it was reported to have been performed on chimps, although it had been extensively investigated in rabbits, hamsters, and mice. The same is true for ICSI. The reproductive use of ICSI was first introduced by Belgian researchers in 1992. Two years later, relying on a two-study review of safety and efficacy, ASRM (The American Society for Reproductive Medicine) declared ICSI to be a "clinical" rather than "experimental" procedure. Yet the first nonhuman primate conceived by ICSI was born only in 1997 and the first successful ICSI procedure in mice was reported in 1995. Absent long-term studies of the children conceived using ICSI or other novel procedures, it is unclear to what extent these alterations in the ART process affect the health and development of the children so conceived. . . .

A. Well-Being of the Child

The central figure in the process of assisted reproduction, directly affected by every action taken but incapable of consenting to such actions, is the child born with the aid of ART. Each intervention or stage in the ART process might affect this child's health and well-being: gamete retrieval and preparation, fertilization, embryo culture, embryo transfer, pregnancy, and of course birth. . . .

There have been very few comprehensive or long-term studies of the health and well-being of children born using ART, although more than 170,000 such children have been born in the United States. The fact that no major investigation or public study has yet been called for in this area might suggest that there is no discernible health crisis in assisted reproduction, as does the fact that demand for ART has grown substantially and continuously since its inception. At the same time, however, our ability to know this with certainty is limited, both because of the absence of major longitudinal studies of the well-being of children born using different assisted reproduction techniques, and because the oldest person conceived through ART is only in her mid-twenties.

Some recent studies have associated various birth defects and developmental difficulties with the uses of various technologies and practices of assisted reproduction. None of these studies provide a causal link between ART and the dysfunctions observed, and some commentators have taken issue with some of the methodologies used. Nevertheless, these findings have raised some concerns. One such study concluded that children conceived by assisted reproduction are twice as likely to suffer major birth defects as children conceived without such assistance. Other recent studies have reached similar conclusions. Additional studies have associated the use of assisted reproduction technologies with a higher incidence of diseases and malformations, including Beckwith-Wiedemann syndrome (BWS), rare urological defects, retinoblastoma, neural tube defects, and Angelman syndrome.

While many are concerned about the increased risk to children suggested by these studies, the overall incidence of such harms is low enough that infertile couples have not been deterred in their efforts to conceive using IVF or ICSI. Indeed, ART clinicians (and in some cases the authors of these studies) advise their patients that such data should not dissuade them from pursuing infertility treatment.

ICSI has raised concerns among some observers largely for the very reasons that it has proven so successful as a means of fertilization: ICSI circumvents the ovum's natural barrier against sperm otherwise incapable of insemination. Some suspect that removing this barrier may permit a damaged sperm (for example, aneuploid or with damaged DNA) to fertilize an ovum, resulting in spontaneous abortion or harm to the resulting child. Some male ART patients have a gene mutation or a chromosomal deletion that renders them infertile. Yet, if a sperm can be retrieved from these patients, they may be able to conceive a child via ICSI, possibly passing along the genetic abnormality to the resulting child. . . .

It is a matter of concern that there have been few longitudinal studies analyzing the long-term effects of ICSI on the children born with its aid. The Belgian group that pioneered ICSI has collected a database that details neonatal outcome and congenital malformations in children conceived through ICSI. But there do not seem to be any ongoing or published studies of this kind investigating the long-term effects of ICSI beyond the neonatal stage. . . .

Multiple gestations, far more common in the context of assisted reproduction than in natural conception, have a higher incidence of adverse impacts on the health of the children born. Such pregnancies greatly increase the risk of prenatal death. Multiple pregnancies are also more likely to lead to premature birth; and prematurity is associated with myriad health problems including serious infection, respiratory distress syndrome, and heart defects. One in ten children born following high-order pregnancies dies before one year of age. Children born following a multiple pregnancy are at greater risk for such disabilities as blindness, respiratory dysfunction, and brain damage. Moreover, infants born following such a pregnancy tend to have an extremely low birthweight, which is itself associated with a number of health problems, including some that manifest themselves only later in life, such as hypertension, cardiac disease, stroke, and osteoporosis in middle age. Interestingly, the higher incidence of low birthweight may not be limited to infants born from multiple pregnancies. According to recent studies, singletons born with the aid of ART tend to have an abnormally high incidence of prematurity and low birthweight. . . .

. . . [T]he significance of these various studies is uncertain. They raise a broad range of concerns, but the scale of the research has been limited. In many cases, there are observed correlations between ART and a higher incidence of certain health problems in the resulting children. But in most studies, there is no demonstrable causal relationship between a particular facet of ART and the undesirable health effect. Infertile individuals seeking assisted reproduction may be disproportionately afflicted with heritable disorders, and these may in part account for the higher incidence of birth and developmental abnormalities in ART children compared to those conceived in vivo. The results are therefore still preliminary. The need seems clear for more data to determine what risks, if any, different assisted reproduction techniques present to the well-being of the future child. Moreover, in cases where ART is the only available means for individuals or couples to conceive a biologically related child, it is an important ethical and social question what level of increased risk can be privately justified by patients and doctors, and what level of increased risk should be publicly justified by society as a whole, especially should the society bear the costs of caring for any resulting health problems.

B. Well-Being of Women in the ART Process

Another concern is for the well-being of the women who participate directly in the process of assisted reproduction.

Aside from the discomforts and burdens of ovarian stimulation and monitoring, there are also some risks attached to hormonal stimulation. One such risk is "ovarian hyperstimulation syndrome," characterized by dramatic enlargement of the ovaries and fluid imbalances that can be (in extreme cases) life threatening. Complications can include rupture of the ovaries, cysts, and cancers. The reported incidence of severe ovarian hyperstimulation syndrome is between 0.5 and 5.0 percent. Additionally, adverse side effects of the hormones administered during superovulation have included memory loss, neurological dysfunction, cardiac disorders, and even sudden death. There do not appear to be any studies on the incidence of such side effects. . . .

Multiple pregnancies are far more common following ART, owing especially to the practice of transferring multiple embryos but also to the higher incidence of spontaneous twinning with any single embryo. Multiple pregnancies pose greater risks to mothers than do singleton pregnancies. A woman carrying multiple fetuses has a greater chance of suffering from high blood pressure, anemia, or pre-eclampsia. Because multiple-gestation pregnancies are generally more taxing on the mother's body, they are likelier to aggravate pre-existing medical conditions. Moreover, such pregnancies expose the woman to higher risks of uterine rupture, placenta previa, or abruption. . . .

The Presumptive Primacy of Procreative Liberty

John A. Robertson

John A. Robertson, Professor of Law and the Vinson and Elkins Chair at The University of Texas School of Law at Austin, obtained his J.D. from Harvard University. He has written and lectured widely on law and bioethical issues. In this chapter from *Children of Choice: Freedom and the New Reproductive Technologies*, Robertson argues that procreative liberty, the freedom either to have children or to avoid having them, is a negative right against public or private interference. This moral right to reproduce is respected because of the centrality of reproduction to personal identity, dignity, and the meaning of one's life. He argues that restricting procreative liberty would require a justification that reproduction would create a substantive harm. Moreover, he suggests that this argument applies to coital reproduction as well as the use of reproductive technologies.

What is Procreative Liberty?

At the most general level, procreative liberty is the freedom either to have children or to avoid having them. Although often expressed or realized in the context of a couple, it is first and foremost an individual interest. It is to be distinguished from freedom in the ancillary aspects of reproduction, such as liberty in the conduct of pregnancy or choice of place or mode of childbirth.

The concept of reproduction, however, has a certain ambiguity contained within it. In a strict sense, reproduction is always genetic. It occurs by provision of one's gametes to a new person, and thus includes having or producing offspring. While female reproduction has traditionally included gestation, in vitro fertilization (IVF) now allows female genetic and gestational reproduction to be separated. Thus a woman who has provided the egg that is carried by another has reproduced, even if she has not gestated and does not rear resulting offspring. Because of the close link between gestation and female reproduction, a woman who gestates the embryo of another may also reasonably be viewed as having a reproductive experience, even though she does not reproduce genetically.

In any case, reproduction in the genetic or gestational sense is to be distinguished from child rearing. Although reproduction is highly valued in part because it usually leads to child rearing, one can produce offspring without rearing

Excerpted from John A. Robertson, *Children of Choice: Freedom and the New Reproductive Technologies* (Princeton University Press, 1994), chapter 2. Notes for this chapter are omitted. Reproduced with permission of Princeton University Press. Copyright © Princeton University Press, 1994.

them and rear children without reproduction. One who rears an adopted child has not reproduced, while one who has genetic progeny but does not rear them has.

In this [chapter] . . . the terms "procreative liberty" and "reproductive freedom" will mean the freedom to reproduce or not to reproduce in the genetic sense, which may also include rearing or not, as intended by the parties. Those terms will also include female gestation whether or not there is a genetic connection to the resulting child.

. . . Two . . . qualifications on the meaning of procreative liberty should be noted. One is that "liberty" as used in procreative liberty is a negative right. It means that a person violates no moral duty in making a procreative choice, and that other persons have a duty not to interfere with that choice. However, the negative right to procreate or not does not imply the duty of others to provide the resources or services necessary to exercise one's procreative liberty despite plausible moral arguments for governmental assistance.

. . . The second qualification is that not everything that occurs in and around procreation falls within liberty interests that are distinctively procreative. Thus whether the father may be present during childbirth, whether midwives may assist birth, or whether childbirth may occur at home rather than in a hospital may be important for the parties involved, but they do not implicate the freedom to reproduce (unless one could show that the place or mode of birth would determine whether birth occurs at all). Similarly, questions about a pregnant woman's drug use or other conduct during pregnancy . . . implicates liberty in the course of reproduction but not procreative liberty in the basic sense. . . .

The Importance of Procreative Liberty

Procreative liberty should enjoy presumptive primacy when conflicts about its exercise arise because control over whether one reproduces or not is central to personal identity, to dignity, and to the meaning of one's life. For example, deprivation of the ability to avoid reproduction determines one's self-definition in the most basic sense. It affects women's bodies in a direct and substantial way. It also centrally affects one's psychological and social identity and one's social and moral responsibilities. The resulting burdens are especially onerous for women, but they affect men in significant ways as well.

On the other hand, being deprived of the ability to reproduce prevents one from an experience that is central to individual identity and meaning in life. Although the desire to reproduce is in part socially constructed, at the most basic level transmission of one's genes through reproduction is an animal or species urge closely linked to the sex drive. In connecting us with nature and future generations, reproduction gives solace in the face of death. As Shakespeare noted, "nothing 'gainst Time's scythe can make defense/save breed." For many people "breed"—reproduction and the parenting that usually accompanies it—is a central part of their life plan, and the most satisfying and meaningful experience they have. It also has primary importance as an expression of a couple's love or unity. For many persons, reproduction also has religious significance and is ex-perienced as a "gift from God." Its denial—through infertility or governmental restriction—is experienced as a great loss, even if one has already had children or will have little or no rearing role with them.

Decisions to have or to avoid having children are thus personal decisions of great import that determine the shape and meaning of one's life. The person di-rectly involved is best situated to determine whether that meaning should or should not occur. An ethic of personal autonomy as well as ethics of community or family should then recognize a presumption in favor of most personal repro-ductive choices. Such a presumption does not mean that reproductive choices are without consequence to others, nor that they should never be limited. Rather, it means that those who would limit procreative choice have the burden of showing that the reproductive actions at issue would create such substantial harm that they could justifiably be limited. . . .

A closely related reason for protecting reproductive choice is to avoid the highly intrusive measures that governmental control of reproduction usually en-tails. State interference with reproductive choice may extend beyond exhorta-tion and penalties to Gestapo and police state tactics. Margaret Atwood's powerful futuristic novel *The Handmaid's Tale* expresses this danger by creating a world where fertile women are forcibly impregnated by the ruling powers and their pregnancies monitored to replenish a decimated population.

Equally frightening scenarios have occurred in recent years when repressive governments have interfered with reproductive choice. In Romania and China, men and women have had their most private activities scrutinized in the service of state reproductive goals. In Ceausescu's Romania, where contraception and abortion were strictly forbidden, women's menstrual cycles were routinely mon-itored to see if they were pregnant. Women who did not become pregnant or who had abortions were severely punished. Many women nevertheless sought il-legal abortions and died, leaving their children orphaned and subject to sale to Westerners seeking children for adoption.

In China, forcible abortion and sterilization have occurred in the service of a one-child-per-family population policy. Village cadres have seized pregnant women in their homes and forced them to have abortions. A campaign of forcible sterilization in India in 1977 was seen as an "attack on women and chil-dren" and brought Indira Gandhi's government down. In the United States, state-imposed sterilization of "mental defectives," sanctioned in 1927 by the United States Supreme Court in *Buck v. Bell*, resulted in 60,000 sterilizations over a forty-year period. Many mentally normal people were sterilized by mis-take, and mentally retarded persons who posed little risk of harm to others were subjected to surgery. . . .

Two Types of Procreative Liberty

. . . An essential distinction is between the freedom to avoid reproduction and the freedom to reproduce. When people talk of reproductive rights, they usually have one or the other aspect in mind. Because different interests and justifications underlie each and countervailing interests for limiting each aspect vary, recognition of one aspect does not necessarily mean that the other will also be respected; nor does limitation of one mean that the other can also be denied.

However, there is a mirroring or reciprocal relationship here. Denial of one type of reproductive liberty necessarily implicates the other. If a woman is not able to avoid reproduction through contraception or abortion, she may end up reproducing, with all the burdens that unwanted reproduction entails. Similarly, if one is denied the liberty to reproduce through forcible sterilization, one is forced to avoid reproduction, thus experiencing the loss that absence of progeny brings. By extending reproductive options, new reproductive technologies present challenges to both aspects of procreative choice.

The Freedom to Procreate

In addition to freedom to avoid procreation, procreative liberty also includes the freedom to procreate—the freedom to beget and bear children if one chooses. As with avoiding reproduction, the right to reproduce is a negative right against public or private interference, not a positive right to the services or the resources needed to reproduce. It is an important freedom that is widely accepted as a basic, human right. But its various components and dimensions have never been fully analyzed, as technologies of conception and selection now force us to do.

As with avoiding reproduction, the freedom to procreate involves the freedom to engage in a series of actions that eventuate in reproduction and usually in child rearing. One must be free to marry or find a willing partner, engage in sexual intercourse, achieve conception and pregnancy, carry a pregnancy to term, and rear offspring. Social and natural barriers to reproduction would involve the unavailability of willing or suitable partners, impotence or infertility, and lack of medical and childcare resources. State barriers to marriage, to sexual intercourse, to conception, to infertility treatment, to carrying pregnancies to term, and to certain child-rearing arrangements would also limit the freedom to procreate. The most commonly asserted reasons for limiting coital reproduction are overpopulation, unfitness of parents, harm to offspring, and costs to the state or others. Technologies that treat infertility raise additional concerns that are discussed below.

The moral right to reproduce is respected because of the centrality of reproduction to personal identity, meaning, and dignity. This importance makes the liberty to procreate an important moral right, both for an ethic of individual autonomy and for ethics of community or family that view the purpose of marriage and sexual union as the reproduction and rearing of offspring. Because of this importance, the right to reproduce is widely recognized as a prima facie moral right that cannot be limited except for very good reason.

Recognition of the primacy of procreation does not mean that all reproduction is morally blameless, much less that reproduction is always responsible and praiseworthy and can never be limited. However, the presumptive primacy of procreative liberty sets a very high standard for limiting those rights, tilting the balance in favor of reproducing but not totally determining its acceptability. A two-step process of analysis is envisaged here. The first question is whether a distinctively procreative interest is involved. If so, the question then is whether the harm threatened by reproduction satisfies the strict standard for overriding this liberty interest.

The personal importance of procreation helps answer questions about who holds procreative rights and about the circumstances under which the right to reproduce may be limited. A person's capacity to find significance in reproduction should determine whether one holds the presumptive right, though this question is often discussed in terms of whether persons with such a capacity are fit parents. To have a liberty interest in procreating, one should at a minimum have the mental capacity to understand or appreciate the meanings associated with reproduction. This minimum would exclude severely retarded persons from having reproductive interests, though it would not remove their right to bodily integrity. However, being unmarried, homosexual, physically disabled, infected with HIV, or imprisoned would not disqualify one from having reproductive interests, though they might affect one's ability to rear offspring. Whether those characteristics justify limitations on reproduction is discussed later. Nor would already having reproduced negate a person's interest in reproducing again, though at a certain point the marginal value to a person of additional offspring diminishes.

What kinds of interests or harms make reproduction unduly selfish or irresponsible and thus could justifiably limit the presumptive right to procreate? To answer this question, we must distinguish coital and noncoital reproduction. Surprisingly, there is a widespread reluctance to speak of coital reproduction as irresponsible, much less to urge public action to prevent irresponsible coital reproduction from occurring. If such a conversation did occur, reasons for limiting coital reproduction would involve the heavy costs that it imposed on others—costs that outweighed whatever personal meaning or satisfaction the person(s) reproducing experienced. With coital reproduction, such costs might arise if there were severe overpopulation, if the persons reproducing were unfit parents, if reproduction would harm offspring, or if significant medical or social costs were imposed on others.

Because the United States does not face the severe overpopulation of some countries, the main grounds for claiming that reproduction is irresponsible is where the person(s) reproducing lack the financial means to raise offspring or will otherwise harm their children. As later discussions will show, both grounds are seriously inadequate as justifications for interfering with procreative choice. Imposing rearing costs on others may not rise to the level of harm that justifies depriving a person of a fundamental moral right. Moreover, protection of offspring from unfit parenting requires that unfit parents not rear, not that they not reproduce. Offspring could be protected by having others rear them without interfering with parental reproduction.

A further problem, if coital reproduction were found to be unjustified, concerns what action should then be taken. Exhortation or moral condemnation might be acceptable, but more stringent or coercive measures would act on the body of the person deemed irresponsible. Past experience with forced sterilization of retarded persons and the inevitable focus on the poor and minorities as targets of coercive policies make such proposals highly unappealing. Because of these doubts, there have been surprisingly few attempts to restrict coital reproduction in the United States since the era of eugenic sterilization, even though some instances of reproduction—for example, teenage pregnancy, inability to care for offspring—appear to be socially irresponsible.

An entirely different set of concerns arises with noncoital reproductive techniques. Charges that noncoital reproduction is unethical or irresponsible arise because of its expense, its highly technological character, its decomposition of parenthood into genetic, gestational, and social components, and its potential effects on women and offspring. To assess whether these effects justify moral condemnation or public limitation, we must first determine whether noncoital reproduction implicates important aspects of procreative liberty.

The Right to Reproduce and Noncoital Technology

If the moral right to reproduce presumptively protects coital reproduction, then it should protect noncoital reproduction as well. The moral right of the coitally infertile to reproduce is based on the same desire for offspring that the coitally fertile have. They too wish to replicate themselves, transmit genes, gestate, and rear children biologically related to them. Their infertility should no more disqualify them from reproductive experiences than physical disability should disqualify persons from walking with mechanical assistance. The unique risks posed by noncoital reproduction may provide independent justifications for limiting its use, but neither the noncoital nature of the means used nor the infertility of their beneficiaries means that the presumptively protected moral interest in reproduction is not present.

A major question about this position, however, is whether the noncoital or collaborative nature of the means used truly implicates reproductive interests. For example, what if only one aspect of reproduction-genetic transfer, gestation, or rearing occurs, as happens with gamete donors or surrogates who play no rearing role? Is a person's procreative liberty substantially implicated in such partial reproductive roles? The answer will depend on the value attributed to the particular collaborative contribution and on whether the collaborative enterprise is viewed from the donor's or recipient's perspective.

Gamete donors and surrogates are clearly reproducing even though they have no intention to rear. Because reproduction *tout court* may seem less important than reproduction with intent to rear, the donor's reproductive interest may appear less important. However, more experience with these practices is needed to determine the inherent value of "partial" reproductive experiences to donors and surrogates. Experience may show that it is independently meaningful, regardless of their contact with offspring. If not, then countervailing interests would more easily override their right to enter these roles.

Viewed from the recipient's perspective, however, the donor or surrogate's reproduction *tout court* does not lessen the reproductive importance of her contribution. A woman who receives an egg or embryo donation has no genetic connection with offspring but has a gestational relation of great personal significance. In addition, gamete donors and surrogates enable one or both rearing partners to have a biological relation with offspring. If one of them has no biological connection at all, they will still have a strong interest in rearing their partner's biologic offspring. Whether viewed singly through the eyes of the partner who is reproducing, or jointly as an endeavor of a couple seeking to rear children who are biologically related to at least one of the two, a significant reproductive interest is at stake. If so, noncoital, collaborative treatments for infertility should be respected to the same extent as coital reproduction is.

Questions about the core meaning of reproduction will also arise in the temporal dislocations that cryopreservation of sperm and embryos make possible. For example, embryo freezing allows siblings to be conceived at the same time, but born years apart and to different gestational mothers. Twins could be created by splitting one embryo into two. If one half is frozen for later use, identical twins could be born at widely different times. Sperm, egg, and embryo freezing also make posthumous reproduction possible.

Such temporally dislocative practices clearly implicate core reproductive interests when the ultimate recipient has no alternative means of reproduction. However, if the procreative interests of the recipient couple are not directly implicated, we must ask whether those whose gametes are used have an independent procreative interest, as might occur if they directed that gametes or embryos be thawed after their death for purposes of posthumous reproduction. In that case the question is whether the expectancy of posthumous reproduction is so central to an individual's procreative identity or life-plan that it should receive the same respect that one's reproduction when alive receives. The answer to such a question will be important in devising policy for storing and posthumously disposing of gametes and embryos. The answer will also affect inheritance questions and have implications for management of pregnant women who are irreversibly comatose or brain dead.

The problem of determining whether technology implicates a major reproductive interest also arises with technologies that select offspring characteristics. Some degree of quality control would seem logically to fall within the realm of procreative liberty. For many couples the decision whether to procreate depends on the ability to have healthy children. Without some guarantee or protection against the risk of handicapped children, they might not reproduce at all.

Thus viewed, quality control devices become part of the liberty interest in procreating or in avoiding procreation, and arguably should receive the same degree of protection. If so, genetic screening and selective abortion, as well as the right to select a mate or a source for donated eggs, sperm, or embryos should be protected as part of procreative liberty. The same arguments would apply to positive interventions to cure disease at the fetal or embryo stage. However, futuristic practices such as nontherapeutic enhancement, cloning, or intentional diminishment of offspring characteristics may so deviate from the core interests that make reproduction meaningful as to fall outside the protective canopy of procreative liberty.

Finally, technology will present questions of whether one may use one's reproductive capacity to produce gametes, embryos, and fetuses for nonreproductive uses in research or therapy. Here the purpose is not to have children to rear, but to get material for research or transplant. Are such uses of reproductive capacity tied closely enough to the values and interests that underlie procreative freedom to warrant similar respect? Even if procreative choice is not directly involved, other liberties may protect the activity.

Are Noncoital Technologies Unethical?

If this analysis is accepted, then procreative liberty would include the right to use noncoital and other technologies to form a family and shape the characteristics of offspring. Neither infertility nor the fact that one will only partially reproduce eliminates the existence of a *prima facie* reproductive experience for someone. However, judgments about the proximity of these partial reproductive experiences to the core meanings of reproduction will be required in balancing those claims against competing moral concerns.

Judgment about the reproductive importance of noncoital technologies is crucial because many people have serious ethical reservations about them, and are more than willing to restrict their use. The concerns here are not the fears of overpopulation, parental unfitness, and societal costs that arise with allegedly irresponsible coital reproduction. Instead, they include reduction of demand for hard-to-adopt children, the coercive or exploitive bargains that will be offered to poor women, the commodification of both children and reproductive collaborators, the objectification of women as reproductive vessels, and the undermining of the nuclear family.

However, often the harms feared are deontological in character. In some cases they stem from a religious or moral conception of the unity of sex and reproduction or the definition of family. Such a view characterizes the Vatican's strong opposition to IVF, donor sperm, and other noncoital and collaborative techniques. Other deontological concerns derive from a particular conception of the proper reproductive role of women. Many persons, for example, oppose paid surrogate motherhood because of a judgment about the wrongness of a woman's willingness to sever the mother-child bond for the sake of money. They also insist that the gestational mother is always morally entitled to rear, despite her preconception promise to the contrary. Closely related are dignitary objections to allowing any reproductive factors to be purchased, or to having offspring selected on the basis of their genes.

Finally, there is a broader concern that noncoital reproduction will under-mine the deeper community interest in having a clear social framework to de-fine boundaries of families, sexuality, and reproduction. The traditional family provides a container for the narcissism and irrationality that often drives human reproduction. This container assures commitments to the identifications and taboos that protect children from various types of abuse. The technical ability to disaggregate and recombine genetic, gestational, and rearing connections and to control the genes of *offspring* may thus undermine essential protections for off-spring, couples, families, and society.

These criticisms are powerful ones that explain much of the ambivalence that surrounds the use of certain reproductive technologies. They call into ques-tion the wisdom of individual decisions to use them, and the willingness of so-ciety to promote or facilitate their use. Unless one is operating out of a specific religious or deontological ethic, however, they do not show that all individual uses of these techniques are immoral, much less that public policy should restrict or discourage their use.

. . . [T]hese criticisms seldom meet the high standard necessary to limit pro-creative choice. Many of them are mere hypothetical or speculative possibilities. Others reflect moralisms concerning a "right" view of reproduction, which in-dividuals in a pluralistic society hold or reject to varying degrees. In any event, without a clear showing of substantial harm to the tangible interests of others, speculation or mere moral objections alone should not override the moral right of infertile couples to use those techniques to form families. Given the primacy of procreative liberty, the use of these techniques should be accorded the same high protection granted to coital reproduction.

Resolving Disputes over Procreative Liberty

As this brief survey shows, new reproductive technologies will generate ethical and legal disputes about the meaning and scope of procreative liberty. Because procreative liberty has never been fully elaborated, the importance of procre-ative choice in many novel settings will be a question of first impression. The ul-timate decision reached will reflect the value assigned to the procreative interest at stake in light of the *effects* causing concern. In an important sense, the mean-ing of procreative liberty will be created or constituted for society in the process of resolving such disputes.

If procreative liberty is taken seriously, a strong presumption in favor of using technologies that centrally implicate reproductive interests should be rec-ognized. Although procreative rights are not absolute, those who would limit procreative choice should have the burden of establishing substantial harm. This is the standard used in ethical and legal analyses of restrictions on traditional re-productive decisions. Because the same procreative goals are involved, the same standard of scrutiny should be used for assessing moral or governmental restric-tions on novel reproductive techniques.

In arbitrating these disputes, one has to come to terms with the importance of procreative interests relative to other concerns. The precise procreative interest at stake must be identified and weighed against the core values of reproduction. As noted, this will raise novel and unique questions when the technology deviates from the model of two-person coital reproduction, or otherwise disaggregates or alters ordinary reproductive practices. However, if an important reproductive interest exists, then use of the technology should be presumptively permitted. Only substantial harm to tangible interests of others should then justify restriction.

In determining whether such harm exists, it will be necessary to distinguish between harms to individuals and harms to personal conceptions of morality, right order, or offense, discounted by their probability of occurrence. As previously noted, many objections to reproductive technology rest on differing views of what "proper" or "right" reproduction is aside from tangible effects on others. For example, concerns about the decomposition of parenthood, through the use of donors and surrogates, about the temporal alteration of conception, gestation and birth, about the alienation or commercialization of gestational capacity, and about selection and control of offspring characteristics do not directly affect persons so much as they affect notions of right behavior. Disputes over early abortion and discard or manipulation of IVF-created embryos also exemplify this distinction, if we grant that the embryo/previable fetus is not a person or entity with rights in itself.

At issue in these cases is the symbolic or constitutive meaning of actions regarding prenatal life, family, maternal gestation, and respect for persons over which people in a secular, pluralistic society often differ. A majoritarian view of "right" reproduction or "right" valuation of prenatal life, family, or the role of women should not suffice to restrict actions based on differing individual views of such preeminently personal issues. At a certain point, however, a practice such as cloning, enhancement, or intentional diminishment of offspring may be so far removed from even pluralistic notions of reproductive meaning that they leave the realm of protected reproductive choice. People may differ over where that point is, but it will not easily exclude most reproductive technologies of current interest.

To take procreative liberty seriously, then, is to allow it to have presumptive priority in an individual's life. This will give persons directly involved the final say about use of a particular technology, unless tangible harm to the interests of others can be shown. Of course, people may differ over whether an important procreative interest is at stake or over how serious the harm posed from use of the reproductive technology is. Such a focused debate, however, is legitimate and ultimately essential in developing ethical standards and public policy for use of new reproductive technologies.

The Limits of Procreative Liberty

The emphasis on procreative liberty that informs this book provides a useful but by no means complete or final perspective on the technologies in question. Theological, social, psychological, economic, and feminist perspectives would emphasize different aspects of reproductive technology, and might be much less sanguine about potential benefits and risks. Such perspectives might also offer better guidance in how to use these technologies to protect offspring, respect women, and maintain other important values.

A strong rights perspective has other limitations as well. Recognition of procreative liberty, whether in traditional or in new technological settings, does not guarantee that people will achieve their reproductive goals, much less that they will be happy with what they do achieve. Nature may be recalcitrant to the latest technology. Individuals may lack the will, the perseverance, or the resources to use effective technologies. Even if they do succeed, the results may be less satisfying than envisaged. In addition, many individual instances of procreative choice may cumulate into larger social changes that from our current vantage point seem highly undesirable. But these are the hazards and limitations of any scheme of individual rights.

Recognition of procreative liberty will protect the right of persons to use technology in pursuing their reproductive goals, but it will not eliminate the ambivalence that such technologies engender. Societal ambivalence about reproductive technology is recapitulated at the individual level, as individuals and couples struggle with whether to use the technologies in question. Thus recognition of procreative liberty will not eliminate the dilemmas of personal choice and responsibility that reproductive choice entails. The freedom to act does not mean that we will act wisely, yet denying that freedom may be even more unwise, for it denies individuals' respect in the most fundamental choices of their lives.

Feminist Ethics and In Vitro Fertilization

Susan Sherwin

Susan Sherwin received her Ph.D. in philosophy from Stanford University and is a Professor of Philosophy and Women's Studies at Dalhousie University in Halifax, Nova Scotia, Canada. Most of her research and teaching efforts are concentrated in the areas of health care ethics and feminist theory. Although access to in vitro fertilization, or IVF, has increased since this article was published (available now through some private insurers) and the safety concerns she identifies have not materialized to the degree that she asserts, her arguments are worth hearing. In this article she suggests that rather than increase reproductive freedom, IVF may decrease women's freedom. She argues that IVF cannot be considered as a technology in isolation; it needs to be considered as it is used within the social and political structures that have maintained power relationships to the disadvantage of women and people of lower socioeconomic status. This line of argument is heard not only with regard to IVF; it is also adopted by some individuals to argue against the use of stem cells.

New technology in human reproduction has provoked wide-ranging arguments about the desirability and moral justifiability of many of these efforts. Authors of biomedical ethics have ventured into the field to offer the insight of moral theory to these complex moral problems of contemporary life. I believe, however, that the moral theories most widely endorsed today are problematic, and that a new approach to ethics is necessary if we are to address the concerns and perspectives identified by feminist theorists in our considerations of such topics. Hence, I propose to look at a particular technique in the growing repertoire of reproductive technologies, in vitro fertilization (IVF), in order to consider the insight which the mainstream approaches to moral theory have offered to this debate, and to see the difference made by a feminist approach to ethics.

I have argued elsewhere that the most widely accepted moral theories of our time are inadequate for addressing many of the moral issues we encounter in our lives, since they focus entirely on such abstract qualities of moral agents as autonomy or quantities of happiness, and they are addressed to agents who are conceived of as independent . . . individuals. In contrast, I claimed, we need a theory which places the locus of ethical concerns in a complex social network of interrelated persons who are involved in special sorts of relations with one another. Such a theory, as I envision it, would be influenced by the insights and concerns of feminist theory, and hence, I have called it feminist ethics.

In this [essay], I propose to explore the differences between a feminist approach to ethics and other, more traditional approaches in examining the propriety of developing and implementing in vitro fertilization and related technologies. This is a complicated task, since each sort of ethical theory admits of a variety of interpretations and hence of a variety of conclusions on concrete ethical issues. Nonetheless, certain themes and trends can be seen to emerge. Feminist thinking is also ambivalent in application, for feminists are quite torn about their response to this sort of technology. It is my hope that a systematic theoretic evaluation of IVF from the point of view of a feminist ethical theory will help feminists like myself sort through our uncertainty on these matters.

. . . Let us turn now to the responses that philosophers working within the traditional approaches to ethics have offered on this subject. A review of the literature in bioethics identifies a variety of concerns with this technology. Philosophers who adopt a theological perspective tend to object that such technology is wrong because it is not "natural" and undermines God's plan for the family. . . . Leon Kass[,] . . . in "'Making Babies' Revisited[,]"[1] . . . worries that our conception of humanness will not survive the technological permutations before us, and that we will treat these new artificially conceived embryos more as objects than as subjects; he also fears that we will be unable to track traditional human categories of parenthood and lineage, and that this loss would cause us to lose track of important aspects of our identity.

The . . . position paper of the Catholic Church on reproductive technology reflects related concerns:

> It is through the secure and recognized relationship to his [sic] own parents that the child can discover his own identity and achieve his own proper human development. . . .
>
> Heterologous artificial fertilization violates the rights of the child; it deprives him of his filial relationship with his parental origins and can hinder the maturing of his personal identity.[2]

Philosophers partial to utilitarianism prefer a more scientific approach; they treat these sorts of concerns as sheer superstition. They carefully explain to their theological colleagues that there is no clear sense of "natural" and certainly no sense that demands special moral status. All medical activity, and perhaps all human activity, can be seen in some sense as being "interference with nature," but that is hardly grounds for avoiding such action. "Humanness," too, is a concept that admits of many interpretations generally; it does not provide satisfactory grounds for moral distinctions. Further, it is no longer thought appropriate to focus too strictly on questions of lineage and strict biological parentage, and, they note, most theories of personal identity do not rely on such matters.

Where some theologians object that "fertilization achieved outside the bodies of the couple remains by this very fact deprived of the meanings of the values which are expressed in the language of the body and the union of human persons,"[3] utilitarians quickly dismiss the objection against reproduction without sexuality in a properly sanctified marriage. See, for instance, Michael Bayles in *Reproductive Ethics:* ". . . even if reproduction should occur only within a context of marital love, the point of that requirement is the nurturance of offspring. Such nurturance does not depend on the sexual act itself. The argument confuses the biological act with the familial context."[4]

Another area of disagreement between theological ethicists and their philosophical critics is the significance of the wedge argument to the debate about IVF. IVF is already a complex technology involving research on superovulation, "harvesting" of ova, fertilization, and embryo implants. It is readily adaptable to technology involving the transfer of ova and embryos, and hence their donation or sale, as well as to the rental of womb space; it also contributes to an increasing ability to foster fetal growth outside of the womb and, potentially, to the development of artificial wombs covering the whole period of gestation. It is already sometimes combined with artificial insemination and is frequently used to produce surplus fertilized eggs to be frozen for later use. Theological ethicists worry that such activity, and further reproductive developments we can anticipate (such as human cloning), violate God's plan for human reproduction. They worry about the cultural shift involved in viewing reproduction as a scientific enterprise, rather than the "miracle of love" which religious proponents prefer: "[He] cannot be desired or conceived as the product of an intervention of medical or biological techniques; that would be equivalent to reducing him to an object of scientific technology."[5] And, worse, they note, we cannot anticipate the ultimate outcome of this rapidly expanding technology.

The where-will-it-all-end hand-wringing that comes with this sort of religious futurology is rejected by most analytical philosophers; they urge us to realize that few slopes are as slippery as the pessimists would have us believe, that scientists are moral people and quite capable of evaluating each new form of technology on its own merits, and that IVF must be judged by its own consequences and not the possible result of some future technology with which it may be linked. Samuel Gorovitz is typical:

> It is not enough to show that disaster awaits if the process is not controlled. A man walking East in Omaha will drown in the Atlantic—if he does not stop. The argument must also rest on the evidence about the likelihood that judgment and control will be exercised responsibly. . . . Collectively we have significant capacity to exercise judgment and control . . . our record has been rather good in regard to medical treatment and research.[6]

The question of the moral status of the fertilized eggs is more controversial. Since the superovulation involved in producing eggs for collection tends to produce several at once, and the process of collecting eggs is so difficult, and since the odds against conception on any given attempt are so slim, several eggs are usually collected and fertilized at once. A number of these fertilized eggs will be introduced to the womb with the hope that at least one will implant and gestation will begin, but there are frequently some "extras." Moral problems arise as to what should be done with the surplus eggs. They can be frozen for future use (since odds are against the first attempt "taking"), or they can be used as research material, or simply discarded. Canadian clinics get around the awkwardness of their ambivalence on the moral status of these cells by putting them all into the woman's womb. This poses the devastating threat of six or eight "successfully" implanting, and a woman being put into the position of carrying a litter; something, we might note, her body is not constructed to do.

Those who take a hard line against abortion and argue that the embryo is a person from the moment of conception object to all these procedures, and, hence, they argue, there is no morally acceptable means of conducting IVF. To this line, utilitarians offer the standard responses. Personhood involves moral, not biological categories. A being neither sentient nor conscious is not a person in any meaningful sense. For example, Gorovitz argues, "Surely the concept of person involves in some fundamental way the capacity for sentience, or an awareness of sensations at the least."[7] Bayles says, "For fetuses to have moral status they must be capable of good or bad in their lives . . . What happens to them must make a difference to them. Consequently some form of awareness is necessary for moral status."[8] (Apparently, clinicians in the field have been trying to avoid this whole issue by coining a new term in the hopes of identifying a new ontological category, that of the "pre-embryo.")

Many bioethicists have agreed here, as they have in the abortion debate, that the principal moral question of IVF is the moral status and rights of the embryo. Once they resolve that question, they can, like Engelhardt, conclude that since fetuses are not persons, and since reproductive processes occurring outside a human body pose no special moral problems, "there will be no sustainable moral arguments in principle . . . against in vitro fertilization."[9] He argues,

> in vitro fertilization and techniques that will allow us to study and control human reproduction are morally neutral instruments for the realization of profoundly important human goals, which are bound up with the realization of the good of others: children for infertile parents and greater health for the children that will be born.[10]

Moral theorists also express worries about the safety of the process, and by that they tend to mean the safety to fetuses that may result from this technique. Those fears have largely been put to rest in the years since the first IVF baby was born in 1978, for the . . . infants reportedly produced by this technique to date seem no more prone to apparent birth defects than the population at large, and, in fact, there seems to be evidence that birth defects may be less common in this group—presumably because of better monitoring and pre- and post-natal care. (There is concern expressed, however, in some circles outside of the bioethical literature about the long-term effect of some of the hormones involved, in light of our belated discoveries of the effect of DES usage on offspring. This concern is aggravated by the chemical similarity of Clomid, one of the hormones used in IVF, to DES.[11])

Most of the literature tends to omit comment on the uncertainties associated with the effect of drugs inducing superovulation in the woman concerned, or with the dangers posed by the general anaesthetic required for the laparoscopy procedure; the emotional costs associated with this therapy are also overlooked, even though there is evidence that it is extremely stressful in the 85–90% of the attempts that fail, and that those who succeed have difficulty in dealing with common parental feelings of anger and frustration with a child they tried so hard to get. Nonetheless, utilitarian theory could readily accommodate such concerns, should the philosophers involved think to look for them. In principle, no new moral theory is yet called for, although a widening of perspective (to include the effects on the women involved) would certainly be appropriate.

The easiest solution to the IVF question seems to be available to ethicists of a deontological orientation who are keen on autonomy and rights and free of religious prejudice. For them, IVF is simply a private matter, to be decided by the couple concerned together with a medical specialist. The desire to have and raise children is a very common one and generally thought to be a paradigm case of a purely private matter. Couples seeking this technology face medical complications that require the assistance of a third party, and it is thought, "it would be unfair to make infertile couples pass up the joys of rearing infants or suffer the burdens of rearing handicapped children."[12] Certainly, meeting individuals' desires/needs is the most widely accepted argument in favour of the use of this technology.

What is left, then, in the more traditional ethical discussions, is usually some hand waving about costs. This is an extremely expensive procedure. . . . Gorovitz says, for instance, "there is the question of the distribution of costs, a question that has heightened impact if we consider the use of public funds to pay for medical treatment."[13] Debate tends to end here in the mystery of how to balance soaring medical costs of various sorts and a comment that no new ethical problems are posed.

Feminists share many of these concerns, but they find many other moral issues involved in the development and use of such technology and note the silence of the standard moral approaches in addressing these matters. Further, feminism does not identify the issues just cited as the primary areas of moral concern. Nonetheless, IVF is a difficult issue for feminists.

On the one hand, most feminists share the concern for autonomy held by most moral theorists, and they are interested in allowing women freedom of choice in reproductive matters. This freedom is most widely discussed in connection with access to safe and effective contraception and, when necessary, to abortion services. For women who are unable to conceive because of blocked fallopian tubes, or certain fertility problems of their partners, IVF provides the technology to permit pregnancy which is otherwise impossible. Certainly most of the women seeking IVF perceive it to be technology that increases their reproductive freedom of choice. So, it would seem that feminists should support this sort of technology as part of our general concern to foster the degree of re-productive control women may have over their own bodies. Some feminists have chosen this route. But feminists must also note that IVF as practiced does not altogether satisfy the motivation of fostering individual autonomy.

It is, after all, the sort of technology that requires medical intervention, and hence it is not really controlled by the women seeking it, but rather by the med-ical staff providing this "service." IVF is not available to every woman who is medically suitable, but only to those who are judged to be worthy by the med-ical specialists concerned. To be a candidate for this procedure, a woman must have a husband and an apparently stable marriage. She must satisfy those spe-cialists that she and her husband have appropriate resources to support any chil-dren produced by this arrangement (in addition, of course, to the funds required to purchase the treatment in the first place), and that they generally "deserve" this support. IVF is not available to single women, lesbian women, or women not securely placed in the middle class or beyond. Nor is it available to women whom the controlling medical practitioners judge to be deviant with respect to their norms of who makes a good mother. The supposed freedom of choice, then, is provided only to selected women who have been screened by the per-sonal values of those administering the technology.

Further, even for these women, the record on their degree of choice is un-clear. Consider, for instance, that this treatment has always been very experi-mental: it was introduced without the prior primate studies which are required for most new forms of medical technology, and it continues to be carried out under constantly shifting protocols, with little empirical testing, as clinics try to raise their very poor success rates. Moreover, consent forms are perceived by pa-tients to be quite restrictive procedures and women seeking this technology are not in a particularly strong position to bargain to revise the terms; there is no al-ternate clinic down the street to choose if a woman dislikes her treatment at some clinic, but there are usually many other women waiting for access to her place in the clinic should she choose to withdraw.

Some recent studies indicate that few of the women participating in current programs really know how low the success rates are.[14] And it is not apparent that participants are encouraged to ponder the medical unknowns associated with various aspects of the technique, such as the long-term consequences of super-ovulation and the use of hormones chemically similar to DES. Nor is it the case that the consent procedure involves consultation on how to handle the disposal of "surplus" zygotes. It is doubtful that the women concerned have much real choice about which procedure is followed with the eggs they will not need. These policy decisions are usually made at the level of the clinic. It should be noted . . . that at least one feminist argues that neither the woman, nor the doc-tors have the right to choose to destroy these embryos: ". . . because no one, not even its parents, owns the embryo/fetus, no one has the right to destroy it, even at a very early development stage . . . to destroy an embryo is not an automatic entitlement held by anyone, including its genetic parents."[15]

Moreover, some participants reflect deep-seated ambivalence on the part of many women about the procedure—they indicate that their marriage and status depends on a determination to do whatever is possible in pursuit of their 'natural' childbearing function, and they are not helped to work through the seeming imponderables associated with their long-term well being. Thus, IVF as practiced involves significant limits on the degree of autonomy deontologists insist on in other medical contexts, though the non-feminist literature is insensitive to this anomaly.

From the perspective of consequentialism, feminists take a long view and try to see IVF in the context of the burgeoning range of techniques in the area of human reproductive technology. While some of this technology seems to hold the potential of benefiting women generally—by leading to better understanding of conception and contraception, for instance—there is a wary suspicion that this research will help foster new techniques and products such as human cloning and the development of artificial wombs which can, in principle, make the majority of women superfluous. . . .

Many authors from all traditions consider it necessary to ask why it is that some couples seek this technology so desperately. Why is it so important to so many people to produce their "own" child? On this question, theorists in the analytic tradition seem to shift to previously rejected ground and suggest that this is a natural, or at least a proper, desire. Engelhardt, for example, says, "The use of technology in the fashioning of children is integral to the goal of rendering the world congenial to persons."[16] Bayles more cautiously observes that, "A desire to beget for its own sake . . . is probably irrational"; nonetheless, he immediately concludes, "these techniques for fulfilling that desire have been found ethically permissible."[17] R. G. Edwards and David Sharpe state the case most strongly: "the desire to have children must be among the most basic of human instincts, and denying it can lead to considerable psychological and social difficulties."[18] Interestingly, although the recent pronouncement of the Catholic Church assumes that "the desire for a child is natural,"[19] it denies that a couple has a right to a child: "The child is not an object to which one has a right."[20]

Here, I believe, it becomes clear why we need a deeper sort of feminist analysis. We must look at the sort of social arrangements and cultural values that underlie the drive to assume such risks for the sake of biological parenthood. We find that the capitalism, racism, sexism, and elitism of our culture have combined to create a set of attitudes which views children as commodities whose value is derived from their possession of parental chromosomes. Children are valued as privatized commodities, reflecting the virility and heredity of their parents. They are also viewed as the responsibility of their parents and are not seen as the social treasure and burden that they are. Parents must tend their needs on pain of prosecution, and, in return, they get to keep complete control over them. Other adults are inhibited from having warm, stable interactions with the children of others—it is as suspect to try to hug and talk regularly with a child who is not one's own as it is to fondle and hang longingly about a car or a bicycle which belongs to someone else—so those who wish to know children well often find they must have their own. Women are persuaded that their most important purpose in life is to bear and raise children; they are told repeatedly that their life is incomplete, that they are lacking in fulfillment if they do not have children. And, in fact, many women do face a barren existence without children. Few women have access to meaningful, satisfying jobs. Most do not find themselves in the centre of the romantic personal relationships which the culture pretends is the norm for heterosexual couples. And they have been socialized to be fearful of close friendships with others—they are taught to distrust other women, and to avoid the danger of friendship with men other than their husbands. Children remain the one hope for real intimacy and for the sense of accomplishment which comes from doing work one judges to be valuable.

To be sure, children can provide that sense of self worth, although for many women (and probably for all mothers at some times) motherhood is not the romanticized satisfaction they are led to expect. But there is something very wrong with a culture where childrearing is the only outlet available to most women in which to pursue fulfillment. Moreover, there is something wrong with the ownership theory of children that keeps other adults at a distance from children. There ought to be a variety of close relationships possible between children and adults so that we all recognize that we have a stake in the well-being of the young, and we all benefit from contact with their view of the world.

In such a world, it would not be necessary to spend the huge sums on designer children which IVF requires while millions of other children starve to death each year. Adults who enjoyed children could be involved in caring for them whether or not they produced them biologically. And, if the institution of marriage survives, women and men would marry because they wished to share their lives together, not because the men needed someone to produce heirs for them and women needed financial support for their children. That would be a world in which we might have reproductive freedom of choice. The world we now live in has so limited women's options and self-esteem, it is legitimate to question the freedom behind women's demand for this technology, for it may well be largely a reflection of constraining social perspectives.

Nonetheless, I must acknowledge that some couples today genuinely mourn their incapacity to produce children without IVF, and there are very significant and unique joys which can be found in producing and raising one's own [biological] children which are not accessible to persons in infertile relationships. We must sympathize with these people. None of us shall live to see the implementation of the ideal cultural values outlined above which would make the demand for IVF less severe. It is with real concern that some feminists suggest that the personal wishes of couples with fertility difficulties may not be compatible with the overall interests of women and children.

Feminist thought, then, helps us to focus on different dimensions of the problem than do other sorts of approaches. But, with this perspective, we still have difficulty in reaching a final conclusion on whether to encourage, tolerate, modify, or restrict this sort of reproductive technology. I suggest that we turn to the developing theories of feminist ethics for guidance in resolving this question.

In my view, a feminist ethics is a moral theory that focuses on relations among persons as well as on individuals. It has as a model an inter-connected social fabric, rather than the familiar one of isolated, independent atoms; and it gives primacy to bonds among people rather than to rights to independence. It is a theory that focuses on concrete situations and persons and not on free-floating abstract actions. Although many details have yet to be worked out, we can see some of its implications in particular problem areas such as this.

It is a theory that is explicitly conscious of the social, political, and economic relations that exist among persons; in particular, as a feminist theory, it attends to the implications of actions or policies on the status of women. Hence, it is necessary to ask questions from the perspective of feminist ethics in addition to those which are normally asked from the perspective of mainstream ethical theories. We must view issues such as this one in the context of the social and political realities in which they arise, and resist the attempt to evaluate actions or practices in isolation (as traditional responses in biomedical ethics often do). Thus, we cannot just address the question of IVF per se without asking how IVF contributes to general patterns of women's oppression. As Kathryn Payne Addleson has argued about abortion,[21] a feminist perspective raises questions that are inadmissible within the traditional ethical frameworks, and yet, for women in a patriarchal society, they are value questions of greater urgency. In particular, a feminist ethics, in contrast to other approaches in biomedical ethics, would take seriously the concerns just reviewed which are part of the debate in the feminist literature.

A feminist ethics would also include components of theories that have been developed as feminine ethics sketched out by the empirical work of Carol Gilligan.[22] (The best example of such a theory is the work of Nel Noddings in her influential book, Caring.[23]) In other words, it would be a theory that gives primacy to interpersonal relationships and woman-centered values such as nurturing, empathy, and cooperation. Hence, in the case of IVF, we must care for the women and men who are so despairing about their infertility as to want to spend the vast sums and risk the associated physical and emotional costs of the treatment, in pursuit of "their own children." That is, we should, in Noddings' terms, see their reality as our own and address their very real sense of loss. In so doing, however, we must also consider the implications of this sort of solution to their difficulty. While meeting the perceived desires of some women—desires which are problematic in themselves, since they are so compatible with the values of a culture deeply oppressive to women—this technology threatens to further entrench those values which are responsible for that oppression. A larger vision suggests that the technology offered may, in reality, reduce women's freedom and, if so, it should be avoided.

A feminist ethics will not support a wholly negative response, however, for that would not address our obligation to care for those suffering from infertility; it is the responsibility of those who oppose further implementation of this technology to work towards the changes in the social arrangements that will lead to reduction of the sense of need for this sort of solution. On the medical front, research and treatment ought to be stepped up to reduce the rates of [puerperal] sepsis and gonorrhea which often result in tubal blockage, more attention should be directed at the causes and possible cures for male infertility, and we should pursue techniques that will permit safe reversible sterilization providing women with better alternatives to ligation as a means of fertility control; these sorts of technology would increase the control of many women over their own fertility and would be compatible with feminist objectives. On the social front, we must continue the social pressure to change the status of women and children in our society from that of breeder and possession, respectively; hence, we must develop a vision of society as community where all participants are valued members, regardless of age or gender. And we must challenge the notion that having one's wife produce a child with his own genes is sufficient cause for the wives of men with low sperm counts to be expected to undergo the physical and emotional assault such technology involves.

Further, a feminist ethics will attend to the nature of the relationships among those concerned. Annette Baier has eloquently argued for the importance of developing an ethics of trust,[24] and I believe a feminist ethics must address the question of the degree of trust appropriate to the relationships involved. Feminists have noted that women have little reason to trust the medical specialists who offer to respond to their reproductive desires, for commonly women's interests have not come first from the medical point of view. In fact, it is accurate to perceive feminist attacks on reproductive technology as expressions of the lack of trust feminists have in those who control the technology. Few feminists object to reproductive technology per se; rather they express concern about who controls it and how it can be used to further exploit women. The problem with reproductive technology is that it concentrates power in reproductive matters in the hands of those who are not directly involved in the actual bearing and rearing of the child; i.e., in men who relate to their clients in a technical, professional, authoritarian manner. It is a further step in the medicalization of pregnancy and birth which, in North America, is marked by relationships between pregnant women and their doctors which are very different from the traditional relationships between pregnant women and midwives. The latter relationships fostered an atmosphere of mutual trust which is impossible to replicate in hospital deliveries today. In fact, current approaches to pregnancy, labour, and birth tend to view the mother as a threat to the fetus who must be coerced to comply with medical procedures designed to ensure delivery of healthy babies at whatever cost necessary to the mother. Frequently, the fetus-mother relationship is medically characterized as adversarial and the physicians choose to foster a sense of alienation and passivity in the role they permit the mother. However well IVF may serve the interests of the few women with access to it, it more clearly serves the interests (be they commercial, professional, scholarly, or purely patriarchal) of those who control it.

Questions such as these are a puzzle to those engaged in the traditional approaches to ethics, for they always urge us to separate the question of evaluating the morality of various forms of reproductive technology in themselves, from questions about particular uses of that technology. From the perspective of a feminist ethics, however, no such distinction can be meaningfully made. Reproductive technology is not an abstract activity; it is an activity done in particular contexts and it is those contexts which must be addressed.

Feminist concerns cited earlier made clear the difficulties we have with some of our traditional ethical concepts; hence, feminist ethics directs us to rethink our basic ethical notions. Autonomy, or freedom of choice, is not a matter to be determined in isolated instances, as is commonly assumed in many approaches to applied ethics. Rather it is a matter that involves reflection on one's whole life situation. The freedom of choice feminists appeal to in the abortion situation is freedom to define one's status as a childbearer, given the social, economic, and political significance of reproduction for women. A feminist perspective permits us to understand that reproductive freedom includes control of one's sexuality, protection against coerced sterilization (or iatrogenic sterilization, e.g., as caused by the Dalkon shield), and the existence of a social and economic network of support for the children we may choose to bear. It is the freedom to redefine our roles in society according to our concerns and needs as women.

In contrast, the consumer freedom to purchase technology, allowed only to a few couples of the privileged classes (in traditionally approved relationships), seems to entrench further the patriarchal notions of woman's role as childbearer and of heterosexual monogamy as the only acceptable intimate relationship. In other words, this sort of choice does not seem to foster autonomy for women on the broad scale. IVF is a practice which seems to reinforce sexist, classist, and often racist assumptions of our culture; therefore, on our revised understanding of freedom, the contribution of this technology to the general autonomy of women is largely negative.

We can now see the advantage of a feminist ethics over mainstream ethical theories, for a feminist analysis explicitly accepts the need for a political component to our understanding of ethical issues. In this, it differs from traditional ethical theories and it also differs from a simply feminine ethics approach, such as the one Noddings offers, for Noddings seems to rely on individual relations exclusively and is deeply suspicious of political alliances as potential threats to the pure relation of caring. Yet, a full understanding of both the threat of IVF, and the alternative action necessary should we decide to reject IVF, is possible only if it includes a political dimension reflecting on the role of women in society.

From the point of view of feminist ethics, the primary question to consider is whether this and other forms of reproductive technology threaten to reinforce the lack of autonomy which women now experience in our culture—even as they appear, in the short run, to be increasing freedom. We must recognize that the interconnections among the social forces oppressive to women underlie feminists' mistrust of this technology which advertises itself as increasing women's autonomy. The political perspective which directs us to look at how this technology fits in with general patterns of treatment for women is not readily accessible to traditional moral theories, for it involves categories of concern not accounted for in those theories—e.g., the complexity of issues which makes it inappropriate to study them in isolation from one another, the role of oppression in shaping individual desires, and potential differences in moral status which are connected with differences in treatment.

It is the set of connections constituting women's continued oppression in our society which inspires feminists to resurrect the old slippery slope arguments to warn against IVF. We must recognize that women's existing lack of control in reproductive matters begins the debate on a pretty steep incline. Technology with the potential to further remove control of reproduction from women makes the slope very slippery indeed. This new technology, though offered under the guise of increasing reproductive freedom, threatens to result, in fact, in a significant decrease in freedom, especially since it is a technology that will always include the active involvement of designated specialists and will not ever be a private matter for the couple or women concerned.

Ethics ought not to direct us to evaluate individual cases without also looking at the implications of our decisions from a wide perspective. My argument is that a theory of feminist ethics provides that wider perspective, for its different sort of methodology is sensitive to both the personal and the social dimensions of issues. For that reason, I believe it is the only ethical perspective suitable for evaluating issues of this sort.

Notes

1. Leon Kass, "'Making Babies' Revisited," *The Public Interest* 54 (Winter 1979), 32–60.
2. Joseph Cardinal Ratzinger and Alberto Bovone, "Instruction on Respect for Human Life in its Origin and on the Dignity of Procreation: Replies to Certain Questions of the Day" (Vatican City: Vatican Polyglot Press 1987), 23–24.
3. Ibid., 28.
4. Michael Bayles, *Reproductive Ethics* (Englewood Cliffs, Prentice-Hall 1984). 15.
5. Ratzinger and Bovone, 28.
6. Samuel Gorovitz, *Doctors' Dilemmas: Moral Conflict or Medical Care* (New York: Oxford University Press 1982), 168.
7. Ibid., 173.
8. Bayles, 66.
9. H. Tristram Engelhardt, *The Foundations of Bioethics* (Oxford: Oxford University Press 1986), 237.
10. Ibid., 241.

11. Anita Direcks, "Has the Lesson Been Learned?" DES Action Voice 28 (Spring 1986), 1–4; and Nikita A. Crook, "Clomid." I DES Action/Toronto Factsheet #442 (available from 60 Grosvenor Toronto, MSS 1136).
12. Bayles, 32. Though Bayles is not a deontologist, he concisely expresses a deontological concern here.
13. Gorovitz, 177.
14. Michael Soules, "The In Vitro Fertilization Pregnancy I: Let's Be Honest with One Another," *Fertility and Sterility* 43, 511–13.
15. Christine Overall, *Ethics and Human Reproduction: A Feminist Analysis* (Allen and Unwin, 1987).
16. Engelhardt, 239.
17. Bayles, 31.
18. Robert G. Edwards and David J. Sharpe, "Social Value Research in Human Embryology," *Nature* 231 (May 14, 1971).
19. Ratzinger and Bovone, 33.
20. Ibid., 34.
21. Kathryn Payne Addelson, "Moral Revolution," in M. Pearsall, ed., *Women and Values* (Belmont, CA: Wadsworth) 291–309.
22. Carol Gilligan, *In a Different Voice* (Cambridge, MA: Harvard University Press 1982).
23. Nel Noddings, *Caring* (Berkeley: University of California Press 1984).
24. Annette Baier, "What Do Women Want in a Moral Theory?" *Nous* 19 (March 1985), 53–64; and "Trust and Antitrust," *Ethics* 96 (January 1986), 231–60.

GENETICS AND REPRODUCTION

Genetics and Reproductive Risk:
Can Having Children Be Immoral?

Laura M. Purdy

Laura Purdy received her Ph.D. in philosophy from Stanford University and is Professor of Philosophy and Ruth and Albert Koch Professor of Humanities at Wells College, New York. In this chapter she argues that it is morally wrong to reproduce when we know there is a high risk of transmitting a serious disease or defect that would prevent a child's ability to enjoy a minimally satisfying life. She focuses her discussion on Huntington's Disease, but the arguments are relevant to many other genetic diseases.

Is it morally permissible for me to have children? A decision to procreate is surely one of the most significant decisions a person can make. So it would seem that it ought not to be made without some moral soul-searching.

There are many reasons why one might hesitate to bring children into this world if one is concerned about their welfare. Some are rather general, like the deteriorating environment or the prospect of poverty. Others have a narrower focus, like continuing civil war in Ireland, or the lack of essential social support for child rearing persons in the United States. Still others may be relevant only to individuals at risk of passing harmful diseases to their offspring.

There are many causes of misery in this world, and most of them are unrelated to genetic disease. In the general scheme of things, human misery is most efficiently reduced by concentrating on noxious social and political arrangements. Nonetheless, we shouldn't ignore preventable harm just because it is confined to a relatively small corner of life. So the question arises: can it be wrong to have a child because of genetic risk factors?

Unsurprisingly, most of the debate about this issue has focused on prenatal screening and abortion: much useful information about a given fetus can be made available by recourse to prenatal testing. This fact has meant that moral questions about reproduction have become entwined with abortion politics, to the detriment of both. The abortion connection has made it especially difficult to think about whether it is wrong to prevent a child from coming into being since doing so might involve what many people see as wrongful killing; yet there is no necessary link between the two. Clearly, the existence of genetically compromised children can be prevented not only by aborting already existing fetuses but also by preventing conception in the first place.

Worse yet, many discussions simply assume a particular view of abortion, without any recognition of other possible positions and the difference they make in how people understand the issues. For example, those who object to aborting fetuses with genetic problems often argue that doing so would undermine our conviction that all humans are in some important sense equal. However, this position rests on the assumption that conception marks the point at which humans are endowed with a right to life. So aborting fetuses with genetic problems looks morally the same as killing "imperfect" people without their consent.

This position raises two separate issues. One pertains to the legitimacy of different views on abortion. Despite the conviction of many abortion activists to the contrary, I believe that ethically respectable views can be found on different sides of the debate, including one that sees fetuses as developing humans without any serious moral claim on continued life. There is no space here to address the details, and doing so would be once again to fall into the trap of letting the abortion question swallow up all others. Fortunately, this issue need not be resolved here. However, opponents of abortion need to face the fact that many thoughtful individuals do not *see* fetuses as moral persons. It follows that their reasoning process and hence the implications of their decisions are radically different from those envisioned by opponents of prenatal screening and abortion. So where the latter see genetic abortion as murdering people who just don't measure up, the former see it as a way to prevent the development of persons who are more likely to live miserable lives. This is consistent with a world view that values persons equally and holds that each deserves high quality life. Some of those who object to genetic abortion appear to be oblivious to these psychological and logical facts. It follows that the nightmare scenarios they paint for us are beside the point: many people simply do not share the assumptions that make them plausible.

How are these points relevant to my discussion? My primary concern here is to argue that conception can sometimes be morally wrong on grounds of genetic risk, although this judgment will not apply to those who accept the moral legitimacy of abortion and are willing to employ prenatal screening and selective abortion. If my case is solid, then those who oppose abortion must be especially careful not to conceive in certain cases.... Those ... who do not see abortion as murder have more ways to prevent birth.

Huntington's Disease

There is always some possibility that reproduction will result in a child with a serious disease or handicap. Genetic counselors can help individuals determine whether they are at unusual risk and, as the Human Genome Project rolls on, their knowledge will increase by quantum leaps. As this knowledge becomes available, I believe we ought to use it to determine whether possible children are at risk *before* they are conceived.

I want in this paper to defend the thesis that it is morally wrong to reproduce when we know there is a high risk of transmitting a serious disease or defect. This thesis holds that some reproductive acts are wrong, and my argument puts the burden of proof on those who disagree with it to show why its conclusions can be overridden. Hence it denies that people should be free to reproduce mindless of the consequences. However, as moral argument, it should be taken as a proposal for further debate and discussion. It is not, by itself, an argument in favor of legal prohibitions of reproduction.

There is a huge range of genetic diseases. Some are quickly lethal; others kill more slowly, if at all. Some are mainly physical, some mainly mental; others impair both kinds of function. Some interfere tremendously with normal functioning, others less. Some are painful, some are not. There seems to be considerable agreement that rapidly lethal diseases, especially those, like Tay-Sachs, accompanied by painful deterioration, should be prevented even at the cost of abortion. Conversely, there seems to be substantial agreement that relatively trivial problems, especially cosmetic ones, would not be legitimate grounds for abortion. In short, there are cases ranging from low risk of mild disease or disability to high risk of serious disease or disability. Although it is difficult to decide where the duty to refrain from procreation becomes compelling, I believe that there are some clear cases. I have chosen to focus on Huntington's Disease to illustrate the kinds of concrete issues such decisions entail. However, the arguments presented here are also relevant to many other genetic diseases.

The symptoms of Huntington's Disease usually begin between the ages of thirty and fifty. It happens this way:

> Onset is insidious. Personality changes (obstinacy, moodiness, lack of initiative) frequently antedate or accompany the involuntary choreic movements. These usually appear first in the face, neck, and arms, and are jerky, irregular, and stretching in character. Contractions of the facial muscles result in grimaces, those of the respiratory muscles, lips, and tongue lead to hesitating, explosive speech. Irregular movements of the trunk are present; the gait *is* shuffling and dancing. Tendon reflexes are increased. . . . Some patients display a fatuous euphoria; others are spiteful, irascible, destructive, and violent. Paranoid reactions are common. Poverty of thought and impairment of attention, memory, and judgment occur. As the disease progresses, walking becomes impossible, swallowing difficult, and dementia profound. Suicide is not uncommon.[1]

The illness lasts about fifteen years, terminating in death.

Huntington's Disease is an autosomal dominant disease, meaning that it is caused by a single defective gene located on a non-sex chromosome. It is passed from one generation to the next via affected individuals. Each child of such an affected person has a fifty percent risk of inheriting the gene and thus of eventually developing the disease, even if he or she was born before the parent's disease was evident.

Until recently, Huntington's Disease was especially problematic because most affected individuals did not know whether they had the gene for the disease until well into their childbearing years. So they had to decide about childbearing before knowing whether they could transmit the disease or not. If, in time, they did not develop symptoms of the disease, then their children could know they were not at risk for the disease. If unfortunately they did develop symptoms, then each of their children could know there was a fifty percent chance that they, too, had inherited the gene. In both cases, the children faced a period of prolonged anxiety as to whether they would develop the disease. Then, in the 1980s, thanks in part to an energetic campaign by Nancy Wexler, a genetic marker was found that, in certain circumstances, could tell people with a relatively high degree of probability whether or not they had the gene for the disease. Finally, in March 1993, the defective gene itself was discovered. Now individuals can find out whether they carry the gene for the disease, and prenatal screening can tell us whether a given fetus has inherited it. These technological developments change the moral scene substantially.

How serious are the risks involved in Huntington's Disease? Geneticists often think a ten percent risk is high. But risk assessment also depends on what is at stake: the worse the possible outcome the more undesirable an otherwise small risk seems. In medicine, as elsewhere, people may regard the same result quite differently. But for devastating diseases like Huntington's this part of the judgment should be unproblematic: no one wants a loved one to suffer in this way.

There may still be considerable disagreement about the acceptability of a given risk. So it would be difficult in many circumstances to say how we should respond to a particular risk. Nevertheless, there are good grounds for a conservative approach, for it is reasonable to take special precautions to avoid very bad consequences, even if the risk is small. But the possible consequences here are very bad: a child who may inherit Huntington's Disease has a much greater than average chance of being subjected to severe and prolonged suffering. And it is one thing to risk one's own welfare, but quite another to do so for others and without their consent.

Is this judgment about Huntington's Disease really defensible? People appear to have quite different opinions. Optimists argue that a child born into a family afflicted with Huntington's Disease has a reasonable chance of living a satisfactory life. After all, even children born of an afflicted parent still have a fifty percent chance of escaping the disease. And even if afflicted themselves, such people will probably enjoy some thirty years of healthy life before symptoms appear. It is also possible, although not at all likely, that some might not mind the symptoms caused by the disease. Optimists can point to diseased persons who have lived fruitful lives, as well as those who seem genuinely glad to be alive. One is Rick Donohue, a sufferer from the Joseph family disease: "You know, if my mom hadn't had me, I wouldn't be here for the life I have had. So there is a good possibility I will have children."[2] Optimists therefore conclude that it would be a shame if these persons had not lived.

Pessimists concede some of these facts, but take a less sanguine view of them. They think a fifty percent risk of serious disease like Huntington's appallingly high. They suspect that many children born into afflicted families are liable to spend their youth in dreadful anticipation and fear of the disease. They expect that the disease, if it appears, will be perceived as a tragic and painful end to a blighted life. They point out that Rick Donohue is still young, and has not experienced the full horror of his sickness. It is also well-known that some young persons have such a dilated sense of time that they can hardly envision themselves at thirty or forty, so the prospect of pain at that age is unreal to them.

More empirical research on the psychology and life history of sufferers and potential sufferers is clearly needed to decide whether optimists or pessimists have a more accurate picture of the experiences of individuals at risk. But given that some will surely realize pessimists' worst fears, it seems unfair to conclude that the pleasures of those who deal best with the situation simply cancel out the suffering of those others when that suffering could be avoided altogether

I think that these points indicate that the morality of procreation in situations like this demands further investigation. I propose to do this by looking first at the position of the possible child, then at that of the potential parent.

Possible Children and Potential Parents

The first task in treating the problem from the child's point of view is to find a way of referring to possible future offspring without seeming to confer some sort of morally significant existence upon them. I will follow the convention of calling children who might be born in the future but who are not now conceived "possible" children, offspring, individuals, or persons.

Now, what claims about children or possible children are relevant to the morality of childbearing in the circumstances being considered? Of primary importance is the judgment that we ought to try to provide every child with something like a minimally satisfying life. I am not altogether sure how best to formulate this standard but I want clearly to reject the view that it is morally permissible to conceive individuals so long as we do not expect them to be so miserable that they wish they were dead.[3] I believe that this kind of moral minimalism is thoroughly unsatisfactory and that not many people would really want to live in a world where it was the prevailing standard. Its lure is that it puts few demands on us, but its price is the scant attention it pays to human well-being.

How might the judgment that we have a duty to try to provide a minimally satisfying life for our children be justified? It could, I think, be derived fairly straightforwardly from either utilitarian or contractarian theories of justice, although there is no space here for discussion of the details. The net result of such analysis would be the conclusion that neglecting this duty would create unnecessary unhappiness or unfair disadvantage for some persons.

Of course, this line of reasoning confronts us with the need to spell out what is meant by "minimally satisfying" and what a standard basal on this concept would require of us. Conceptions of a minimally satisfying life vary tremendously among societies and also within them. *De rigeur* in some circles are private music lessons and trips to Europe, while in others providing eight years of schooling is a major accomplishment. But there is no need to consider this complication at length here since we are concerned only with health as a prerequisite for a minimally satisfying life. Thus, as we draw out what such a standard might require of us, it seems reasonable to retreat to the more limited claim that parents should try to ensure something like normal health for their children. It might be thought that even this moderate claim is unsatisfactory since in some places debilitating conditions are the norm, but one could circumvent this objection by saying that parents ought to try to provide for their children health normal for that culture, even though it may be inadequate if measured by some outside standard. This conservative position would still justify efforts to avoid birth of children at risk for Huntington's Disease and other serious genetic diseases in virtually all societies.

This view is reinforced by the following considerations. Given that possible children do not presently exist as actual individuals, they do not have a right to be brought into existence, and hence no one is maltreated by measures to avoid the conception of a possible person. Therefore, the conservative course that avoids the conception of those who would not be expected to enjoy a minimally satisfying life is at present the only fair course of action. The alternative is a laissez-faire approach which brings into existence the lucky, but only at the expense of the unlucky. Notice that attempting to avoid the creation of the unlucky does not necessarily lead to fewer people being brought into being: the question boils down to taking steps to bring those with better prospects into existence, instead of those with worse ones.

I have so far argued that if people with Huntington's Disease are unlikely to live minimally satisfying lives, then those who might pass it on should not have genetically related children. This is consonant with the principle that the greater the danger of serious problems, the stronger the duty to avoid them. But this principle is in conflict with what people think of as the right to reproduce. How might one decide which should take precedence?

Expecting people to forego having genetically related children might seem to demand too great a sacrifice of them. But before reaching that conclusion we need to ask what is really at stake. One reason for wanting children is to experience family life, including love, companionship, watching kids grow, sharing their pains and triumphs, and helping to form members of the next generation. Other reasons emphasize the validation of parents as individuals within a continuous family line, children as a source of immortality, or perhaps even the gratification of producing partial replicas of oneself. Children may also be desired in an effort to prove that one is an adult, to try to cement a marriage or to benefit parents economically.

Are there alternative ways of satisfying these desires? Adoption or new reproductive technologies can fulfill many of them without passing on known genetic defects. Replacements for sperm have been available for many years via artificial insemination by donor. More recently, egg donation, sometimes in combination with contract pregnancy, has been used to provide eggs for women who prefer not to use their own. Eventually it may be possible to clone individual humans, although that now seems a long way off. All of these approaches to avoiding the use of particular genetic material are controversial and have generated much debate. I believe that tenable moral versions of each do exist.

None of these methods permits people to extend both genetic lines or realize the desire for immortality or for children who resemble both parents; nor is it clear that such alternatives will necessarily succeed in proving that one is an adult, cementing a marriage, or providing economic benefits. Yet, many people feel these desires strongly. Now, I am sympathetic to William James's dictum regarding desires: "Take any demand, however slight, which any creature, however weak, may make. Ought it not, for its own sole sake be satisfied? If not, prove why not."[4] Thus a world where more desires are satisfied is generally better than one where fewer are. However, not all desires can be legitimately satisfied since, as James suggests, there may be good reasons—such as the conflict of duty and desire—why some should be overruled.

Fortunately, further scrutiny of the situation reveals that there are good reasons why people should attempt—with appropriate social support—to talk themselves out of the desires in question or to consider novel ways of fulfilling them. Wanting to see the genetic line continued is not particularly rational when it brings a sinister legacy of illness and death. The desire for immortality cannot really be satisfied anyway, and people need to face the fact that what really matters is how they behave in their own lifetime. And finally, the desire for children who physically resemble one is understandable, but basically narcissistic, and its fulfillment cannot be guaranteed even by normal reproduction. There are other ways of proving one is an adult, and other ways of cementing marriages—and children don't necessarily do either. Children, especially prematurely ill children, may not provide the expected economic benefits anyway. Nongenetically related children may also provide benefits similar to those that would have been provided by genetically related ones, and expected economic benefit is, in many cases, a morally questionable reason for having children.

Before the advent of reliable genetic testing, the options of people in Huntington's families were cruelly limited. On the one hand, they could have children, but at the risk of eventual crippling illness and death for them. On the other, they could refrain from childbearing, sparing their possible children from significant risk of inheriting this disease, perhaps frustrating intense desires to procreate—only to discover, in some cases, that their sacrifice was unnecessary because they did not develop the disease. Or they could attempt to adopt or try new reproductive approaches.

Reliable genetic testing has opened up new possibilities. Those at risk who wish to have children can get tested. If they test positive, they know their possible children are at risk. Those who are opposed to abortion must be especially careful to avoid conception if they are to behave responsibly. Those not opposed to abortion can responsibly conceive children, but only if they are willing to test each fetus and abort those who carry the gene. If individuals at risk test negative, they are home free.

What about those who cannot face the test for themselves? They can do prenatal testing and abort fetuses who carry the defective gene. A clearly positive test also implies that the parent is affected, although negative tests do not rule out that possibility. Prenatal testing can thus bring knowledge that enables one to avoid passing the disease to others, but only, in some cases, at the cost of coming to know with certainty that one will indeed develop the disease. This situation raises with peculiar force the question of whether parental responsibility requires people to get tested.

Some people think that we should recognize a right "not to know." It seems to me that such a right could be defended only where ignorance does not put others at serious risk. So if people are prepared to forego genetically related children, they need not get tested. But if they want genetically related children then they must do whatever is necessary to ensure that affected babies are not the result. There is, after all, something inconsistent about the claim that one has a right to be shielded from the truth, even if the price is to risk inflicting on one's children the same dread disease one cannot even face in oneself.

In sum, until we can be assured that Huntington's Disease does not prevent people from living a minimally satisfying life, individuals at risk for the disease have a moral duty to try not to bring affected babies into this world. There are now enough options available so that this duty needn't frustrate their reasonable desires. Society has a corresponding duty to facilitate moral behavior on the part of individuals. Such support ranges from the narrow and concrete (like making sure that medical testing and counseling is available to all) to the more general social environment that guarantees that all pregnancies are voluntary, that pronatalism is eradicated, and that women are treated with respect regardless of the reproductive options they choose.

Notes

1. *The Merck Manual* (Railway, NJ: Merck, 1972), pp. 1363, 1346.
2. *The New York Times*, September 30, 1975, p. 1, col. 6. The Joseph family disease is similar to Huntington's Disease except that symptoms start appearing in the twenties. Rick Donohue was in his early twenties at the time he made this statement.
3. The view I am rejecting has been forcefully articulated by Derek Parfit. *Reasons and Persons* (Oxford: Oxford University Press, 1984).
4. *Essays in Pragmatism*, ed. A. Castell (New York, 1948), p. 73.

Reproductive Freedom and the Prevention of Harm

Allen Buchanan, Dan W. Brock, Norman Daniels, and Daniel Wikler

These authors worked together in the 1980s on the President's Commission for the Study of Ethical Problems in Medicine and Biomedical and Behavioral Research. Allen Buchanan is the James B. Duke Professor of Philosophy and Public Policy Studies at Duke University. Dan Brock is the Frances Glessner Lee Professor of Medical Ethics in the Department of Social Medicine, the Director of the Division of Medical Ethics at the Harvard Medical School, and the Director of the Harvard University Program in Ethics and Health. Norman Daniels is Professor of Public Health at Harvard University. Dan Wikler is the Mary B. Saltonstall Professor of Population Ethics in the Department of Population and International Health, School of Public Health, at Harvard University. In this chapter the authors first characterize the general perception that conceiving or bringing a child to term when the result is a life not worth living is wrong. They then identify the challenging problem of comparing a life not worth living with nonexistence. This line of reasoning is extended to lives that are worth living, albeit disabled. The authors then discuss the moral dimensions of these types of decisions in different contexts—known preventable risk before conception, known treatable risk during pregnancy, and known treatable risk after birth.

Public Policy and Wrongful Life Issues

It is perhaps fortunate that, despite the great expansion of genetic information that will be available in the future both from pre-conception testing for genetic risks to potential offspring and from prenatal diagnosis of the genetic condition of a fetus, public policy may be able largely to avoid the most contentious and intractable wrongful life issues for at least two reasons. First, only a very small proportion of genetic abnormalities and diseases are both compatible with life and also so severe as to result in the affected child having a life not worth living. Second, courts and legislatures are likely to continue to be reluctant to permit wrongful life legal suits, both because damages covering the child's medical and extra care expenses can usually be obtained by a suit brought in the name of the parents instead of a wrongful life suit in the name of the child, and because uncertainty exists about how to assess damages for wrongful life. But regardless of what occurs in the courts, moral choices about whether conceiving a child or carrying a pregnancy to term would constitute an action of wrongful life will be increasingly faced in the future by parents or would-be parents.

A complicating factor is that the woman or couple making the choice will often face only a risk, not a certainty, that the child will not have a life worth living and that risk can vary from very low to approaching certainty. Whether it is morally wrong to conceive in the face of such risks will depend in part on the woman's willingness and intention to do appropriate prenatal genetic testing and to abort her fetus if it is found to have a disease or condition incompatible with a worthwhile life.

As noted before, pursuing the moral complexities of abortion would take us too far afield here. Nevertheless, suppose, as the authors of this book believe, that the fetus at least through the first two trimesters is not a person and so aborting it then is morally permissible. Aborting a fetus found during the first two trimesters to have a disease that would make life a burden to the child prevents the creation of a person with a life not worth living; no wrongful life then occurs, so there is no question of moral wrong-doing. Even conceiving when there is a relatively high risk of genetic transmission of a disease incompatible with a life worth living could be morally acceptable so long as the woman firmly intends to test the fetus for the disease and to abort it if the disease is present.

On the other hand, a woman may intend not to test her fetus and abort it if such a disease is present, either because she considers abortion morally wrong or for other reasons. In that case, the higher the risk that her child will have a genetic disease or condition incompatible with a life worth living, the stronger the moral case that she does a serious moral wrong to that child in conceiving it and carrying it to term.

If a mother or anyone else knowingly and responsibly caused harm to an already born child so serious as to make its life no longer worth living, that would constitute extremely serious child abuse and be an extremely serious moral and legal wrong. In that case, however, the child would have had a worthwhile life that was taken away by whomever was guilty of the child abuse; the wrong to the child then is depriving it of a worthwhile life that it otherwise would have had. That is a different and arguably more serious wrong than wrongful life, where the alternative to the life not worth living is never having a life at all, and so not having a worthwhile life taken away. The wrong in nearly all cases of wrongful life is bringing into existence a child who will have a short life dominated by severe and unremitting suffering—that is, being caused to undergo that suffering without compensating benefits.

How high must the risk be of a child having a genetic disease incompatible with a worthwhile life for it to be morally wrong for the parents to conceive it and allow it to be born? There is, of course, no precise probability at which the risk of the harm makes it morally wrong to conceive or not to abort; different cases fall along a spectrum in the degree to which undertaking the risk is morally justified. How seriously wrong, if a wrong at all, it is to risk the conception and birth of a child with such a life will depend on several factors. How bad is the child's life, and in particular how severe and unremitting is its suffering? How high is the probability of the child having a genetic disease incompatible with a worthwhile life? How weighty are its parents' interests in having the child? For example, is this likely its parents' only opportunity to become parents, or are they already parents seeking to have additional children? How significant is the possibility of the parents having an unaffected child if this pregnancy is terminated and another conception pursued? How willing and able are the parents to support and care for the child while it lives?

These factors, and no doubt others unique to specific cases, will determine how strong the moral case is against individuals risking having a child who will not have a life worth living. It is worth underlining that any case for the wrongness of parents conceiving and bringing to term such a child depends on their having reasonable access to genetic testing, contraception, and abortion services, and this can require public provision and funding of these services for those who otherwise cannot afford them.

We hope that our analysis so far makes it clear why we believe that there are some cases, albeit very few, in which it would be clearly and seriously morally wrong for individuals to risk conceiving and having such a child. However, use of government power to force an abortion on an unwilling woman would be so deeply invasive of her reproductive freedom, bodily integrity, and right to decide about her own health care as to be virtually never morally justified. Allowing the child to be born and then withholding life support even over its parents' objections would probably be morally preferable. The government's doing this forcibly and over the parents' objections would be extraordinarily controversial, both morally and legally, but in true cases of wrongful life, the wrong done is sufficiently serious as to possibly justify doing so in an individual case. However, at the present time and as a practical matter, the common and strong bias in favor of life, even in the face of serious suffering, makes it nearly inconceivable that public policy might authorize the government forcibly to take an infant from its parents, not for the purpose of securing beneficial treatment for it, but instead to allow it to die because it could not have a worthwhile life. Moreover, the risk of abuse of such a governmental power to intervene forcibly in reproductive choices to prevent wrongful life is too great to warrant granting that power.

There is a stronger moral case for the use of government coercive power to prevent conception in some wrongful life cases. Similar power is now exercised by government over severely mentally disabled people who are sterilized to prevent them from conceiving. In such cases, the individual sterilized is typically deemed incompetent to make a responsible decision about conception, as well as unable to raise a child. Forced sterilization of a competent individual is more serious morally, but the harm to be prevented of wrongful life is more serious than the harm prevented in typical involuntary sterilization cases where the child would have a worthwhile life if raised by others. Nevertheless, the historical abuses of "eugenic sterilizations" . . . are enough to warrant not giving government the coercive power to prevent wrongful life conceptions unless their occurrence was very common and widespread. Wrongful life conceptions are sufficiently uncommon, and practical and moral difficulties in using the coercive power of government to prevent them sufficiently great, to rule out policies that prevent people from conceiving wrongful lives. Coercive government intrusion into reproductive freedom to prevent wrongful life would be wrong.

Pre-Conception Interventions to Prevent Conditions Compatible with a Worthwhile Life

The Human Genome Project and related research will produce information permitting genetic screening for an increasing number of genetically transmitted diseases, or susceptibilities to diseases and other harmful conditions. In the foreseeable future, our capacities for pre-conception and prenatal screening for these diseases and conditions will almost certainly far outstrip our capacities for genetic or other therapy to correct for the harmful genes and their effects. The vast majority of decisions faced by prospective parents, consequently, will not be whether to pursue genetic or other therapy for their fetus or child, but instead whether to test for particular genetic risks and/or conditions and, when they are found to be present, whether to avoid conception or to terminate a pregnancy. Moreover, the vast majority of genetic risks that will be subject to testing will not be for conditions incompatible with a life worth living—the wrongful life cases—but rather for less severe conditions compatible with having a life well worth living.

These genetically transmitted conditions and diseases will take different forms. Sometimes their disabling features will be manifest during much of the individual's life, but still will permit a worthwhile life, as with most cases of Down syndrome, which is caused by a chromosomal abnormality. Sometimes the disease or condition will result in significant disability and a significantly shorter than normal life span, but not so disabling or short as to make the life not worth living, as with cystic fibrosis. Sometimes the disease or condition, although devastating in its effects on the afflicted individual's quality of life, will only manifest symptoms after a substantial period of normal life and function, as with Huntington's chorea and Alzheimer's dementia.

When the genetically transmitted conditions could and should have been prevented, they will constitute what we have called cases of wrongful disability. But in which cases will the failure to prevent a genetically transmitted disability be morally wrong? Again, different cases fall along a spectrum in the degree of moral justification for undertaking or not undertaking to prevent the disability.

Whether failure to prevent a disability is wrong in specific cases will typically depend on many features of that case. For example, what is the relative seriousness of the disability for the child's well-being and opportunities? What measures are available to the child's parents to prevent the condition—such as abortion, artificial insemination by donor, or oocyte donation—and how acceptable are these means to the prospective parents? Is it possible, and if so how likely, that they can conceive another child without the disabling condition, or will any child they conceive have or be highly likely to have the condition? If the disability can only be prevented by not conceiving at all, do the couple have alternative means, such as adoption, of becoming parents? When the condition can be prevented or its adverse impact compensated for, what means are necessary to do so?

These and other considerations can all bear on the threshold question: Is the severity of a genetically transmitted disability great enough that particular parents are morally obligated to prevent it, given the specific means necessary for them to do so; or is it sufficiently limited and minor that it need not be prevented, but is instead a condition that the child can reasonably be expected to live with?

Different prospective parents will answer this threshold question differently because they differ about such matters as the burdensomeness and undesirability of particular alternative methods of reproduction that may be necessary to prevent the disability, the seriousness of the impact of the particular disability on a person's well-being and opportunity, aspects of reproductive freedom such as the importance of having children and of having biologically related as opposed to adopted or only partially biologically related children, the extent of society's obligation and efforts to make special accommodations to eliminate or ameliorate the disability, and their willingness to assume the burdens of raising a child with the disability in question.

Because there are these multiple sources of reasonable disagreement bearing on the threshold question, and because the aspects of reproductive freedom at stake will usually be of substantial importance, public policy should usually permit prospective parents to make and act on their own judgments about whether they morally ought to prevent particular genetically transmitted disabilities for the sake of their child. But there is a systematic objection to all preconception wrongful disability cases that must be met in order to clear the way for individual judgments about specific cases.

To fix attention on the general problem in question, which is not restricted to cases of genetically transmitted disease, let us imagine a case, call it P1, in which a woman is told by her physician that she should not attempt to become pregnant now because she has a condition that is highly likely to result in moderate mental retardation in her child. Her condition is easily and fully treatable by taking a quite safe medication for one month. If she takes the medication and delays becoming pregnant for a month, the risk to her child will be eliminated and there is every reason to expect that she will have a normal child. Because the delay would interfere with her vacation travel plans, however, she does not take the medication, gets pregnant now, and gives birth to a child who is moderately retarded.

According to commonsense moral views, this woman acts wrongly, and in particular, wrongs her child by not preventing its disability for such a morally trivial reason, even if for pragmatic reasons many people would oppose government intrusion into her decision. According to commonsense morality, her action is no different morally than if she failed to take the medicine in a case, P2, in which the condition is discovered, and so the medicine must be taken, after conception and when she is already pregnant. Nor is it different morally than if she failed to provide a similar medication to her born child, in a case, P3, if doing so is again necessary to prevent moderate mental retardation in her child. It is worth noting that in most states in this country, her action in P3 would probably constitute medical neglect, and governmental child protection agencies could use coercive measures, if necessary, to ensure the child's treatment.

This suggests that it might only be because her reproductive freedom and her right to decide about her own health care are also involved in P1 that we are reluctant to coerce her decision, if necessary, there as well. On what Derek Parfit has called the "no difference" view, the view of commonsense morality, her failure to use the medication to prevent her child's mental retardation would be equally seriously wrong, and for the same reason, in each of the three cases.[1] But her action in P1, which is analogous in relevant respects to preconception genetic screening to prevent disabilities, has a special feature that makes it not so easily shown to be wrong as commonsense morality might suppose.

What is the philosophical problem at the heart of wrongful disability cases like P1? As with wrongful life cases, in which the necessary comparison of life with nonexistence is thought to create both philosophical and policy problems, so also in wrongful disability cases do the philosophical and policy problems arise from having to compare a disabled existence with not having existed at all. But the nature of the philosophical problems in wrongful life and wrongful disability cases are in fact quite different. The philosophical objections we considered to wrongful life cases centered on whether it is coherent to compare an individual's quality of life with never having existed at all—that is, with nonexistence—and whether merely possible persons can have moral rights or be owed moral obligations. In wrongful disability cases, a person's disability uncontroversially leaves him or her with a worthwhile life. The philosophical problem, as noted earlier, is how this is compatible with the commonsense view that it would be wrong not to prevent the disability.

The special difficulty in wrongful disability cases, which Derek Parfit has called the "nonidentity problem," is that it would not be better for the person with the disability to have had it prevented, since that could only be done by preventing him or her from ever having existed at all. Preventing the disability would deny the disabled individual a worthwhile, although disabled, life. That is because the disability could only have been prevented either by conceiving at a different time and/or under different circumstances (in which case a different child would have been conceived) or by terminating the pregnancy (in which case this child also never would have been born, although a different child may or may not have been conceived instead). None of these possible means of preventing the disability would be better for the child with the disability—all would deny him or her a worthwhile life.

But if the mother's failure to prevent the disability did not make her child worse off than he or she would have been without the intervention, then her failure to prevent it seems not to harm her child. And if she did not harm her child by not preventing its disability, then why does she wrong her child morally by failing to do so? How could making her child better off, or at least not worse off, by giving it a life worth living, albeit a life with a significant disability, wrong it? A wrong action must be bad for someone, but her choice to create her child with its disability is bad for no one, so she does no wrong. Of course, there is a sense in which it is bad for her child to have the disability, in comparison with being without it, but there is nothing the mother could have done to enable that child to be born without the disability, and so nothing she does or omits doing is bad for her child.

So actions whose harmful effects would constitute seriously wrongful child abuse if done to an existing child are no harm, and so not wrong, if their harmful effects on a child are inextricable from the act of bringing that child into existence with a worthwhile life! This argument threatens to undermine common and firmly held moral judgments, as well as public policy measures, concerning prevention of such disabilities for children.

Actual versus Possible Persons

David Heyd has accepted the implications of this argument and concludes that in all of what he calls "genesis" choices—that is, choices that inextricably involve whether a particular individual will be brought into existence—only the interests of actual persons, not those of possible persons such as the disabled child in case P1, are relevant to the choice.[2] So in case P1, the effects on the parents and the broader society, such as the greater childrearing costs and burdens of having the moderately retarded child instead of taking the medication and having a normal child a month later are relevant to the decision. But the effects on and interests of the child who would be moderately retarded are not relevant. In cases P2 and P3, on the other hand, Heyd presumably would share the commonsense moral view that the fundamental reason the woman's action would be wrong is the easily preventable harm that she allows her child to suffer. In these situations, the preventable harm to her child is the basis of the moral wrong she does her child.

In Parfit's "no difference" view, the woman's action in P1 is equally wrong, and for the same reason, as her action in P2 and P3. We share with Parfit, in opposition to Heyd, the position that the woman's action in P1 is wrong because of the easily preventable effect on her child. But we do not accept the "no difference" thesis. We will suggest a reason why her action in P1 may not be as seriously wrong as in P2 car P3, and also suggest that the reason her action is wrong in P1 is similar to but nevertheless importantly different from the reason it is wrong in P2 and P3.

As Parfit notes, the difficulty is identifying and formulating a moral principle that implies that the woman's action in P1 is seriously wrong, but does not have unacceptable implications for other cases. Before proceeding further, we must emphasize that we cannot explore this difficulty fully here. The issues are extraordinarily complex and involve testing the implications of such a principle in a wide variety of cases outside of the genetic context that is our concern here (e.g., in population policy contexts and, in particular, avoiding what Parfit calls the "Repugnant Conclusion" and explaining what he calls the "Asymmetry"). Its relationship to other principles and features of a moral theory must also be explained, including that to the principle applicable to P2 and P3.[1]

The apparent failure to account for common and firmly held moral views in the genetics cases of wrongful disabilities like P1 constitutes one of the most important practical limitations (problems of population policy are another) of traditional ethical theories and of their principles of beneficence—doing good—and nonmaleficence—not causing or preventing harm. Where the commonsense moral judgment about cases like P1 is that the woman is morally wrong to go ahead and have the disabled child instead of waiting and having a normal child, the principles of traditional ethical theories apparently fail to support that judgment. New or revised moral principles appear to be needed. What alternatives and resources, either within or beyond traditional moral principles or theories, could account for and explain the wrong done in wrongful disability cases?

Person-Affecting Moral Principles

Perhaps the most natural way to account for the moral wrong in wrongful disability cases like P1 is to abandon the specific feature of typical moral principles about obligations to prevent or not cause harm which generates difficulty when we move from standard cases of prevention of harm to existing persons, as in P3, to harm prevention in genesis cases like P1. That feature is what philosophers have called the "person-affecting" property of principles of beneficence and nonmaleficence. Recall that earlier we appealed to principle M: Those individuals responsible for a child's, or other dependent person's, welfare are morally required not to let her suffer a serious harm or disability or a serious loss of happiness or good that they could have prevented without imposing substantial burdens or costs or loss of benefits on themselves or others.

The person-affecting feature of M is that the persons who will suffer the harm if it is not prevented and not suffer it if it is prevented must be one and the same distinct individual. If M is violated, a distinct child or dependent person is harmed without good reason, and so the moral wrong is done to that person. Since harms to persons must always be harms to some person, it may seem that there is no alternative to principles that are person-affecting, but that is not so. The alternative is clearest if we follow Derek Parfit by distinguishing "same person" from "same number" choices.

In same person choices, the same persons exist in each of the different alternative courses of action from which an agent chooses. Cases P2 and P3 above were same person choices (assuming in P2 that the fetus is or will become a person, though that is not essential to the point)—the harm of moderate retardation prevented is to the woman's fetus or born child. In same number choices, the same number of persons exist in each of the alternative courses of action from which an agent chooses, but the identities of some of the persons—that is, who exists in those alternatives—are affected by the choice. P1 is a same number but not a same person choice—the woman's choice affects which child will exist. If the woman does not take the medication and wait to conceive, she gives birth to a moderately retarded child, whereas if she takes the medication and waits to conceive, she gives birth to a different child who is not moderately retarded.

The concept of "harm," arguably, is necessarily comparative, and so the concept of "harm prevention" may seem necessarily person affecting; this is why harm prevention principles seem not to apply to same number, different person choices like P1. But it would be a mistake to think that non-person-affecting principles, even harm prevention principles, are not coherent. Suppose for simplicity that the harm in question in P1 from the moderate retardation is suffering and limited opportunity. Then in P1, if the woman chooses to have the moderately retarded child, she causes suffering and limited opportunity to exist that would be prevented and not exist if she chooses to take the medication and wait to conceive a different normal child. An example of a non-person-affecting principle that applies to P1 is:

> N: Individuals are morally required not to let any child or other dependent person for whose welfare they are responsible experience serious suffering or limited opportunity or serious loss of happiness or good, if they can act so that, without affecting the number of persons who will exist and without imposing substantial burdens or costs or loss of benefits on themselves or others, no child or other dependent person for whose welfare they are responsible will experience serious suffering or limited opportunity or serious loss of happiness or good.

Any suffering and limited opportunity must, of course, be experienced by some person—they cannot exist in disembodied form—and so in that sense N remains person-affecting. But N does not require that the same individuals who experience suffering and limited opportunity in one alternative exist without the suffering and limited opportunity in the other alternative; it is a same number, not a same person, principle. N allows the child who does not experience the suffering and limited opportunity to be a different person from the child who does; that is why the woman's action in P1 is morally wrong according to N, but not according to M. If the woman in P1 does take the medication and wait to conceive a normal child, she acts so as to make the suffering and limited opportunity "avoidable by substitution."

A different way of making the same point is to say that this principle for the prevention of suffering applies not to distinct individuals, so that the prevention of suffering must make a distinct individual better off than he or she would have been, as M requires, but to the classes of individuals who will exist if the suffering is or is not prevented, as N does. Assessing the prevention of suffering by the effect on classes of persons, as opposed to distinct individuals, also allows for avoidability by substitution—an individual who does not suffer if one choice is made is substituted for a different person who does suffer if the other choice is made. A principle applied to the classes of all persons who will exist in each of two or more alternative courses of action will be a non-person-affecting principle. The preceding discussion referred only to the prevention of harm or loss of opportunity because that is the focus of this chapter. However, it should be noted that N allows, for the same reasons as does M, the weighing of securing happiness or good against preventing suffering and loss of opportunity. If it did not, but required only preventing serious suffering, then N would require not creating a child who would experience serious suffering, but also great happiness and good, in favor of creating a child who would suffer less, but experience no compensating happiness or good, even though the latter child on balance would have a substantially worse life. We note as well that we have not defined "serious" as it functions in either M or N; it is difficult to do so in a sufficiently general yet precise way to make the application of the principle simple and straightforward for a wide range of cases. Judgment must be used in applying N. The seriousness of suffering and loss of opportunity, or loss of happiness and good, that could be prevented must be assessed principally in light of their potential impact on the child's life, the probability of that impact, and the possibility and probability of compensatory measures to mitigate that impact. Applying N requires judgment as well regarding what are "substantial burdens or costs, or loss of benefits, on themselves or others." For example, how serious are possible moral objections by the parents to the use of abortion, and how great are the financial costs or medical risks of having an alternative child using assisted reproduction technologies, and so forth?

We do not claim that all moral principles concerning obligations to prevent harm, or of beneficence and nonmaleficence more generally, are non-person-affecting, and so we do not reject principle M. In typical cases of harm where a distinct individual is made worse off, the moral principles most straightforwardly applicable to them are person-affecting. Our claim is only that an adequate moral theory should include as well non-person-affecting principles like N. How these principles are related, as well as what principles apply to different number cases in a comprehensive moral theory, involves deep difficulties in moral theory that we cannot pursue here. In this respect, we do not propose a full solution to the nonidentity problem.

References

1. Derek Parfit, *Reasons and Persons*. New York: Oxford University Press, 1984.
2. David Heyd, *Genethics: Moral Issues in the Creation of People*. Berkeley, CA: University of California Press, 1992.

The Case Against Perfection:
What's Wrong with Designer Children, Bionic Athletes, and Genetic Engineering

Michael J. Sandel

Michael J. Sandel received his doctorate in philosophy from Oxford University in 1981. He is a professor of government at Harvard University, where he has taught political philosophy for the last 30 years. Sandel served on the President's Council on Bioethics, a national body appointed by President George W. Bush to examine the ethical implications of new biomedical technologies. Well known for his critique of Rawls's Theory of Justice, he has become popular to a mass audience through public television and Internet access to his popular course on justice. In this article Sandel provides examples of proposed enhancements such as reversing muscle loss, increasing memory, enhancing height, and sex selection of children, and presents arguments against genetic enhancement. Most importantly, he argues that "genetic engineering represents a one-sided triumph of willfulness over giftedness, of dominion over reverence, of molding over beholding"—all of which undermine the moral sentiments that social solidarity requires.

Breakthroughs in genetics present us with a promise and a predicament. The promise is that we may soon be able to treat and prevent a host of debilitating diseases. The predicament is that our newfound genetic knowledge may also enable us to manipulate our own nature—to enhance our muscles, memories, and moods; to choose the sex, height, and other genetic traits of our children; to make ourselves "better than well." When science moves faster than moral understanding, as it does today, men and women struggle to articulate their unease. In liberal societies they reach first for the language of autonomy, fairness, and individual rights. But this part of our moral vocabulary is ill equipped to address the hardest questions posed by genetic engineering. The genomic revolution has induced a kind of moral vertigo.

Consider cloning. The birth of Dolly the cloned sheep, in 1997, brought a torrent of concern about the prospect of cloned human beings. There are good medical reasons to worry. Most scientists agree that cloning is unsafe, likely to produce offspring with serious abnormalities. (Dolly recently died a premature death.) But suppose technology improved to the point where clones were at no greater risk than naturally conceived offspring. Would human cloning still be objectionable? Should our hesitation be moral as well as medical? What, exactly, is wrong with creating a child who is a genetic twin of one parent, or of an older sibling who has tragically died—or, for that matter, of an admired scientist, sports star, or celebrity?

Some say cloning is wrong because it violates the right to autonomy: by choosing a child's genetic makeup in advance, parents deny the child's right to an open future. A similar objection can be raised against any form of bioengineering that allows parents to select or reject genetic characteristics. According to this argument, genetic enhancements for musical talent, say, or athletic prowess, would point children toward particular choices, and so designer children would never be fully free.

At first glance the autonomy argument seems to capture what is troubling about human cloning and other forms of genetic engineering. It is not persuasive, for two reasons. First, it wrongly implies that absent a designing parent, children are free to choose their characteristics for themselves. But none of us chooses his genetic inheritance. The alternative to a cloned or genetically enhanced child is not one whose future is unbound by particular talents but one at the mercy of the genetic lottery.

Second, even if a concern for autonomy explains some of our worries about made-to-order children, it cannot explain our moral hesitation about people who seek genetic remedies or enhancements for themselves. Gene therapy on somatic (that is, nonreproductive) cells, such as muscle cells and brain cells, repairs or replaces defective genes. The moral quandary arises when people use such therapy not to cure a disease but to reach beyond health, to enhance their physical or cognitive capacities, to lift themselves above the norm.

Like cosmetic surgery, genetic enhancement employs medical means for nonmedical ends—ends unrelated to curing or preventing disease or repairing injury. But unlike cosmetic surgery, genetic enhancement is more than skin deep. If we are ambivalent about surgery or Botox injections for sagging chins and furrowed brows, we are all the more troubled by genetic engineering for stronger bodies, sharper memories, greater intelligence, and happier moods. The question is whether we are right to be troubled, and if so, on what grounds.

In order to grapple with the ethics of enhancement, we need to confront questions largely lost from view—questions about the moral status of nature, and about the proper stance of human beings toward the given world. Since these questions verge on theology, modern philosophers and political theorists tend to shrink from them. But our new powers of biotechnology make them unavoidable. To see why this is so, consider four examples already on the horizon: muscle enhancement, memory enhancement, growth hormone treatment, and reproductive technologies that enable parents to choose the sex and some genetic traits of their children. In each case what began as an attempt to treat a disease or prevent a genetic disorder now beckons as an instrument of improvement and consumer choice.

Muscles   Everyone would welcome a gene therapy to alleviate muscular dystrophy and to reverse the debilitating muscle loss that comes with old age. But what if the same therapy were used to improve athletic performance? Researchers have developed a synthetic gene that, when injected into the muscle cells of mice, prevents and even reverses natural muscle deterioration. The gene not only repairs wasted or injured muscles but also strengthens healthy ones. This success bodes well for human applications. H. Lee Sweeney, of the University of Pennsylvania, who leads the research, hopes his discovery will cure the immobility that afflicts the elderly. But Sweeney's bulked-up mice have already attracted the attention of athletes seeking a competitive edge. Although the therapy is not yet approved for human use, the prospect of genetically enhanced weight lifters, home-run sluggers, linebackers, and sprinters is easy to imagine. The widespread use of steroids and other performance-improving drugs in professional sports suggests that many athletes will be eager to avail themselves of genetic enhancement.

Suppose for the sake of argument that muscle-enhancing gene therapy, unlike steroids, turned out to be safe—or at least no riskier than a rigorous weight-training regimen. Would there be a reason to ban its use in sports? There is something unsettling about the image of genetically altered athletes lifting SUVs or hitting 650-foot home runs or running a three-minute mile. But what, exactly, is troubling about it? Is it simply that we find such superhuman spectacles too bizarre to contemplate? Or does our unease point to something of ethical significance?

It might be argued that a genetically enhanced athlete, like a drug-enhanced athlete, would have an unfair advantage over his unenhanced competitors. But the fairness argument against enhancement has a fatal flaw: it has always been the case that some athletes are better endowed genetically than others, and yet we do not consider this to undermine the fairness of competitive sports. From the standpoint of fairness, enhanced genetic differences would be no worse than natural ones, assuming they were safe and made available to all. If genetic enhancement in sports is morally objectionable, it must be for reasons other than fairness.

Memory   Genetic enhancement is possible for brains as well as brawn. In the mid-1990s scientists managed to manipulate a memory-linked gene in fruit flies, creating flies with photographic memories. More recently researchers have produced smart mice by inserting extra copies of a memory-related gene into mouse embryos. The altered mice learn more quickly and remember things longer than normal mice. The extra copies were programmed to remain active even in old age, and the improvement was passed on to offspring.

Human memory is more complicated, but biotech companies, including Memory Pharmaceuticals, are in hot pursuit of memory-enhancing drugs, or "cognition enhancers," for human beings. The obvious market for such drugs consists of those who suffer from Alzheimer's and other serious memory disorders. The companies also have their sights on a bigger market: the 81 million Americans over 50, who are beginning to encounter the memory loss that comes naturally with age. A drug that reversed age-related memory loss would be a bonanza for the pharmaceutical industry: a Viagra for the brain. Such use would straddle the line between remedy and enhancement. Unlike a treatment for Alzheimer's, it would cure no disease; but insofar as it restored capacities a person once possessed, it would have a remedial aspect. It could also have purely nonmedical uses: for example, by a lawyer cramming to memorize facts for an upcoming trial, or by a business executive eager to learn Mandarin on the eve of his departure for Shanghai.

Some who worry about the ethics of cognitive enhancement point to the danger of creating two classes of human beings: those with access to enhancement technologies, and those who must make do with their natural capacities. And if the enhancements could be passed down the generations, the two classes might eventually become subspecies—the enhanced and the merely natural. But worry about access ignores the moral status of enhancement itself. Is the scenario troubling because the unenhanced poor would be denied the benefits of bioengineering, or because the enhanced affluent would somehow be dehumanized? As with muscles, so with memory: the fundamental question is not how to ensure equal access to enhancement but whether we should aspire to it in the first place.

Height   Pediatricians already struggle with the ethics of enhancement when confronted by parents who want to make their children taller. Since the 1980s human growth hormone has been approved for children with a hormone deficiency that makes them much shorter than average. But the treatment also increases the height of healthy children. Some parents of healthy children who are unhappy with their stature (typically boys) ask why it should make a difference whether a child is short because of a hormone deficiency or because his parents happen to be short. Whatever the cause, the social consequences are the same.

In the face of this argument some doctors began prescribing hormone treatments for children whose short stature was unrelated to any medical problem. By 1996 such "off-label" use accounted for 40 percent of human-growth-hormone prescriptions. Although it is legal to prescribe drugs for purposes not approved by the Food and Drug Administration, pharmaceutical companies cannot promote such use. Seeking to expand its market, Eli Lilly & Co. recently persuaded the FDA to approve its human growth hormone for healthy children whose projected adult height is in the bottom one percentile—under five feet three inches for boys and four feet eleven inches for girls. This concession raises a large question about the ethics of enhancement: If hormone treatments need not be limited to those with hormone deficiencies, why should they be available only to very short children? Why shouldn't all shorter-than-average children be able to seek treatment? And what about a child of average height who wants to be taller so that he can make the basketball team?

Some oppose height enhancement on the grounds that it is collectively self-defeating; as some become taller, others become shorter relative to the norm. Except in Lake Wobegon, not every child can be above average. As the un-enhanced began to feel shorter, they, too, might seek treatment, leading to a hormonal arms race that left everyone worse off, especially those who couldn't afford to buy their way up from shortness.

But the arms-race objection is not decisive on its own. Like the fairness objection to bioengineered muscles and memory, it leaves unexamined the attitudes and dispositions that prompt the drive for enhancement. If we were bothered only by the injustice of adding shortness to the problems of the poor, we could remedy that unfairness by publicly subsidizing height enhancements. As for the relative height deprivation suffered by innocent bystanders, we could compensate them by taxing those who buy their way to greater height. The real question is whether we want to live in a society where parents feel compelled to spend a fortune to make perfectly healthy kids a few inches taller.

Sex selection   Perhaps the most inevitable nonmedical use of bioengineering is sex selection. For centuries parents have been trying to choose the sex of their children. Today biotech succeeds where folk remedies failed.

One technique for sex selection arose with prenatal tests using amniocentesis and ultrasound. These medical technologies were developed to detect genetic abnormalities such as spina bifida and Down syndrome. But they can also reveal the sex of the fetus—allowing for the abortion of a fetus of an undesired sex. Even among those who favor abortion rights, few advocate abortion simply because the parents do not want a girl. Nevertheless, in traditional societies with a powerful cultural preference for boys, this practice has become widespread.

Sex selection need not involve abortion, however. For couples undergoing in vitro fertilization (IVF), it is possible to choose the sex of the child before the fertilized egg is implanted in the womb. One method makes use of pre-implantation genetic diagnosis (PGD), a procedure developed to screen for genetic diseases. Several eggs are fertilized in a petri dish and grown to the eight-cell stage (about 3 days). At that point the embryos are tested to determine their sex. Those of the desired sex are implanted; the others are typically discarded. Although few couples are likely to undergo the difficulty and expense of IVF simply to choose the sex of their child, embryo screening is a highly reliable means of sex selection. And as our genetic knowledge increases, it may be possible to use PGD to cull embryos carrying undesired genes, such as those associated with obesity, height, and skin color. The science fiction movie *Gattaca* depicts a future in which parents routinely screen embryos for sex, height, immunity to disease, and even IQ. There is something troubling about the *Gattaca* scenario, but it is not easy to identify what exactly is wrong with screening embryos to choose the sex of our children.

One line of objection draws on arguments familiar from the abortion debate. Those who believe that an embryo is a person, reject embryo screening for the same reasons they reject abortion. If an eight-cell embryo growing in a petri dish is morally equivalent to a fully developed human being, then discarding it is no better than aborting a fetus, and both practices are equivalent to infanticide. Whatever its merits, however, this "pro-life" objection is not an argument against sex selection as such.

The latest technology poses the question of sex selection unclouded by the matter of an embryo's moral status. The Genetics & IVF Institute, a for-profit infertility clinic in Fairfax, Virginia, now offers a sperm-sorting technique that makes it possible to choose the sex of one's child before it is conceived. X-bearing speml, which produce girls, carry more DNA than Y-bearing sperm, which produce boys; a device called a flow cytometer can separate them. The process, called MicroSort, has a high rate of success.

If sex selection by sperm sorting is objectionable, it must be for reasons that go beyond the debate about the moral status of the embryo. One such reason is that sex selection is an instrument of sex discrimination—typically against girls, as illustrated by the chilling sex ratios in India and China. Some speculate that societies with substantially more men than women will be less stable, more violent, and more prone to crime or war. These are legitimate worries—but the sperm-sorting company has a clever way of addressing them. It offers MicroSort only to couples who want to choose the sex of a child for purposes of "family balancing." Those with more sons than daughters may choose a girl, and vice versa. But customers may not use the technology to stock up on children of the same sex, or even to choose the sex of their firstborn child. (So far the majority of MicroSort clients have chosen girls.) Under restrictions of this kind, do any ethical issues remain that should give us pause?

The case of MicroSort helps us isolate the moral objections that would persist if muscle-enhancement, memory-enhancement, and height-enhancement technologies were safe and available to all.

It is commonly said that genetic enhancements undermine our humanity by threatening our capacity to act freely, to succeed by our own efforts, and to consider ourselves responsible—worthy of praise or blame—for the things we do and for the way we are. It is one thing to hit 70 home runs as the result of disciplined training and effort, and something else, something less, to hit them with the help of steroids or genetically enhanced muscles. Of course, the roles of effort and enhancement will be a matter of degree. But as the role of enhancement increases, our admiration for the achievement fades—or, rather, our admiration for the achievement shifts from the player to his pharmacist. This suggests that our moral response to enhancement is a response to the diminished agency of the person whose achievement is enhanced.

Though there is much to be said for this argument, I do not think the main problem with enhancement and genetic engineering is that they undermine effort and erode human agency. The deeper danger is that they represent a kind of hyperagency—a Promethean aspiration to remake nature, including human nature, to serve our purposes and satisfy our desires. The problem is not the drift to mechanism but the drive to mastery. And what the drive to mastery misses and may even destroy is an appreciation of the gifted character of human powers and achievements.

To acknowledge the giftedness of life is to recognize that our talents and powers are not wholly our own doing, despite the effort we expend to develop and to exercise them. It is also to recognize that not everything in the world is open to whatever use we may desire or devise. Appreciating the gifted quality of life constrains the Promethean project and conduces to a certain humility. It is in part a religious sensibility. But its resonance reaches beyond religion.

It is difficult to account for what we admire about human activity and achievement without drawing upon some version of this idea. Consider two types of athletic achievement. We appreciate players like Pete Rose, who are not blessed with great natural gifts but who manage, through striving, grit, and determination, to excel in their sport. But we also admire players like Joe DiMaggio, who display natural gifts with grace and effortlessness. Now, suppose we learned that both players took performance-enhancing drugs. Whose turn to drugs would we find more deeply disillusioning? Which aspect of the athletic ideal—effort or gift— would be more deeply offended?

Some might say effort: the problem with drugs is that they provide a short-cut, a way to win without striving. But striving is not the point of sports; excellence is. And excellence consists at least partly in the display of natural talents and gifts that are no doing of the athlete who possesses them. This is an uncomfortable fact for democratic societies. We want to believe that success, in sports and in life, is something we earn, not something we inherit. Natural gifts, and the admiration they inspire, embarrass the meritocratic faith; they cast doubt on the conviction that praise and rewards flow from effort alone. In the face of this embarrassment we inflate the moral significance of striving, and depreciate giftedness. This distortion can be seen, for example, in network television coverage of the Olympics, which focuses less on the feats the athletes perform than on heartrending stories of the hardships they have overcome and the struggles they have waged to triumph over an injury or a difficult upbringing or political turmoil in their native land.

But effort isn't everything. No one believes that a mediocre basketball player who works and trains even harder than Michael Jordan deserves greater acclaim or a bigger contract. The real problem with genetically altered athletes is that they corrupt athletic competition as a human activity that honors the cultivation and display of natural talents. From this standpoint, enhancement can be seen as the ultimate expression of the ethic of effort and willfulness—a kind of high-tech striving. The ethic of willfulness and the biotechnological powers it now enlists are arrayed against the claims of giftedness.

The ethic of giftedness, under siege in sports, persists in the practice of parenting. But here, too, bioengineering and genetic enhancement threaten to dislodge it. To appreciate children as gifts is to accept them as they come, not as objects of our design or products of our will or instruments of our ambition. Parental love is not contingent on the talents and attributes a child happens to have. We choose our friends and spouses at least partly on the basis of qualities we find attractive. But we do not choose our children. Their qualities are unpredictable, and even the most conscientious parents cannot be held wholly responsible for the kind of children they have. That is why parenthood, more than other human relationships, teaches what the theologian William F. May calls an "openness to the unbidden."

May's resonant phrase helps us see that the deepest moral objection to enhancement lies less in the perfection it seeks than in the human disposition it expresses and promotes. The problem is not that parents usurp the autonomy of a child they design. The problem lies in the hubris of the designing parents, in their drive to master the mystery of birth. Even if this disposition did not make parents tyrants to their children, it would disfigure the relation between parent and child, and deprive the parent of the humility and enlarged human sympathies that an openness to the unbidden can cultivate.

To appreciate children as gifts or blessings is not, of course, to be passive in the face of illness or disease. Medical intervention to cure or prevent illness or restore the injured to health does not desecrate nature but honors it. Healing sickness or injury does not override a child's natural capacities but permits them to flourish.

Nor does the sense of life as a gift mean that parents must shrink from shaping and directing the development of their child. Just as athletes and artists have an obligation to cultivate their talents, so parents have an obligation to cultivate their children, to help them discover and develop their talents and gifts. As May points out, parents give their children two kinds of love: accepting love and transforming love. Accepting love affirms the being of the child, whereas transforming love seeks the well-being of the child. Each aspect corrects the excesses of the other, he writes: "Attachment becomes too quietistic if it slackens into mere acceptance of the child as he is." Parents have a duty to promote their children's excellence.

These days, however, overly ambitious parents are prone to get carried away with transforming love—promoting and demanding all manner of accomplishments from their children, seeking perfection. "Parents find it difficult to maintain an equilibrium between the two sides of love," May observes. "Accepting love, without transforming love, slides into indulgence and finally neglect. Transforming love, without accepting love, badgers and finally rejects." May finds in these competing impulses a parallel with modern science: it, too, engages us in beholding the given world, studying and savoring it, and also in molding the world, transforming and perfecting it.

The mandate to mold our children, to cultivate and improve them, complicates the case against enhancement. We usually admire parents who seek the best for their children, who spare no effort to help them achieve happiness and success. Some parents confer advantages on their children by enrolling them in expensive schools, hiring private tutors, sending them to tennis camp, providing them with piano lessons, ballet lessons, swimming lessons, SAT prep courses, and so on. If it is permissible and even admirable for parents to help their children in these ways, why isn't it equally admirable for parents to use whatever genetic technologies may emerge (provided they are safe) to enhance their children's intelligence, musical ability, or athletic prowess?

The defenders of enhancement are right to this extent: improving children through genetic engineering is similar in spirit to the heavily managed, high-pressure child-rearing that is now common. But this similarity does not vindicate genetic enhancement. On the contrary, it highlights a problem with the trend toward hyperparenting. One conspicuous example of this trend is sports-crazed parents bent on making champions of their children. Another is the frenzied drive of overbearing parents to mold and manage their children's academic careers.

As the pressure for performance increases, so does the need to help distractible children concentrate on the task at hand. This may be why diagnoses of attention deficit and hyperactivity disorder have increased so sharply. Lawrence Diller, a pediatrician and the author of *Running on Ritalin*, estimates that 5 to 6 percent of American children under 18 (a total of four to five million kids) are currently prescribed Ritalin, Adderall, and other stimulants, the treatment of choice for ADHD. (Stimulants counteract hyperactivity by making it easier to focus and sustain attention.) The number of Ritalin prescriptions for children and adolescents has tripled over the past decade, but not all users suffer from attention disorders or hyperactivity. High school and college students have learned that prescription stimulants improve concentration for those with normal attention spans, and some buy or borrow their classmates' drugs to enhance their performance on the SAT or other exams. Since stimulants work for both medical and nonmedical purposes, they raise the same moral questions posed by other technologies of enhancement.

However those questions are resolved, the debate reveals the cultural distance we have traveled since the debate over marijuana, LSD, and other drugs a generation ago. Unlike the drugs of the 1960s and 1970s, Ritalin and Adderall are not for checking out but for buckling down, not for beholding the world and taking it in but for molding the world and fitting in. We used to speak of nonmedical drug use as "recreational." That term no longer applies. The steroids and stimulants that figure in the enhancement debate are not a source of recreation but a bid for compliance—a way of answering a competitive society's demand to improve our performance and perfect our nature. This demand for performance and perfection animates the impulse to rail against the given. It is the deepest source of the moral trouble with enhancement.

Some see a clear line between genetic enhancement and other ways that people seek improvement in their children and themselves. Genetic manipulation seems somehow worse—more intrusive, more sinister—than other ways of enhancing performance and seeking success. But morally speaking, the difference is less significant than it seems. Bioengineering gives us reason to question the low-tech, high-pressure child-rearing practices we commonly accept. The hyperparenting familiar in our time represents an anxious excess of mastery and dominion that misses the sense of life as a gift. This draws it disturbingly close to eugenics.

The shadow of eugenics hangs over today's debates about genetic engineering and enhancement. Critics of genetic engineering argue that human cloning, enhancement, and the quest for designer children are nothing more than "privatized" or "free market" eugenics. Defenders of enhancement reply that genetic choices freely made are not really eugenic—at least not in the pejorative sense. To remove the coercion, they argue, is to remove the very thing that makes eugenic policies repugnant.

Sorting out the lesson of eugenics is another way of wrestling with the ethics of enhancement. The Nazis gave eugenics a bad name. But what, precisely, was wrong with it? Was the old eugenics objectionable only insofar as it was coercive? Or is there something inherently wrong with the resolve to deliberately design our progeny's traits?

James Watson, the biologist who, with Francis Crick, discovered the structure of DNA, sees nothing wrong with genetic engineering and enhancement, provided they are freely chosen rather than state-imposed. And yet Watson's language contains more than a whiff of the old eugenic sensibility. "If you really are stupid, I would call that a disease," he recently told *The Times* of London. "The lower 10 percent who really have difficulty, even in elementary school, what's the cause of it? A lot of people would like to say, 'Well, poverty, things like that.' It probably isn't. So I'd like to get rid of that, to help the lower 10 percent." A few years ago Watson stirred controversy by saying that if a gene for homosexuality were discovered, a woman should be free to abort a fetus that carried it. When his remark provoked an uproar, he replied that he was not singling out gays but asserting a principle: women should be free to abort fetuses for any reason of genetic preference—for example, if the child would be dyslexic, or lacking musical talent, or too short to play basketball.

Watson's scenarios are clearly objectionable to those for whom all abortion is an unspeakable crime. But for those who do not subscribe to the pro-life position, these scenarios raise a hard question: If it is morally troubling to contemplate abortion to avoid a gay child or a dyslexic one, doesn't this suggest that something is wrong with acting on any eugenic preference, even when no state coercion is involved?

Consider the market in eggs and sperm. The advent of artificial insemination allows prospective parents to shop for gametes with the genetic traits they desire in their offspring. It is a less predictable way to design children than cloning or preimplantation genetic screening, but it offers a good example of a procreative practice in which the old eugenics meets the new consumerism. A few years ago some Ivy League newspapers ran an ad seeking an egg from a woman who was at least five feet ten inches tall and athletic, had no major family medical problems, and had a combined SAT score of 1400 or above. The ad offered $50,000 for an egg from a donor with these traits. More recently a Web site was launched claiming to auction eggs from fashion models whose photos appeared on the site, at starting bids of $15,000 to $150,000.

On what grounds, if any, is the egg market morally objectionable? Since no one is forced to buy or sell, it cannot be wrong for reasons of coercion. Some might worry that hefty prices would exploit poor women by presenting them with an offer they couldn't refuse. But the designer eggs that fetch the highest prices are likely to be sought from the privileged, not the poor. If the market for premium eggs gives us moral qualms, this, too, shows that concerns about eugenics are not put to rest by freedom of choice.

A tale of two sperm banks helps explain why. The Repository for Germinal Choice, one of America's first sperm banks, was not a commercial enterprise. It was opened in 1980 by Robert Graham, a philanthropist dedicated to improving the world's "germ plasm" and counteracting the rise of "retrograde humans." His plan was to collect the sperm of Nobel Prize–winning scientists and make it available to women of high intelligence, in hopes of breeding supersmart babies. But Graham had trouble persuading Nobel laureates to donate their sperm for his bizarre scheme, and so settled for sperm from young scientists of high promise. His sperm bank closed in 1999.

In contrast, California Cryobank, one of the world's leading sperm banks, is a for-profit company with no overt eugenic mission. Cappy Rothman, M.D., a co-founder of the firm, has nothing but disdain for Graham's eugenics, although the standards Cryobank imposes on the sperm it recruits are exacting. Cryobank has offices in Cambridge, Massachusetts, between Harvard and MIT, and in Palo Alto, California, near Stanford. It advertises for donors in campus newspapers (compensation up to $900 a month), and accepts less than 5 percent of the men who apply. Cryobank's marketing materials play up the prestigious source of its sperm. Its catalogue provides detailed information about the physical characteristics of each donor, along with his ethnic origin and college major. For an extra fee prospective customers can buy the results of a test that assesses the donor's temperament and character type. Rothman reports that Cryobank's ideal sperm donor is six feet tall, with brown eyes, blond hair, and dimples, and has a college degree—not because the company wants to propagate those traits, but because those are the traits his customers want: "If our customers wanted high school dropouts, we would give them high school dropouts."

Not everyone objects to marketing sperm. But anyone who is troubled by the eugenic aspect of the Nobel Prize sperm bank should be equally troubled by Cryobank, consumer-driven though it be. What, after all, is the moral difference between designing children according to an explicit eugenic purpose and designing children according to the dictates of the market? Whether the aim is to improve humanity's "germ plasm" or to cater to consumer preferences, both practices are eugenic insofar as both make children into products of deliberate design.

A number of political philosophers call for a new "liberal eugenics." They argue that a moral distinction can be drawn between the old eugenic policies and genetic enhancements that do not restrict the autonomy of the child. "While old-fashioned authoritarian eugenicists sought to produce citizens out of a single centrally designed mould," writes Nicholas Agar, "the distinguishing mark of the new liberal eugenics is state neutrality." Government may not tell parents what sort of children to design, and parents may engineer in their children only those traits that improve their capacities without biasing their choice of life plans. A recent text on genetics and justice, written by the bioethicists Allen

Buchanan, Dan W. Brock, Norman Daniels, and Daniel Wikler, offers a similar view. The "bad reputation of eugenics," they write, is due to practices that "might be avoidable in a future eugenic program." The problem with the old eugenics was that its burdens fell disproportionately on the weak and the poor, who were unjustly sterilized and segregated. But provided that the benefits and burdens of genetic improvement are fairly distributed, these bioethicists argue, eugenic measures are unobjectionable and may even be morally required.

The libertarian philosopher Robert Nozick proposed a "genetic super-market" that would enable parents to order children by design without imposing a single design on the society as a whole: "This supermarket system has the great virtue that it involves no centralized decision fixing the future human type(s)."

Even the leading philosopher of American liberalism, John Rawls, in his classic *A Theory of Justice* (1971), offered a brief endorsement of noncoercive eugenics. Even in a society that agrees to share the benefits and burdens of the genetic lottery, it is "in the interest of each to have greater natural assets," Rawls wrote. "This enables him to pursue a preferred plan of life." The parties to the social contract "want to insure for their descendants the best genetic endow-ment (assuming their own to be fixed)." Eugenic policies are therefore not only permissible but required as a matter of justice. "Thus over time a society is to take steps at least to preserve the general level of natural abilities and to prevent the diffusion of serious defects."

But removing the coercion does not vindicate eugenics. The problem with eugenics and genetic engineering is that they represent the one-sided triumph of willfulness over giftedness, of dominion over reverence, of molding over beholding. Why, we may wonder, should we worry about this triumph? Why not shake off our unease about genetic enhancement as so much superstition? What would be lost if biotechnology dissolved our sense of giftedness?

From a religious standpoint the answer is clear: To believe that our talents and powers are wholly our own doing is to misunderstand our place in creation, to confuse our role with God's. Religion is not the only source of reasons to care about giftedness, however. The moral stakes can also be described in secular terms. If bioengineering made the myth of the "self-made man" come true, it would be difficult to view our talents as gifts for which we are indebted, rather than as achievements for which we are responsible. This would transform three key features of our moral landscape: humility, responsibility, and solidarity.

In a social world that prizes mastery and control, parenthood is a school for humility. That we care deeply about our children and yet cannot choose the kind we want teaches parents to be open to the unbidden. Such openness is a disposi-tion worth affirming, not only within families but in the wider world as well. It invites us to abide the unexpected, to live with dissonance, to rein in the impulse to control. A Gattaca-like world in which parents became accustomed to specify-ing the sex and genetic traits of their children would be a world inhospitable to the unbidden, a gated community writ large. The awareness that our talents and abilities are not wholly our own doing restrains our tendency toward hubris.

Though some maintain that genetic enhancement erodes human agency by overriding effort, the real problem is the explosion, not the erosion, of responsibility. As humility gives way, responsibility expands to daunting proportions. We attribute less to chance and more to choice. Parents become responsible for choosing, or failing to choose, the right traits for their children. Athletes become responsible for acquiring, or failing to acquire, the talents that will help their teams win.

One of the blessings of seeing ourselves as creatures of nature, God, or fortune is that we are not wholly responsible for the way we are. The more we become masters of our genetic endowments, the greater the burden we bear for the talents we have and the way we perform. Today when a basketball player misses a rebound, his coach can blame him for being out of position. Tomorrow the coach may blame him for being too short. Even now the use of performance-enhancing drugs in professional sports is subtly transforming the expectations players have for one another; on some teams players who take the field free from amphetamines or other stimulants are criticized for "playing naked."

The more alive we are to the chanced nature of our lot, the more reason we have to share our fate with others. Consider insurance. Since people do not know whether or when various ills will befall them, they pool their risk by buying health insurance and life insurance. As life plays itself out, the healthy wind up subsidizing the unhealthy, and those who live to a ripe old age wind up subsidizing the families of those who die before their time. Even without a sense of mutual obligation, people pool their risks and resources and share one another's fate.

But insurance markets mimic solidarity only insofar as people do not know or control their own risk factors. Suppose genetic testing advanced to the point where it could reliably predict each person's medical future and life expectancy. Those confident of good health and long life would opt out of the pool, causing other people's premiums to skyrocket. The solidarity of insurance would disappear as those with good genes fled the actuarial company of those with bad ones.

The fear that insurance companies would use genetic data to assess risks and set premiums recently led the Senate to vote to prohibit genetic discrimination in health insurance. But the bigger danger, admittedly more speculative, is that genetic enhancement, if routinely practiced, would make it harder to foster the moral sentiments that social solidarity requires.

Why, after all, do the successful owe anything to the least-advantaged members of society? The best answer to this question leans heavily on the notion of giftedness. The natural talents that enable the successful to flourish are not their own doing but, rather, their good fortune—a result of the genetic lottery. If our genetic endowments are gifts, rather than achievements for which we can claim credit, it is a mistake and a conceit to assume that we are entitled to the full measure of the bounty they reap in a market economy. We therefore have an obligation to share this bounty with those who, through no fault of their own, lack comparable gifts.

A lively sense of the contingency of our gifts—a consciousness that none of us is wholly responsible for his or her success—saves a meritocratic society from sliding into the smug assumption that the rich are rich because they are more deserving than the poor. Without this, the successful would become even more likely than they are now to view themselves as self-made and self-sufficient, and hence wholly responsible for their success. Those at the bottom of society would be viewed not as disadvantaged, and thus worthy of a measure of compensation, but as simply unfit, and thus worthy of eugenic repair. The meritocracy, less chastened by chance, would become harder, less forgiving. As perfect genetic knowledge would end the simulacrum of solidarity in insurance markets, so perfect genetic control would erode the actual solidarity that arises when men and women reflect on the contingency of their talents and fortunes.

Thirty-five years ago Robert L. Sinsheimer, a molecular biologist at the California Institute of Technology, glimpsed the shape of things to come. In an article titled "The Prospect of Designed Genetic Change" he argued that freedom of choice would vindicate the new genetics, and set it apart from the discredited eugenics of old.

> To implement the older eugenics . . . would have required a massive social program carried out over many generations. Such a program could not have been initiated without the consent and co-operation of a major fraction of the population, and would have been continuously subject to social control. In contrast, the new eugenics could, at least in principle, be implemented on a quite individual basis, in one generation, and subject to no existing restrictions.

According to Sinsheimer, the new eugenics would be voluntary rather than coerced, and also more humane. Rather than segregating and eliminating the unfit, it would improve them. "The old eugenics would have required a continual selection for breeding of the fit, and a culling of the unfit," he wrote. "The new eugenics would permit in principle the conversion of all the unfit to the highest genetic level."

Sinsheimer's paean to genetic engineering caught the heady, Promethean self-image of the age. He wrote hopefully of rescuing "the losers in that chromosomal lottery that so firmly channels our human destinies," including not only those born with genetic defects but also "the 50,000,000 'normal' Americans with an IQ of less than 90." But he also saw that something bigger than improving on nature's "mindless, age-old throw of dice" was at stake. Implicit in technologies of genetic intervention was a more exalted place for human beings in the cosmos. "As we enlarge man's freedom, we diminish his constraints and that which he must accept as given," he wrote. Copernicus and Darwin had "demoted man from his bright glory at the focal point of the universe," but the new biology would restore his central role. In the mirror of our genetic knowledge we would see ourselves as more than a link in the chain of evolution: "We can be the agent of transition to a whole new pitch of evolution. This is a cosmic event."

There is something appealing, even intoxicating, about a vision of human freedom unfettered by the given. It may even be the case that the allure of that vision played a part in summoning the genomic age into being. It is often assumed that the powers of enhancement we now possess arose as an inadvertent by-product of biomedical progress—the genetic revolution came, so to speak, to cure disease, and stayed to tempt us with the prospect of enhancing our performance, designing our children, and perfecting our nature. That may have the story backwards. It is more plausible to view genetic engineering as the ultimate expression of our resolve to see ourselves astride the world, the masters of our nature. But that promise of mastery is flawed. It threatens to banish our appreciation of life as a gift, and to leave us with nothing to affirm or behold outside our own will.

Human Nature and Enhancement

Allen Buchanan

Allen Buchanan received his doctorate in philosophy from the University of North Carolina at Chapel Hill in 1975. He taught at the University of Arizona and the University of Wisconsin–Madison before becoming a professor of philosophy at Duke University. Buchanan served as staff philosopher for the President's [Council] on Medical Ethics in 1983. He served on the Advisory Council for the National Human Genome Research Institute from 1996 to 2000. He writes extensively on the ethics of medical enhancements and topics germane to political philosophy and bioethics. In this article Buchanan argues that "appeals to human nature tend to obscure rather than illuminate the debate over the ethics of enhancements." He argues that "there is nothing wrong, *per se*, with altering or destroying human nature," and that modifications in human nature need not "result in the loss of our ability to make judgments about the good."

Appeals to the idea of human nature are frequent in the voluminous literature on the ethics of enhancing human beings through biotechnology.[1] Two chief concerns about the impact of enhancements on human nature have been voiced. The first is that enhancement may alter human nature. The second is that if enhancement alters or destroys human nature, this will undercut our ability to ascertain the good because, for us, the good is determined by our nature.[2] The first concern assumes that altering human nature is in itself a bad thing. The second concern assumes that human nature provides a normative perspective without which we cannot make coherent, defensible judgments about what is good.

I aim to show that neither of these concerns is cogent. I will argue (i) that there is nothing wrong, *per se*, with altering human nature, because, on plausible understandings of what human nature is, it contains bad as well as good characteristics and there is no reason to believe that in every case eliminating some of the bad characteristics would so imperil the good ones as to make the elimination of the bad impermissible, and (ii) that altering human nature need not result in the loss of our ability to make judgments about the good, because we possess a conception of the good by which we can and do evaluate human nature. I will also argue (iii) that appeals to human nature tend to obscure, rather than illuminate the debate over the ethics of enhancement and can be eliminated in favor of more cogent considerations.

Allen Buchanan, "Human Nature and Enhancement," *Bioethics* 23(3), 2009: 141–150. Reprinted with permission from John Wiley and Sons, www.interscience.wiley.com.

In Section I, I distinguish several conceptions of human nature, articulate five different roles that appeals to human nature can play in ethics, and explain their implications for the enhancement debate. In Section II, I focus on the one appeal to human nature that has been especially prominent in the enhancement debate, the view that reflection on our nature can supply substantive moral rules, including a prohibition on enhancements that would alter our nature. I argue that this latter view, which I call normative essentialism, is irreparably flawed. In Section III, I examine the idea of human nature as a whole, a set of complex interdependencies. I argue that the worry ... that biomedical enhancements might disrupt complex interdependencies can be more fruitfully express[ed] without recourse to the concept of human nature.

I. Different Roles for the Concept of Human Nature in Ethics

Conceptions of Human Nature

At least since Aristotle, the dominant philosophical conception of human nature is that of a set of characteristics that are common to all humans and that distinguish humans from other kinds of beings. If one adds the assumption that there are natural kinds, then these characteristics are thought of as essential, rather than merely contingent. From this it follows that if any of these characteristics disappeared, we would no longer be human beings.

There are less philosophically sophisticated conceptions of human nature as well, both secular and theological. According to what might be called folk (or colloquial) conceptions, human nature consists of a set of dispositions that all (or at least most) humans have and that shape behavior across a wide range of human activities, regardless of cultural context, throughout human history. There is much disagreement about which characteristics manifested by human beings fit this description. It has been said that human beings are selfish by nature, or that they tend to have biases against members of "outgroups," that it is human nature to seek the transcendent, to love one's offspring and to be prepared to make sacrifices for them, to be prone to fall foolishly in love, to rationalize about one's failings, to be sociable, to be capable of knowing the laws of reason, to be able to make moral judgments, etc. Some contemporary evolutionary biologists and neuroscientists would say that if the concept of human nature has any value, it is as shorthand for those "hard-wired"[3,4] characteristics that most humans now have as a result of their common evolutionary development. Common to both folk and evolutionary conceptions is the idea that if something is part of human nature it is recalcitrant to alteration by acculturation, education, or indoctrination, and stable across a wide variety of environments. Neither folk nor evolutionary conceptions of human nature need include the assumption that what is part of human nature is unique to humans, as the example of sociability indicates (some non-human animals are said to be social as well).

One thing that most religious conceptions, folk conceptions, and evolutionary conceptions have in common is that they do not restrict human nature to good characteristics. Folk conceptions sometimes include the idea that human nature has a dark side. Religious conceptions often hold that human nature includes sinfulness. Evolutionary conceptions employ the concept of fitness, not goodness, but they are compatible with judgments about the goodness or badness o[f] those traits that are said to be part of human nature. For example, if certain dispositions are now part of our nature because they evolved during what evolutionary theorists call the ancestral evolutionary environment, they may be not only maladaptive in our current, quite different environment, but also bad from the standpoint of our moral values.

On any of these quite different conceptions of human nature, altering human nature would presumably require at least this much: large-scale changes to one or more of the characteristics that have been common to all or most normal human beings. The qualifier "large-scale" is intended to indicate that if such changes occurred only in some but not most human beings, we would probably say that *they* have become something other than human, rather than that human nature has changed.[5]

To begin to evaluate concerns about the impact of enhancement on human nature, it will prove useful to distinguish different roles that the idea of human nature (HN) has played in ethics. Five different roles may be distinguished: (1) HN as a condition of moral agency, (2) HN as a feasibility constraint on morality, (3) HN as a constraint on the good for humans, (4) HN as a source of substantive moral rules, and (5) HN as a whole—a complex set of interdependent characteristics—that can be seriously damaged by efforts to improve it. Each of these roles will be explained, and in each case the implications for the enhancement debate will be explained. The discussion that follows makes no pretence to completeness; that would be hubristic, given the many different ways in which the term "human nature" has been used. But it should suffice to show that objections to enhancement based on appeals to human nature are either otiose or ineffectual.

1. HN as Including a Precondition for Moral Agency

Practical rationality, understood as the capacity to recognize and act on reasons or, in Kantian terms, to be motivated to act by the belief that one ought to do something, has often been regarded as an important constituent of human nature. On some views, including Kant's, to destroy our capacity for practical rationality would be perhaps the deepest wrong. But no one advocating the moral permissibility of enhancements is suggesting that the destruction of our capacity for practical rationality would be an enhancement. So, the only relevance of this first sort [of] appeal to human nature in Ethics is simply to warn us to beware of one especially bad possible unintended consequence of efforts to enhance, namely, the inadvertent destruction or impairment of our capacity for practical rationality. Saying that we should take care not to damage our capacity for practical rationality in the pursuit of enhancements sheds very limited light, however, on the question of whether we ought to undertake any particular enhancement that is likely to be proposed. Moreover, the concern about damaging our practical rationality can be expressed without recourse to the concept of human nature.

2. HN as a Feasibility Constraint on Morality

Some moral philosophers, including Hume and contemporary practitioners of "neuroethics," emphasize that a properly realistic understanding of morality must take into account the cognitive and motivational limitations of human beings (in neuroethics jargon, the "hard-wiring" we happen to have as a result of evolution). If, for example, by virtue of our evolved biological makeup, we have a limited capacity to act altruistically toward strangers or toward those who are not members of our "primary group," then a plausible account of our moral obligations to others must take this into account.

There is an obvious reason why this sort of appeal to human nature sheds scant light on the ethics of enhancement: included among the enhancements regarding which we seek moral guidance are those that would involve the removal or relaxing of the very sorts of limitations that Hume and "neuro-ethicists" emphasize. Suppose it becomes possible, for example, to administer drugs that increase our ability to empathize with strangers and hence to be motivated to act altruistically toward them. What had been a limitation on the human capacity for altruism would to that extent be relaxed.

Regardless of whether a given limitation thought to be part of human nature is alterable or not, this second sort of appeal to human nature, like the first, can do little work in the ethics of enhancement. On the one hand, if a current motivational or cognitive limitation *can* be removed or relaxed by some biomedical technique, then the question is whether it would be a good idea, all things considered, to remove or relax it. The fact that until now we have been subject to this limitation tells us nothing about whether we should continue to be so. This point is especially clear and obvious if we think of human nature in the way evolutionary biologists do: the most we can say about any characteristic that is part of human nature is that it, or some characteristic to which it is tied in the processes of human biological development, *was* adaptive at some point in the development of our species. We cannot assume that it is still adaptive at present, much less that it will be adaptive in the future or that it is inextricably tied to something that is or will be adaptive. On the other hand, if a particular limitation is unalterable, then there is no issue of whether altering it is permissible and the idea of that it is a part of human nature adds nothing to the claim that it is unalterable. In either case, the appeal to human nature as a set of limitations on human motivation and cognition and hence on morality does not advance our understanding of the ethics of enhancement; nor does it lend support to the thesis that it is wrong *per se* to change human nature.[6] What is more, the very technologies that have prompted the enhancement debate also call into question the usefulness of the concept of human nature, insofar as it includes the idea of inalterability. Because these technologies challenge the idea that we have a fixed core of characteristics, an appeal to the idea of a fixed core cannot help resolve the debate about whether we should employ them.

3. HN as a Constraint on the Good for Us

In one plausible interpretation, Aristotle held that a being's nature determines its good, but only in a rather minimal sense: by constituting a constraint on what can count as a good life for that kind of being. According to this view, if human beings are by nature rational, then the good life for human beings, whatever else it is like, must include significant scope for the exercise of reason. Similarly, if human beings are by nature sociable, then the good life for them must include ample scope for social interactions.

This third sort of appeal to human nature, like the preceding two, is of meager value in wrestling with the ethics of enhancement, for two reasons. First, it neither forbids nor condones enhancements that would alter our nature, because it merely says that in seeking the good, we should remember that a being's nature constrains its good. By itself, this reminder is silent on the question of whether we should continue to live under the constraint that our nature imposes. Consider this analogy: if we are limited to a particular canvas, we can only create a painting that fits within its boundaries and we should take that into account in deciding what to paint—*on it*. But if we have the option of using *a different* canvas, then there will be other possibilities (and other constraints, as well). Recognizing that a given canvas limits the artistic good we can achieve does not imply that we should refrain from changing canvases; nor does it imply that we *should* change canvases. But it does raise the question of whether there might be reasons for using a different canvas, if we can. In principle, we might come to think that if we changed our nature in a certain way, we would come to be capable of goods that are not available to us and that they would be worth pursuing.

Perhaps some aspects of our nature constrain our good in unfortunate ways; to recur to an earlier example, our limited altruism may cut us off from forms of sociability that would greatly improve the quality of our lives. If such were the case, the question would be whether this sort of enhancement could be safely achieved and whether it would be good, all things considered. As I have already noted, in attempting to answer this question we would have to take the risk of unforeseen bad consequences very seriously.[7]

Knowing whether an intervention would be a good thing, all things considered, is a daunting task; but it is not made easier by recourse to the notion of human nature. Nor can an appeal to the concept of human nature help us decide whether, in the face of such uncertainties, it would be best to adopt some version of a Precautionary Principle or some other set of cautionary maxims that would function to counterbalance tendencies to underestimate risks in the pursuit of benefits.

Second, suppose that we see no reason to try to "paint on a new canvas." Suppose also that we agree that sociability, rationality, compassion, etc., are aspects of our nature which, given our resolution to seek the good within the constraints of our nature, we should do nothing to imperil. What follows about whether we should undertake this or that biomedical intervention for the sake of enhancement? Not much, for the simple reason that the plausibility of saying that any such characteristic is part of our nature depends upon characterizing it in a very abstract fashion. Consider sociability, for example. Sociability is a good candidate for being something for which all normal humans have the capacity and which is deeply ingrained if not innate; in that sense one might say that sociability is part of human nature. But a moment's attention to cultural diversity indicates that sociability can take many forms. There is a dilemma, then. Either one characterizes the features of human nature that are to constrain our pursuit of the good very abstractly, in order to make more plausible the claim that they are in fact features of human nature and not merely characteristics that some humans have, in which case they can provide few limitations on the sorts of enhancements we might opt for. Or, one characterizes them in more determinate way, so that the notion that they are to be preserved can significantly constrain our options regarding enhancement, but at the price of rendering implausible the claim that they are really features of human nature, rather than characteristics that some humans have.

Nevertheless, it is conceivable that a biomedical enhancement of some human capacity would damage some human capacity we rightly value, for example, our capacity for empathizing with others. If this damage occurred, then our capacity to achieve the good, to live well, might be seriously compromised, but not because the capacity for empathy is part of our nature; rather, because the capacity for empathy is either itself an important component of our good or instrumental for other goods, or both.[8]

Beyond being an unnecessary circumlocution for an admonition to avoid unwittingly destroying capacities upon which our good depends, then, the idea that human nature is a constraint on the good for us cannot supply much content for an ethics of enhancement. Nor can it provide an argument for the claim that we should not change our nature, because it only tells us that if we have a certain nature, that should be taken into account when we try to determine what our good is, not that we should persist with that nature and the constraints on goodness that it entails.

4. HN as a Source of Substantive Moral Rules: Normative Essentialism

Some of the harshest critics of enhancement, including Leon Kass and the President's Council on Bioethics, which [Kass] chaired for a time, embrace what might be called *normative essentialism:* they believe it is possible to derive substantive moral rules from reflection on human nature. More specifically, they believe it is possible to derive moral rules that include a prohibition on enhancement and on cloning as well. For example, the Council's *Report on Human Cloning and Human Dignity* solemnly declares that "human reproduction is sexual" (meaning that it involves the combining of genetic material from a male and a female) and then proceeds as if this is a strong and even conclusive reason against cloning, which is a type of asexual reproduction.[9] It is important to understand that this declaration is not offered as a descriptive statement that would have been more accurately framed in the past tense: the Council's claim is *not* that so far human reproduction has been sexual, that is, has involved the uniting of a sperm and an egg. That statement, though true (virgin births aside) would by itself give us no reason whatsoever to oppose asexual reproduction, including cloning. The Council clearly advances the claim that human reproduction is sexual as a *normative* claim, as the claim that sexual reproduction is the only form of reproduction that is fitting for human beings or in keeping with the dignity that their nature bestows. So, on the Council's view, any attempt at enhancement that involved cloning would be impermissible. At least in the Council's view, the connection between the "unnaturalness" of cloning and of enhancement is clear: the Council Report *Beyond Therapy* sees cloning as opening the door to a "new eugenics" that would involve not only "selecting out" undesirable traits, but also engineering desirable ones, and not necessarily just those falling in the normal range.[10]

The Council's claim is not simply that the engineering of human beings would produce bad results, but that it would alter our nature, "distorting" human relationships. According to this view, human nature is comprised not only of certain capacities that individual human beings have, but also of certain kinds of relationships among human beings. The fear is that some kinds of relationships that might emerge from the widespread use of biomedical enhancements would be "unnatural," that is, contrary to human nature and damaging to it.

The key point is that the adjective "human" in "human reproduction is sexual" is being used by the Council as a *normative* term. The clear implication is that asexual reproduction and hence cloning is debasing, demeaning, unnatural, or even perverse—in a word, less than human. Similarly, the Council states that in procreation "a man and woman give themselves in love to each other, setting their projects aside in order to do just that."[11] Taken as a descriptive claim, this is surely false as a generalization, unless one simply equates love with sex (which the Council would never do), because human procreation sometimes has nothing to do with love. Taken simply as a descriptive claim, it rules out by definitional fiat the possibility of same sex partners procreating. So if biomedical technology eventually made it possible to create an individual by combining DNA from partners who were of the same sex, this would not be human procreation, according to the Council's stipulative definition of human creation as sexual. As a normative claim, it implies that if same sex partners could procreate this would be inhuman or less than human, and that the same is true when men and women procreate without the procreative act being an act of love. Notice that even if it is true that procreation that involves love is better (other things being equal) than procreation that does not, it does not follow that procreation which is not an act of love between a man and woman is less than human, incompatible with the fundamental dignity of humanity.

The [Council's] appeal to normative essentialism is perhaps clearest in its criticism of cloning, but it is also implicit in the critique of enhancement in the report *Beyond Therapy*. This is not surprising, since, as I have already noted, the [Council] explicitly links cloning to enhancement, viewing it as the doorway to a "new eugenics" that would "alter the nature" of human procreation.[12] In its discussion of the possibility of genetic interventions in human embryos to enhance normal human characteristics, the Commission states that "The salient fact about human procreation in its natural context is that children are not *made* but *begotten*. By this we mean that children are the issue of our love, not the product of our wills."[13] Taken as a descriptive generalization, this last statement is clearly false: in many cases children are not the issue of love (they are often the result of careless sex and sometimes the result of rape) and in some cases they are the product of human willing, as when people deliberately try to conceive an heir or to produce another child to take care of them in old age. If the Commission's point is simply that the practice of genetically engineering our offspring would involve a new connection between willing and procreating—namely, the choosing of a child's genotype—that is certainly true, but adding that this would be a departure from "the natural context" does nothing to show that it would be wrong. Similarly, the Commission claims (without offering any data) that many people find repulsive the idea of 70-year-olds having children and suggests that this reaction of repugnance indicates that such an extension of the normal period of human reproductive activity would be contrary to what is "naturally human," with the clear implication that this counts against such a development.[14]

The Commission, however, does not rely entirely on the implication that genetically designing children would be "unnatural" in its condemnation of enhancement. Instead, it goes on to suggest that genetically designing children is either equivalent to or will lead to the replacement of the "natural" activity of procreating with a process of manufacturing children. The description of current procreation as "natural" in this claim, however, does no argumentative work: the issue is whether genetically designing children is or is likely to lead to parents regarding their offspring as manufactured items. Reference to human nature or the natural can be eliminated without loss.

We can now see that the problem with normative essentialism is not merely that it disguises normative claims as descriptive claims. If that were the only problem, it could be solved by reformulating the claims in a way that makes their normativity obvious. There is a deeper flaw. Normative essentialist claims confuse two quite different kinds of moral judgments, and fail to provide justifications for either: judgments about what is best for human beings and judgments about what is compatible with human beings' fundamental moral status or dignity. It is one thing to claim that a certain way of procreating is best, or even that other ways of procreating are in some way defective. Such judgments are controversial enough, and normative essentialists typically do not take up the burden of providing plausible justifications for them. It is even more problematic to assert that anything other than this particular way of procreating is *less than human, incompatible with human dignity*.

The history of prejudice and persecution is replete with normative essentialist claims: homosexuality is unnatural, marriages between the races are unnatural, social equality for inferior and superior types of humans (Aryans and non-Aryans, Nietzschean "higher-types" and "lower-types") is unnatural, demeaning (to the superior), etc. That alone should make one suspicious of normative essentialist claims and prompt an insistence that they be backed up with evidence or argument. What sort of backing do Kass and the President's Council provide for normative essentialist claims like "human reproduction is sexual" or that "in procreation a man and a woman give themselves to each other in love"?

Most who are skeptical of such claims have focused on Kass et al.'s appeal to "the wisdom of repugnance." Although Kass et al. have acknowledged the simple but powerful point that feelings of repugnance can be not just unreliable, but counter-reliable (as when a racist feels repugnance upon seeing a White and a Black kissing), they have not begun to take up the burden of articulating a moral epistemology that would enable us to distinguish between those feelings of repugnance that reveal moral truth and those that do not. And they have certainly not articulated a normative account of human nature that would either explain which feelings of repugnance are veridical, or make the appeal to such feelings unnecessary for determining what is permitted and what is not.

The history of Ethics suggests that such an account is not likely to be forthcoming. Reflection on the concept of human nature can provide some constraints on the good for humans, but it is not a source of substantive moral rules that could decide controversial moral issues, at least not rules prohibiting homosexuality, mixed marriages, or asexual reproduction.

The normative essentialist faces a destructive dilemma. If the concept of human nature from which controversial substantive moral rules (prohibiting asexual reproduction or enhancement, or procreation by same sex partners, etc.) are supposed to be derived is itself normatively rich enough to ground those rules, then that highly normative concept of human nature will itself be equally controversial and no argumentative leverage will be gained. But if the concept of human nature the normative essentialist invokes is thin enough to be plausible to those who do not already accept the substantive moral rules that are supposed to be derived from it, then it will be too thin to ground those rules. In either case, the appeal to human nature does not help us resolve enhancement issues.

It would be unfair to Kass et al. to let the matter rest there, however. They might concede that feelings of repugnance and reflection on the concept of human nature are insufficient to ground the very strong and determinate moral prohibitions on cloning an[d] enhancement they advance. Instead, they might recast their view roughly as follows. "A good human life is one in which there are certain relationships—between parent and child, between man and women, among siblings, between older and younger generations. The goods that flow from these relationships are deeply interdependent; altering any of them not only entails a loss of the good that it involves, but also threatens to undermine other goods. Sexual reproduction (i.e., reproduction involving a genetic contribution from a male and a female) is one important aspect of the good life for human beings; it is valuable in its own right but is also connected, through various interdependencies, to other important human goods. So, abandoning sexual reproduction in favor of reproduction by cloning or abandoning traditional parenting in favor of parenting that includes the designing of embryos would imperil the good life for humans."

This version of the Council's message avoids normative essentialism: it does not attempt to resolve moral controversies by deriving substantive moral rules from reflection on the concept of human nature, and it seems to dispense with unsupported and question-begging pronouncements about what is fitting for human beings or in keeping with their dignity. Instead, it appeals to the idea of a good life and to the interdependency of goods within it. Call this The Good Life Argument, to distinguish it from normative essentialism.

The first thing to notice about this new argument is that it is not an attempt to show that cloning or enhancement or anything else is wrong *because it would alter or destroy human nature.* Thus it avoids the problems of normative essentialism.

This gain comes at an exorbitant price, however. Proponents of The Good Life Argument must bear an enormous burden of empirical evidence to make plausible their sweeping claims about the relevant causal interdependencies. They have to show, in effect, that the good life is a seamless web; that severing one fiber is likely to result in the whole thing unraveling. Empirical evidence, not armchair speculation, must be marshaled to show that the various aspects of the good human life are so thoroughly interdependent. In particular, there must be credible evidence for the claim that if asexual reproduction or procreation by same sex partners or the genetic enhancement of children become common, this will undermine various relationships that the argument assumes to be of great value.

Notice that it will not do to state, as the [*Beyond Therapy*] report does, that cloning or the designing of children will "confound" relationships between generations. If "confound" is an evaluative term, meaning roughly "distort" or "derange," then this claim presupposes just what is at issue, namely, that the fitting, proper, true, or natural relationship between individuals is precisely the one that has obtained until now, before the possibility of the biomedical intervention in question, and the argument collapses back into naked normative essentialism. If, instead, the term "confound" is shorthand for the assertion that certain biomedical interventions will in fact disrupt certain causal dependencies, then this is an empirical claim that needs evidence.[15]

The facts of cross-cultural diversity suggest that this daunting burden of empirical evidence is not likely to be borne. Some societies now recognize same-sex marriages and homosexuality and all include reproduction that has more to do with sex or economics than love. (To borrow the Council's biblical phrasing: many people are not "begotten.") Yet it appears that good human lives are possible within such societies, even for those who depart from the Council's standards of "naturalness," and this suggests either that the behavior in question is not bad or that if it is bad, the goodness in human life is much more independent of it and more resilient than the seamless web metaphor implies.

Further, Kass et al. must argue, not assume, that the forms of human relationships they believe are endangered by cloning or enhancement are not only goods (objectively speaking, not merely things *they* prefer), but also goods of such overwhelming value that they can never be reasonably compromised for other goods. Simply to declare that certain kinds of relationships or activities (male–female parenting, sexual reproduction, parenting that does not involve genetic intervention in embryos, etc.) are so valuable that they must never be imperiled for the sake of any other good is to beg the question in an argumentative context in which their opponents do not agree with the Council's particular perfectionist theory of the good.

The threads of my argument so far can now be pulled together. The type of argument against enhancement (and cloning) that the Council advances either relies on the concept of human nature and is an unpersuasive normative essentialist argument, an attempt to derive substantive and controversial moral rules from the concept of human nature; or instead presupposes a very controversial, perfectionist theory of the good and an unarticulated and ambitious empirical social theory to support very strong claims about the causal interdependence of the elements of a good human life, neither of which its proponents have begun to articulate, much less defend.

II. Human Nature as a Complex Whole

The Assumption of Strong Interdependency

Sometimes the term "nature" or "the natural" is used to convey the idea of a complex whole, a web of harmonious interdependencies, as when those concerned about the effects of human activity on the environment speak of the disruption of Nature. In the enhancement context, the worry is that if we change any part of human nature, even an undesirable part, we endanger the good parts as well. This worry rests on a very strong empirical claim about the interdependency of the various parts or aspects of human nature.

Strong interdependency as sweeping empirical claim   It is unlikely that an appeal to empirical evidence of interdependencies could yield a prescription to eschew enhancements altogether. Instead, it would ground a counsel of prudence to go slowly and to take seriously the possibility that we may unwittingly do damage to ourselves in the pursuit of betterment. One plausible heuristic for reducing the risk of unwittingly disrupting benign interdependencies would be to favor enhancement interventions on biological subsystems that are relatively self-contained or "modular."

It is important to see that even if there were good empirical evidence for a claim about interdependency that was strong enough to ground an absolute prohibition on efforts to enhance, the result would not be a sound argument from human nature to an anti-enhancement conclusion but rather the abandonment of any role for an appeal to human nature in the argument. What matters is whether proposed enhancements are likely to produce bad unintended consequences because of our ignorance of interdependencies that include what we value; whether the things that are interdependent are parts of our nature is irrelevant. Once again, the appeal to human nature is eliminable without loss.

Focusing on interdependency, not human nature   In brief, if the appeal to human nature is simply shorthand for an appeal to the fragility of wholeness—to the dangers of meddling with complex interdependencies—it is shorthand we can do without. Moreover, relying on this kind of shorthand is not only unnecessary; it is pernicious to the extent that it encourages the delusion that reflection on human nature can yield substantive moral rules capable of resolving controversies about enhancement. Having to grapple with the ambiguity of the notion of human nature and to resolve long-standing disputes about what is and is not included in human nature simply distracts attention from the issue of complex interdependencies.

III. Human Nature and the Ability to Judge What Is Good

Once we appreciate the implausibility of normative essentialism, we are in a better position to evaluate the charge that enhancement could, by altering or destroying human nature, undermine our ability to make judgments about the good. The idea here is that human nature provides a standard for the good in the following sense: to know whether something is good (for us) we need to know whether it conforms to or is consistent with or "fits" our nature. I have already suggested one reason to reject the claim that if we alter our nature we will lose the capacity to make judgments about the good. I have argued that the role that the appeal to our nature plays in making judgments about the good is that of providing a constraint on what can be good for us, *so long as that is our nature.* Therefore, altering our nature need not result in the inability to judge what the good is. An altered nature could simply supply new constraints.

There is a second, more important reason to reject the claim that an alteration in our nature would rob us of the ability to make judgments about the good: we already make what appear to be perfectly sound judgments about the goodness of human nature, judgments that can supply reasons for altering that nature. For as I have already observed, human nature is often thought to include bad as well as good characteristics. It is said to be human nature to be selfish, inclined to excessive partiality or sinful, for example. The idea that human nature is a mixed bag is perhaps most credible in the case of evolutionary conceptions of human nature: there is no reason to assume that all of the human traits that have evolved are good.

Whether or not the claim that some aspects of human nature are bad is true, there seems to be nothing incoherent about it. But if that is so, then it appears that we have a standard of goodness that is somewhat independent of our concept of human nature. In principle, then, there seems to be nothing wrong with the idea of changing those parts of our nature that are bad, if this can be done without imperiling the good parts. To assume otherwise is to beg the question at hand, to assume, rather than to argue, that changing human nature is itself a wrong. Further, we can give reasons—just the sort of reasons that we use to support judgments about the good in other cases—for why it would be good, other things being equal, to alter some of the more unsavory traits which, on some views, are part of human nature. For example, suppose, as some evolutionary biologists claim, that it is part of human nature—a widespread trait due to our evolutionary past—that we have a bias toward negative evaluations of those we regard as alien, as "not one of us." Perhaps if this propensity were reduced, there would be fewer wars and a reduction of the miseries that wars bring. The reasons we would have for changing this putative aspect of human nature are familiar reasons, reasons having to do with what is bad for us, with what tends to undercut our well-being. So to that extent, we have an evaluative perspective that is independent of our nature, and changing some aspect of our nature need not result in the inability to make well-grounded judgments about the good.

Nor is there reason to assume that if such changes were made, we would no longer be able to make coherent judgments about the good. Even if our capacity to make evaluative judgments is in some way dependent upon our nature, it does not follow that it is dependent on each and every aspect of our nature.

It might be replied that the idea of human nature is nonetheless needed to flesh out an adequate conception of *our* well-being, to distinguish it from the well-being of other animals. For example, it could be argued that our well-being requires that we have the capacity for complex and self-conscious forms of sociability, for a degree of autonomy (the capacity to lead a life, not just to live), etc.

All of that is no doubt true, but it can be said without invoking the idea of human nature, where this means a set of characteristics that is universal in all humans and unique to them. Presumably if there are intelligent extraterrestrials, then the capacity for complex, self-conscious forms of sociability and for autonomy will be important for their well-being, too, even if they were not human beings. If we say that certain capacities that we believe (perhaps wrongly) are peculiar to human nature are important for our well-being, all the normative work is being done by the idea that they are important for well-being, not by the claim that they are part of our nature.

Consider one last interpretation of the claim that our capacity to judge the good depends upon human nature and hence could be imperiled by efforts to alter human nature through the application of biomedical technologies. Suppose that the capacity to make judgments about goodness is part of human nature, where human nature is understood, not as the problematic metaphysical notion of essence, but simply as a bundle of properties that most normal, mature members of our biological species have and that is not possessed (in toto) by any other creatures on earth. Depending upon how robustly we construe the capacity for judging goodness, it may be plausible to say that this capacity is part of human nature in this metaphysically lean sense. If a biomedical intervention had the unintended consequence of destroying the capacity for judging goodness, then it would follow trivially that *this* alteration of human nature undercut our capacity for judging goodness. But it would not follow that *any* alteration of our nature would undercut that capacity. So, even if it is true that the capacity for judging goodness is part of human nature, it does not follow that by altering human nature we undercut our ability to judge goodness. Here, as elsewhere, we can eliminate the appeal to human nature and focus instead on the commonsensical admonition to try to avoid enhancement efforts that may have unintended bad consequences.

Conclusion

I have argued for three theses. (i) The fact that an enhancement would alter or destroy human nature is in itself not a reason to forgo the enhancement. Setting aside the difficulties of determining which of our characteristics are part of human nature (and in which sense of that ambiguous term), the question is whether we have good reason to try to change a particular characteristic. The answer to that question will depend upon a number of factors, including whether the characteristic is undesirable on balance and whether in attempting to change it we would imperil things we rightly value. Whether it is part of our nature is irrelevant, unless, by stipulative definition we say that what is part of our nature is impossible to change, in which case the appeal to human nature is nothing more than the admonition not to try to do what cannot be done. (ii) Reasonable worries about enhancement that are sometimes expressed in the language of human nature—such as the concern about unintended consequences due to unnoticed interdependencies between what we wish to change and what we wish to preserve—can be more clearly expressed without appealing to human nature. Because appeals to human nature in this context are not only unnecessary, but also run the risk of degenerating into the errors of normative essentialism, they are best avoided. (iii) It is not the case that in altering human nature we would thereby undermine our ability to make judgments about the good. We already possess standards of evaluation that are independent of our nature in the sense that we can and do make coherent judgments about the defective aspects of human nature, and if those defects were remedied this need not affect our ability to judge what is good.

Finally, a disclaimer is in order. I have not tried to make a case in favor of enhancement. To be "in favor of enhancement" would reveal a failure to appreciate the diversity of changes that might be labeled enhancements or oblivious to the risks that some of them undoubtedly carry. Nor am I denying that some of the concerns that critics of enhancement have formulated in terms of human nature are serious. My conclusion, rather, is that given that genuine concerns about enhancement can be formulated without reference to human nature and given that normative essentialist appeals to human nature ought to be rejected, the enhancement debate is more fruitfully pursued in other terms.[16]

References

1. Enhancements of physical, cognitive, or motivational characteristics might be brought about through genetic engineering of embryos or gametes, implantation of laboratory-grown tissues and organs, human/machine interfacing (including nanotechnologies), and pharmacological interventions.
2. For some relevant literature, see: Harold W. Baillie and Timothy K. Casey, editors (2004) *Is human nature obsolete? Genetics, bioengineering, and the future*

of the human condition (Cambridge, MA: MIT Press); Leon Kass (2000) "The Wisdom of Repugnance," in *The Human Cloning Debate*, 2nd edition, ed. Glenn McGee, pp. 68–106. (Berkeley, CA: Berkeley Hills Books); and Erik Parens, editor (1998) *Enhancing Human Traits: Ethical and Social Implications* (Washington, DC: Georgetown University Press).
3. Two prominent examples include Jurgen Habermas (2003) *The Future of Human Nature* (Cambridge: Polity) and Francis Fukuyama (2002) *Our Post-Human Future: Consequences of the Biotechnology Revolution* (New York: Farrar, Straus & Giroux).
4. "Hard-wired" here does not mean deterministic. Rather, it is used to convey either the idea that the dispositions in question are recalcitrant to modification by training, education, and acculturation (hence the adjective "hard") and also that their existence is more innate than learned.
5. Norman Daniels (forthcoming) "Can anyone really be talking about ethically modifying human nature?" In Julian Savulescu and Nick Bostrom, editors, *Enhancement of Human Beings* (Oxford: Oxford University Press).
6. Some critics of enhancement, including Michael Sandel and Erik Parens, contend that enhancement is objectionable precisely because it involves the removal of limitations on what human beings can do, because they believe that there are irreplaceable goods that depend upon our having limitations. For a systematic critique of this view, see Allen Buchanan, "Human Development and Human Enhancement," *Kennedy Institute Journal of Ethics*, forthcoming. See Erik Parens (1995) "The Goodness of Fragility: On the Prospect of Genetic Technologies Aimed at the Enhancement of Human Capacities," *Kennedy Institute of Ethics Journal* 5(2): 141–53, and Michael Sandel (2004) "The Case Against Perfection," *The Atlantic Monthly* 293(3): 50–62.
7. Notice that it would be a mistake to assume that an enhancement that altered an individual's nature—that made her no longer a human—would result in the loss of that individual's identity. Personal identity could be preserved through such a transformation, regardless of whether one assumes that personal identity requires only continuity of psychological states or that plus the persistence of the body. Such a change would result in the loss of the individual's identity only if it were true that [persons] are essentially human.
8. I am grateful to Tom Douglas for this point.
9. President's Council on Bioethics (2002) *Human Cloning and Human Dignity: An Ethical Inquiry* (Washington, DC: President's Council on Bioethics).
10. President's Council on Bioethics, *Beyond Therapy*, (2002) p. 70. Here the Council says that cloning and eugenic interventions might "alter the very nature" of human procreation.
11. President's Council on Bioethics, *Human Cloning and Human Dignity* (2002), p. 99.
12. President's Council on Bioethics, *Beyond Therapy*, p. 70.
13. *Ibid.*, p. 70.
14. *Ibid.*, p. 287.

15. Perhaps the claim that cloning would "confound" relationships among generations is a clumsy shorthand for the empirical psychological prediction that if reproduction by cloning were pervasive, people would be confused and disturbed in their thinking about the relationships among generations. The Council provides no evidence for this prediction and no indication of how serious or pervasive the alleged psychological distress would be. Nor do they address the fact that there is evidence that people can adjust, rather rapidly, to new reproductive technologies, for example, in vitro fertilization. Despite some initial speculation that "test-tube babies" would be regarded as freaks, most people seem to have taken IVF in stride.

16. Research for this paper was done when the author was Scholar in Residence at the Duke University Institute for Genome Sciences and Policy.

Ethical Topics at the End of Life

Withholding and Withdrawing Medical Treatment

૨૱

The Terri Schiavo Case:
Legal, Ethical, and Medical Perspectives

Joshua E. Perry, Larry R. Churchill, and Howard S. Kirshner

These authors from Vanderbilt University represent backgrounds in law (Joshua Perry, research fellow at the Center for Biomedical Ethics philosophy), religion (Larry Churchill, the
Ann Geddes Stahlman Chair in Medical Ethics and Co-director of the Center for Biomedical
Ethics), and neurology (Howard Kirshner, Professor of Neurology and Vice-Chairman of that
department). In this article they review the current understanding of persistent vegetative state
(PVS) and chronicle the clinical and legal history of the case. They argue that the judicial
processes functioned well in protecting Ms. Schiavo's individual liberty and privacy interests.
Moreover, the authors contend that using self-determination as the primary framework for addressing the challenges in this case is better than relying on beneficence, quality of life, or
sanctity of life judgments.

On 25 February 1990, Terri Schiavo, 26 years of age, collapsed in the hall of her apartment and experienced severe hypoxia for several minutes. She had not executed a living will or a durable power of attorney. Four months after her

Condensed from *Annals of Internal Medicine* 143 (10), November 2005. Reproduced with
permission of the American College of Physicians. Copyright © 2005 American College of
Physicians.

injury, Mrs. Schiavo was judged incompetent and her husband, Michael Schiavo, was appointed her legal guardian without objection from her parents, Robert and Mary Schindler. Because she was unable to swallow, Mrs. Schiavo underwent placement of a percutaneous endoscopic gastrostomy (PEG) tube. By late 1990, Mrs. Schiavo was determined to be in a persistent vegetative state.

Understanding the Persistent Vegetative State

The Schiavo case rests critically on the concept of the persistent vegetative state and the certainty of the prediction that a patient in this state will have no meaningful recovery. The persistent vegetative state is distinguished from several other states of reduced consciousness. Brain death implies the loss of not only all higher brain functions but also all brainstem functions, including pupillary light reflexes, reflex eye movements, respirations, and gag and corneal reflexes. Determination of brain death is straightforward and is generally accepted as a criterion for death. Coma is a complete state of unresponsiveness to stimuli, although the patient may have brainstem reflexes. Stupor and obtundation refer to states of reduced consciousness in which meaningful responses are still possible, if the patient receives enough stimulation. Finally, the "locked-in syndrome" denotes the condition of a patient who is paralyzed and cannot move or speak but is completely awake. Such patients can often communicate by blinking their eyes or looking up and down.

The American Academy of Neurology, along with representatives of the American Neurological Association, the Child Neurology Society, the American Association of Neurological Surgeons, and the American Academy of Pediatrics, set up a Multi-Society Task Force to establish criteria for diagnosing the persistent vegetative state (Table 1). In 1994, the Task Force published its findings,[1-3] which have been adopted as a practice guideline by the American Academy of Neurology (Table 2). The Task Force estimated that 10,000 to 25,000 adults and 6000 to 10,000 children in the United States are in the persistent vegetative state. The criteria for this diagnosis must be met at least 1 year after traumatic brain injury in young patients and at least 3 months after nontraumatic illnesses.

The Task Force reviewed case series from the literature, which included 434 adults and 106 children with traumatic brain injury and 169 adults and 45 chil-

TABLE 1   Requirements for the Examination of Persistent Vegetative State[4]

No evidence of awareness of self or environment, no interaction with others
No meaningful response to stimuli
No receptive or expressive language
Return of sleep-wake cycles, arousal, even smiling, frowning, yawning
Preserved brain stem/hypothalamic autonomic functions to permit survival
Bowel and bladder incontinence
Variably preserved cranial nerve and spinal reflexes

TABLE 2 American Academy of Neurology Practice Guidelines[7]

"The decision to discontinue fluid and nutrition should be made in the same manner as other medical treatment decisions"

"Artificial provision of nutrition and hydration is analogous to other forms of life-sustaining treatment, such as . . . a respirator"

"Administration of fluids and nutrition by medical means, such as a gastrostomy tube, is a medical procedure"

"Treatments which provide no benefit to the patient or the family may be discontinued. . . . Medical treatment provides no benefit to patients in a persistent vegetative state, once the diagnosis is established"

dren with nontraumatic injuries, mostly related to hypoxia. Of the patients in the persistent vegetative state for more than 3 months after nontraumatic injuries, the probability of moderate disability or good recovery was 1% (99% CI, 0% to 4%), but for patients still in the persistent vegetative state at 6 months, this probability was 0%. No patient, even those with traumatic brain injury, has been reported to recover after a full year of being in the persistent vegetative state. Delayed recoveries after traumatic brain injury are more common than with nontraumatic brain injuries. Certainly, no patient has recovered after 15 years, the period during which Terri Schiavo survived in this state. The criteria make clear that the patient can have periods of sleep alternating with periods of an awake-like state, in which his or her eyes are open and may move about, and the patient may breathe, yawn, and open his or her mouth, but not interact meaningfully with others. . . .

The Legal History and Commentary

Throughout the early 1990s, Michael Schiavo and the Schindlers worked together coordinating extensive rehabilitation efforts for Mrs. Schiavo, including regular and aggressive physical, occupational, and speech therapies. Despite their best efforts and explorations of all potentially viable treatments, her condition failed to improve.

By the mid-1990s, Mr. Schiavo's relationship with the Schindlers had chilled. Increasingly, Mr. Schiavo understood that his wife's condition was irreversible. In addition, he insisted that he knew his wife would not want to live in a persistent vegetative state. The Schindlers, however, insisted that Mrs. Schiavo be kept alive through artificial nutrition and hydration. Unable to reach a consensus with his wife's parents, Mr. Schiavo decided not to request withdrawal of the PEG tube. Instead, he petitioned the guardianship court "to function as the proxy" and asked the court "to make an independent determination of Mrs. Schiavo's terminal condition and to make the decision to continue or discontinue life-prolonging procedures."[4]

In January 2000, Judge George Greer held a trial to resolve the dispute over the extent of Terri Schiavo's neurologic devastation and to determine how she would exercise her right of privacy, or liberty interest, to forgo life-supporting medical treatment, if she were able to communicate. The proceeding was adversarial in nature, with both Mr. Schiavo and the Schindlers presenting witnesses and making arguments pursuant to the rules of evidence and civil procedure.

It is important to note that end-of-life guardianship controversies, such as those that emerged in this case, are resolved on the basis of state law, which varies throughout the United States. In Florida, both statutes and case law recognize the fundamental right of every individual to control his or her person, regardless of the capacity to communicate.[5] In Terri Schiavo's case, where the inability to communicate directly was established and no written medical directive existed, Florida law considers previously made oral declarations, as long as they satisfy the standard of "clear and convincing evidence."[6]

At the 2000 trial, the opposing parties presented Judge Greer with Mrs. Schiavo's entire medical history, numerous exhibits, and the testimony of 18 witnesses, including the parties, 2 physicians, and various family members.

On the first issue of her medical condition and whether Mrs. Schiavo had the ability or potential to communicate her wishes directly, the court found that she met the statutory definition of the persistent vegetative state and that there was no hope of her regaining consciousness or the ability to communicate. The court's analysis then shifted to the issue of exercising Mrs. Schiavo's autonomy rights, independently of state, family, or public opinion. After hearing from all of the witnesses and observing their cross examinations, Judge Greer determined that testimony from the 3 witnesses recalling Mrs. Schiavo's prior oral declarations that "she would not want to live like that" satisfied the clear-and-convincing evidence standard.[7] Accordingly, the court ordered that the PEG tube be removed.

The Schindlers appealed Judge Greer's decision. The appellate court reviewed both the substantive evidence and procedural formalities at the 2000 trial and concluded that after 10 years in a persistent vegetative state with no hope of recovery, the ultimate issue was whether Mrs. Schiavo "would choose to continue the constant nursing care and the supporting tubes ... or whether she would wish to permit a natural death process to take its course."[8] The appellate court affirmed Judge Greer's conclusion that the clear and convincing evidence required removal of the PEG tube.

Judge Greer ordered cessation of Terri Schiavo's artificial nutrition and hydration on 24 April 2001. Two days later, however, the treatment resumed when the Schindlers presented a different trial court judge with 7 affidavits from licensed physicians. Although none of the physicians had physically examined Mrs. Schiavo, each contended that she was neither in a persistent vegetative state nor beyond the aid of medical treatment. In 2002, after numerous legal proceedings, the Schindlers successfully convinced the appellate court "to permit discovery and conduct an evidentiary hearing only for the purpose of assessing Mrs. Schiavo's current medical condition, the nature of the new medical treatments described in the affidavits and their acceptance in the relevant scientific community . . ."[8]

At the 2002 hearing, again in the context of an adversary proceeding, Judge Greer heard testimony from 5 board-certified experts whose examinations of Mrs. Schiavo were videotaped. Mr. Schiavo's 2 experts, as well as the independent, court-appointed expert, were all neurologists who unanimously concluded that Mrs. Schiavo had irreversible neurologic damage and was undoubtedly in a persistent vegetative state. The Schindlers' experts, 1 radiologist and 1 neurologist, questioned whether Mrs. Schiavo was in such a state and advocated for the use of hyperbaric oxygen and vasodilation therapies.

Judge Greer found the Schindlers' experts unconvincing and their rehabilitation theories unsupported by the mainstream neurologic community. "Viewing all of the evidence as a whole, and acknowledging that medicine is not a precise science," the court found "the credible evidence overwhelmingly support[ed] the view that Terri Schiavo remain[ed] in a persistent vegetative state."[9] Accordingly, the court once again ordered the withdrawal of the feeding tube.

On appeal, Judge Greer's decision was again scrutinized for both procedural and substantive error. The appellate court affirmed Judge Greer's decision, emphasizing that the proceedings in 2000 and again in 2002 had been undertaken with care, objectivity, and according to "a cautious legal standard designed to promote the value of life." In conclusion, the appellate court stated that the "value of life" could be adequately promoted "if all people are . . . entitled to a personalized decision about life-prolonging procedures."[10]

With Terri Schiavo's "personalized decision" determined at the 2000 trial and her medical condition conclusively confirmed at the 2002 hearing, artificial hydration and nutrition were stopped for the second time, on 15 October 2003, pursuant to court order. This order was superseded 6 days later when the Florida legislature passed and Governor Jeb Bush signed Terri's Law, which provided the governor unfettered discretion to order resumption of Mrs. Schiavo's medical treatment.

Eleven months later, the Florida Supreme Court held Terri's Law unconstitutional as a violation of separation of powers,[11] and a third date was set for removal of the PEG tube. On 18 March 2005, the tube was removed for the final time, and despite the intervention of Congress and President George W. Bush, Terri Schiavo's saga ended with her death on 31 March 2005.

On 1 April 2005, an autopsy was performed, revealing that Mrs. Schiavo's brain was "grossly abnormal and weighed only 615 grams"—less than half of the expected weight for an adult her age—and showing neuronal loss in her occipital lobes consistent with cortical blindness.[12] There was also extensive loss of neurons in the basal ganglia and in the hippocampus, and loss of Purkinje cells in the cerebellum. In short, the findings from the autopsy were fully consistent with the expectation of widespread, hypoxic-ischemic brain damage. In addition, the presence of an implanted thalamic stimulator explained why an MRI scan was never done.

In our opinion, the law did not fail Terri Schiavo. In fact, no end-of-life guardianship case in U.S. history has generated as much high-quality evidence, judicial attention, or legal scrutiny as the Terri Schiavo case. Throughout a lengthy trial and evidentiary hearing, countless motions, oral arguments, and numerous appeals to every available state and federal court, this case shows that the judicial process works at the end of life.

The Florida guardianship law was clear, and the law was followed. The judiciary was charged with 2 questions: 1) What was Terri Schiavo's medical condition? 2) In such a condition, what would she choose to do? In the midst of an intense and intractable family dispute, amid dizzying media attention and unprecedented political intervention, the judicial process produced 2 answers. The process and the resulting answers were reviewed repeatedly by cautious, nonpartisan judges who demonstrated restraint and care in adjudicating Mrs. Schiavo's case pursuant to her individual liberty and privacy interests.

Competing Ethical Frameworks

Expressed as an ethical principle, the central focus in this case is respect for autonomy. Many have argued, however, that the proper moral framework is not autonomy at all, but rather sanctity of life, discrimination toward the disabled community, or the moral character of those empowered to decide. Examining these alternative frameworks is important, both to discern why respect for autonomy is the key ethical feature of this case and to display the weaknesses of the alternatives.

Several commentators argued that the central question of the Terri Schiavo case is a struggle between sanctity of life versus quality of life. For example, columnist David Brooks advanced this thesis, neatly dividing the contending parties into "social conservatives," who believe in the "intrinsic value" of all life and "social liberals," who emphasize "quality."[13] However, this rendering of the ethical issues is too neat, and it compounds descriptive vagueness with stereotypes, thus ensuring that the debate remains at the superficial level of moral slogans. Unless one adopts the position that sheer biological existence is what is sacred about human life, considerations of sanctity inevitably involve judgments of quality. More important, this dichotomous rendering of the issues begs the essential question of whose notion of "sanctity" and "quality" counts. By taking the autonomy and liberty interests of patients as the central question, the courts preserved the prerogative of individuals to decide according to their own values, even after they have lost the ability to speak for themselves.

Others felt passionately that discontinuing Terri Schiavo's artificial nutrition and hydration would initiate a cascading disregard for disabled persons or others who are judged to have poor quality of life.[14] The fundamental flaw with this approach is overgeneralization. It is descriptively inaccurate to refer to a person in the persistent vegetative state as being disabled. For example, there are far more differences than similarities between Terri Schiavo and persons who have paralysis, such as the late Christopher Reeve. A related kind of overgeneralization is the idea that the action taken in the Terri Schiavo case has wide-ranging public policy implications. The wisdom of deciding this case within the moral framework of individual autonomy and liberty interests is that it portends nothing as a general policy for what should be done to and for other persons in the persistent vegetative state. In fact, stressing liberty as the fundamental issue means precisely that individuals will be able to choose for themselves and that a variety of differing living wills and legitimate proxy decisions can be honored.

Finally, some argued that Michael Schiavo should not be allowed a voice in decisions concerning his wife's treatment because over the 15-year period of caring for her, he developed a relationship with another woman. Those who espoused this perspective seemed ignorant of Mr. Schiavo's aggressive attempts to rehabilitate his wife during the 1990s. However, the more general flaw in reasoning here is an elementary one in ethics—seeking to discredit the person making the decision instead of considering the merits of the decision itself. The courts correctly avoided this ad hominem fallacy, although many media commentators did not.

Securing individual liberties and honoring self-determination for oneself and others were the impetus for the founding of the United States and have been a prominent part of the American sensibility ever since. However, protecting the liberty interests of those who have lost competence was not always the favored framework for either law or medical ethics. It has become so for both fields through a series of momentous court decisions and through changes in medical ethics that both prefigured the court decisions and became codified in the wake of them. For instance, the cases of Karen Ann Quinlan in 1976,[15] Paul Brophy in 1986,[16] and Nancy Cruzan in 1990[17] all involved patients in the persistent vegetative state who did not have written advance directives. In each of these cases, the courts deemed the liberty and privacy interests of the patient as paramount.

Likewise, medical codes of ethics echo this legal consensus when they frame end-of-life issues in terms of autonomy and affirm the right of patients to informed consent and respect for their wishes, even beyond their loss of competence, when it can be reasonably determined what they would have wanted. Of course, to assert that respect for self-determination is paramount does not mean that it is the only principle at work in this and similar cases—it also does not mean that discerning someone's preferences for medical care from the accounts of others is simple or without problems. Rather, our claim is that using self-determination as the primary framework is simply the best we can do and is far less problematic than relying on medical or parental beneficence, quality of life, or sanctity of life judgments. Relying fundamentally on these other norms would jeopardize the rights of Terri Schiavo and patients like her to receive or not to receive continuing interventions and would threaten this basic principle of U.S. law and American medical ethics.

Conclusion

The Terri Schiavo case will probably remain controversial, with some people feeling strongly that the wrong decision was made. We have argued that, to the contrary, this case is an example of good standards and processes in medicine, law, and ethics. It exemplifies the use of medical consensus to create standards to diagnose persistent vegetative state, it is characterized by careful proceedings and review in keeping with a long tradition of legal procedure, and it embodies respect for self-determination as a fundamental U.S. principle that honors both individual preferences and a wide moral pluralism.

References

1. The Multi-Society Task Force on PVS. Medical aspects of the persistent vegetative state (1). *N Engl J Med*. 1994;330:1499–508. [PMID: 781863311].
2. The Multi-Society Task Force on PVS. Medical aspects of the persistent vegetative state (2). *N Engl J Med*. 1994;330:1572–9. [PMID: 8177248].
3. American Academy of Neurology. Practice parameters: assessment and management of patients in the persistent vegetative states. Report of the Quality Standards Subcommittee of the American Academy of Neurology. *Neurology* 1995;45:1015–8. [PMID: 7746375].
4. *Schindler v Schiavo*, 792 So 2d 551, 557 (Fla Dist Ct App 2001).
5. *John F Kennedy Memorial Hospital Inc v Bludworth*, 452 So 2d 921, 932 (Fla 1984).
6. *In re Browning*, 568 So 2d 4 (Fla 1990).
7. American Academy of Neurology. Position of the American Academy of Neurology on certain aspects of the care and management of the persistent vegetative state patient. *Neurology* 1989;39:125–126.
8. *Schindler v Schiavo*, 800 So 2d 640, 647 (Fla Dist Ct App 2001).
9. *In re Schiavo*, 90-2908GB-003 (Fla Cir Ct, Pinellas Co, 22 November 2002).
10. *Schindler v Schiavo*, 800 So 2d 640, 647 (Fla Dist Ct App 2001).
11. *Bush V. Schiavo*, 885 So 2d 321, 324 (Fla, 2004).
12. Thogmartin JR. Chief Medical Examiner Report of Autopsy, Theresa Schiavo. Case 5050429, 13 June 2005.
13. Brooks D. Morality and reality. *New York Times*. 2005;26 Mar:A13.
14. Lytle T. Rep. Dave Weldon's efforts raise questions among some of his colleagues. *Orlando Sentinel*. 2005;20 Apr:A18.
15. *In re Quinlan*, 355 A 2d 647 (NJ 1976).
16. *Brophy v New England Sinai Hospital Inc*, 497 NE 2d 626 (Mass 1986).
17. *Cruzan v Director*, 110 S Ct 2841, 2851 (1990).

Nutrition and Hydration:
Moral and Pastoral Reflections

U.S. Bishops' Pro-Life Committee

The U.S. Bishops' Pro-Life Committee prepared this document to assist Catholics and others in making treatment decisions. The paper outlines relevant moral principles and then addresses the question of artificial nutrition and hydration. The Bishops express concern that withdrawal of all life-sustaining treatment, including artificial nutrition and hydration, not be viewed as appropriate or automatically indicated for an entire class of patients with persistent vegetative state simply because of their irreversible state of reduced consciousness. They are worried about a society that might dismiss patients without apparent mental faculties as nonpersons or as undeserving of human care and concern.

Introduction

Modern medical technology seems to confront us with many questions not faced even a decade ago. Corresponding changes in medical practice have benefited many, but have also prompted fears by some that they will be aggressively treated against their will or denied the kind of care that is their due as human persons with inherent dignity. Current debates about life-sustaining treatment suggest that our society's moral reflection is having difficulty keeping pace with its technological progress.

A religious view of life has an important contribution to make to these modern debates. Our Catholic tradition has developed a rich body of thought on these questions, which affirms a duty to preserve human life but recognizes limits to that duty.

Our first goal in making this statement is to reaffirm some basic principles of our moral tradition, to assist Catholics and others in making treatment decisions in accord with respect for God's gift of life. These principles do not provide clear and final answers to all moral questions that arise as individuals make difficult decisions. Catholic theologians may differ on how best to apply moral principles to some questions not explicitly resolved by the church's teaching authority. Likewise, we understand that those who must make serious healthcare decisions for themselves or for others face a complexity of issues, circumstances, thoughts and emotions in each unique case. This is the case with some questions involving the medically assisted provision of nutrition and hydration to helpless patients— those who are seriously ill, disabled or persistently unconscious. These questions have been made more urgent by widely publicized court cases and the public debate to which they have given rise. Our second purpose in issuing this statement, then, is to provide some clarification of the moral issues involved in decisions about medically assisted nutrition and hydration. We are fully aware that such guidance is not necessarily final, because there are many unresolved medical and ethical questions related to these issues and the continuing development of medical technology will necessitate ongoing reflection. But these decisions already confront patients, families and health-care personnel every day. They arise whenever competent patients make decisions about medically assisted nutrition and hydration for their own present situation, when they consider signing an advance directive such as a "living will" or health-care proxy document, and when families or other proxy decision-makers make decisions about those entrusted to their care. We offer guidance to those who, facing these issues, might be confused by opinions that at times threaten to deny the inherent dignity of human life. We therefore address our reflections first to those who share our Judeo-Christian traditions, and secondly to others concerned about the dignity and value of human life who seek guidance in making their own moral decisions.

Moral Principles

The Judeo-Christian moral tradition celebrates life as the gift of a loving God and respects the life of each human being because each is made in the image and likeness of God. As Christians we also believe we are redeemed by Christ and called to share eternal life with him. From these roots the Catholic tradition has developed a distinctive approach to fostering and sustaining human life. Our church views life as a sacred trust, a gift over which we are given stewardship and not absolute dominion. The church thus opposes all direct attacks on innocent life. As conscientious stewards we have a duty to preserve life, while recognizing certain limits to that duty:

(1) Because human life is the foundation for all other human goods, it has a special value and significance. Life is "the first right of the human person" and "the condition of all the others."

(2) All crimes against life, including "euthanasia or willful suicide," must be opposed. Euthanasia is "an action or an omission which of itself or by intention causes death, in order that all suffering may in this way be eliminated." Its terms of reference are to be found "in the intention of the will and in the methods used." Thus defined, euthanasia is an attack on life which no one has a right to make or request, and which no government or other human authority can legitimately recommend or permit. Although individual guilt may be reduced or absent because of

suffering or emotional factors that cloud the conscience, this does not change the objective wrongfulness of the act. It should also be recognized that an apparent plea for death may really be a plea for help and love.

(3) Suffering is a fact of human life, and has special significance for the Christian as an opportunity to share in Christ's redemptive suffering. Nevertheless there is nothing wrong in trying to relieve someone's suffering; in fact it is a positive good to do so, as long as one does not intentionally cause death or interfere with other moral and religious duties.

(4) Everyone has the duty to care for his or her own life and health and to seek necessary medical care from others, but this does not mean that all possible remedies must be used in all circumstances. One is not obliged to use either "extraordinary" means or "disproportionate" means of preserving life—that is, means which are understood as offering no reasonable hope of benefit or as involving excessive burdens. Decisions regarding such means are complex, and should ordinarily be made by the patient in consultation with his or her family chaplain or pastor and physician when that is possible.

(5) In the final stage of dying one is not obliged to prolong the life of a patient by every possible means: "When inevitable death is imminent in spite of the means used, it is permitted in conscience to take the decision to refuse forms of treatment that would only secure a precarious and burdensome prolongation of life, so long as the normal care due to the sick person in similar cases is not interrupted."

(6) While affirming life as a gift of God, the church recognizes that death is unavoidable and that it can open the door to eternal life. Thus, "without in any way hastening the hour of death," the dying person should accept its reality and prepare for it emotionally and spiritually.

(7) Decisions regarding human life must respect the demands of justice, viewing each human being as our neighbor and avoiding all discrimination based on age or dependency. A human being has "a unique dignity and an independent value, from the moment of conception and in every stage of development, whatever his or her physical condition." In particular, "the disabled person (whether the disability be the result of a congenital handicap, chronic illness or accident, or from mental or physical deficiency, and whatever the severity of the disability) is a fully human subject, with the corresponding innate, sacred and inviolable rights." First among these is "the fundamental and inalienable right to life."

(8) The dignity and value of the human person, which lie at the foundation of the church's teaching on the right to life, also provide a basis for any just social order. Not only to become more Christian, but to become more truly human, society should protect the right to life through its laws and other policies.

While these principles grow out of a specific religious tradition, they appeal to a common respect for the dignity of the human person. We commend them to all people of good will.

Questions about Medically Assisted Nutrition and Hydration

In what follows we apply these well-established moral principles to the difficult issue of providing medically assisted nutrition and hydration to persons who are seriously ill, disabled or persistently unconscious. We recognize the complexity involved in applying these principles to individual cases and acknowledge that, at this time and on this particular issue, our applications do not have the same authority as the principles themselves.

1. Is the withholding or withdrawing of medically assisted nutrition and hydration always a direct killing?

In answering this question one should avoid two extremes.

First, it is wrong to say that this could not be a matter of killing simply because it involves an omission rather than a positive action. In fact a deliberate omission may be an effective and certain way to kill, especially to kill someone weakened by illness. Catholic teaching condemns as euthanasia "an action or an omission which of itself or by intention causes death, in order that all suffering may in this way be eliminated." Thus "euthanasia includes not only active mercy killing but also the omission of treatment when the purpose of the omission is to kill the patient."

Second, we should not assume that all or most decisions to withhold or withdraw medically assisted nutrition and hydration are attempts to cause death. To be sure, any patient will die if all nutrition and hydration are withheld. But sometimes other causes are at work—for example, the patient may be imminently dying, whether feeding takes place or not, from an already existing terminal condition. At other times, although the shortening of the patient's life is one foreseeable result of an omission, the real purpose of the omission was to relieve the patient of a particular procedure that was of limited usefulness to the patient or unreasonably burdensome for the patient and the patient's family or caregivers. This kind of decision should not be equated with a decision to kill or with suicide.

The harsh reality is that some who propose withdrawal of nutrition and hydration from certain patients do directly intend to bring about a patient's death and would even prefer a change in the law to allow for what they see as more "quick and painless" means to cause death. In other words, nutrition and hydration (whether orally administered or medically assisted) are sometimes withdrawn not because a patient is dying, but precisely because a patient is not dying (or not dying quickly) and someone believes it would be better if he or she did, generally because the patient is perceived as having an unacceptably low "quality of life" or as imposing burdens on others.

When deciding whether to withhold or withdraw medically assisted nutrition and hydration, or other forms of life support, we are called by our moral tradition to ask ourselves: What will my decision do for this patient? And what am I trying to achieve by doing it? We must be sure that it is not our intent to cause the patient's death either for its own sake or as a means to achieving some other goal such as the relief of suffering.

2. Is medically assisted nutrition and hydration a form of "treatment" or "care"?

Catholic teaching provides that a person in the final stages of dying need not accept "forms of treatment that would only secure a precarious and burdensome prolongation of life," but should still receive "the normal care due to the sick person in similar cases." . . . But the teaching of the church has not resolved the question whether medically assisted nutrition and hydration should always be seen as a form of normal care.

Almost everyone agrees that oral feeding, when it can be accepted and assimilated by a patient, is a form of care owed to all helpless people. . . . But our obligations become less clear when adequate nutrition and hydration require the skills of trained medical personnel and the use of technologies that may be perceived as very burdensome—that is, as intrusive, painful or repugnant. Such factors vary from one type of feeding procedure to another, and from one patient to another, making it difficult to classify all feeding procedures as either "care" or "treatment."

Perhaps this dilemma should be viewed in a broader context. Even medical "treatments" are morally obligatory when they are "ordinary" means—that is, if they provide a reasonable hope of benefit and do not involve excessive burdens. Therefore we believe people should make decisions in light of a simple and fundamental insight: Out of respect for the dignity of the human person we are obliged to preserve our own lives and help others preserve theirs, by the use of means that have a reasonable hope of sustaining life without imposing unreasonable burdens on those we seek to help, that is, on the patient and his or her family and community.

We must therefore address the question of benefits and burdens next, recognizing that a full moral analysis is only possible when one knows the effects of a given procedure on a particular patient.

3. What are the benefits of medically assisted nutrition and hydration?

Nutrition and hydration, whether provided in the usual way or with medical assistance . . . benefit patients in several ways. First, for all patients who can assimilate them, suitable food and fluids sustain life, and providing them normally expresses loving concern and solidarity with the helpless. Second, for patients being treated with the hope of a cure, appropriate food and fluids are an important element of sound health care. Third, even for patients who are imminently dying and incurable, food and fluids can prevent the *suffering* that may arise from dehydration, hunger and thirst.

. . . But sometimes even food and fluids are no longer effective in providing this benefit because a patient has entered the final stage of a terminal condition. At such times we should make the dying person as comfortable as possible and provide nursing care and proper hygiene as well as companionship and appropriate spiritual aid. Such a person may lose all desire for food and drink and even be unable to ingest them. Initiating medically assisted feeding or intravenous fluids in this case may increase the patient's discomfort while providing no real benefit; ice chips or sips of water may instead be appropriate to provide comfort and counteract the adverse effects of dehydration. Even in the case of the imminently dying patient, of course, any action or omission that of itself or by intention causes death is to be absolutely rejected. . . .

4. What are the burdens of medically assisted nutrition and hydration?

The risks and objective complications of medically assisted nutrition and hydration will depend on the procedure used and the condition of the patient. In a given case a feeding procedure may become harmful or even life-threatening. . . .

If the risks and burdens of a particular feeding procedure are deemed serious enough to warrant withdrawing it, we should not automatically deprive the patient of all nutrition and hydration but should ask whether another procedure is feasible that would be less burdensome. We say this because some helpless patients, including some in a "persistent vegetative state," receive tube feedings not because they cannot swallow food at all but because tube feeding is less costly and *difficult* for healthcare personnel. . . .

Many people see feeding tubes as frightening or even as bodily violations. Assessments of such burdens are necessarily subjective; they should not be dismissed on that account, but we offer some practical cautions to help prevent abuse.

First, in keeping with our moral teaching against the intentional causing of death by omission, one should distinguish between repugnance to a particular procedure and repugnance to life itself. The latter may occur when a patient views a life of helplessness and dependency on others as itself a heavy burden, leading him or her to wish or even to pray for death. Especially in our achievement-oriented society, the burden of living in such a condition may seem to outweigh any possible benefit of medical treatment and even lead a person to despair. But we should not assume that the burdens in such a case always outweigh the benefits; for the sufferer, given good counseling and spiritual support, may be brought again to appreciate the precious gift of life.

Second, our tradition recognizes that when treatment decisions are made, "account will have to be taken of the reasonable wishes of the patient and the patient's family, as also of the advice of the doctors who are specially competent in the matter." . . .

Third, we should not assume that a feeding procedure is inherently repugnant to all patients without specific evidence. In contrast to Americans' general distaste for the idea of being supported by "tubes and machines," some studies indicate surprisingly favorable views of medically assisted nutrition and hydration among patients and families with actual experience of such procedures. . . .

While some balk at the idea, in principle cost can be a valid factor in decisions about life support. For example, money spent on expensive treatment for one family member may be money otherwise needed for food, housing and other necessities for the rest of the family. Here, also, we offer some cautions. . . . Even for altruistic reasons a patient should not directly intend his or her own death by malnutrition or dehydration, but may accept an earlier death as a consequence of his or her refusal of an unreasonably expensive treatment.

. . . Individual decisions about medically assisted nutrition and hydration should [not] be determined by macroeconomic concerns such as national budget priorities and the high cost of health care. These social problems are serious, but it is by no means established that they require depriving chronically ill and helpless patients of effective and easily tolerated measures that they need to survive.

. . . [T]ube feeding alone is generally not very expensive and may cost no more than oral feeding. What is seen by many as a grave financial and emotional burden on caregivers is the total long-term care of severely debilitated patients, who may survive for many years with no life support except medically assisted nutrition and hydration and nursing care. . . .

In the context of official church teaching, it is not yet clear to what extent we may assess the burden of a patient's total care rather than the burden of a particular treatment when we seek to refuse "burdensome" life support. On a practical level, those seeking to make good decisions might assure themselves of their own intentions by asking: Does my decision aim at relieving the patient of a particularly grave burden imposed by medically assisted nutrition and hydration? Or does it aim to avoid the total burden of caring for the patient? If so, does it achieve this aim by deliberately bringing about his or her death?

Rather than leaving families to confront such dilemmas alone, society and government should improve their assistance to families whose financial and emotional resources are strained by long-term care of loved ones.

5. What role should "Quality of Life" play in our decisions?

Financial and emotional burdens are willingly endured by most families to raise their children or to care for mentally aware but weak and elderly family members. It is sometimes argued that we need not endure comparable burdens to feed and care for persons with severe mental and physical disabilities, because their low "quality of life" makes it unnecessary or pointless to preserve their lives.

But this argument—even when it seems motivated by a humanitarian concern to reduce suffering and hardship—ignores the equal dignity and sanctity of all human life. Its key assumption—that people with disabilities necessarily enjoy life less than others or lack the potential to lead meaningful lives—is also mistaken. Where suffering does exist, society's response should not be to neglect or eliminate the lives of people with disabilities, but to help correct their inadequate living conditions. Very often the worst threat to a good "quality of life" for these people is not the disability itself, but the prejudicial attitudes of others—attitudes based on the idea that a life with serious disabilities is not worth living.

This being said, our moral tradition allows for three ways in which the "quality of life" of a seriously ill patient is relevant to treatment decisions:

(1) Consistent with respect for the inherent sanctity of life, we should relieve needless suffering and support morally acceptable ways of improving each patient's quality of life.

(2) One may legitimately refuse a treatment because it would itself create an impairment imposing new serious burdens or risks on the patient. This decision to avoid the new burdens or risks created by a treatment is not the same as directly intending to end life in order to avoid the burden of living in a disabled state.

(3) Sometimes a disabling condition may directly influence the benefits and burdens of a specific treatment for a particular patient. For example, a confused or demented patient may find medically assisted nutrition and hydration more frightening and burdensome than other patients do because he or she cannot understand what it is. The patient may even repeatedly pull out feeding tubes, requiring burdensome physical restraints if this form of feeding is to be continued. In such cases, ways of alleviating such special burdens should be explored before concluding that they justify withholding all food and fluids needed to sustain life.

These humane considerations are quite different from a "quality of life" ethic that would judge individuals with disabilities or limited potential as not worthy of care or respect. It is one thing to withhold a procedure because it would impose new disabilities on a patient, and quite another thing to say that patients who already have such disabilities should not have their lives preserved. A means considered ordinary or proportionate for other patients should not be considered extraordinary or disproportionate for severely impaired patients solely because of a judgment that their lives are not worth living.

In short, while considerations regarding a person's quality of life have some validity in weighing the burdens and benefits of medical treatment, at the present time in our society judgments about the quality of life are sometimes used to promote euthanasia. The Church must emphasize the sanctity of life of each person as a fundamental principle in all moral decisionmaking.

6. Do persistently unconscious patients represent a special case?

Even Catholics who accept the same basic moral principles may strongly disagree on how to apply them to patients who appear to be persistently unconscious—that is, those who are in a permanent coma or a "persistent vegetative state" (PVS). Some moral questions in this area have not been explicitly resolved by the church's teaching authority.

On some points there is wide agreement among Catholic theologians:

(1) An unconscious patient must be treated as a living human person with inherent dignity and value. Direct killing of such a patient is as morally reprehensible as the direct killing of anyone else. Even the medical terminology used to describe these patients as "vegetative" unfortunately tends to obscure this vitally important point, inviting speculation that a patient in this state is a "vegetable" or a subhuman animal.

(2) The area of legitimate controversy does not concern patients with conditions like mental retardation, senility, dementia or even temporary unconsciousness. Where serious disagreement begins is with the patient who has been diagnosed as completely and permanently unconscious after careful testing over a period of weeks or months.

Some moral theologians argue that a particular form of care or treatment is morally obligatory only when its benefits outweigh its burdens to a patient or the care providers. In weighing burdens, they say, the total burden of a procedure and the consequent requirements of care must be taken into account. If no benefit can be demonstrated, the procedure, whatever its burdens, cannot be obligatory. These moralists also hold that the chief criterion to determine the benefit of a procedure cannot be merely that it prolongs physical life, since physical life is not an absolute good but is relative to the spiritual good of the person. They assert that the spiritual good of the person is union with God, which can be advanced only by human acts, i.e., conscious, free acts. Since the best current medical opinion holds that persons in the persistent vegetative state (PVS) are incapable now or in the future of conscious, free human acts, these moralists conclude that, when careful diagnosis verifies this condition, it is not obligatory to prolong life by such interventions as a respirator, antibiotics or medically assisted hydration and nutrition. To decide to omit non-obligatory care, therefore, is not to intend the patient's death, but only to avoid the burden of the procedure. Hence, though foreseen, the patient's death is to be attributed to the patient's pathological condition and not to the omission of care. Therefore, these theologians conclude, while it is always wrong directly to intend or cause the death of such patients, the natural dying process which would have occurred without these interventions may be permitted to proceed.

While this rationale is convincing to some, it is not theologically conclusive and we are not persuaded by it. In fact, other theologians argue cogently that theological inquiry could lead one to a more carefully limited conclusion.

These moral theologians argue that while particular treatments can be judged useless or burdensome, it is morally questionable and would create a dangerous precedent to imply that any human life is not a positive good or "benefit." They emphasize that while life is not the highest good, it is always and everywhere a basic good of the human person and not merely a means to other goods. They further assert that if the "burden" one is trying to relieve by discontinuing medically assisted nutrition and hydration is the burden of remaining alive in the allegedly undignified condition of PVS, such a decision is unacceptable because one's intent is only achieved by deliberately ensuring the patient's death from malnutrition or dehydration. Finally, these moralists suggest that PVS is best seen as an extreme form of mental and physical disability, one whose causes, nature and prognosis are as yet imperfectly understood—and not as a terminal illness or fatal pathology from which patients should generally be allowed to die. Because the patient's life can often be sustained indefinitely by medically assisted nutrition and hydration that is not unreasonably risky or burdensome for that patient, they say, we are not dealing here with a case where "inevitable death is imminent in spite of the means used." Rather, because the patient will die in a few days if medically assisted nutrition and hydration are discontinued, but can often live a long time if they are provided, the inherent dignity and worth of the human person obligates us to provide this patient with care and support.

Further complicating this debate is a disagreement over what responsible Catholics should do in the absence of a final resolution of this question. Some point to our moral tradition of probabilism, which would allow individuals to follow the appropriate moral analysis that they find persuasive. Others point to the principle that in cases where one might risk unjustly depriving someone of life, we should take the safer course.

In the face of the uncertainties and unresolved medical and theological issues, it is important to defend and preserve important values. On the one hand, there is a concern that patients and families should not be subjected to unnecessary burdens, ineffective treatments and indignities when death is approaching. On the other hand, it is important to ensure that the inherent dignity of human persons, even those who are persistently unconscious, is respected and that no one is deprived of nutrition and hydration with the intent of bringing on his or her death.

It is not easy to arrive at a single answer to some of the real and personal dilemmas involved in this issue. In study, prayer, and compassion, we continue to reflect on this issue and hope to discover additional information that will lead to its ultimate resolution.

In the meantime, at a practical level we are concerned that withdrawal of all life support, including nutrition and hydration, not be viewed as appropriate or automatically indicated for the entire class of PVS patients simply because of a judgment that they are beyond the reach of medical treatment that would restore consciousness. We note the current absence of conclusive scientific data on the causes and implications of different degrees of brain damage, on the PVS patient's ability to experience pain and on the reliability of prognoses for many such patients. We do know that many of these patients have a good prognosis for long-term survival when given medically assisted nutrition and hydration, and a certain prognosis for death otherwise—and we know that many in our society view such an early death as a positive good for a patient in this condition. Therefore we are gravely concerned about current attitudes and policy trends in our society that would too easily dismiss patients without apparent mental faculties as non-persons or as undeserving of human care and concern. In this climate, even legitimate moral arguments intended to have a careful and limited application can easily be misinterpreted, broadened and abused by others to erode respect for the lives of some of our society's most helpless members. . . .

Medical Futility:

Its Meaning and Ethical Implications

Lawrence J. Schneiderman, Nancy S. Jecker, and Albert R. Jonsen

Larry Schneiderman is an internist (M.D. from Harvard University) and an emeritus professor in the Department of Family and Preventive Medicine at the University of California at
San Diego. Nancy Jecker (Ph.D. in philosophy from the University of Washington) is a pro-
fessor in the Department of Medical History and Ethics at the University of Washington,
School of Medicine. Al Jonsen (Ph.D. in religious studies from Yale University) is an emeritus
professor in the Department of Medical History and Ethics at the University of Washington,
School of Medicine, where he served as chair from 1987 to 1999. He is also Co-Director and
Senior Ethics Scholar at the Program in Medicine and Human Values, California Pacific Medical Center, San Francisco.
In this paper the authors characterize two forms of medical futil-
ity: quantitative (e.g., less than one percent likelihood of success) and qualitative (e.g., living
totally dependent on intensive medical care). Importantly, they propose standards for evaluating medical futility and invite further examination of these proposed standards. The authors
argue that when an intervention is clearly medically futile, based not on the judgment of an individual physician but on professional standards of care, the intervention should not be presented as a medical option to the patient or family.

A 62-year-old man with irreversible respiratory disease is in the intensive care
unit. He is severely obtunded. During 3 weeks in the unit, repeated efforts
to wean him from ventilatory support have been unsuccessful. There is general
agreement among his physicians that he could not survive outside of an intensive care setting. They debate whether therapy should include cardiopulmonary
resuscitation if the patient has a cardiac arrest or antibiotics if he develops infection. The patient gave no previous indication of his wishes nor executed an advance directive. Some physicians argue that a "do not resuscitate" order may be
written without consulting the family, because resuscitation would be futile.
Other physicians object, pointing out that resuscitation cannot be withheld on
grounds of medical futility, because the patient could survive indefinitely in the
intensive care unit. They agree to consult the family on this matter. At first there
is considerable disagreement within the family, until a son asks whether there is

Condensed from *Annals of Internal Medicine* 112, 1990: 949–954. Many notes are omitted
from the original. Reproduced with permission of the American College of Physicians. Copyright © American College of Physicians, 1990.

any hope at all that his father might recover. The physicians look at each other. There is always hope. This unites the family. They insist that if the situation is not hopeless, the physicians should continue all measures including resuscitation.

How should these physicians deal with this family's demands? The answer depends on both how the physicians define futility and the weight they give it when patients or surrogates strongly express treatment preferences. Are these issues perhaps too complex or ambiguous to resolve? We submit that they are not, and we offer both a theoretical and practical approach to the concept of futility, an approach that we believe serves in this case and more generally in similar cases by restoring a common sense notion of medical duty. We recognize that if futility is held to be nothing more than a vague notion of physician discretion, it is subject to abuse: therefore, we propose specific standards by which this idea can be appropriately invoked. In our view, judgments of futility emerge from either quantitative or qualitative evaluations of clinical situations. Such evaluations determine whether physicians are obligated to offer an intervention. If an intervention is judged to be futile, the duty to present the intervention as an option to the patient or the patient's family is mitigated or eliminated. We recognize—indeed invite—examination and challenge of our proposal.

The Glare of Autonomy

Less than a few decades ago, the practice of medicine was characterized by a paternalism exemplified in the expression, "doctor's orders." Physicians determined by themselves or in consultation with colleagues the usefulness of courses of treatment. The art of medicine was considered to include selectively withholding as well as disclosing information in order to maintain control over therapy. The dramatic shift toward patient self-determination that has taken place in recent decades almost certainly received much of its momentum from society's backlash to this paternalism. In addition, philosophical and political concerns about the rights of individuals and respect for persons elevated the principle of autonomy to a position in ethics that it had not previously held. Today, ethics and the law give primacy to patient autonomy, defined as the right to be a fully informed participant in all aspects of medical decision making and the right to refuse unwanted, even recommended and life-saving, medical care. So powerful has this notion of autonomy become that its glare often blinds physicians (and ethicists) to the validity of earlier maxims that had long defined the range of physicians' moral obligations toward patients. Among these was the maxim, respected in ethics and law, that futile treatments are not obligatory. No ethical principle or law has ever required physicians to offer or accede to demands for treatments that are futile.[1] Even the so-called Baby Doe regulations, notorious for their advocacy of aggressive medical intervention, permit physicians to withhold treatment that is "futile in terms of the survival of the infant" or "virtually futile."[2] Even when this maxim is accepted in theory, however, physicians frequently practice as though every available medical measure, including absurd and overzealous interventions, must be used to prolong life unless patients give definitive directions to the contrary. Some physicians allow patient surrogates to decide when a treatment is futile, thereby overriding medical judgment and potentially allowing the patient surrogate to demand treatment that offers no benefit.

Comparison of Effect and Benefit

In the early nineteenth century, all medications were, by definition, effective: They inevitably brought about the effect that their names described. Emetics could be counted on to cause vomiting; purgatives to cause taxation; sudorifics, sweating; and so on. These effects, given the medical theories of the times, were presumed always to be beneficial. Failure to heal was a defect of nature, not of the physician or the treatment. However, one advance of modern medicine, particularly with the introduction of controlled clinical trials, was to clarify by empiric methods the important distinction between effect and benefit. In examining the notion of futility, physicians sometimes fail to keep this distinction in mind.

For example, a recent discussion of futility includes the following: "[Physicians] may acknowledge that therapy is effective, in a limited sense, but believe that the goals that can be achieved are not desirable, as when considering prolonged nutritional support for patients in a persistent vegetative state. Physicians should acknowledge that, in such situations, potentially achievable goals exist. Therapy is not, strictly speaking, futile."[3] On the contrary, we believe that the goal of medical treatment is not merely to cause an effect on some portion of the patient's anatomy, physiology, or chemistry, but to benefit the patient as a whole. No physician would feel obligated to yield to a patient's demand to treat pneumonia with insulin. The physician would rightly argue that (in the absence of insulin-requiring diabetes) such treatment is inappropriate; insulin might have a physiologic effect on the patient's blood sugar, but would offer no benefit to the patient with respect to the pneumonia. Similarly, nutritional support could effectively preserve a host of organ systems in a patient in persistent vegetative state, but fail to restore a conscious and sapient life. Is such nutritional treatment futile or not? We argue that it is futile for the simple reason that the ultimate goal of any treatment should be improvement of the patient's prognosis, comfort, well-being, or general state of health. A treatment that fails to provide such a benefit—even though it produces a measurable effect—should be considered futile.

Approaching a Definition

The word futility comes from the Latin word meaning leaky (*futilis*). According to the *Oxford English Dictionary*, a futile action is "leaky, hence untrustworthy, vain, failing of the desired end through intrinsic defect." In Greek mythology, the daughters of Danaus were condemned in Hades to draw water in leaky sieves. Needless to say, their labors went for naught. The story conveys in all its fullness the meaning of the term: A futile action is one that cannot achieve the goals of the action, no matter how often repeated. The likelihood of failure may be predictable because it is inherent in the nature of the action proposed, and it may become immediately obvious or may become apparent only after many failed attempts.

This concept should be distinguished from etymologic neighbors. Futility should not be used to refer to an act that is, in fact, impossible to do. Attempting to walk to the moon or restore cardiac function in an exsanguinated patient would not be futile acts; they would be physically and logically impossible. Nor should futility be confused with acts that are so complex that, although theoretically possible, they are implausible. The production of a human infant entirely outside the womb, from in-vitro combination of sperm and egg to physiologic viability, may be theoretically possible but, with current technology, is implausible.

Further, futile, because the term is not merely descriptive, but also operational, denoting an action that will fail and that ought not be attempted, implies something more than simply rare, uncommon, or unusual. Some processes that are quite well understood and quite probable may occur only occasionally, perhaps because of their complexity and the need for many circumstances to concur at the same time. For example, successful restoration to health of a drug addict with bacterial endocarditis might require a combination of medical, psychological, social, and educational efforts. These interventions could work but, due to various factors (including limited societal resources), they rarely work. However, they are not futile.

Futility should also be distinguished from hopelessness. Futility refers to the objective quality of an action; hopelessness describes a subjective attitude. Hope and hopelessness bear more relation to desire, faith, denial, and other psychological responses than to the objective possibility or probability that the actions being contemplated will be successful. Indeed, as the chance for success diminishes, hope may increase and replace reasonable expectation. Something plausible is hardly ever hopeless, because hope is what human beings summon up to seek a miracle against overwhelming odds. It is possible then to say in the same breath, "I know this is futile, but I have hope." Such a statement expresses two facts, one about the objective properties of the situation, the other about the speaker's psychological state.

Futility refers to an expectation of success that is either predictably or empirically so unlikely that its exact probability is often incalculable. Without specific data, one might predict futility from closely analogous experience. For example, one might avoid a trial of an extended experience as the source of their conclusions. Here, specialty practice contributes an essential element; for example, an intensive care pulmonary specialist who sees several hundred patients who have similar disease conditions and receive similar therapy can often group together "futility characteristics" better than a generalist who does not see cases in so focused a manner.

Without systematic knowledge of the various factors that cause a therapy to have less than a 1% chance of success—knowledge that would allow the physician to address these factors—we regard it as unreasonable to require that the physicians offer such therapy. To do so forces the physician to offer any therapy that may have seemed to work or that may conceivably work. In effect, it obligates the physician to offer a placebo. Only when empirically observed (though not understood) outcomes rise to a level higher than that expected by any placebo effect, can a specific therapy be considered to be "possibly helpful" in rare or occasional cases and its appropriateness evaluated according to rules of decision analysis. In the clinical setting, such judgments also would be influenced, of course, by considering such tradeoffs as how cheap and simple the intervention is and how serious or potentially fatal the disease (see Exceptions and Cautions).

Although our proposed selection of proportions of success is admittedly arbitrary, it seems to comport reasonably well with ideas actually held by physicians. For example, Murphy and colleagues[4] invoked the notion of futility in their series of patients when survival after cardiopulmonary resuscitation was no better than 2% (upper limit of 95% CI as calculated by authors), and Lantos and colleagues[5] when survival was no better than 7% (upper limit of 95% CI as calculated by authors).

Obviously, as medical data on specific situations are gathered under appropriate experimental conditions, empiric uncertainty can be replaced with empiric confidence. Admittedly, some disorders may be too rare to provide sufficient experience for a confident judgment of futility, even when efforts are made to pool data. We acknowledge this difficulty but adhere to our conservative standard to prevent arbitrary abuse of power. In judging futility, as in other matters, physicians should admit uncertainty rather than impose unsubstantiated claims of certainty. Therefore, our view of futility should be considered as encouraging rather than opposing well conducted clinical trials. Important examples of such work in progress include studies of survival after cardiopulmonary resuscitation[6-13] and use of prognostic measures in patients requiring intensive medical care.[14,15]

Already, data on burn patients[16] and on patients in persistent vegetative state with abnormal neuroophthalmic signs[17] are sufficient to help with decision making. The latter group of patients present a particular challenge to presently confused notions of futility, perhaps accounting in part for why an estimated 5000 to 10,000 patients in persistent vegetative state are now being maintained in medical institutions.[18] The mythologic power of the coma patient who "wakes up" apparently overrides the rarity of documented confirmation of such miraculous recoveries (which have resulted, moreover, in incapacitating mental impairment or total dependence). This point bears on the frequently heard excuse for pushing ahead with futile therapies: "It is only by so doing that progress is made and the once futile becomes efficacious. Remember the futility of treating childhood leukemia or Hodgkin lymphoma." These statements hide a fallacy. It is not through repeated futility that progress is made, but through careful analysis of the elements of the "futile case," followed by well designed studies, that advances knowledge. We also point out that our proposal is intended for recognized illness in the acute clinical setting. It does not apply to preventive treatments, such as immunizations, estrogen prophylaxis for hip fractures, or penicillin prophylaxis for rheumatic heart disease and infectious endocarditis, all of which appear to have lower rates of efficacy because they are purposely administered to large groups of persons, many of whom will never be at risk for or identified with the particular diseases that their treatments are intended to prevent.

Qualitative Aspects

In keeping with the qualitative notion of futility we propose that any treatment that merely preserves permanent unconsciousness or that fails to end total dependence on intensive medical care should be regarded as nonbeneficial and, therefore, futile. We do not regard futility as "an elusive concept."[19] It is elusive only when effects on the patient are confused with benefits to the patient or when the term is stretched to include either considerations of 5-year survival in patients with cancer (not at all pertinent to the notion of futility) or the "symbolic" value to society of treating handicapped newborns or patients in persistent vegetative state (which rides roughshod over patient-centered decision making).[20]

Here is the crux of the matter. If futility is qualitative, why should the patient not always decide whether the quality achieved is satisfactory or not? Why should qualitatively "futile" results not be offered to the patient as an option? We believe a distinction is in order. Some qualitatively poor results should indeed be the patient's option, and the patient should know that they may be attainable. We believe, however, that other sorts of qualitatively poor results fall outside the range of the patient's autonomy and need not be offered as options. The clearest of these qualitatively poor results is continued biologic life without conscious autonomy. The patient has no right to be sustained in a state in which he or she has no purpose other than mere vegetative survival; the physician has no obligation to offer this option or services to achieve it. Other qualitatively poor results are conditions requiring constant monitoring, ventilatory support, and intensive care nursing (such as in the example at the beginning of our paper) or conditions associated with overwhelmingly suffering for a predictably brief time. Admittedly, these kinds of cases fall along a continuum, and there are well known examples of the most remarkable achievements of life goals despite the most burdensome handicaps. However, if survival requires the patient's entire preoccupation with intensive medical treatment, to the extent that he or she cannot achieve any other life goals (thus obviating the goal of medical care), the treatment is effective but not beneficial; it need not be offered to the patient, and the patient's family has no right to demand it.

Specifically excluded from our concept of futility is medical care for patients for whom such care offers the opportunity to achieve life goals, however limited. Thus, patients whose illnesses are severe enough to require frequent hospitalization, patients confined to nursing homes, or patients with severe physical or mental handicaps are not, in themselves, objects of futile treatments. Such patients (or their surrogates) have the right to receive or reject any medical treatment according to their own perceptions of benefits compared with burdens.

Some observers might object, as a matter of principle, to excluding patient input from assessments of qualitative futility. Others might be concerned that such exclusion invites abuse, neglect, and a retreat to the paternalistic "silent world" of the past in which doctors avoided communication with their patients.[21] In response to the latter objection, we acknowledge that potential for abuse is present and share this concern. We would deplore the use of our proposal to excuse physicians from engaging patients in ongoing informed dialogue. Nonetheless, the alternative is also subject to abuse (for example, when legal threats made by patients and surrogates cow hospitals into providing excessive care). We reiterate that the distinction between medical benefit and effect justifies excluding patients from determination of qualitative futility. Physicians are required only to provide medical benefits to patients. Physicians are permitted, but not obligated, to offer other, non-medical benefits. For example, a physician is not obligated to keep a patient alive in an irreversible vegetative state, because doing so does not medically benefit the patient. However, as noted below, a physician may do so on compassionate grounds, when temporary continuance of biologic life achieves goals of the patient or family.

Exceptions and Cautions

We have attempted to provide a working definition of futility. We also have drawn attention to the ethical notion that futility is a professional judgment that takes precedence over patient autonomy and permits physicians to withhold or withdraw care deemed to be inappropriate without subjecting such a decision to patient approval. Thus, we regard our proposal as representing the ordinary duties of physicians, duties that are applicable where there is medical agreement that the described standards of futility are met. We recognize, however, that the physician's duty to serve the best interests of the patient may require that exceptions to our approaches be made under special circumstances.

An exception could well be made out of compassion for the patient with terminal metastatic cancer who requests resuscitation in the event of cardiac arrest to survive long enough to see a son or daughter who has not yet arrived from afar to pay last respects. Such an exception could also be justified to facilitate coping and grieving by family members, a goal the patient might support. Although resuscitation may be clearly futile (that is, would keep the patient alive in the intensive care unit for only 1 or 2 more days), complying with the patient's wishes would be appropriate, provided such exceptions do not impose undue burdens on other patients, health care providers, and the institution, by directly threatening the health care of others. We hasten to add, however, that our notion of futility does not arise from considerations of scarce resources. Arguments for limiting treatments on grounds of resource allocation should proceed by an entirely different route and with great caution in our present open system of medical care, as there is no universally accepted value system for allocation and no guarantee that any limits a physician imposes on his or her patients will be equitably shared by other physicians and patients in the same circumstances.

Admittedly, in cases in which treatment has begun already, there may be an emotional bias to continue, rather than withdraw, futile measures. If greater attention is paid at the outset to indicating futile treatments, these situations would occur less frequently: however, the futility of a given treatment may not become clear until it has been implemented. We submit that physicians are entitled to cease futile measures in such cases, but should do so in a manner sensitive to the emotional investments and concerns of caretakers.

What if a hospitalized patient with advanced cancer demands a certain medication (for example, a particular vitamin), a treatment that the physician believes to be futile? Several aspects of this demand support its overriding the physician's invocation of futility. Certain death is expected and, although an objective goal such as saving the patient's life or even releasing the patient from the hospital might be unachievable, the subjective goal of patient well-being might be enhanced (a placebo induced benefit). In this particular situation, the effort and resources invested to achieve this goal impose a negligible burden on the health care system and do not threaten the health care of others. Thus, although physicians are not obligated to offer a placebo, they occasionally do. For example, Imbus and Zawacki[22] allowed burn patients to opt for treatment even when survival was unprecedented. In this clinical situation, compassionate yielding imposes no undue burden. because survival with or without treatment is measured in days. In contrast, yielding to a surrogate's demand for unlimited life-support for a patient in persistent vegetative state may lead to decades of institutional care.

References

1. President's Commission for the Study of Ethical Problems in Medicine and Biomedical and Behavioral Research. *Deciding to Forego Life-Sustaining Treatment: A Report on the Ethical, Medical, and Legal Issues in Treatment Decisions*. Washington, D.C.: U.S. Government Printing Office: 1983:60–89.
2. *Child Abuse and Neglect Prevention and Treatment*. Washington, DC: U.S. Department of Health and Human Services. Office of Human Development Services: 1985 *Federal Register* 50:14887–14878.
3. Lantos JD, Singer PA, Walker RM, et al. The illusion of futility in clinical practice. *Am J Med* 1989;87:81–84.
4. Murphy DJ, Murray AM, Robinson BE, Campion EW. Outcomes of cardiopulmonary resuscitation in the elderly. *Ann Intern Med* 1989;111:199–205.

5. Lantos JD, Miles SH, Silverstein MD, Stocking CB. Survival after cardiopulmonary resuscitation in babies of very low birth weight. *N Engl J Med* 1988;318:91–95.

6. Murphy DJ, Murray AM, Robinson BE, Campion EW. Outcomes of cardiopulmonary resuscitation in the elderly. *Ann Intern Med* 1989;111:199–205.

7. Lantos JD, Miles SH, Silverstein MD, Stocking CB. Survival after cardiopulmonary resuscitation in babies of very low birth weight. *N Engl J Med* 1988;318:91–95.

8. Freiman JA, Chalmers TC, Smith H Jr, Kuebler RR. The importance of beta, the type 11 error and sample size in the design and interpretation of the randomized control trial Survey of 71 "negative" trials. *N Engl J Med* 1978;299:690–694.

9. Bedell SE, Delbanco TL, Cook EF, Epstein *FH*. Survival after cardiopulmonary resuscitation in the hospital *N Engl J Med* 1983;309:569–576.

10. Gordon M, Hurowitz E. Cardiopulmonary resuscitation of the elderly. *J Am Geriatr Soc* 1984;32:930–934.

11. *Life-Sustaining Technologies and the Elderly*, Washington, D.C. U. S. Congress, Office of Technology Assessment: 1987: publication OTA-BA-306. 167.201.

12. Johnson AL, Tanser PH, Ulan RA, Wood TE. Results of cardiac resuscitation in 552 patients. *Am J Cardiol* 1967;20:831–835.

13. Taffet GE, Teasdale TA, Luchl RJ. In-hospital cardiopulmonary resuscitation. *JAMA* 1988;260:2069–2072.

14. Knaus WA, Draper EA, Wagner DP, Zimmerman JE. APACHE 11: a severity of disease classification system. *Crit Care Med* 1985;13:818–829.

15. Knaus WA, Draper EA, Wagner DP, Zimmerman JE. An evaluation of outcome from intensive care in major medical centers. *Ann Intern Med* 1986;104:410–418.

16. Imbus SH, Zawacki BE. Autonomy for burned patients when survival is unprecedented. *N Engl J Med* 1977;297:308–11.

17. Plum F, Posner JB. *The Diagnosis of Stupor and Coma*. 3rd ed. Philadelphia: F. A. Davis: 1980.

18. Cranford RE. The persistent vegetative state as medical reality (getting the facts straight). *Hastings Cent Rep.* 1988;18:27–32.

19. Lantos JD, Singer PA, Walker RM, et al. The illusion of futility in clinical practice. *Am J Med* 1989;87:81–84.

20. Lantos JD, Singer PA, Walker RM, et al. The illusion of futility in clinical practice. *Am J Med* 1989;87:81–84.

21. Katz J. *The Silent World of Doctor and Patient*. New York: Free Press: 1984

22. Imbus SH, Zawacki BE. Autonomy for burned patients when survival is unprecedented. *N Engl J Med* 1977;297:308–311.

ADVANCE CARE PLANNING AND SURROGATE DECISION MAKING

Natural Death Act

The State of California

The California Natural Death Act was passed in 1976. This Act became the precedent and template for other states interested in sanctioning living wills. The language of this first Living Will statute embodies the thinking of the time. This living will was intended for people with terminal illnesses and death anticipated imminently. It was meant to avoid unwanted life-sustaining treatment that would merely prolong the moment of death. More recent and generic advance directives and supplementary comments attached to Durable Power of Attorney for Health Care documents expand the scope of circumstances in which a mentally incapacitated person may forgo life-sustaining treatment.

The people of the State of California do enact as follows:

Directive to Physicians

Directive made this _____ day of _____ (month, year).

I _____ , being of sound mind, willfully, and voluntarily make known my desire that my life shall not be artificially prolonged under the circumstances set forth below, do hereby declare:

1. If at any time I should have an incurable injury, disease, or illness certified to be a terminal condition by two physicians, and where the application of life-sustaining procedures would serve only to artificially prolong the moment of my death and where my physician determines that my death is imminent whether or not life-sustaining procedures are utilized, I direct that such procedures be withheld or withdrawn, and that I be permitted to die naturally.

Reprinted from "California Natural Death Act," from *California Health and Safety Code*, Part I, Division 7, Chapter 3.9, Section 7188. Approved by the Governor on the 30th of September, 1976.

2. In the absence of my ability to give directions regarding the use of such life-sustaining procedures, it is my intention that this directive shall be honored by my family and physician(s) as the final expression of my legal right to refuse medical or surgical treatment and accept the consequences from such refusal.

3. If I have been diagnosed as pregnant and that diagnosis is known to my physician, this directive shall have no force or effect during the course of my pregnancy.

4. I have been diagnosed and notified at least 14 days ago as having a terminal condition by _____ M.D., whose address is _____ and whose telephone number is _____. I understand that if I have not filled in the physician's name and address, it shall be presumed that I did not have a terminal condition when I made out this directive.

5. This directive shall have no force or effect five years from the date filled in above.

6. I understand the full import of this directive and I am emotionally and mentally competent to make this directive.

Signed _____

City, County and State of Residence_____

The declarant has been personally known to me and I believe him or her to be of sound mind.

Witness _____

Witness _____

Integrating Preferences for Life-Sustaining Treatments and Health States Ratings into Meaningful Advance Care Discussions

Robert A. Pearlman, Helene Starks, Kevin C. Cain,
William G. Cole, Donald L. Patrick, and Richard F. Uhlmann

Robert Pearlman is Professor of Medicine at the University of Washington and Chief of the Ethics Evaluation Service for the National Center for Ethics in Health Care. Helene Starks is an assistant professor in the Department of Medical History and Ethics, University of Washington School of Medicine. Kevin Cain is a research associate professor at the University of Washington School of Nursing. William Cole is a private consultant in medical informatics. Donald Patrick is Professor of Health Services at the University of Washington School of Public Health. Richard Uhlmann is a geriatrician and gastroenterologist in private practice in Idaho.
In this paper the authors describe the results of an empirical study about preferences for life-sustaining treatments. When individuals decline noninvasive treatments, they usually decline more invasive treatments, and when they want to receive invasive treatments, they usually accept less invasive ones. Similarly, when individuals decline life-sustaining treatment in a good state of health, they usually decline the same treatment in worse health states. Conversely, when individuals want treatment in a poor health state, they usually want the same treatment in a better health state. These logically consistent findings may help health care providers frame advance care planning discussions with patients.

A dvance care planning has received attention as an important means to enhance end-of-life care. The major goal of advance care planning is to extend a patient's right to self-determination into the period when he or she becomes decisionally incapacitated. This is supposed to occur by ensuring that medical decisions made on behalf of patients without decisional capacity are based on either their previous wishes or their best interests. Advance care planning aims to accomplish this by having (1) the patient's wishes, expressed during a period of prior decisional capacity, serve as an action guide, and/or

(2) the patient specify a surrogate decision-maker who will represent him or her in making decisions.

It is important to differentiate advance care planning from advance directives. Advance care planning is a process that involves four steps: (1) thinking about one's values and preferences for medical care if one is unable to communicate, (2) communicating these values and preferences to loved ones and health care providers, (3) documenting values and preferences, and (4) ensuring that these documents are accessible and up-to-date. Advance directives represent only one part of this process: they are the mechanisms used to document patients' wishes or appoint surrogate decision-makers.

It is hoped that advance care planning will serve several additional functions: (1) reduce the risk of over-treatment and under-treatment, (2) minimize the conflicts among family members and between clinicians and family members, and (3) reduce the burden of surrogate decision-making that is placed on family members. Unfortunately, there are limited data that support the effectiveness of advance directives. Before indicting the use of advance directives, it is prudent to recognize the barriers to effective advance care planning. These are outlined below.

- Reimbursement mechanisms for advance care planning discussions are uncertain.
- Efforts to promote efficiency in the outpatient setting have reduced the length of provider-patient visits.
- In spite of patient interest, physicians often wait too long or never initiate advance care planning discussions.
- When discussions occur, they are often superficial (for instance, cardiopulmonary resuscitation (CPR) is often discussed without reference to the need for mechanical ventilation or likelihood of failure).
- Advance directives often are written using vague language or are restricted to terminal illness or permanent vegetative states. This inhibits clinical applicability.
- Clinicians frequently are inadequately educated and trained to conduct advance care planning discussions.
- Clinicians and surrogates lack good understanding of patients' wishes.

These barriers led us to investigate preferences for life-sustaining treatment and attitudes about health states with a diverse sample of volunteers. Participants were provided with detailed descriptions of health states and treatments to facilitate more informed decision-making. Once informed choices were elicited, the relationships between assessments of health states and treatment preferences were characterized as well as the relationships between different treatment preferences within a health state and across health states. These data provide a valid profile of attitudes and preferences that could form the basis for meaningful advance care planning discussions. This in turn could result in better discussions, more meaningful advance directives, and increased utility in clinical settings. The specific study questions addressed in this research are as follows:

1. When people consider life in a particular circumstance as "worse than death," what is the likelihood that they will refuse life-sustaining treatments in that circumstance?
2. How well does a person's preference for one treatment in a specific health state predict that person's preferences for other treatments in the same health state?
3. How well does a person's preference for one treatment in a specific health state predict that person's preferences for the same treatment in other health states?
4. If an advance care planning discussion is organized based on the results from this data set, how would it be structured, and why?

Methods

Overview

The research findings reviewed in this chapter are derived from a longitudinal study conducted between 1991 and 1995 in which preferences for life-sustaining treatments were elicited under a variety of conditions. Participants also rated their current health state and two hypothetical states depicting severe dementia and permanent coma.

Patient Population

The study participants were volunteers from seven groups in the Seattle area: younger well adults age 21 to 65 (n = 50), older well adults over age 65 (n = 49), older adults (over 65 years of age) with at least one chronic illness (n = 49), persons with cancer and a physician-estimated life expectancy of 6–24 months (n = 49), persons with AIDS or class IV HIV infection (n = 50), survivors of a stroke that occurred within the last ten years and resulted in residual impairment (n = 45), and nursing home residents who were expected to remain in the nursing home for at least six months (n = 50). Participants had to be at least 21 years of age, have no major vision or hearing impairments, show cognitive ability according to the Telephone Interview for Cognitive Status, and speak English. Well adults could not have any health condition that had lasted longer than one year, be receiving regular treatment by a health care provider or be taking medications more than twice monthly.

Well adults were recruited by sending letters to addresses that were randomly selected from the telephone directory. Eligible patients were identified with the help of community and university-affiliated physicians and social service intermediaries. Potential participants were sent or given information statements about the project. If an individual was interested in learning more about the study, he or she could contact the study office. All persons who contacted the study office were screened. Informed consent occurred at the time of the interview. A total of 342 persons participated in the study.

Questionnaire Description

Treatment preferences, health state ratings, and health status data were collected during in-person interviews. Preferences for antibiotics, long-term mechanical ventilation (with tracheostomy), long-term hemodialysis, long-term jejunal tube feeding, short-term mechanical ventilation, and CPR were elicited in each participant's current health and two hypothetical states representing severe dementia and permanent coma. Two versions of a visual aid to facilitate decision-making were used. That visual aid showed that the outcome of choosing treatment would result in a 100% chance of returning to the baseline state. Treatment preferences were elicited after reviewing the visual aid with the simple question, "Do you want to receive treatment?".

The health states were characterized in four domains: (1) thinking, remembering and talking; (2) walking and mobility; (3) self care; and (4) pain and discomfort. For each domain, three to four levels were described with examples. For the current health situation, the participant selected the appropriate level of function for each domain. The dementia state was characterized as "think, remember, and talk with great difficulty; get around with great difficulty; perform self care with some difficulty; and are in no physical pain or discomfort." The permanent coma situation was described as "do not think, remember, or communicate in any way; are confined to a bed; do not perform self care activities; and are in no physical pain or discomfort." These descriptions were complemented by examples written in everyday language. For example, "get around with great difficulty" was further characterized with "walk or use a cane, walker, or wheelchair but are limited to the house." . . . Health state ratings were elicited after reviewing the four domain descriptions with their corresponding examples. The ratings were elicited with the simple question, "How would you rate this health state?"

Measurement

Treatment preferences were indicated on a five-point scale: "definitely no," "probably no," "not sure," "probably yes," and "definitely yes." Health states were rated on a seven-point scale: "much worse than death," "somewhat worse than death," "a little worse than death," "neither better nor worse than death," "a little better than death," "somewhat better than death," and "much better than death."

Analytic Strategies

To facilitate analyses, the five-point treatment preference scale was collapsed into three clinically based categories: forego treatment (representing "definitely no" and "probably no"), accept treatment (including "definitely yes" and "probably yes"), and not sure. The health state rating scale also was collapsed into three categories: worse than death, neither better nor worse, and better than death. To address the first study question, the percentage of treatment decisions that were refused when the health states were rated as "worse than death" was calculated. To address the second and third study questions, positive and negative predictive values were used to assess the relationship between treatment preferences. Positive predictive value is the conditional probability that a person will want one treatment given a preference in favor of a different treatment. Negative predictive value is the conditional probability that a person will forego one treatment given a preference to forego a different treatment.

Results (Previously Reported)

Distribution of Health State Ratings and Treatment Preferences

Nearly all participants rated their current health as better than death. In contrast, 52% rated permanent coma as worse than death and 27% rated severe dementia as worse than death. Table 1 shows the distribution of treatment refusals in the three health states.

Relationship Between "Worse than Death" Health State Ratings and Treatment Preferences

When health states were rated as worse than death, participants chose to forego life-sustaining treatments 85% of the time. The 15% of decisions in which participants accepted treatment in health states rated as "worse than death" had several explanations.

First, many people wanted antibiotic treatment, viewing it as relatively simple and short-term. Second, some people used the "worse than death" language to connote an undesirable state, rather than a literal interpretation in which death would be preferred to continued existence in that situation. Another reason given was that people wanted to respect the wishes of family members that they continue living.

TABLE 1   Percentage of Treatment Refusals* in Each of the Three Health States

Treatment	Current Health	Dementia	Coma
Antibiotics	5	20	62
Short-term mechanical ventilation	12	44	71
Cardiopulmonary resuscitation	23	60	85
Long-term dialysis	25	56	86
Long-term feeding tube	41	64	86
Long-term mechanical ventilation	58	77	86

*Includes "probably no" and "definitely no" treatment preference ratings.
Preferences for treatments differed across health states for every treatment ($P < 0.0001$) and across treatments for every health state ($P < 0.0001$).
Reprinted with permission from the *Annals of Internal Medicine*.

Predictive Values Between Treatment Preferences in Current Health

In general, the preference to receive more invasive treatments had high positive predictive value for less invasive treatments in current health (range, 0.86–1.0). For example, if a participant was willing to accept long-term treatment with a feeding tube, there was a greater than an 88% probability that s/he also would accept CPR, short-term mechanical ventilation, or intravenous antibiotics. In addition, preferences to forego less invasive treatments had moderately high negative predictive value for more invasive treatments (range, 0.61–0.94). For example, saying "no" to intravenous antibiotics generalized to saying "no" to all other life-sustaining treatments with greater than a 77% probability.

Predictive Values of Treatment Preferences Between Health States

The positive predictive values of treatments in permanent coma for the same treatment in severe dementia were high (range, 0.8–1.0). In contrast, wanting a treatment in current health did not generalize well to wanting the same treatment in the dementia situation (range, 0.44–0.77). Weak positive predictive values also were seen when trying to generalize treatment preferences from the dementia situation to the permanent coma situation (range, 0.27–0.51).

The negative predictive values of treatment preferences from current health to severe dementia were high (range, 0.88–0.97), excluding antibiotics which had a negative predictive value of 0.72. There were high negative predictive values when generalizing treatment preferences from the severe dementia situation to the permanent coma situation (range, 0.92–0.99). In contrast, weak negative predictive values were found generalizing from the permanent coma situation to the severe dementia situation (range, 0.20–0.77).

Results (Not Previously Reported)

Predictive Values Between Treatments Within Hypothetical Health States

As shown in Table 2A, the preference to receive more invasive treatments in the severe dementia situation had moderate to high positive predictive values for less invasive treatments, especially for intravenous antibiotics. For example, if a participant wanted long-term mechanical ventilation, there was a greater than a 78% probability that s/he also would want all other treatments. However, the converse was not true: no preference for any other life-sustaining treatment generalized well to long-term mechanical ventilation. Preferences for CPR did generalize well to antibiotics, but only moderately well to short-term mechanical ventilation, long-term dialysis and long-term feeding tubes, and rather poorly to long-term mechanical ventilation.

Table 2B shows that preferences to forego less invasive treatments in the severe dementia situation had moderate to high negative predictive values for more invasive treatments. For example, if a participant wanted to forego antibi-

TABLE 2A   Positive Predictive Value of Treatments in Dementia for Each Other

If a person said "yes" to:	n*	ABX	SMV	CPR	DYL	LFT	LMV
		Then, the probability of saying "yes" to this treatment is:					
Antibiotics (ABX)	249	—	.51	.41	.37	.33	.18
Short-term mechanical ventilation (SMV)	134	.96	—	.60	.58	.49	.33
Cardiopulmonary resuscitation (CPR)	106	.97	.75	—	.64	.59	.37
Long-term dialysis (DYL)	100	.92	.78	.68	—	.63	.41
Long-term feeding tube (LFT)	82	.99	.79	.77	.77	—	.45
Long-term mechanical ventilation (LMV)	47	.96	.94	.83	.87	.79	—
Overall rate = yes**	342	.73	.39	.31	.29	.24	.14

TABLE 2B   Negative Predictive Value of Treatments in Dementia for Each Other

If a person said "no" to:	n*	ABX	LMV	CPR	DYL	LFT	LMV
		Then, the probability of saying "no" to this treatment is:					
Antibiotics (ABX)	67	—	-91	.99	.90	.96	1.00
Short-term mechanical ventilation (SMV)	150	.41	—	.85	.83	.91	.95
Cardiopulmonary resuscitation (CPR)	206	.32	.62	—	.76	.85	.93
Long-term dialysis (DYL)	192	.31	.65	.82	—	.89	.96
Long-term feeding tube (LFT)	219	.29	.63	.80	.78	—	.93
Long-term mechanical ventilation (LMV)	262	.26	.54	.73	.70	.77	—
Overall rate = no**	342	.20	.44	.60	.56	.64	.77

*The n refers to the number of respondents who said "yes" (Table 2A) or "no" (Table 2B) to the treatments in the rows.
**The overall rates indicate how often participants wanted (Table 2A) or did not want (Table 2B) each of the treatments shown in the columns. All predictive values are significantly different from the overall rates ($P < 0.001$).

otics, there was greater than an 89% probability that s/he would want to forego all other treatments. Preferences to forego most treatments generalized well to foregoing long-term feeding tubes and long-term mechanical ventilation. However, saying "no" to long-term mechanical ventilation had moderate to poor negative predictive value for other treatments.

Tables 3A and 3B show results for predictive values between treatments in the permanent coma situation. There are similar patterns as with the dementia

Table 3A   Positive Predictive Values (PPV) for Treatments in Coma for Each Other

If a person says "yes" to:	n*	ABX	SMV	CPR	DYL	LFT	LMV
Antibiotics (ABX)	89	—	.52	.36	.26	.20	Value?
Short-term mechanical ventilation (SMV)	68	.68	—	.44	.37	.34	Value?
Cardiopulmonary resuscitation (CPR)	36	.89	.83	—	.61	.50	Value?
Long-term dialysis (DYL)	33	.70	.76	.67	—	.61	Value?
Long-term feeding tube (LFT)	26	.69	.88	.69	.77	—	Value?
Long-term mechanical ventilation (LMV)	30	.83	.80	.70	.63	.53	—
Overall rate = yes**	342	.26	.20	.11	.10	.08	.09

Table 3B   Negative Predictive Values (NPV) of Treatments in Coma for Each Other

If a person says "no" to:	n*	ABX	SMV	CPR	DYL	LFT	LMV
Antibiotics (ABX)	Value?	—	.91	.97	.95	.96	.98
Short-term mechanical ventilation (SMV)	Value?	.79	—	.98	.98	.99	.98
Cardiopulmonary resuscitation (CPR)	Value?	.70	.82	—	.95	.95	.95
Long-term dialysis (DYL)	Value?	.68	.81	.93	—	.97	.95
Long-term feeding tube (LFT)	Value?	.68	.82	.93	.97	—	.94
Long-term mechanical ventilation (LMV)	Value?	.70	.80	.93	.94	.94	—
Overall rate = no**	342	.61	.71	.84	.86	.86	.86

*The n refers to the number of respondents who said "yes" (Table 3A) or "no" (Table 3B) to the treatments in the rows, considering the predicting health state to the left of the arrows.
**The rates reflect the fraction of participants who wanted (Table 3A) or did not want (Table 3B) the treatments in the rows, considering the second health state to the right of the arrows. These predictive values do not differ significantly at the $P = 0.01$ level. All other values of PPV and NPV are significantly different from the rates ($P < 0.001$).

situation, but with lower positive predictive values and higher negative predictive values. . . .

Discussion

In this research, the relationship between treatment preferences and ratings of health states as well as the predictive values of life-sustaining treatment preferences were examined. The data further support earlier research indicating that quality of life and perceptions of states worse than death motivate the desire to forego life-sustaining treatment. The relationships between treatment preferences affirm earlier findings showing that (1) when patients decline noninvasive treatments, they usually decline more invasive treatments, and (2) when they want to receive invasive treatments they usually accept less invasive ones.

The data also suggest an empirically derived, organizing sequence in which to order treatments. The treatments as listed in Tables 2 and 3 (antibiotics, short-term mechanical ventilation, CPR, long-term dialysis, long-term feeding tube, and long-term mechanical ventilation) represent degrees of "aggressiveness" that incorporate invasiveness and duration of treatment. . . .

Advance care planning discussions that pertain to treatment preferences and health state ratings can be organized by systematically reviewing the results from these analyses. Practice would suggest that a good place to begin a discussion of advance care planning is to inquire about who would be the best person to speak on the patient's behalf. Following this, the clinician can ask about life-sustaining treatment preferences in current health. If the patient says she "wants nothing" in current health, the clinician should probe for an explanation that provides the context for these preferences. Patients who "want nothing" in their current health are very likely not to want treatment under any circumstances. This interpretation can and should be verified directly with the patient.

If, however, a patient is interested in receiving life-sustaining treatment in current health, the next set of questions should determine whether the patient is interested in receiving life-sustaining treatment in all circumstances (including long-term coma and/or terminal illness). Since only a small minority of individuals desire life-sustaining treatment in all circumstances, identifying them quickly may streamline the discussion. The most common situation, however, is that patients have a mix of preferences. Thus, the next set of questions should have two goals: (1) to characterize two thresholds of unacceptability: one for health states and the other for treatments, and (2) to understand why the person would not want treatments under certain circumstances.

To achieve the first goal, a clinician should inquire about whether living under specified situations, such as severe dementia or permanent coma, would be considered a "fate worse than death." A question that introduces this topic is, "What kinds of situations do you fear the most?" If the patient identifies one or more of these situations, the clinician should explore the patient's reasons. Afterwards, the clinician should confirm that the patient would not want life-sustaining treatment if faced with a life-threatening illness in the situation(s). Asking about a few specific treatments should verify the inference that life-sustaining treatments should be withheld or withdrawn under these unacceptable circumstances. Occasionally a patient will indicate that a health state would be unacceptable and yet she would want one or more life-sustaining treatments. In these circumstances, asking the "why" question should illuminate other important and clinically relevant values or concerns that have bearing on advance care planning.

When a patient indicates that a situation is acceptable, follow-up questions about preferences for a few treatments should illuminate the patient's threshold for treatment acceptability. For example, if the patient is asked about long-term use of a mechanical ventilator and would desire such treatment, then the chances are good that she will want all other treatments in a particular state of health. Conversely, if a patient is asked and says no to treatment with antibiotics, she would likely not want other treatments.

There are three important clinical caveats that derive from these results. First, eliciting preferences only for CPR, as is often done, is not enough to understand a patient's overall preferences for life-sustaining treatment. CPR generalizes poorly to other life-sustaining treatments that are perceived to be more invasive or long-term. Second, wanting treatment in one's current health does not generalize well to wanting treatment in more impaired functional health states. Third, refusing treatment in a severely impaired state of health (severe dementia or long-term coma) does not generalize well to refusing treatment in less impaired states of health (for instance, current health).

It is known that at present, physicians spend a very limited amount of time discussing advance care planning with their patients and often do not develop a shared understanding of their patients' values or preferences. The extrapolation of these data into an approach to advance care planning discussions may help clinicians, as it is organized, balanced, and straightforward. Moreover, by asking the "why" questions after eliciting preferences and listening to the responses, the discussion will stay patient-focused. In addition, many patients have diagnoses with predictable prognoses. In these situations, the advance care planning discussions can be streamlined further by focusing on the anticipated circumstances for the particular patient.

The proposed advance care planning questions do not address the important challenge that patients need to understand the health states and treatments that are raised in any discussion. Without rich descriptions, patients may be unable to visualize the treatments and health states, and therefore may be poorly prepared to formulate preferences that reflect their values and interests. . . .

Some patients may prefer to discuss general values or goals of care rather than specific treatment preferences. Unfortunately, reliance on general values has shown limited generalizability to treatment preferences. Similarly, treatment goals that rely on general statements, such as "attempt cure" or "consider quality of life," appear to have limited ability to translate consistently into treatment preferences. Other patients may prefer to have family members decide what is best when the situation arises. Reliance on the family may prevent over-interpretation of directives and is supported by social custom. However, relying on the family to make decisions for decisionally incapacitated patients does not lessen the value of explicit discussions between patients and their family members before the need arises.

A major study limitation is that the people who agreed to participate differ from the general population in the United States because (1) they were predominantly white and well educated, and (2) they were willing to think about these issues. Another limitation is that participants were asked to assume three things that make the decisions less realistic than they would be in actual practice when formulating their treatment preferences. These included considering the hypothetical health states to be permanent, accepting the stated probabilities of treatment success, and that the decisions would not have economic implications.

Despite these reservations, we believe these data and the resultant approach to advance care planning discussions may help clinicians, patients, and their family members. Prior to asking the recommended questions, however, clinicians should decide how they plan to (1) introduce the topic, (2) address the emotional content of the discussions, (3) facilitate communication between the patient and family or surrogate decision-maker, (4) ensure that patients understand what they are talking about, and (5) follow up either with regard to further deliberation or developing an advance directive.

Enough: The Failure of the Living Will

Angela Fagerlin and Carl E. Schneider

Angela Fagerlin received her Ph.D. from Kent State University in research psychology and is a Research Scientist in Internal Medicine at the University of Michigan. Carl Schneider received his J.D. from the University of Michigan and is the Chauncey Stillman Professor for Ethics, Morality, and the Practice of Law and Professor of Internal Medicine at the University of Michigan.
In this article the authors provide a comprehensive review of the conceptual and practical issues that undermine the goals, use, and value of the living will. The authors' indictment does not extend to proxy directives such as the durable power of attorney for health care. Moreover, the failure of the living will is not tantamount to a failure of repeated advance care planning discussions between patients and their future surrogate decision makers.

. . . Living wills are a bioethical idea that has passed from controversy to conventional wisdom, from the counsel of academic journals to the commands of law books, from professors' proposal to professional practice. Advance directives generally are embodied in federal policy by the Patient Self-Determination Act, which requires medical institutions to give patients information about their state's advance directives. In turn, the law of every state provides for advance directives, almost all states provide for living wills, and most states "have at least two statutes, one establishing a living will type directive, the other establishing a proxy or durable power of attorney for health care."[1] Not only are all these statutes very much in effect, but new legislative activity is constant. . . .

Although some sophisticated observers have long doubted the wisdom of living wills,[2] proponents have tended to respond in one of three ways, all of which preserve an important role for living wills. First, proponents have supposed that the principal problem with living wills is that people just won't sign them. These proponents have persevered in the struggle to find ways of getting more people to sign up.[3]

Second, proponents have reasserted the usefulness of the living wills. For example, Norman Cantor, distinguished advocate of living wills, acknowledges that "(s)ome commentators doubt the utility or efficacy of advance directives" (by which he means the living will), but he concludes that "these objections don't obviate the importance of advance directives."[4] Other proponents are daunted by the criticisms of living wills but offer new justifications for them.

Condensed from *Hastings Center Report* 34 (2), March/April 2004: 30–42. Reproduced with permission of The Hastings Center. Copyright © 2004 The Hastings Center.

Linda Emanuel, another eminent exponent of living wills, writes that "living wills can help doctors and patients talk about dying" and can thereby "open the door to a positive, caring approach to death."[5]

Third, some proponents concede the weaknesses of the living will and the advantages of the durable power of attorney and then propose a durable power of attorney that incorporates a living will. That is, the forms they propose for establishing a durable power of attorney invite their authors to provide the kinds of instructions formerly confined to living wills.[6]

None of these responses fully grapples with the whole range of difficulties that confound the policy promoting living wills. In fairness, this is partly because the case against that policy has been made piecemeal and not in a full-fledged and full-throated analysis of the empirical literature on living wills.

. . . The time has come to investigate . . . [the] polic[y] . . . systematically. . . . We ask an obvious but unasked question: What would it take for a regime of living wills to function as their advocates hope? First, people must have living wills. Second, they must decide what treatment they would want if incompetent. Third, they must accurately and lucidly state that preference. Fourth, their living wills must be available to people making decisions for a patient. Fifth, those people must grasp and heed the living will's instructions. These conditions are unmet and largely unmeetable.

Do People Have Living Wills?

At the level of principle, living wills have triumphed among the public as among the princes of medicine. People widely say they want a living will, and living wills have so much become conventional medical wisdom "that involvement in the process is being portrayed as a duty to physicians and others."[7] Despite this, and despite decades of urging, most Americans lack them.[8] While most of us who need one have a property will, roughly 18 percent have living wills.[9] The chronically or terminally ill are likelier to prepare living wills than the healthy, but even they do so fitfully.[10] In one study of dialysis patients, for instance, only 35 percent had a living will, even though all of them thought living wills a "good idea."[11]

Why do people flout the conventional wisdom? The flouters advance many explanations.[12] They don't know enough about living wills,[13] they think living wills hard to execute,[14] they procrastinate,[15] they hesitate to broach the topic to their doctors (as their doctors likewise hesitate).[16] Some patients doubt they need a living will. Some think living wills are for the elderly or infirm and count themselves in neither group.[17] Others suspect that living wills do not change the treatment people receive; 91 percent of the veterans in one study shared that suspicion.[18] Many patients are content or even anxious to delegate decisions to their families,[19] often because they care less what decisions are made than that they are made by people they trust. Some patients find living wills incompatible with their cultural traditions.[20] Thus in the large SUPPORT and HELP studies, most patients preferred to leave final resuscitation decisions to their family and physician instead of having their own preferences expressly followed (70.8% in HELP and 78.0% in SUPPORT). "This result is so striking that it is worth restating: not even a third of the HELP patients and hardly more than a fifth of the SUPPORT patients would want their own preferences followed."[21]

If people lacked living wills only because of ignorance, living wills might proliferate with education. But studies seem not to "support the speculations found in the literature that the low level of advance directives use is due primarily to a lack of information and encouragement from health care professionals and family members."[22] Rather, there is considerable evidence "that the elderly's action of delaying execution of advance directives and deferring to others is a deliberate, if not an explicit, refusal to participate in the advance directives process."[23]

The federal government has sought to propagate living wills through the Patient Self-Determination Act,[24] which essentially requires medical institutions to inform patients about advance directives. However, "empirical studies demonstrate that: the PSDA has generally failed to foster a significant increase in advance directives use; it is being implemented by medical institutions and their personnel in a passive manner; and the involvement of physicians in its implementation is lacking."[25] One commentator even thinks "the PSDAs legal requirements have become a ceiling instead of a floor."[26]

In short, people have reasons, often substantial and estimable reasons, for eschewing living wills, reasons unlikely to be overcome by persuasion. . . .

Do People Know What They Will Want?

Suppose, counterfactually, that people executed living wills. For those documents to work, people would have to predict their preferences accurately. This is an ambitious demand. Even patients making contemporary decisions about contemporary illnesses are regularly daunted by the decisions' difficulty. They are human. We humans falter in gathering information, misunderstand and ignore what we gather, lack well-considered preferences to guide decisions, and rush headlong to choice.[27] How much harder, then, is it to conjure up preferences for an unspecifiable future confronted with unidentifiable maladies with unpredictable treatments?

. . . To take one example from many, people grossly overestimate the effectiveness of CPR and in fact hardly know what it is.[28] For such information, people must rely on doctors. But doctors convey that information wretchedly even to competent patients making contemporaneous decisions. Living wills can be executed without even consulting a doctor,[29] and when doctors are consulted, the conversations are ordinarily short, vague, and tendentious. In the Tulsky study, for example, doctors only described either "dire scenarios . . . in which few people, terminally ill or otherwise, would want treatment" or "situations in which patients could recover with proper treatment."[30]

. . . Not only do people regularly know too little when they sign a living will, but often (again, we're human) they analyze their choices only superficially before placing them in the time capsule. An ocean of evidence affirms that answers are shaped by the way questions are asked. Preferences about treatments are influenced by factors like whether success or failure rates are used,[31] the level of detail employed,[32] and whether long or short-term consequences are explained first.[33] Thus in one study, "201 elderly subjects opted for the intervention 12% of the time when it was presented negatively, 18% of the time when it was phrased as in an advance directive already in use, and 30% of the time when it was phrased positively. Seventy-seven percent of the subjects changed their minds at least once when given the same case scenario but a different description of the intervention."[34]

If patients have trouble with contemporaneous decisions, how much more trouble must they have with prospective ones. For such decisions to be "true," patients' preferences must be reasonably stable. Surprisingly often, they are not. A famous study of eighteen women in a "natural childbirth" class found that preferences about anesthesia and avoiding pain were relatively stable before childbirth, but at "the beginning of active labor (4–5 cm dilation) there was a shift in the preference toward avoiding labor pains. . . . During the transition phase of labor (8–10 cm) the values remained relatively stable, but then . . . the mothers' preferences shifted again at postpartum toward avoiding the use of anesthesia during the delivery of her next child."[35] And not only are preferences surprisingly labile, but people have trouble recognizing that their views have changed.[36] This makes it less likely they will amend their living wills as their opinions develop and more likely that their living wills will treasonously misrepresent their wishes.

. . . At least sixteen studies have investigated the stability of people's preferences for life-sustaining treatment.[37] A meta-analysis of eleven of these studies found that the stability of patients' preferences was 71 percent (the range was 57 percent to 89 percent).[38] Although stability depended on numerous factors (including the illness, the treatment, and demographic variables), the bottom line is that, over periods as short as two years, almost one-third of preferences for life-sustaining medical treatment changed. More particularly, illness and hospitalization change people's preferences for life-sustaining treatments.[39] In a prospective study, the desire for life-sustaining treatment declined significantly after hospitalization but returned almost to its original level three to six months later.[40] Another study concluded that the "will to live is highly unstable among terminally ill cancer patients."[41] The authors thought their findings "perhaps not surprising, given that only 10–14% of individuals who survive a suicide attempt commit suicide during the next 10 years, which suggests that a desire to die is inherently changeable."

The consistent finding that interest in life-sustaining treatment shifts over time and across contexts coincides tellingly with research charting people's struggles to predict their own tastes, behavior, and emotions even over short periods and under familiar circumstances.[42] People mispredict what poster they will like,[43] how much they will buy at the grocery store,[44] how sublimely they will enjoy an ice cream,[45] and how they will adjust to tenure decisions.[46] And people "miswant" for numerous reasons.[47] They imagine a different event from the one that actually occurs, nurture inaccurate theories about what gives them pleasure,[48] forget they might outwit misery, concentrate on salient negative events and ignore offsetting happier ones,[49] and misgauge the effect of physiological sensations like pain.[50] Given this rich stew of research on people's missteps in predicting their tastes generally, we should expect misapprehensions about end-of-life preferences. Indeed, those preferences should be especially volatile, since people lack experience deciding to die.

Can People Articulate What They Want?

Suppose, arguendo, that patients regularly made sound choices about future treatments and write living wills. Can they articulate their choices accurately? This question is crucially unrealistic, of course, because the assumption is false. People have trouble reaching well-considered decisions, and you cannot state clearly on paper what is muddled in your mind. And indeed people do, for instance, issue mutually inconsistent instructions in living wills.[51]

But assume this difficulty away and the problem of articulation persists. In one sense, the best way to divine patients' preferences is to have them write their own living wills to give surrogates the patient's gloriously unmediated voice. This is not a practical policy. Too many people are functionally illiterate,[52] and most of the literate cannot express themselves clearly in writing. It's hard, even for the expert writer. Furthermore, most people know too little about their choices to cover all the relevant subjects. Hence living wills are generally forms that demand little writing. But the forms have failed. For example, "several studies suggest that even those patients who have completed AD forms . . . may not fully understand the function of the form or its language."[53] Living wills routinely baffle patients with their

> "syntactic complexity, concept density, abstractness, organization, coherence, sequence of ideas, page format, length of line of print, length of paragraph, punctuation, illustrations, color, and reader interest." Unfortunately, most advance directive forms . . . often have neither a reasonable scope nor depth. They do not ask all the right questions and they do not ask those questions in a manner that elicits clear responses.[54]

Doctors and lawyers who believe their clients are all above average should ask them what their living will says. One of us (CES) has tried the experiment. The modal answer is, in its entirety: "It says I don't want to be a vegetable."

No doubt the forms could be improved, but not enough to matter. The world abounds in dreadfully drafted forms because writing complex instructions for the future is crushingly difficult. Statutes read horribly because their authors are struggling to (1) work out exactly what rule they want, (2) imagine all the circumstances in which it might apply, and (3) find language to specify all those but only those circumstances. Each task is ultimately impossible, which is why statutes explicitly or implicitly confide their enforcers with some discretion and why courts must interpret—rewrite?—statutes. However, these skills and resources are not available to physicians or surrogates.

... The lamentable history of the living will demonstrates just how recalcitrant these problems are. There have been, essentially, three generations of living wills. At first, they stated fatuously general desires in absurdly general terms. As the vacuity of over-generality became clear, advocates of living wills did the obvious: Were living wills too general? Make them specific. Were they "one size fits all"? Make them elaborate questionnaires. Were they uncritically signed? "Require" probing discussions between doctor and patient. However, the demand for specificity forced patients to address more questions than they could comprehend. So, generalities were insufficiently specific and insufficiently considered. Specifics were insufficiently general and perhaps still insufficiently considered. What was a doctor—or lawyer—to do? Behold the "values history," a disquisition on the patient's supposed overarching beliefs from which to infer answers to specific questions.[55] That patients can be induced to trek through these interminable and imponderable documents is unproved and unlikely. That useful conclusions can be drawn from the platitudes they evoke is false. As Justice Holmes knew, "General propositions do not decide concrete cases."[56]

The lessons of this story are that drafting instructions is harder than proponents of living wills seem to believe and that when you move toward one blessing in structuring these documents, you walk away from another. The failure to devise workable forms is not a failure of effort or intelligence. It is a consequence of attempting the impossible.

Where Is the Living Will?

Suppose that, mirabile dictu, people executed living wills, knew what they will want, and could say it. That will not matter unless the living will reaches the people responsible for the incompetent patient. Often, it does not. This should be no surprise, for long can be the road from the drafters chair to the ICU bed.

First, the living will may be signed years before it is used, and its existence and location may vanish in the mists of time.[57] Roughly half of all living wills are drawn up by lawyers and must somehow reach the hospital, and 62 percent of patients do not give their living will to their physician.[58] On admission to the hospital, patients can be too assailed and anxious to recall and mention their advance directives.[59] Admission clerks can be harried, neglectful, and loath to ask patients awkward questions.

Thus when a team of researchers reviewed the charts of 182 patients who had completed a living will before being hospitalized, they found that only 26 percent of the charts accurately recorded information about those directives,[60] and only 16 percent of the charts contained the form. And in another study only 35 percent of the nursing home patients who were transferred to the hospital had their living wills with them.[61]

Will Proxies Read It Accurately?

Suppose, *per impossibile*, that patients wrote living wills, correctly anticipated their preferences, articulated their desires lucidly, and conveyed their document to its interpreters. How acutely will the interpreters analyze their instructions? Living wills are not self-executing: someone must decide whether the patient is incompetent, whether a medical situation described in the living will has arisen, and what the living will then commands.

Usually, the patient's intimates will be central among a living will's interpreters. We might hope that intimates already know the patient's mind, so that only modest demands need be made on their interpreting skills. But many studies have asked such surrogates to predict what treatment the patient would choose.[62] Across these studies, approximately 70 percent of the predictions were correct—not inspiring success for life and death decisions.

Do living wills help? We know of only one study that addresses that question. In a randomized trial, researchers asked elderly patients to complete a disease- and treatment-based or a value-based living will.[63] A control group of elderly patients completed no living will. The surrogates were generally spouses or children who had known the patient for decades. Surrogates who were not able to consult their loved one's living will predicted patients' preferences about 70 percent of the time. Strikingly, surrogates who consulted the living will did no better than surrogates denied it. Nor were surrogates more successful when they discussed living wills with patients just before their prediction.

What is more, a similar study found that primary care physicians' predictions were similarly unimproved by providing them with patients' advance directives.[64] On the other hand, emergency room doctors (complete strangers) given a living will more accurately predicted patients' preferences than ER doctors without one.[65]

Do Living Wills Alter Patient Care?

Our survey of the mounting empirical evidence shows that none of the five requisites to making living wills successful social policy is met now or is likely to be. The program has failed, and indeed is impossible.

That impossibility is confirmed by studies of how living wills are implemented, which show that living wills seem not to affect patients' treatments. For instance, one study concluded that living wills "do not influence the level of medical care overall. This finding was manifested in the quantitatively equal use of diagnostic testing, operations, and invasive hemodynamic monitoring among patients with and without advance directives. Hospital and ICU lengths of stay, as well as health care costs, were also similar for patients with and without advance directive statements."[66] Another study found that in thirty of thirty-nine cases in which a patient was incompetent and the living will was in the patient's medical record, the surrogate decision-maker was not the person the patient had appointed.[67] In yet a third study, a quarter of the patients received care that was inconsistent with their living will.[68]

But all this is normal. Harry Truman rightly predicted that his successor would "sit here, and he'll say, 'Do this! Do that!' And nothing will happen. Poor Ike—it won't be a bit like the army. He'll find it very frustrating." (Of course, the army isn't like the army either, as Col. Truman surely knew.) Indeed, the whole law of bioethics often seems a whited sepulchre for slaughtered hopes, for its policies have repeatedly fallen woefully short of their purposes. Informed consent is a "fairytale."[69] Programs to increase organ donation have persistently disappointed. Laws regulating DNR orders are hardly better. Legal definitions of brain death are misunderstood by astonishing numbers of doctors and nurses. And so on.[70]

But why don't living wills affect care?[71] Joan Teno and colleagues saw no evidence "that a physician unilaterally decided to ignore or disregard an AD." Rather, there was "a complex interaction of . . . three themes." First (as we have emphasized), "the contents of ADs were vague and difficult to apply to current clinical situations." The imprecision of living wills not only stymies interpreters, it exacerbates their natural tendency to read documents in light of their own preferences. Thus "(e)ven with the therapy-specific AD accompanied by designation of a proxy and prior patient-physician discussion, the proportion of physicians who were willing to withhold therapies was quite variable: cardiopulmonary resuscitation, 100%; administration of artificial nutrition and hydration, 82%; administration of antibiotics, 80%; simple tests, 70%; and administration of pain medication, 13%."[72]

Second, the Teno team found that "patients were not seen as 'absolutely, hopelessly ill,' and thus, it was never considered the time to invoke the AD." Living wills typically operate when patients become terminally ill, but neither doctors nor families lightly conclude patients are dying, especially when that means ending treatment. And understandably. For instance, "on the day before death, the median prognosis for patients with heart failure is still a 50% chance to live 6 more months because patients with heart failure typically die quickly from an unpredictable complication like arrhythmia or infection."[73] So by the time doctors and families finally conclude the patient is dying, the patient's condition is already so dire that treatment looks pointless quite apart from any living will. "In all cases in which life-sustaining treatment was withheld or withdrawn, this decision was made after a trial of life-sustaining treatment and at a time when the patient was seen as 'absolutely, hopelessly ill' or 'actively dying.' Until patients crossed this threshold, ADs were not seen as applicable." Thus "it is not surprising that our previous research has shown that those with ADs did not differ in timing of DNR orders or patterns of resource utilization from those without ADs."[74]

Third, "family members or the surrogate designated in a [durable power of attorney] were not available, were ineffectual, or were overwhelmed with their own concerns and did not effectively advocate for the patient." Family members are crucial surrogates because they should be: patients commonly want them to be; they commonly want to be; they specially cherish the patient's interests. Doctors ordinarily assume families know the patient's situation and preferences and may not relish responsibility for life-and-death decisions, and doctors intent on avoiding litigation may realize that the only plausible plaintiffs are families. The family, however, may not direct attention to the advance directive and may not insist on its enforcement. In fact, surrogates may be guided by either their own treatment preferences or an urgent desire to keep their beloved alive.[75]

In sum, not only are we awash in evidence that the prerequisites for a successful living wills policy are unachievable, but there is direct evidence that living wills regularly fail to have their intended effect. That failure is confirmed by the numerous convincing explanations for it. And if living wills do not affect treatment, they do not work. . . .

What Is To Be Done?

Living wills attempt what undertakers like to call "pre-need planning," and on inspection they are as otiose as the mortuary version. Critically, empiricists cannot show that advance directives affect care. This is damning, but were it our only evidence, perhaps we might not be weary in well doing: for in due season we might reap, if we faint not. However, our survey of the evidence suggests that living wills fail not for want of effort, or education, or intelligence, or good will, but because of stubborn traits of human psychology and persistent features of social organization.

Thus when we reviewed the five conditions for a successful program of living wills, we encountered evidence that not one condition has been achieved or, we think, can be. First, despite the millions of dollars lavished on propaganda, most people do not have living wills. And they often have considered and considerable reasons for their choice. Second, people who sign living wills have generally not thought through its instructions in a way we should want for life-and-death decisions. Nor can we expect people to make thoughtful and stable decisions about so complex a question so far in the future. Third, drafters of living wills have failed to offer people the means to articulate their preferences accurately. And the fault lies primarily not with the drafters; it lies with the inherent impossibility of living wills' task. Fourth, living wills too often do not reach the people actually making decisions for incompetent patients. This is the most remediable of the five problems, but it is remediable only with unsustainable effort and unjustifiable expense. Fifth, living wills seem not to increase the accuracy with which surrogates identify patients' preferences. And the reasons we surveyed when we explained why living wills do not affect patients' care suggest that these problems are insurmountable.

The cost-benefit analysis here is simple: If living wills lack detectable benefits, they cannot justify any cost, much less the considerable costs they now exact. Any attempt to increase their incidence and their availability to surrogates must be expensive. And the evidence suggests that broader use of living wills can actually disserve rather than promote patients' autonomy: If, as we have argued, patients sign living wills without adequate reflection, lack necessary information, and have fluctuating preferences anyway, then living wills will not lead surrogates to make the choices patients would have wanted. Thus, as Pope suggests, the "PSDA, rather than promoting autonomy has 'done a disservice to most real patients and their families and caregivers.' It has promoted the execution of un-informed and under-informed advance directives, and has undermined, not pro-tected, self-determination."[76]

If living wills have failed, we must say so. We must say so to patients. If we believe our declamations about truth-telling, we should frankly warn patients how faint is the chance that living wills can have their intended effect. More broadly, we should abjure programs intended to cajole everyone into signing liv-ing wills. We should also repeal the PSDA, which was passed with arrant and ar-rogant indifference to its effectiveness and its costs and which today imposes accumulating paperwork and administrative expense for paltry rewards.[77]

Of course we recognize the problems presented by the decisions that must be made for incompetent patients, and our counsel is not wholly negative. Pa-tients anxious to control future medical decisions should be told about durable powers of attorney. These surely do not guarantee patients that their wishes will blossom into fact, but nothing does. What matters is that powers of attorney have advantages over living wills. First, the choices that powers of attorney de-mand of patients are relatively few, familiar, and simple. Second, a regime of powers of attorney requires little change from current practice, in which family members ordinarily act informally for incompetent patients. Third, powers of attorney probably improve decisions for patients, since surrogates know more at the time of the decision than patients can know in advance. Fourth, powers of attorney are cheap; they require only a simple form easily filled out with little advice. Fifth, powers of attorney can be supplemented by legislation (already in force in some states) akin to statutes of intestacy. These statutes specify who is to act for incompetent patients who have not specified a surrogate. In short, durable powers of attorney are—as these things go—simple, direct, modest, straightforward, and thrifty.

In social policy as in medicine, plausible notions can turn out to be bad ideas. Bad ideas should be renounced. Bloodletting once seemed plausible, but when it demonstrably failed, the course of wisdom was to abandon it, not to in-sist on its virtues and to scrounge for alternative justifications for it. Living wills were praised and peddled before they were fully developed, much less studied. They have now failed repeated tests of practice. It is time to say, "enough."

References

1. C.P. Sabatino, "End-of-Life Legal Trends," ABA Commission on Legal Problems of the Elderly 2, (2000).
2. R. Dresser, "Recommitment: A Misguided Strategy for securing Death with Dignity," Texas Law Review 81 (2003): 1823–1847.
3. A.R. Eiser and M.D. Weiss, "The Underachieving Advance Directive: Rec-ommendations for Increasing Advance Directive Completion," American Journal of Bioethics 1 (2001): 1–5.

4. N.L. Cantor, "Twenty-five Years after Quinlan: A Review of the jurisprudence of Death and Dying," Journal of Law, Medicine &-Ethics 29 (2001): 182–96.

5. L. Emanuel, "Living Wills Can Help Doctors and Patients Talk about Dying," Western Journal of Medicine 173 (2000): 368.

6. For example, the form provided by a consortium of the American Bar Association, the American Medical Association, and the American Association of Retired Persons "combines and expands the traditional Living Will and Health Care Power of Attorney into a single, comprehensive document" (http://www.ama-assn.org/public/booklets/livgwill.htm).

7. D.M. High, "Why Are Elderly People Not Using Advance Directives?" Journal of Aging and Health 5, no. 4 (1993): 497–515.

8. L.L. Emanuel, "Advance Directives for Medical Care; Reply." NEJM 325 (1991): 1256; N.L. Cantor, "Making Advance Directives Meaningful," Psychology, Public Policy, and Law 4, no. 3 (1998): 629–52; D.M. Cox and G.A. Sachs, "Advance Directives and the Patient Self-Determination Act," Clinics in Geriatric Medicine 10 (1994): 431–43; G.A.D. Havens, "Differences in the Execution/Nonexecution of Advance Directives by Community Dwelling Adults," Research in Nursing and Health 23 (2000): 319–33; D.M. High, "Advance Directives and the Elderly: A Study of Intervention Strategies to Increase Use," Gerontologist 33, no. 3 (1993): 342–49; S.H. Miles, R. Koepp, and E.P. Weber, "Advance End-of-Life Treatment Planning: A Research Review," Archives of Internal Medicine 156, no. 10 (1996): 1062–1068; S.R. Steiber, "Right to Die: Public Balks at Deciding for Others," Hospitals 61 (1987): 572; J. Teno et al., "Do Advance Directives Provide Instructions that Direct Care? SUPPORT Investigators. Study to Understand Prognoses and Preferences for Outcomes and Risks of Treatment," Journal of the American Geriatrics Society 45, no. 4 (1997): 508–512.

9. Emanuel, "Advance Directives for Medical Care; Reply."

10. Miles, Koepp, and Weber, "Advance End-of-Life Treatment Planning"; J.L. Hoiley et al, "Factors Influencing Dialysis Patients' Completion of Advance Directives," American Journal of Kidney Diseases 30, no. 3 (1997): 356–60; L.C. Hanson and E. Rodgman, "The Use of Living Wills at the End of Life: A National Study," Archives of Internal Medicine 156, no. 9 (1996): 1018–1022; J.M. Teno et al., "Do Advance Directives Provide Instructions that Direct Care? SUPPORT Investigators. Study to Understand Prognoses and Preferences for Outcomes and Risks of Treatment," Journal of the American Geriatrics Society 45, no. 4 (1997): 508–512.

11. Holley et al., "Factors Influencing Dialysis Patients' Completion of Advance Directives."

12. Cox and Sachs, "Advance Directives and the Patient Self-Determination Act"; Miles, Koepp, and Weber, "Advance End-of-Life Treatment Planning"; D.M. High, "All in the Family: Extended Autonomy and Expecta-

tions in Surrogate Health Care Decision-Making," Gerontologist 28 (suppl) (1988): 46–51.

13. L.L. Emanuel and E.J. Emanuel, "The Medical Directive: A New Comprehensive Advance Care Document," JAMA 261 (1989): 3288–93.

14. High, "Advance Directives and the Elderly"; J.M. Roe et al., "Durable Power of Attorney for Health Care: A Survey of Senior Center Participants," Archives of Internal Medicine 152 (1992): 292–96.

15. High, "Why Are Elderly People Not Using Advance Directives?"; Roe et al., "Durable Power of Attorney for Health Care."

16. High, "Why Are Elderly People Not Using Advance Directives?"; Roe et al., "Durable Power of Attorney for Health care"; E.J. Emanuel, L.L. Emanuel, and D. Orentlicher, "Advance Directives," JAMA 266 (1991): 2563–63; G.A. Sachs, C.B. Stocking, and S.H. Miles, "Empowerment of the older patient? A Randomized, Controlled Trial to Increase Discussion and Use of Advance Directives," Journal of the American Geriatrics Society 40, no. 3 (1992): 269–73; L.L. Brunetti, S.D. Carperos, and R.E. Westlund, "Physicians' Attitudes towards Living Wills and Cardiopulmonary Resuscitation," Journal of General Internal Medicine 6 (1991): 323–29; T.E. Finucane et al, "Planning with Elderly Outpatients for Contingencies of Severe Illness: A Survey and Clinical Trial," Journal of General Internal Medicine 3, no. 4 (1988): 322–25; B. Lo, G.A. McLeod, and G. Saika, "Patient Attitudes to Discussing Life-sustaining Treatment," Archives of Internal Medicine 146, no. 8 (1986): 1613–15; R. Yamada et al., "A Multimedia Intervention on Cardiopulmonary Resuscitation and Advance Directives," Journal of General Internal Medicine 14 (1999): 559–63.

17. Cox and Sachs, "Advance Directives and the Patient Self-Determination Act"; L.L. Emanuel and E. Emanuel, "Advance Directives," Annals of Internal Medicine 116 (1992): 348–49; B.B. Ott, "Advance Directives: The Emerging Body of Research," American Journal of Critical Care 8 (1999): 514–19.

18. J. Sugarman, M. Weinberger, and G. Samsa, "Factors Associated with Veterans' Decisions about Living Wills," Archives of Internal Medicine 152 (1992): 343–47.

19. Cox and Sachs, "Advance Directives and the Patient Self-Determination Act"; Holley et al., "Factors Influencing Dialysis Patients' Completion of Advance Directives," High, "All in the Family"; Roe et al, "Durable Power of Attorney for Health Care"; Ott, "Advance Directives"; N.A. Hawkins et al., "Do Patients Want to Micro-manage Their Own Deaths? Process Preferences, Values and Goals in End-of-Life Medical Decision Making," Unpublished manuscript. R.B. Terry et al., "End-of-Life Decision Making: When Patients and Surrogates Disagree," Journal of Clinical Ethics 10, no. 4 (1999): 286–93.

20. J. Carrese and L. Rhodes, "Western Bioethics on the Navajo Reservation: Benefit or Harm?" JAMA 274 (1995): 826–29; L.J. Blackball et al., "Ethnicity and Attitudes toward Patient Autonomy," JAMA 274 (1995): 820–25.

21. C.M. Puchalski et al., Patients Who Want their Family and Physician to Make Resuscitation Decisions for Them: Observations from SUPPORT and HELP; JAGS 48 (2000): S84.
22. High, "Why Are Elderly People Not Using Advance Directives?"
23. Ibid.
24. Patient Self-Determination Act of 1990 of the Omnibus Reconciliation Act of 1990.
25. J.L. Yates and H.R. Click, "The Failed Patient Self-Determination Act and Policy Alternatives for the Right to Die," Journal of Aging and Social Policy 29 (1997): 29, 31.
26. M.T. Pope, "The Maladaptation of Miranda to Advance Directives: A Critique of the Implementation of the Patient Self-Determination Act," Health Matrix 9 (1999): 139.
27. C.E. Schneider, The Practice of Autonomy: Patients, Doctors, and Medical Decisions (New York: Oxford University Press, 1998).
28. Yamada et al., "A Multimedia Intervention on Cardiopulmonary Resuscitation and Advance Directives"; S.H. Miles, "Advanced Directives to Limit Treatment: The Need for Portability," Journal of the American Geriatrics Society 35, no. 1 (1987): 7476; K.M. Coppola et al., "Perceived Benefits and Burdens of Life-Sustaining Treatments: Differences among Elderly Adults, Physicians, and Young Adults," Journal of Ethics, Law, and Aging 4, no. 1 (1998): 3–13.
29. Roe et al., "Durable Power of Attorney for Health Care."
30. J.A. Tulsky et al., "Opening the Black Box: How Do Physicians Communicate about Advance Directives?" Annals of Internal Medicine 129 (1998): 441, 444.
31. B.J. McNeil et al., "On the Elicitation of Preferences for Alternative Therapies," NEJM 306 (1982): 1259–62.
32. T.R. Malloy et al., "The Influence of Treatment Descriptions on Advance Medical Directive Decisions," Journal of the American Geriatrics Society 40, no. 12 (1992): 1255–60; D.J. Mazur and D.H. Hickman, "Patient Preferences: Survival versus Quality-of-Life Considerations," Journal of General Internal Medicine 8, no. 7 (1993): 374–77; D.J. Mazur and J.F. Merz, "How Age, Outcome Severity, and Scale Influence General Medicine Clinic Patients' Interpretations of Verbal Probability Terms" (See comments), Journal of General Internal Medicine 9 (1994): 268–71.
33. Miles, Koepp, and Weber, "Advance End-of-Life Treatment Planning."
34. Ott, "Advance Directives." pp. 514, 517.
35. J.J. Christenscn-Szalanski, "Discount Functions and the Measurement of Patients' Values: Women's Decisions during Childbirth," Medical Decision Making 4, no. 1 (1984): 47–58.
36. R.M. Gready et al., "Actual and Perceived Stability of Preferences for Life-Sustaining Treatment" Journal of Clinical Ethics 11, no. 4 (2000): 334–46.
37. Gready et al., "Actual and Perceived Stability of Preferences for Life-Sustaining Treatment"; J.T. Berger and D. Majerovitz, "Stability of Prefer-

ences for Treatment among Nursing Home Residents," Gerontologist 28, no. 2 (1998): 217–23; S. Carmel and E. Mutran, "Stability of Elderly Persons' Expressed Preferences regarding the Use of Life-Sustaining Treatments," Social Science and Medicine 49, no. 3 (1999): 303–311; M. Danis et al., "Stability of Choices about Life-Sustaining Treatments," Annals of Internal Medicine 120, no. 7 (1994): 56773; R.H. Ditto et al., "A Prospective Study of the Effects of Hospitalization on Life-Sustaining Treatment Preferences: Context Changes Choices," Unpublished manuscript; E.H. Ditto et al, "The Stability of Older Adults' Preferences for Life-Sustaining Medical Treatment," Unpublished manuscript; E.J. Emanuel, "Commentary on Discussions about Life-Sustaining Treatments," Journal of Clinical Ethics 5, no. 3 (1994): 250–51; L.L. Emanuel et al., "Advance Directives: Stability of Patients' Treatment Choices," Archives of Internal Medicine 154 (1994): 209–217; M.A. Everhart and R.A. Pearlman, "Stability of Patient Preferences regarding Life-Sustaining Treatments," GW 97 (1990): 159–64; L. Ganzini et al., "The Effect of Depression Treatment on Elderly Patients' Preferences for Life-Sustaining Medical Therapy," American Journal of Psychiatry 151, no. 11 (1994): 1631–36; N. Kohut et al., "Stability of Treatment Preferences: Although Most Preferences Do Not Change, Most People Change Some of their Preferences," Journal of Clinical Ethics 8, no. 2 (1997): 124–35; M.D. Silverstein et al., "Amyotrophic Lateral Sclerosis and Life-Sustaining Therapy: Patients' Desires for Information, Participation in Decision Making, and Life-Sustaining Therapy," Mayo Clinic Proceedings 66 (1991): 906–913; J.S. Weissman et al., "The Stability of Preferences for Life-Sustaining Care among Persons with AIDS in the Boston Health Study," Medical Decision Making 19 (1999): 16–26; K.M. Coppola et al., "Are Life-Sustaining Treatment Preferences Stable over Time? An Analysis of the Literature," unpublished manuscript.

38. Coppola et al., "Are Life-Sustaining Treatment Preferences Stable over Time?"

39. Danis et al., "Stability of Choices about Life-Sustaining Treatments"; Ditto et al., "A Prospective Study of the Effects of Hospitalization"; Weissman et al., "The Stability of Preferences for Life-Sustaining Care."

40. Ditto et al., "A Prospective Study of the Effects of Hospitalization."

41. H.M. Chochinov et al., "Will to Live in the Terminally 111," Lancet 354 (1999): 816, 818.

42. D.T. Gilbert and T.D. Wilson, "Miswanting: Some Problems in the Forecasting of Future Affective States," in Feeling and Thinking: The Role of Affect in Social Cognition, ed. J.P. Forgas (New York: Cambridge University Press, 2000): 178–97; C.H. Griffith 3rd et al., "Knowledge and Experience with Alzheimer's Disease: Relationship to Resuscitation Preference," Archives of Family Medicine 4, no. 9 (1995): 780–84; T.M. Osbergand, J.S. Shrauger, "Self-prediction: Exploring the Parameters of Accuracy," Journal of Personality and Social Psychology 51, no. 5 (1986): 1044–57.

43. Griffith 3rd et al., "Knowledge and Experience with Alzheimer's Disease."
44. Gilbert and Wilson, "Miswanting."
45. D. Kahneman and J. Snell, "Predicting a Changing Taste: Do People Know What They Will Like?" Journal of Behavioral Decision Making 1, no. 3 (1992): 187–200.
46. Gilbert and Wilson, "Miswanting."
47. Ibid.
48. G. Loewenstein and D. Schkade, "Wouldn't It be Nice? Predicting Future Feelings," in Hedonic Psychology: Scientific Approaches to Enjoyment, Suffering and Well-being, ed. N. Schwartz and D. Kahneman (New York: Russell Sage Foundation, 1997).
49. D. Schkade, "Does Living in California Make People Happy? A Focusing Illusion in Judgments of Life Satisfaction," Psychological Science 9 (1998): 340–46.
50. Loewenstein and Schkade, "Wouldn't It be Nice?"
51. A.S. Brett, "Limitations of Listing Specific Medical Interventions in Advance Directives," JAMA 266 (1991): 825–28.
52. I.S. Kirsch et al., Adult Literacy in America: A First Look at the Results of the National Adult Literacy Survey, U.S. Department of Education; August 1993; NCES 93275.
53. Cox and Sachs, "Advance Directives and the Patient Self-Determination Act"; Miles, Koepp, and Weber, "Advance End-of-Life Treatment Planning"; Silveira et al., "Patient's Knowledge of Options at the End of Life"; Coppola et al., "Perceived Benefits and Burdens of Life-Sustaining Treatments."
54. Pope, "The Maladaptation of Miranda to Advance Directives." pp. 139, 165–66.
55. D.J. Doukas and L.B. McCullough, "The Values History: The Evaluation of the Patient's Values and Advance Directives," Journal of Family Practice 32, no. 2 (1991): 145–53.
56. Lochner v. New York N. 198 U.S. 45: Supreme Court of the United States; 1905.
57. H.J. Silverman et al., "Implementation of the Patient Self-Determination Act in a Hospital Setting: An Initial Evaluation," Archives of Internal Medicine 155, no. 5 (1995): 502–510.
58. Roe et al., "Durable Power of Attorney for Health Care."
59. R.S. Morrison et al., "The Inaccessibility of Advance Directives on Transfer from Ambulatory to Acute Care Settings," JAMA 274 (1995): keep together 478–82.
60. Ibid.
61. M. Danis et al., "A Prospective Study of the Impact of Patient Preferences on Life-Sustaining Treatment and Hospital Cost," Critical Care Medicine 24, 11 (1996): 181117.
62. J.A. Druley et al., "Physicians' Predictions of Elderly Outpatients' Preferences for Life-Sustaining Treatment," Journal of Family Practice 37 (1993):

469–75; J. Hare, C. Pratt, and C. Nelson, "Agreement between Patients and Their Self-Selected Surrogates on Difficult Medical Decisions," Archives of Internal Medicine 152, no. 5 (1992): 1049–1054; P.M. Layde et al, "Surrogates' Predictions of Seriously 111 Patients' Resuscitation Preferences," Archives of Family Medicine 4, no. 6 (1995): 518–23; J.G. Ouslander, A.J. Tymchuk, and B. Rahbar, "Health Care Decisions among Elderly Long-term Care Residents and Their Potential Proxies," Archives of Internal Medicine 149 no. 6 (1989): 1367–72; A.B. Seckler et al., "Substituted judgment: How Accurate Are Proxy Predictions?" Annals of Internal Medicine 115 (1991): 92–98; O.P. Sulmasy et al., "The Accuracy of Substituted judgments in Patients with Terminal Diagnoses," Annals of Internal Medicine 128, no. 8 (1998): 621–29; R.F. Uhlmann, R.A. Pearlman, and K.C. Cain, "Physicians' and Spouses' Predictions of Elderly Patients' Resuscitation Preferences," Journal of Gerontology 43, no. 5 (1988): M115–M121; R.F. Uhlmann, R.A. Pearlman, and K.C. Cain, "Understanding of Elderly Patients' Resuscitation Preferences by Physicians and Nurses," Western Journal of Medicine 150 (1989): 705–707; N.R. Zweibel and C.K. Cassell, "Treatment Choices at the End of Life: A Comparison of Decisions by Older Patients and Their Physician-Selected Proxies," Gerontologist 29, no. 5 (1989): 615–21.

63. L. Emanuel, "The Health Care Directive: Learning How to Draft Advance Care Documents," Journal of the American Geriatrics Society 39, no. 12 (1991): 1221–28; P.H. Ditto et al., "Fates Worse than Death: The Role of Valued Life Activities in Health-State Evaluations," Health Psychology 15, no. 5 (1996): 332–43.

64. K.M. Coppola et al., "Accuracy of Primary Care and Hospital-based Physicians' Predictions of Elderly Outpatients' Treatment Preferences with and without Advance Directives," Archives of Internal Medicine 161, no. 3 (2001): 431–40.

65. Ibid.

66. M.D. Goodman, M. Tarnoff, and G.J. Slotman, "Effect of Advance Directives on the Management of Elderly Critically 111 Patients," Critical Care Medicine 26, no. 4 (1998): 701–704.

67. Morrison et al., "The Inaccessibility of Advance Directives."

68. M. Danis and J.M. Garrett, "Advance Directives for Medical Care: Reply," NEJM 325 (1991).

69. J. Katz, "Informed Consent—A Fairy Tale? Law's Vision," University of Pittsburgh Law Review 39, no. 2 (1977): 137–74; C.H. Braddock 3rd et al., "Informed Decision Making in Outpatient Practice: Time to Get Back to Basics," JAMA 282, no. 24 (1999): 2313–20.

70. C.E. Schneider, "The Best-Laid Plans," Hastings Center Report 30, no. 4 (2000): 24–25; C.E. Schneider, "Gang Aft Agley," Hastings Center Report 31, no. 1 (2001): 27–28.

71. Teno et al., "Do Advance Directives Provide Instructions that Direct Care?"

72. W.R. Mower and L.J. Baraff, "Advance Directives: Effect of Type of Directive on Physicians' Therapeutic Decisions," Archives of Internal Medicine 153 (1993): 375, 378.

73. J. Lynn, "Learning to Care for People with Chronic Illness Facing the End of Life," JAMA 284 (2000): 2508–09.

74. J. Teno et al., "The Illusion of End-of-Life Resource Savings with Advance Directives. SUPPORT Investigators Study to Understand Prognoses and Preferences for Outcomes and Risks of Treatment," Journal of the American Geriatrics Society 45, no. 4 (1997): 513–18.

75. A. Fagerlin et al., "Projection in Surrogate Decisions about Life-Sustaining Medical Treatments," Health Psychology 20, no. 3 (2001): 166–75.

76. Pope, "The Maladaptation of Miranda to Advance Directives." pp. 139, 167.

77. Yates and Click, "The Failed Patient Self-Determination Act"; Sugarman et al., "The Cost of Ethics Legislation."

CARE OF THE DYING

Sedation as a Therapy of Last Resort

National Ethics Committee, Veterans Health Administration

The National Ethics Committee is a Veterans Health Administration advisory committee that issues "white papers." These reports are disseminated throughout the VA health care system and often serve as standards in ethical deliberations. In this report the National Ethics Committee addresses palliative sedation as a therapy of last resort. The report concludes that palliative sedation is meaningfully different than physician-assisted death and euthanasia; should be restricted to patients who are imminently dying; should not require that a patient forgo life-sustaining treatment; should not be used for existential suffering in the absence of severe, refractory clinical symptoms; and may be provided to a patient who lacks decision-making capacity with the informed consent of the authorized surrogate decision maker.

For most patients nearing the end of life, there comes a point at which the goals of care evolve from an emphasis on prolonging life and optimizing function to maximizing the quality of remaining life, and palliative care becomes the priority. Providing adequate relief of symptoms for dying patients is one of the hallmarks of good palliative care.[1] Yet for some patients, even aggressive, high-quality palliative care fails to provide relief. For patients who suffer from severe pain, dyspnea, vomiting, or other symptoms that prove refractory to treatment, there is consensus that deep sedation—so-called palliative sedation—is an appropriate intervention of last resort.[2-5] The National Hospice and Palliative Care Organization[6] and the American Academy of Hospice and Palliative Medicine[7] support the use of sedation to treat otherwise unrelievable suffering at the end of life, and the practice has been endorsed by the End-of-Life Care Consensus Panel of the American College of Physicians (ACP)–American Society of Internal Medicine (ASIM),[8] and the American Medical Association.[9]

This report by the National Ethics Committee (NEC), Veterans Health Administration (VHA) examines what is meant by palliative sedation, explores ethical concerns about the practice, reviews the emerging professional consensus regarding the use of palliative sedation for managing severe, refractory symptoms at the end of life, and offers recommendations for ethical practice within the VHA.

What Do We Mean by "Palliative Sedation"?

The literature describes several uses of sedation as a palliative intervention at the end of life, variously referred to as *total, palliative,* or *terminal* sedation.[3,6,8,10–13] Broadly, the practice involves "sedating a patient to the point of unconsciousness to relieve one or more symptoms that are intractable and unrelieved despite aggressive symptom-specific treatments, and maintaining that condition until the patient dies."[14] The intent, thus, is to provide symptom relief for a dying patient when all other efforts have failed.

Palliative sedation is distinct from sedation that normally accompanies therapeutic interventions, such as intubation or treatment of severe burns, when recovery is expected or more likely to occur.[15–17] Intentionally sedating the patient as a palliative intervention is also distinct from the unintended and variable sedative effects of medications administered for pain relief.[18,19] Some scholars and practitioners further distinguish palliative sedation from "respite sedation" for terminally ill patients; that is, time-limited therapy (e.g., 24 to 48 hours) offered in the hope that temporary sedation will break a cycle of pain, anxiety, and distress.[17]

For purposes of this analysis, the NEC defines palliative sedation as:

> The administration of nonopioid drugs to sedate a terminally ill patient to unconsciousness as an intervention of last resort to treat severe, refractory pain or other clinical symptoms that have not been relieved by aggressive, symptom-specific palliation.

There is broad professional agreement that palliative sedation is a clinically and ethically appropriate response when patients who are near death suffer severe, unremitting symptoms.[7,8,10,11,19–22] The following algorithm has been proposed to help clinicians determine when a symptom is truly refractory:

1. Are further interventions capable of providing further relief?
2. Is the anticipated acute or chronic morbidity of the intervention tolerable to the patient?
3. Are the interventions likely to provide relief within a tolerable time frame?[15,23]

If the answer to any of these three questions is "no," then these are refractory symptoms for which palliative sedation may be considered.

Palliative sedation is provided for a wide range of symptoms. One recent review of published studies, for example, found that the primary indications for this intervention included pain, nausea and vomiting, shortness of breath, and agitated delirium.[13,20,24–26] Other indications for which palliative sedation has been reported include urinary retention due to clot formation,[24] gastrointestinal pain, uncontrolled bleeding,[13] and myoclonus.[27] Many also support palliative sedation to relieve severe psychologic distress in a dying patient,[15,20,28] with the important caveat that potentially treatable mental health conditions first be ruled out.[29]

Ethical Concerns About Palliative Sedation

Ethical debate about palliative sedation has been framed largely in terms of five key questions:

1. Is palliative sedation ethically different from physician-assisted suicide and euthanasia?
2. Is palliative sedation ever ethically appropriate for patients who are not imminently dying?
3. Should willingness to forgo life-sustaining treatment be a condition for administering/receiving palliative sedation?
4. Is palliative sedation an ethically appropriate response to "existential" suffering?
5. May palliative sedation be provided to patients who lack decision-making capacity?

1. Is Palliative Sedation Different From Physician-Assisted Suicide and Euthanasia?

Palliative sedation has been widely discussed in the context of debates about physician-assisted suicide and euthanasia.[11,30,31] Indeed, palliative sedation has been proposed as an ethically acceptable alternative to physician aid in dying.[10,11] Yet despite considerable attention to these questions during the past decade, many clinicians remain uncertain or confused about the ethical differences among these practices.[27]

Although debate continues in some quarters, the dominant view in the professional medical and bioethics communities holds that palliative sedation is ethically different from physician-assisted suicide or euthanasia. These analyses focus on intention and proportionality.[8,10,11,31,32] With respect to intention, the primary intention in both physician-assisted suicide and euthanasia is to cause the patient's death; the patient's suffering ends as a result. In contrast, the primary intention in palliative sedation is to relieve the patient's suffering; death occurs as a result of the underlying disease process. (Note that because death occurs as a result of the disease process, palliative sedation shares a critical feature with established ethically accepted practice of forgoing life-sustaining treatment.) Medication is used only in sufficient doses to achieve unconsciousness (not a lethal dose).[10] The limited evidence currently available suggests that deep sedation is unlikely to hasten death.[19,20,33,34] In response to concerns that it is difficult to assess practitioners' intentions objectively,[35] it has been argued that those intentions can be evaluated indirectly in a general way; for example, by observing practitioners' choice and usage of sedating medications.[4]

Proportionality is a second ethically significant factor in distinguishing palliative sedation from physician-assisted suicide and euthanasia. In medicine, the principle of proportionality requires that "the risk of causing harm must bear a direct relationship to the danger and immediacy of the patient's clinical situation and the expected benefit of the intervention."[11] Practitioners are permitted to perform, and patients to undergo, treatments and procedures that carry grave risks when there are commensurate benefits to be gained. Think of the example of surgery for a patient who is seriously injured in a car accident: Administering general anesthesia carries a foreseeable risk of death. Yet the good intended—for example, saving the patient's leg or minimizing brain damage—is usually held to be significant enough to justify taking a substantial risk to obtain it. In palliative sedation, although the means—deep, continuous sedation for a dying patient—are grave, they are proportional to the goal to be achieved: relieving severe, unremitting suffering when all other interventions acceptable to the patient have failed.[8,11,15,36,37]

The distinction between palliative sedation and either physician-assisted suicide or euthanasia recognized in the emerging medical and ethical consensus is also supported in case law. In its 1997 decisions in *Vacco v. Quill* and *Washington v. Glucksberg*,[38] 2 cases that dealt with physician-assisted suicide, the U.S. Supreme Court seemed to distinguish palliative sedation from assisted suicide as legally acceptable practice.[21,39] The court did not explicitly address palliative sedation as such, but did indicate strong support for aggressive symptom relief for dying patients, even to the point of rendering the patient unconscious.

The Committee concludes that there is a meaningful difference between palliative sedation and physician-assisted suicide or euthanasia.

2. Is Palliative Sedation Ever Ethically Appropriate for Patients Who Are Not Imminently Dying?

The professional community is also divided about whether palliative sedation is ethically appropriate for a patient who experiences intolerable, irremediable suffering but who is not imminently dying. If palliative sedation is an ethically appropriate response to severe, intractable suffering, the argument goes, why should it be available only to patients who are on the verge of death? To withhold palliative sedation from patients whose symptoms are severe and refractory to aggressive care solely because they are not expected to die very soon, or because their condition makes it extremely difficult to predict likely time to death with any confidence, imposes an arbitrary constraint and condemns these individuals to endure unrelieved suffering for a potentially long period of time.[6,8,30]

We recognize the ethical salience of this position. In our judgment, however, more compelling concerns are raised by the prospect of permitting palliative sedation for a patient who is expected to survive for months or years. Sedating a patient to unconsciousness carries significant risks, and palliative sedation is understood to be literally an intervention of *last* resort at the end of life.

Allowing palliative sedation when the patient can reasonably be expected to live for months (or longer) risks eroding the distinction between palliative sedation and physician-assisted suicide or euthanasia. Sedating such a patient to relieve suffering while respecting his or her right to forgo artificially administered nutrition and hydration or other indicated life-sustaining treatment will directly and predictably shorten the patient's life, a result clearly contrary to the goal of palliative sedation.[40]

Providing palliative sedation to patients who are not imminently dying also raises slippery slope concerns. Palliative sedation is generally considered appropriate only for patients who are terminally ill, but if not at the threshold of imminent death, at what other point in the trajectory of terminal illness can we draw a sufficiently bright line to distinguish when palliative sedation is and when it is not ethically permissible? Moreover, accepting "terminally ill" alone as a sufficient criterion for palliative sedation instead of the more restrictive "imminently dying" may increase the risk that the practice would some day be extended to individuals who are not terminally ill.[41]

Furthermore, intentionally sedating a patient and maintaining continuous deep sedation while providing life-sustaining treatment for an indefinite, but possibly prolonged span poses its own challenges. Such scenarios are likely to be emotionally distressing for the patient's intimates, and indeed, for staff.[42]

A further concern, originally raised in reference to physician-assisted suicide,[43] may also be cogent with respect to palliative sedation for patients who are not imminently dying; that is, that deep sedation will come to be seen as an alternative to providing appropriate palliative care. High-quality palliative care is an essential condition for the ethical practice of palliative sedation.

We recognize that it is not possible to predict with certainty how long a patient will live. Patients with terminal cancer follow a relatively predictable course to death,[44] but even for these patients, physicians' predictions about the timing of death are not very accurate.[45,46] For patients with other types of life-limiting illness, such as end-stage lung or heart disease, prognostication is even more challenging.[44] Ultimately, the determination that a patient has entered the final phase of dying rests not on precise predictions of survival but on well-considered, informed professional judgment, which argues for the involvement of practitioners with appropriate expertise, including palliative care specialists, in decision-making about palliative sedation.

The Committee concludes that it is ethically appropriate to restrict palliative sedation to patients who are imminently dying.

3. Should Willingness to Forgo Life-Sustaining Treatment Be a Condition for Administering and Receiving Palliative Sedation?

Professional consensus about best practice for palliative sedation clearly establishes that patients who do not have a do-not-resuscitate (DNR) order should not be considered appropriate candidates for palliative sedation.[12,15,16,20,22,23,47–49] The Committee believes this is an appropriate standard consistent with the overall goals of palliative sedation. However, debate continues within the medical community about whether it is ethically appropriate to provide other life-sustaining interventions, such as ventilator support, dialysis, or artificially administered nutrition and hydration to patients who receive palliative sedation.[30,31,40,49]

For most patients who are appropriate candidates for palliative sedation, the question of life-sustaining treatment is not likely to arise. These are patients near death, for whom the overriding goal of care is no longer to optimize function or prolong survival but to provide comfort and symptom relief. Most such patients will already have stopped eating and drinking.[19,33] As a practical matter, most patients who are candidates for palliative sedation will have already decided to forgo all life-sustaining interventions. When this is not the case, the decision to forgo life-sustaining treatment should be clearly distinguished from and made independently of the decision to provide palliative sedation.[50]

Some dying patients who are appropriate candidates for palliative sedation will, however, want *both* palliative sedation *and* life-sustaining treatment. For these patients, the goal of care is twofold: to relieve suffering *and* to prolong life. Many cultural and religious traditions place high moral value on prolonging life and practitioners have a prima facie obligation to respect these views, an obligation that also resonates with core values of medicine as a profession.

Consensus in the professional community is that candidates for palliative sedation should have a DNR order. However, we find no compelling argument to limit other concurrent life-sustaining interventions, such as artificially administered nutrition and hydration or ventilator support, for patients who receive palliative sedation, so long as those interventions are clinically indicated. To require that a patient consent to forgo *all* life-sustaining treatments as a condition for receiving the only intervention that will relieve the patient's intolerable suffering (i.e., palliative sedation) seems to us ethically and professionally insupportable.

We recognize that views are divided on the question. Most members of the Committee would argue that first and foremost, continuing to provide life-sustaining treatment to a patient who receives palliative sedation and who wants life-sustaining treatment(s) other than cardiopulmonary resuscitation upholds the value of respect for patients as moral agents and autonomous decision makers. Some members, however, see it as unnecessarily prolonging dying, a view we realize others may share. We acknowledge that for both family members and health care professionals who hold this latter view, providing life-sustaining treatment concurrent with palliative sedation may create significant distress.

These considerations carry significant implications for decision making about palliative sedation. Practitioners have an obligation to describe as clearly as possible the likely clinical scenarios for a patient who is considering palliative sedation, and should work with patients and families to establish a clear plan of care before sedation is initiated. This should include discussion of what life-sustaining treatments will be withdrawn, continued, or initiated (if clinically indicated) after the patient has been sedated. This will help patients, their surrogates, other family members, and, indeed, the treatment team understand what is expected to happen once the patient has been sedated and better prepare them for the decisions that may need to be made when the patient is no longer conscious.

As potential sources of conflict, diverging views on the question of life-sustaining treatments for patients who receive palliative sedation also highlight the importance of assuring that appropriate mechanisms are in place to assist stakeholders in resolving disagreements if they arise, including ethics consultation.

The Committee concludes that willingness to forgo life-sustaining treatment should not be a condition for the administration of palliative sedation.

4. Is Palliative Sedation Ethically Appropriate When Suffering Is "Existential"?

One of the most deeply contested questions about palliative sedation is whether the practice is ethically appropriate as a response to existential suffering, as distinct from pain or other clinically defined physical or psychiatric symptoms. The debate about existential suffering has evolved around three basic concerns: (1) the difficulty of clearly defining existential suffering and of distinguishing it clinically from treatable psychiatric conditions such as depression; (2) whether relief of existential suffering represents a "proportionate" goal; and (3) whether relief of existential suffering as such is within the goals of medicine, and thus whether providing a pharmacologic intervention for such suffering is appropriate for health care professionals.

Distinguishing existential suffering from psychologic distress. One difficulty is that there is no single, agreed on definition of existential suffering that is sufficiently clear and concrete to offer guidance in clinical contexts. "Making a diagnosis of suffering," it has been argued, "differs from the usual diagnostic process that internists are familiar with because suffering is an affliction of the person, not the body."[51] The suffering experienced by patients near death may reflect concerns about a prolonged dying process, retaining control, the burden their dying imposes on others, and strengthening personal relationships.[51]

It can, moreover, be extremely difficult to draw bright lines among physical, psychologic, and existential suffering.[29,50] Psychologic distress often contributes to pain, dyspnea, and other symptoms, for example, as well as the reverse. Nor is it always easy for practitioners to determine with confidence whether a patient's distress represents a normal, "appropriate" reaction to the prospect of impending death or indicates the presence of a potentially treatable mental health condition.[52] It is even more challenging to assess whether the patient's distress reflects the kind of response to the irremediable losses imposed by illness and assaults to the sense of self that we would call existential suffering.

"Proportionality" and relief of existential suffering. A further concern can be framed as the following question: Is the goal of relieving severe, refractory existential suffering sufficiently grave or "proportionate" to justify sedating the patient into unconsciousness for the time remaining to him or her? Some answer in the affirmative, arguing that existential suffering "can be just as distressful and refractory as physical suffering,"[36] but acknowledge that practitioners may find it difficult to consider palliative sedation when a patient's existential suffering is not associated with significant physiologic deterioration.[29,47] Opponents of palliative sedation for existential suffering argue that permitting practitioners to make necessarily subjective judgments about the existential well-being of patients risks placing health care professionals and patients on a slippery slope at the bottom of which lies abuse of palliative sedation and danger to patients.[37]

Relief of existential suffering and the goals of medicine. Undeniably, for some patients, suffering at the end of life cannot be attributed solely or primarily to refractory clinical symptoms.[53] But although relieving suffering is one of the core goals of medicine,[51,54] questions have been raised about whether attempting to relieve existential suffering through a specifically clinical intervention, such as palliative sedation, is an appropriate activity for health care professionals.[55] Essentially the same concern has been raised with respect to physician-assisted suicide and euthanasia. As the ACP-ASIM noted in its position statement opposing physician-assisted suicide, "one can raise serious questions about whether medicine should arrogate to itself the task of alleviating all human suffering, even at the end of life."[56]

Despite these concerns, there is some degree of support for palliative sedation in response to existential suffering within the professional hospice and palliative care community in the United States. For example, the Hospice and Palliative Care Federation of Massachusetts has provided guidelines for providers, although it has not formally endorsed palliative sedation,[49] and the National Hospice and Palliative Care Organization has cautiously supported the practice in principle.[6]

These are challenging issues on which the Committee finds that as individuals we do not share a uniform perspective. This lack of consensus within the Committee itself recommends to us the wisdom of taking a conservative stance with respect to palliative sedation for existential suffering. Further, in our view, VA's mission and its unique patient population create a special risk that permitting VA practitioners to offer palliative sedation when the patient's suffering cannot be defined in reference to clinical criteria could erode public trust in the agency; therefore, as a committee, we do not endorse this practice. We acknowledge that restricting the availability of palliative sedation in this way may fail to address the needs of some patients whose suffering cannot be relieved by other means. We commend the commitment of health care professionals and other staff throughout VHA to provide open, empathic support even as clinical interventions fall short of alleviating the individual's suffering. We find the conclusion reached by the ACP-ASIM in its position paper on physician-assisted suicide cogent in our context:

> [W]hen the patient's suffering is interpersonal, existential, or spiritual, the tasks of the physician are to remain present, to "suffer with" the patient in compassion, and to enlist the support of clergy, social workers, family, and friends in healing the aspects of suffering that are beyond the legitimate scope of medical care.[56]

The Committee concludes that palliative sedation should not be used to treat existential suffering in the absence of severe, refractory clinical symptoms.

5. May Palliative Sedation Be Provided to Patients Who Lack Decision-Making Capacity?

Because the decision to sedate a patient to unconsciousness and maintain that state until he or she dies is a serious one, some might argue that palliative sedation should be considered only for patients who can consent to it themselves. However, confining palliative sedation to patients who have decision-making capacity risks excluding many patients whose suffering cannot be relieved by other means for whom surrogates are empowered to make all other treatment decisions. Indeed, many patients for whom palliative sedation would be considered will already have lost the capacity to participate in shared decision making because of the progression of their underlying condition, the effects of treatment, or unmanageable symptoms. To deny a patient's surrogate the possibility of consenting to palliative sedation undermines the surrogate's role in shared decision making and in effect undermines the patient's right to choose this intervention.

The Committee concludes that palliative sedation may be provided to patients who lack decision-making capacity with the informed consent of the authorized surrogate decision maker.

Conditions for Ethically Sound Practice

At various points throughout the foregoing discussion we have noted the important role in palliative sedation of professionals from multiple disciplines. We have also stressed that providing high quality palliative care is a prerequisite to decisions about palliative sedation. These fundamental conditions for ethically appropriate practice of palliative sedation are well recognized in the professional community.[11,12,15,19,22,23,47,48] Consultation with practitioners expert in pain and symptom management is essential to assure that a patient's symptoms truly are refractory before palliative sedation is considered,[11,12,15,22,23,48] and to initiate and monitor sedation.[12] Likewise patients must be assured access to expert psychologic and spiritual assessment and support.[11,15,23,47,48,57]

The decision to sedate a dying patient to unconsciousness for the duration of his or her life should be made only after careful clinical evaluation and thoughtful deliberation, and must be implemented with appropriate monitoring and supervision.

Recommendations

Although debate continues about how broadly to define the range of circumstances in which palliative sedation is appropriate, the emerging professional and ethical consensus is clear: Palliative sedation is an ethically appropriate therapy of last resort for patients who are experiencing severe, unremitting, refractory clinical symptoms at the end of life. The National Ethics Committee therefore recommends that VA adopt policy that:

1. Permits the administration of palliative sedation (by definition, as a last resort) only:
   (a) when severe pain or other clinical symptoms (e.g., dyspnea, nausea and vomiting, agitated delirium) is/are not ameliorated by aggressive symptom-specific interventions that are tolerable to the patient;
   (b) for patients who have entered the final stages of the dying process and who have a DNR order;

(c) with the signed informed consent of the patient, or surrogate if the patient lacks decision-making capacity, as required by VA policy for treatments or procedures involving general anesthesia.[58]

2. Establishes safeguards to protect patients' interests and assure consistent, high quality care by:

  (a) providing for consultation with experts in palliative medicine, psychiatry or clinical psychology, and spiritual care as appropriate in the decision-making process;

  (b) clarifying with the patient and/or surrogate the plan of care regarding:

   (i) concurrent life-sustaining treatment (including, but not limited to, artificially administered nutrition and hydration),

   (ii) regular assessment of the patient's clinical status and ongoing eligibility for palliative sedation, and

   (iii) health care professionals' obligation to discontinue deep sedation in the event the patient's status improves;

  (c) assuring the participation of a health care professional with appropriate expertise in palliative care and the administration of palliative sedation;

  (d) assuring that the patient continues to receive appropriate care and hygiene;

  (e) monitoring sedation to assure adequate and continuous unconsciousness while avoiding inappropriate or unnecessary untoward drug effects;

  (f) documenting the rationale for palliative sedation and the informed consent conversation appropriately in the patient's health record; and

  (g) establishing clear procedures for resolving disagreements about treatment plans or specific treatment decisions, including ethics consultation when appropriate.

Acknowledgments

The National Ethics Committee is grateful to the following individuals, who contributed their expertise in reviewing drafts of this report: Robert M. Arnold, MD, University of Pittsburgh, Section of Palliative Care and Medical Ethics; Amos Bailey, MD, Palliative Care Director, Birmingham VAMC; Matthew J. Bair, MD, Indianapolis VAMC; Rigney Cunningham, Executive Director, Hospice & Palliative Care Federation of Massachusetts; Christine Elnitsky, VA National Pain Management Strategy Coordinating Committee; James Hallenbeck, MD, Medical Director, Hospice, Palo Alto VAMC; Stan Hall, FNP, Boise VAMC; George F. Kelly, MA, VA NJ Health Care System; Robert D. Kerns, PhD, VHA National Program Director for Pain Management, VHA; Anna Lythgoe, VHA Office of Quality and Performance; Hugh Maddry, Director, VHA National Chaplain Center; Kenneth Rosenfeld, MD, Director, Veterans Integrated Palliative (VIP) Program, VA Greater Los Angeles Healthcare System; Paul Rousseau, MD, GEC Administrator, Phoenix VAMC; James Tulsky, MD, Ethics Committee Co-Chair,

Durham VAMC; Joan Van Riper, Director, VHA National Patient Advocacy Program; David Weissman, MD, Palliative Care Center, Medical College of Wisconsin.

Appendix: Members of the National Ethics Committee, Veterans Health Administration

Members: Michael D. Cantor, MD, JD (Chair); Lawrence Biro, EdD; Susan Bowers, MBA; Jeanette Chirico-Post, MD; Sharon P. Douglas, MD; Gwendolyn Gillespie, MSN, RN, APN; Kathleen A. Heaphy, JD; Ware Kuschner, MD; Michael McCoy; Richard Mularski, MD, MSHS; Heather Ohrt, MD; Judy Ozuna, ARNP, MN, CNRN; Peter Poon, JD, MA; Cathy Rick, RN, CNAA, FACHE; Randy Taylor, PhD, MBA, CHE

Ex Officio: Ellen Fox, MD, Director, VHA National Center for Ethics in Health Care

Consultant to the Committee: Michael J. O'Rourke, Veterans of Foreign Wars

Staff to the Committee: Bette-Jane Crigger, PhD; Michael Ford, JD

References

1. National Consensus Project for Quality Palliative Care. *Clinical Practice Guidelines for Quality Palliative Care*, 2004. Available at http://www.national consensusproject.org. Accessed December 3, 2004.
2. Chiu TY, Hu WY, Lue BH, Cheng SY, Chen CY. Sedation for refractory symptoms of terminal cancer patients in Taiwan. *J Pain Symptom Manage.* 2001;21:467–472.
3. Morita T, Tsuneto S, Shima Y. Definition of sedation for symptom relief: a systematic literature review and a proposal of operational criteria. *J Pain Symptom Manage.* 2002;24:447–453.
4. Morita T, Chinone Y, Ikenaga M, et al., for the Japan Pain, Palliative Medicine, Rehabilitation, and Psycho-Oncology Study Group. Ethical validity of palliative sedation therapy: A multicenter, prospective, observational study conducted on specialized palliative care units in Japan. *J Pain Symptom Manage.* 2005;30:308–319.
5. Kaldjian LC, Jekel JF, Bernene JL, Rosenthal GE, Vaughan-Sarrazin M, Duffy TP. Internists' attitudes towards terminal sedation in end of life care. *J Med Ethics.* 2004;30:499–503.
6. National Hospice and Palliative Care Organization. *Total Sedation: A Hospice and Palliative Care Resource Guide.* Alexandria, VA: NHPCO; 2000.
7. American Academy of Hospice and Palliative Medicine. *Statement on Sedation at the End-of-Life.* American Academy of Hospice and Palliative Medicine, September 13, 2002. Available at http://www.aahpm.org/positions/sedation. html. Accessed August 30, 2005.

8. Quill TE, Byock IR, for the ACP-ASIM End-of-Life Care Consensus Panel. Responding to intractable terminal suffering: the role of terminal sedation and voluntary refusal of food and fluids. *Ann Intern Med.* 2000; 132:408–414.
9. American Medical Association. Brief of the American Medical Association, the American Nurses Association, and the American Psychiatric Association, et al. as *Amici Curiae* in Support of Petitioners. *Vacco v. Quill et al* (US S. Ct. No. 95-1858 November 12, 1996).
10. Quill TE, Coombs Lee B, Nunn S. Palliative treatments of last resort: choosing the least harmful alternative. *Ann Intern Med.* 2000;132:488–493.
11. Quill TE, Lo B, Brock DW. Palliative options of last resort: a comparison of voluntarily stopping eating and drinking, terminal sedation, physician-assisted suicide, and voluntary active euthanasia [health law and ethics]. *JAMA.* 1997;278:2099–2104.
12. Braun TC, Hagen NA, Clark T. Development of a clinical practice guideline for palliative sedation. *J Palliat Med.* 2003;6:345–350.
13. Muller-Busch HC, Andres I, Jehser T. Sedation in palliative care—a critical analysis of 7 years experience. *BMC Palliat Care.* 2003;2. Available at http://www.biomedcentral.com/1472-584X2/2. Accessed August 30, 2005.
14. Taylor RM. Is terminal sedation really euthanasia? *Med Ethics (Burlingt, Mass).* 2003;10:3, 8.
15. Cherny NI, Portenoy RK. Sedation in the management of refractory symptoms: guidelines for evaluation and treatment. *J Palliat Care.* 1994;10:31–38.
16. Hospice and Palliative Nurses Association. *Palliative Sedation at End of Life.* Hospice and Palliative Nurses Association, June 2003. Available at http://www.hpna.org/pdf/Palliative_Sedation_Position_Statement_PDF.pdf. Accessed August 30, 2005.
17. Rousseau P. Palliative sedation [guest editorial]. *Am J Hosp Palliat Care.* 2002;19:295–297.
18. Goldstein-Shirley J, Jennings B, Rosen E. Total sedation in hospice and palliative care. Unpublished discussion paper prepared for the Ethics Committee of the National Hospice Association, November 23, 1999.
19. Hallenbeck JL. Terminal sedation for intractable distress: not slow euthanasia but a prompt response to suffering. *West J Med.* 1999;171:222–223.
20. Cowan JD, Palmer TW. Practical guide to palliative sedation. *Curr Oncol Rep.* 2002;4:242–249.
21. Burt RA. The Supreme Court speaks—not assisted suicide but a constitutional right to palliative care. *N Engl J Med.* 1997;337:1234–1236.
22. Schuman Z, Lynch M, Abrahm JL. Implementing institutional change: an institutional case study of palliative sedation. *J Palliat Med.* 2005;8:666–676.
23. Salacz M, Weissman DE. *Fast Fact and Concept #106: Controlled Sedation for Refractory Suffering–Part I.* End-of-Life Physician Education Resource Center, 2004. Available at http://www.eperc.mcw.edu. Accessed January 27, 2005.
24. Greene WR, Davis WH. Titrated intravenous barbiturates in the control of symptoms in patients with terminal cancer. *South Med J.* 1991;84:332–7.

25. Fainsinger RL, Waller A, Bercovici M, et al. A multicentre international study of sedation for uncontrolled symptoms in terminally ill patients. *Palliative Med.* 2000;14:257–265.

26. Kohara H, Ueoka H, Takeyama H, et al. Sedation for terminally ill patients with cancer with uncontrollable physical distress. *J Palliat Med.* 2005;8:20–25.

27. Lo B, Rubenfeld G. Palliative sedation in dying patients: "We turn to it when everything else hasn't worked" [Perspectives on care at the close of life]. *JAMA.* 2005;294:1810–16.

28. Rousseau P. Palliative sedation and sleeping before death: a need for clinical guidelines? *J Palliat Med.* 2003;6:425–427.

29. Cherny NI. Commentary: sedation in response to refractory existential distress: walking the fine line. *J Pain Symptom Manage.* 1998;16:404–406.

30. Billings JA, Block SD. Slow euthanasia. *J Palliat Care.* 1996;12:21–30.

31. Kingsbury RJ, Ducharme HM. The debate over total/terminal/palliative sedation. The Center for Bioethics and Human Dignity, January 24, 2002. Available at http://www.cbhd.org/resources/endoflife/kinsbury-ducharme_2002-01-24.htm. Accessed September 26, 2005.

32. Sulmasy DP, Pellegrino ED. The rule of double effect: clearing up the double talk. *Arch Intern Med.* 1999;159:545–549.

33. Lynn J. Terminal sedation [letter to the editor]. *New Engl J Med.* 1998; 338:1230.

34. Sykes N, Thorn A. Sedative use in the last week of life and the implications for end-of-life decision making. *Arch Intern Med.* 2003;163:341–344.

35. Quill TE, Dresser R, Brock DW. The rule of double effect—a critique of its role in end-of-life decision making [sounding board]. *N Engl J Med.* 1997;337:1764–1771.

36. Rousseau P. The ethical validity and clinical experience of palliative sedation. *Mayo Clin Proc.* 2000;75:1064–1069.

37. Jansen LA, Sulmasy DP. Sedation, alimentation, hydration, and equivocation: careful conversation about care at the end of life. *Ann Intern Med.* 2002;136: 845–849.

38. Vacco v. Quill, 521 U.S. 793 (1997); Washington v. Glucksberg, 521 US 702 (1997).

39. Terminal sedation vs PAS: difference just semantics? *Med Ethics Advis.* 2005; 8:91–93.

40. Gillick MR. Terminal sedation: an acceptable exit strategy? *Ann Intern Med.* 2004;141:236–237.

41. Sheldon T. Dutch euthanasia law should apply to patients "suffering through living," report says. *BMJ.* 2005;330:61.

42. Morita T, Ikenaga M, Adachi I, et al; Japan Pain, Rehabilitation, Palliative Medicine, and Psycho-Oncology Study Group. Family experience with palliative sedation therapy for terminally ill cancer patients. *J Pain Symptom Manage.* 2004;28:557–565.

43. Foley KM. Competent care for the dying instead of physician-assisted suicide [editorial]. *N Engl J Med.* 1997;336:53–58.

44. Lunney JR, Lynn J, Foley DJ, Lipson S, Guralnik JM. Patterns of functional decline at the end of life. *JAMA*. 2003;289:2387–2392.
45. Glare P, Virik K, Jones M, et al. A systematic review of physicians' survival predictions in terminally ill cancer patients. *BMJ*. 2003;327:195–200.
46. Lamont EB, Christakis NA. Complexities in prognostication in advanced cancer: "to help them live their lives the way they want to" [Perspectives on care at the close of life]. *JAMA*. 2003;290:98–104.
47. Rousseau P. Existential suffering and palliative sedation: a brief commentary with a proposal for clinical guidelines. *Am J Hosp Palliat Care*. 2001;18: 151–53.
48. Salacz ME, Weissman DE. *Fast Fact and Concept #107: Controlled Sedation for Refractory Suffering—Part II*. End-of-Life Physician Education Resource Center, 2004. Available at http://www.eperc.mcw.edu. Accessed January 27, 2005.
49. Orentilecher D. The Supreme Court and physician-assisted suicide-rejecting assisted suicide but embracing euthanasia. *N Engl J Med*. 1997;337:1236–39.
50. Hallenbeck JL. Terminal sedation: ethical implications in different situations. *J Palliat Med*. 2000;3:313–320.
51. Cassell EJ. Diagnosing suffering: a perspective. *Ann Intern Med*. 1999;131: 531–534.
52. Block SD. Assessing and managing depression in the terminally ill patient. *Ann Intern Med*. 2000;132:210–218.
53. Morita T, Tsunoda J, Inoue S, Chihara S. Terminal sedation for existential distress. *Am J Hosp Palliat Care*. 2000;17:189–195.
54. Jonsen AR, Siegler M, Winslade WJ. *Clinical Ethics*. 2nd ed. New York, NY: Macmillan Publishing Co; 1986.
55. Callahan D. When self-determination runs amok. *Hastings Center Rep*. 1992; 22:52–55.
56. Snyder L, Sulmasy DP, for the Ethics and Human Rights Committee, ACP-ASIM. Physician-assisted suicide [position paper]. *Ann Intern Med*. 2001; 135:209–216.
57. Block SD. Psychological considerations, growth, and transcendence at the end of life: the art of the possible. *JAMA*. 2001;285:2898–2905.
58. U.S. Department of Veterans Affairs. VHA Handbook 1004.1: *Informed Consent for Clinical Treatments and Procedures*. January 2003.

Recognizing Death While Affirming Life:
Can End of Life Reform Uphold a Disabled Person's Interest in Continued Life?

Adrienne Asch

Adrienne Asch received her doctorate in social psychology from Columbia University in 1992. She is a professor of bioethics at Yeshiva University and professor of epidemiology and population health and family and social medicine at Albert Einstein College of Medicine. Her work focuses on the ethical, political, psychological, and social implications of human reproduction and the family.
 In this article Asch characterizes how the common autonomy arguments used to support a patient's right to withdraw or withhold life-sustaining treatment may actually undermine respect for self-determination and human dignity. Instead, she emphasizes that self-determination and human dignity entail a commitment to fostering the activities, experiences, and relationships that enrich an individual's life by finding techniques and resources to use those capacities that remain.

Evolving Views of End of Life and Disability

... In the years since the 1976 case of Karen Ann Quinlan, much greater weight has been given, both in law and the culture at large, to informed consent; to the experiences, views, and needs of patients and families in the medical encounter; to respect for patient autonomy and family decision-making; and to the quality, not merely the preservation, of an individual's life. These beliefs have meshed well with the efforts of feminists and other marginalized groups to equalize the power relations between doctor and patient, and they have also supported twenty-first century cultural norms of self-fulfillment, self-determination, and control over one's destiny. These ideals should have promoted an alliance between end of life reform, the emerging scholarship of disability studies, and the movement for disability rights and equality. Unfortunately, many scholars and practicing health care professionals have failed to grasp crucial insights of disability scholars or activists. Despite the common cause of disability scholars and activists with those in the end of life movement around maximizing self-

---

determination and giving more respect and authority to patients in their encounters with medicine, the end of life movement has sharply differed with disability theorists and activists in understanding how illness and impairment affect quality of life.

Thanks to the sustained efforts of scholars, clinicians, and grassroots citizen groups like Compassion in Dying, both clinical practice and case law recognize that ill or dying patients and their intimates often are concerned about their experiences and relationships during whatever time they have left to live, not merely with how long they might be maintained by medications, feeding tubes, and breathing machines. Disability activist and lobbying groups such as Not Dead Yet or American Disabled for Attendant Programs Today (ADAPT) also espouse the goals of creating and maintaining opportunities for ill, disabled, or dying people to enjoy fulfilling, meaningful relationships, activities, and experiences for however much time they will live. Compassion in Dying and Not Dead Yet differ in their policy and practice goals for two reasons: they focus on different kinds of paradigm cases, and they have profoundly different understandings of how illness and disability affect life's meaning and rewards. The typical case for the misnamed "right to die" movement is an elderly man or woman in the final stages of an inevitably terminal illness, who will soon die regardless of how much medical treatment is invested in his or her last days or weeks. The case that fuels the disability rights movement is that of a relatively young person with a disability, who could live for several years with the condition, but who instead asks to die—as in *Million Dollar Baby*, and as in many real-life cases.

Although mainstream reformers have criticized the way professionals often dealt with patients and their families, the mainstream has too often accepted medicine's view that illness and disability inevitably diminish life's quality. In contrast, disability theorists and activists point to research demonstrating that people with physical, sensory, and cognitive impairments can and do obtain many satisfactions and rewards in their lives. When people with illness and disability report dissatisfaction and unhappiness, they link their distress not to physical pain or to reliance on medications, dialysis, or ventilators, but to those factors that also trouble nondisabled people—problematic relationships, fears about financial security, or difficulties in playing a valued work or other social role.

Disability theorists and activists endorse the growth of hospice, palliative care, pain relief, and greater attention to the psychological and social needs of patients and their loved ones; however, they argue that endorsing treatment withdrawal from people simply because their health or their capacities are impaired undermines the goals of human dignity, patient self-respect, and quality of life. Such goals are best achieved by helping people discover that changed health status and even impaired cognition need not rob life of its value. Respect for self-determination and human dignity entails a commitment to fostering the activities, experiences, and relationships that enrich an individual's life by finding techniques and resources to use those capacities that remain. In the case of Elizabeth Bouvia, a woman disabled by cerebral palsy and painful arthritis who sought aid in dying, the California Court of Appeals supported her request to end her life by focusing on her limitations, pointing to her physical immobility and her need for assistance with tasks like eating and toileting. Although the court described her as "alert" and "feisty," it also characterized her as "subject to the ignominy, embarrassment, humiliation and dehumanizing aspects created by her helplessness." The 1996 court decision that supported physician-assisted suicide in *Washington v. Glucksberg* was filled with similar portrayals of life with impairment: it referred to people who are in a "childlike state" of helplessness, as exemplified by physical immobility or by their use of diapers to deal with incontinence.

The disability critics of the California court decision revealed an entirely different side to the Elizabeth Bouvia story. They focused on her remaining capacities and on the social and economic problems that contributed to her isolation and depression. Educational discrimination had prevented her from using her mind; she had been denied the full amount of personal assistance services that would have enabled her to stay in the community; and her depression, which stemmed from serious family problems, would have been immediately treated in a nondisabled person who had attempted suicide.

Many of the disability theorists and activists who protested the court decisions in the Bouvia case—and in the similar Michigan case of David Rivlin, who became quadriplegic and sought death rather than remaining in a nursing home—have very similar physical conditions but entirely different life circumstances. By recruiting paid or volunteer personal assistants, they live in their own homes by themselves or with family and friends. They are in the community, not in institutions. They hold jobs, engage in volunteer activities, visit friends, go out to dinner or the movies, and generally participate in ordinary family, civic, and social life. Wheelchairs do not confine; they liberate. Voice synthesizers aid communication for people who can no longer speak. Diapers or catheters are akin to eyeglasses. Using the services and skills of a personal assistant who helps them get into and out of bed, eat their meals, or travel to their next appointment is no more shameful or embarrassing than it is for a nondisabled person to work closely with an administrative assistant or to value the expertise of a mechanic, plumber, or the magician who restores data after a computer crash.

Fortunately, some respected mainstream scholars have acknowledged that societal tolerance of death for people who could live for months or years with disabilities stems from misunderstanding, fear, and prejudice. Excerpts from one clinician–philosopher's recent reflections demonstrate a new receptivity to the disability critique of typical end of life practice and policy.

> I am now embarrassed to realize how limited was the basis on which I made my decisions about David Rivlin. . . . [T]here was no medical need for Rivlin to be effectively incarcerated in a nursing home. If Rivlin had been given access to a reasonable amount of community resources, of the sort that other persons with disabilities were making use of at the time, he could have been moved out of the nursing home and probably could have had his own apartment. He could have been much more able to see friends, get outside a bit, and generally have a much more interesting and stimulating life. . . . If we look at a case one way, it seems that the problem is the person's

physical disability. If we shift our view, we realize that the problem is not the disability, but rather the refusal of society to make reasonable and not terribly expensive accommodations to it.[1]

In his 1979 book *Taking Care of Strangers*, Robert Burt exposed the common discomfort of health care professionals in the presence of patients with very significant impairments: "Rules governing doctor-patient relations must rest on the premise that anyone's wish to help a desperately pained, apparently helpless person is intertwined with a wish to hurt that person, to obliterate him from sight." Speaking of a burned and very disfigured patient, Burt contended: "He is a painful, insistent reminder to others of their frailty, an acknowledgement that, in the routine of everyday life, is ordinarily suppressed. Others cannot avoid wishing that he, and his unwanted lesson, would go away. He cannot avoid knowing this of others and wishing it for himself."

Toward Further Change

These insights should prompt clinicians and policymakers to question how truly autonomous is anyone's wish to die when living with changed, feared, and uncertain physical impairments that lead to anguish and to interpersonal struggles with the very professionals, family members, and friends who are assumed to be supports in a time of trouble. The spirit of such observations illustrates the danger of relying on a simple notion of patient autonomy when deciding to withdraw life-sustaining treatment.

Consider this case from the end of life literature, reported by M. Edwards and Susan Tolle: Their patient—conscious, alert, with mobility impairments that had lasted for 40 years—had recently developed breathing problems that necessitated use of a ventilator, which rendered him unable to speak. Finding this increased disability intolerable, he sought death, and family, professionals, and the hospital ethics committee concurred with his autonomous wish. Edwards and Tolle proposed a seven-step procedure to assure themselves that such an aided death is acceptable. Absent from their analysis is any exposure to or contact with people who have more than two weeks of experience living as ventilator users. The case description provides no information on how effectively this patient was communicating (whether by writing, pointing to letters and words, or using a communication technology). It contains no information about whether this man's decision was affected by concerns over how his relationships with family and friends might be changed by his different means of communication. Presumably these clinicians knew that nonvocal but conscious and responsive individuals have been able to interact in family and work settings. One wonders why these clinicians did not urge such means upon this patient before acceding to his pleas for death rather than life without speech. He may have been psychologically abandoned by his family and clinicians when he most needed their energy, resourcefulness, and imagination to help him devise a new way to express himself.

The most recent report on the workings of Oregon's law on physician-assisted suicide offers yet another illustration of social rather than medical issues at work in requests for assisted dying. The most frequently cited reasons for seeking to die stemmed from loss of enjoyable activities, loss of autonomy, and loss of dignity. Yet these were mentally alert individuals who should have been aided by professionals and their own social networks to discern that autonomy and dignity can reside in self-expression, in determining what activities to pursue, and in obtaining the assistance to undertake them. This reframing of autonomy and dignity is urgently necessary as a way to restore self-respect and pride to people who feel shame at needing physical or emotional help from those around them. Have they lost their own ability to provide love, support, friendship, and guidance to their families and friends, and if so, what professional psychological help might let them regain those capacities? Or have they lost their connections to the social world, and so been denied a way to give and to receive help and support?

For people living with disabilities, the data on Oregon's assisted suicides provokes concern. One can respect individual choice but worry that the Oregon data, like the case involving ventilator withdrawal, graphically support Burt's reflections on the ambivalence of health care professionals and families toward people with significant disability. When these data reveal that fear of burdening others is of much greater concern to patients who seek suicide than concerns about finances or physical pain, then how can professionals and families know that the supposedly autonomous wish to end life is not a response to a patient's deep fear that she has become disliked, distasteful to, and resented by the very people from whom she seeks expertise, physical help, and emotional support? And when we learn that divorced and never-married individuals are twice as likely as married or widowed people to use physician-assisted suicide, we must ponder whether a single dying person feels especially alone and abandoned. It is probably the rare friend who has the time, energy, or willingness to make a sustained, reliable, and deep commitment to live through another's illness and death. Once the severely disabled, ill, or dying person is seen as "other"—as different, not quite in the human and moral community, even past friendship and familial bonds—social bonds can diminish. To anyone with the capacity to perceive the difference between warmth, toleration, and coldness in how he or she is treated by others, the thought of days, months, or years of life subject to resentful, duty-filled physical ministrations may be a fate worse than death, akin to imprisonment and solitary confinement. What needs to change is not the patient's physical or cognitive situation, but the emotional and interpersonal environment; that environment can change only when professionals lead the way to supporting the capacities and thereby affirming the humanity of severely ill and imminently dying people.

Once we have understood the disability community's concerns about cases involving alert people with physical, but not cognitive and affective disabilities, we can better understand the reaction to the unfortunate case of Terri Schiavo. By the time her husband sought to withdraw her feeding tube, all the medical experts were certain that she had not even minimal cognitive capacity or consciousness. Schiavo's supporters in the disability community were almost certainly mistaken about her potential for interaction or responsiveness, and they may have done damage to their efforts to join with others seeking to reform treatment of disabled or dying people. Yet the apprehension in the disability community, apprehension about societal indifference and neglect, is more understandable after reviewing a few of the many instances in which law, medicine, bioethics, and government programs failed to help traumatically disabled patients discover the financial, technological, social, and psychological resources that could sustain them and provide the opportunity for rewarding life. When people with relatively intact cognitive and emotional capacities are neglected, neglect is even more likely for those with greatly diminished cognitive and emotional function.

Although the intense court reviews of Schiavo's situation consistently confirmed her PVS diagnosis, professional literature contains scattered information on patients who were misdiagnosed as being in that state and were consequently denied rehabilitation and treatment from which they might have benefited. Some misdiagnosed patients have limited ability to respond meaningfully to others; this diagnostic error cost such patients between one and four years of interaction with people and the world around them. It is rare for courts or scholars to champion continued treatment for cognitively impaired people who might still enjoy some level of life satisfaction and human interaction.[2]

The disability equality perspective on end of life and treatment withdrawal cases described here should demonstrate that the alliance of disability studies and disability rights with the evangelical religious groups is more apparent than real. Disability critics of much health care practice share more with end of life reformers who seek to promote an emphasis on respect for the dignity and capacities of people facing illness, disability, and death. Like these reformers, they seek the means for maintaining dignity and capacity; the aptly named Not Dead Yet strives to convince people with disabilities, their families, and their health care providers that people can still find satisfaction and quality in their lives. The president of Not Dead Yet clearly articulated the ways in which disability opposition to life-ending decisions is truly a quest for quality, rather than sanctity, of life:

> The far right wants to kill us slowly and painfully by cutting the things we need to live, health care, public housing and transportation, etc. The far left wants to kill us quickly and call it compassion, while also saving money for others perhaps deemed more worthy.
>
> [W]e also have an attitude about disability that diverges from the mainstream . . . Frankly, I think that's why we were deliberately excluded from the last decade of policy making conducted off the public radar screen, why the right-wing-left-wing script was so important . . . no matter how untrue and exclusionary.[3]

These comments lead to a case with disturbing implications for mainstream discussions of patient autonomy, family decision-making, and professional obligation. Barbara Howe, a 79-year-old woman with amyotrophic lateral sclerosis, using a ventilator, was being treated at Boston's Massachusetts General Hospital. Howe's daughters and grandchildren visited her consistently. Howe had indicated that she wished to stay alive as long as she could appreciate family visits and had named one of her daughters to serve as health care proxy. In March of 2005 she was thought to be conscious and alert but was unable to speak or to show responses through any facial or bodily movements. Yet the hospital sought to remove Howe's daughter as her health care proxy and to discontinue ventilator support. After legal wrangling, a reportedly reluctant daughter agreed that ventilatory support could be withdrawn on June 30 (Ms. Howe died while still on ventilatory support before that date).

Although the details of Howe's case are not yet and may never become public record, the published reports give considerable basis for concern. If case law and mainstream end of life practice are to continue their adherence to patient autonomy, health care proxy decision-making, and rights to receive as well as rights to forego life-sustaining treatment, they should question the basis on which the hospital staff sought to end treatment in the face of expressed wishes of patient and family to continue that treatment. On what basis did staff feel that the treatment was inhumane since the patient had requested that she be kept alive regardless of pain if she was appreciating her family's visits? Did the hospital staff have reason to believe that it knew the patient better than did her family because the staff was with her for many more hours every day? Was the staff experiencing the kind of pain and ambivalence Burt describes in the presence of a conscious yet unexpressive woman with complete physical paralysis? Did the hospital, like the hospital in the 1991 case of Helga Wanglie, believe that continuing to provide expensive treatment no longer served either the patient or the public good? Was stewardship of resources an unstated but serious concern, and should it become a legitimate public concern? If end of life practice and law answer yes, as they well may need to do, the field will have to rethink its almost unquestioned championing of patient autonomy and family decision-making if those autonomous or proxy decisions are to maintain, rather than to forego, expensive life-sustaining treatment.

The stories of Helga Wanglie and Barbara Howe clearly reveal the need for end of life reform to re-examine the possibility of setting limits to its own commitments to patient autonomy or family decision-making in the face of public resource constraints. This issue could lead to even more division between the mainstream end of life field and the disability theorists and activists who seek both a shift in an understanding of "quality of life" and a distribution of resources to individuals who need physical, medical, and social support to maintain a life with dignity and meaning.

Next Steps

This largely absent disability perspective could profitably enliven the world of end of life reform. The post-Schiavo reaction, with its renewed calls for advance directives for all Medicare patients, should encourage bioethicists to redesign the current forms, which ask people only about which interventions they do and do not want. Instead, the forms should describe the various medical scenarios that might occur in certain situations and encourage people to consider what they would or would not want done in each instance. Which physical and cognitive capacities can they imagine losing and still find life rewarding? What activities do they envision as essential for life satisfaction? These educational documents should help people imagine not only what physical changes may occur, but also what social, technological, and financial resources they might require to maintain themselves after the onset of serious illness and disability. Recognizing how difficult it is for anyone to project herself into a radically different situation, the end of life field has moved away from advance directives and instead endorsed family decision-making and health care proxies. Indeed, many families will accurately gauge their loved one's desires, whether for continuing or ending life-sustaining treatment; nonetheless, widespread discomfort in the face of physical and cognitive changes in a spouse, parent, sibling, or friend suggest that even intimates may fail to appreciate the rewards and satisfactions remaining in their loved one's life. I would therefore suggest that revamped advance directives and drastically revised educational materials continue to be indispensable in helping us out of the end of life care morass.

End of life reform and society generally have never successfully confronted the rationing question; neither has the disability rights movement or the field of disability studies. Groups like Not Dead Yet bring an invaluable perspective on disability to end of life conversation, and they need to be sought out as we search for progress in reforming end of life practice. Activists from Not Dead Yet and ADAPT, as well as disability scholars from philosophy, psychology, health economics, and other disciplines, need to participate regularly in the mainstream conversation; they need to help determine criteria for allocating national resources among all the many health, disability rights, environmental, and social justice problems we face. They also need to be recruited for hospital and hospice ethics committees, and they need to train physicians, nurses, and social workers in new ways of understanding life with disability. The events of this year demonstrate how desperately the disability perspective needs to become part of the conversation rather than being excluded from it.

At the end of life, facing decline and death, these "disability issues" are issues for everyone—learning how to affirm and celebrate what gives life meaning and simultaneously acknowledge loss of capacity and eventually loss of life itself.

Acknowledgment

I would like to acknowledge the assistance of Jenny Dick Bryan and Ari Schick in the preparation of this essay.

References

1. H. Brody, "Bioethicist's Apology," *Health*, October 6, 2004: 6.
2. Exceptions are R. S. Dresser and J. A. Robertson, "Quality of Life and Non-Treatment Decisions for Incompetent Patients," *Law, Medicine and*

Health Care 17 (1989): 234–44, and *In re Storar,* 52 N.Y.2d 363,420 N.E.2d 64, 438 N.Y.S.2d 266 (1981).

3. D. Coleman, testimony before the Subcommittee on Criminal Justice, Drug Policy and Human Resources of the Committee on Government Reform of the U.S. House of Representatives, *Oversight Hearing on Federal Health Programs and Those Who Cannot Care for Themselves: What Are Their Rights, and Our Responsibilities?* April 19, 2005.

PHYSICIAN-ASSISTED DEATH

2009 Summary of Oregon's Death with Dignity Act

Oregon Department of Human Services

The Oregon Department of Human Services (Public Health Division) is required to collect information on compliance and issue annual reports on the Death with Dignity Act. This report presents longitudinal data that describe the infrequent occurrence of physician-assisted death. More specifically, the report presents data about the patients, who died after ingesting a lethal dose of medication, in relationship to sociodemographic characteristics, underlying illness, end-of-life care, end-of-life concerns, physician-assisted dying process, the presence of health care providers at the time of ingestion and death, rate and type of complications, and the timing of the event.

Oregon's Death with Dignity Act (DWDA), which was enacted in late 1997, allows terminally ill adult Oregonians to obtain and use prescriptions from their physicians for self-administered, lethal doses of medications. The Oregon Public Health Division is required by the act to collect information on compliance and to issue an annual report. The key findings from 2009 are listed below. For more detail, please view the figures and tables on our web site at http://oregon.gov/DHS/ph/pas/index.shtml.

- During 2009, 95 prescriptions for lethal medications were written under the provisions of the DWDA compared to 88 during 2008 (Figure 1). Of these, 53 patients took the medications, 30 died of their underlying illness, and 12 were alive at the end of 2009. In addition, six patients with earlier prescriptions died from taking the medications, resulting in a total of 59 DWDA deaths during 2009. This corresponds to an estimated 19.3 DWDA deaths per 10,000 total deaths.
- Fifty-five physicians wrote the 95 prescriptions (range 1–6).
- Since the law was passed in 1997, 460 patients have died from ingesting medications prescribed under the Death with Dignity Act.
- As in prior years, most participants were between 55 and 84 years of age (78.0%), white (98.3%), well educated (48.3% had at least a baccalaureate

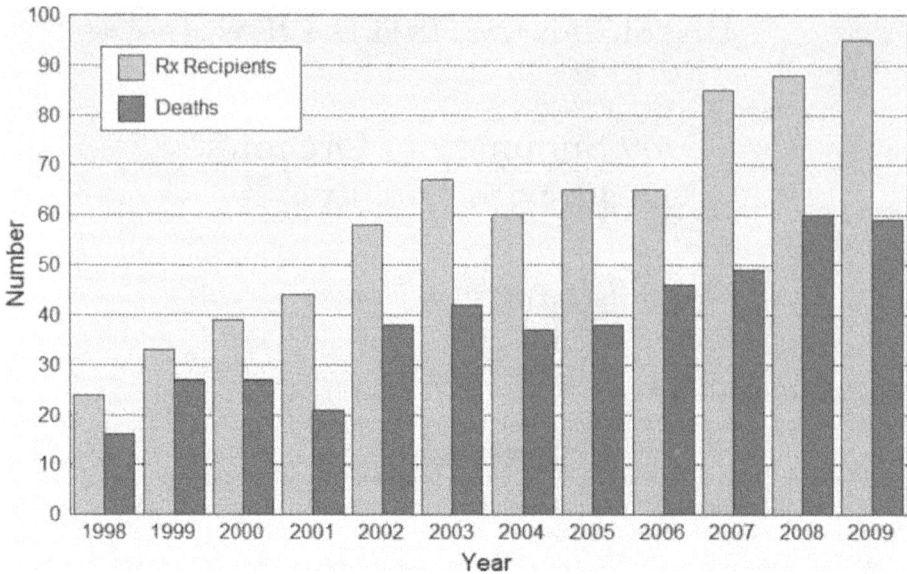

Figure 1 Number of DWDA Prescription Recipients and Deaths, by Year, Oregon, 1998–2009.

degree), and had cancer (79.7%). Patients who died in 2009 were slightly older (median age 76 years) than in previous years (median age 70 years).

- Most patients died at home (98.3%); and most were enrolled in hospice care (91.5%) at time of death.
- In 2009, 98.7% of patients had some form of health care insurance. Compared to previous years, the number of patients who had private insurance (84.7%) was much greater than in previous years (66.8%), and the number of patients who had only Medicare or Medicaid insurance was much less (13.6% compared to 32.0%).
- As in previous years, the most frequently mentioned end-of-life concerns were: loss of autonomy (96.6%), loss of dignity (91.5%), and decreasing ability to participate in activities that made life enjoyable (86.4%).
- In 2009, none of the 59 patients were referred for formal psychiatric or psychological evaluation. Prescribing physicians were present at the time of ingestion for 3 (5.1%) patients compared to 22.6% in previous years; the prescribing physician was present at the time of death for 1 patient. The time from ingestion until death ranged from 2 minutes to 4½ days (104 hours).
- During 2009, one referral was made to the Oregon Medical Board for failing to submit a witnessed written consent form.

Physician-Assisted Death
in the Pacific Northwest

Nancy S. Jecker

Nancy S. Jecker received a Ph.D. in philosophy at the University of Washington, Seattle, in 1986. She has been on the faculty of the University of Washington School of Medicine, Department of Bioethics and Humanities since 1988. Jecker is adjunct professor in the Department of Philosophy and School of Law at the University of Washington. In this article Jecker reviews the experience of Oregon since passage of the Death with Dignity Act, a law that allows terminally ill, competent, adult residents to request and self-administer lethal medication prescribed by a physician. She notes that many of the concerns about physician-assisted death that critics voiced in fact never materialized, and suggests that Washington voters, who recently passed a law similar to Oregon's, may not be the last state to do so.

In 1997, Oregon became the first state to allow physician-assisted suicide. Eleven years later, Oregon's neighbor, Washington State, became the second. Modeled after Oregon's Death with Dignity Act, Washington's Initiative 1000 was approved by 57.91% of voters in November 2008. Washington's new law permits terminally ill, competent, adult Washington residents, who are medically predicted to have 6 months or less to live, to request and self-administer lethal medication prescribed by a physician (Proposed Initiatives to the People 2008).

Initiative 1000 was not Washington's first attempt to pass a citizen initiative allowing physician-assisted suicide. Six years before the 1997 passage of Oregon's Death with Dignity Act, Washington voters successfully petitioned to place Initiative 119 on the ballot. Initiative 119 was a broader measure than either of the Oregon or Washington initiatives that were successfully signed into law. It would have allowed physicians not only to prescribe a lethal dose of medicine on request, but also to administer such medication if the patient could not swallow it, a practice known as "voluntary active euthanasia." Initiative 119 failed in 1991 to gain majority approval, with only 46% of voters supporting it.

History thus suggests that Washington voters draw important distinctions. While Washingtonians said "yes" to physician-assisted suicide (by passing Initiative 1000 in 2008), they said "no" to voluntary active euthanasia (by rejecting Initiative 119 in 1991). Similarly, in the state of Oregon, 11 years after the passage of a citizen initiative allowing physician-assisted suicide (the 1997 Death with Dignity Act), Oregonians have not proposed any measures to broaden the category of

Nancy S. Jecker, "Physician-Assisted Death in the Pacific Northwest," *The American Journal of Bioethics*, 9:3, March 1, 2009, 1–2. Reprinted by permission of Taylor & Francis, Ltd.

eligible patients, or to increase the role of the physicians beyond prescribing medication. This appears to run contrary to the claim that some opponents of physician-assisted suicide make, namely, that allowing physician-assisted suicide is a first step down a slippery slope that will inevitably result in allowing broader categories of physician assistance, including euthanasia.[1]

In 2008, when Washington voters approved legalizing physician-assisted suicide, we had the benefit of over a decade of information from our neighbor to the south. Oregon's law requires physicians to report to Oregon's Department of Human Services (DHS) all prescriptions for lethal medications, and requires DHS to collect information on compliance and issue an annual report. Oregon's annual reports show clearly that many of the negative predictions made by opponents did not materialize (Lindsay 2009).

Contrary to the prediction that vulnerable groups, such as the frail elderly, the uninsured, and people of non-White racial groups, would be disproportionately impacted by the implementation of Oregon's law, the annual reports show that these groups are not represented at a very high rate.[2] During 1998–2007, the overwhelming majority of participants were White (97.4%), college educated (64.2%), and had some form of health insurance (99.1%). Most participants were men (53.7%), with a median age 69 years. Only 9.4% of participants were age 85 years or older. These findings show that rather than targeting the most vulnerable members of society, the law may, ironically, be failing to reach Oregon's most vulnerable citizens.

This possibility is difficult to assess, because we do not know how frequently requests for aid in dying are made by eligible Oregonians without being met. For example, we do not know how frequently requests occur among non-White racial and ethnic groups, such as Asians, American Indians, Hispanics, and African Americans, without follow-up. Oregon's Death with Dignity Act does not require reporting when patients begin the request process but never receive a prescription (Leman and Kohn 2007). For this reason, we cannot say if particular populations are making requests that are disproportionately denied or discouraged. This raises the question of whether all Oregonians have equal access to physician-assisted suicide.

The implementation of Oregon's Death with Dignity Act also did not have the predicted negative effect of cutting patients off from hospice and other alternatives. In fact, from 1998–2007, most participants (85.5%) were enrolled in some form of hospice care.

Armed with this kind of information, Washington voters in 2008 could more confidently support Initiative 1000. From Oregon's experience, we knew that a relatively small number of citizens were likely to end their lives using physician-assisted suicide. In Oregon, 341 patients over an 11-year period were reported to end their lives by this means. We knew that access to prescription medications offered a safe and nonviolent means of hastening death, and that complications were rare. In Oregon, no complications were reported in 94.3% of cases where physician-assisted suicide was used. Finally, we knew that multiple legal efforts to challenge and even repeal Oregon's Death with Dignity Act delayed its implementation, but failed to overturn it. Physician-assisted suicide remains controversial. Yet eligible terminally ill patients who choose to hasten their death in Washington State can now legally receive a physician's help.

Perhaps the Pacific Northwest will be an outpost for physician-assisted suicide, and the practice will not spread to the rest of the country. States such as California, Michigan, and Maine, have considered, but rejected, assisted death measures. Or perhaps other states *will* eventually join the Northwest. Twenty-four states (plus the District of Columbia) offer voters an initiative process, similar to the one used by voters in Washington and Oregon to legalize physician-assisted suicide. With this process, ordinary citizens have the authority to enact new laws (or change existing laws) by petitioning to place proposed legislation on the ballot, with most initiative measures requiring a simple majority vote to become law. I anticipate that Washington will *not* be the last state to allow physician-assisted suicide. Washington's experience going forward will offer additional empirical evidence of the benefits and burdens of this practice. If Washington's experience is similar to Oregon's, many of the empirically based arguments against physician-assisted suicide will further erode, and the arguments supporting this practice will appear all the more persuasive as a result.

Notes

1. This claim is made, for example, by the Coalition Against Physician-Assisted Suicide (2008), and is discussed on their website (http://noassisted suicide.com/i1000claims.html).
2. This objection is made, for example, by the Coalition Against Physician-Assisted Suicide (2008), and is discussed on their website (http://noassisted suicide.com/i1000claims.html).

References

Coalition Against Physician-Assisted Suicide. 2008. Say no to assisted suicide. Protect the vulnerable. Available at: http://noassistedsuicide.com/i1000claims. html (accessed December 15, 2008).

Leman, R., and Kohn, M. A. 2007. Eighth annual report on Oregon's death with dignity act. In *Bioethics: An Introduction to the History, Methods, and Practice*, 2nd edition, eds. N. S. Jecker, A. R. Jonsen, R. A. Pearlman. Sudbury, MA: Jones and Bartlett, 460.

Lindsay, R. A. 2009. Oregon's experience: Evaluating the record. *American Journal of Bioethics* 9(3): 19–27.

Proposed Initiatives to the People. 2008. Available at: http://www.secstate.wa. gov/elections/initiatives/text/i1000.pdf (accessed January 22, 2009).

Death and Dignity:
A Case of Individualized Decision Making

Timothy E. Quill

Timothy Quill received his M.D. from Rochester University. He is a professor in the Departments of Medicine, Psychiatry, and Medical Humanities at Rochester University and director of the local palliative care program. In this article he writes a very personal story of assisting one of his patients with aid-in-dying. Dr. Quill wrote this article because he had decided that it was important to tell Diane's story to the public and to other doctors; in so doing, Quill presents a case-based argument in support of physician-assisted suicide in particular cases similar to Diane's.

Diane was feeling tired and had a rash. A common scenario, though there was something subliminally worrisome that prompted me to check her blood count. Her hematocrit was 22, and the white-cell count was 4.3 with some metamyelocytes and unusual white cells. I wanted it to be viral, trying to deny what was staring me in the face. Perhaps in a repeated count it would disappear. I called Diane and told her it might be more serious than I had initially thought—that the test needed to be repeated and that if she felt worse, we might have to move quickly. When she pressed for the possibilities, I reluctantly opened the door to leukemia. Hearing the word seemed to make it exist. "Oh, shit!" she said. "Don't tell me that." Oh, shit! I thought, I wish I didn't have to.

Diane was no ordinary person (although no one I have ever come to know has been really ordinary). She was raised in an alcoholic family and had felt alone for much of her life. She had vaginal cancer as a young woman. Through much of her adult life, she had struggled with depression and her own alcoholism. I had come to know, respect, and admire her over the previous eight years as she confronted these problems and gradually overcame them. She was an incredibly clear, at times brutally honest, thinker and communicator. As she took control of her life, she developed a strong sense of independence and confidence. In the previous three-and-one-half years, her hard work had paid off. She was completely abstinent from alcohol, she had established much deeper connections

From *New England Journal of Medicine* 324 (10), March 7, 1991: 691–694. Reproduced with permission of the Massachusetts Medical Society. Copyright © 1991, Massachusetts Medical Society.

with her husband, college-age son, and several friends, and her business and her artistic work were blossoming. She felt she was really living fully for the first time.

Not surprisingly, the repeated blood count was abnormal, and detailed examination of the peripheral blood smear showed myelocytes. I advised her to come into the hospital, explaining that we needed to do a bone marrow biopsy and make some decisions relatively rapidly. She came to the hospital knowing what we would find. She was terrified, angry, and sad. Although we knew the odds, we both clung to the thread of possibility that it might be something else.

The bone marrow confirmed the worst: acute myelomonocytic leukemia. In the face of this tragedy, we looked for signs of hope. This is an area of medicine in which technological intervention has been successful, with cures 25 percent of the time long-term cures. As I probed the costs of these cures, I heard about induction chemotherapy (three weeks in the hospital, prolonged neutropenia, probable infectious complications, and hair loss; 75 percent of patients respond, 25 percent do not). For the survivors, this is followed by consolidation chemotherapy (with similar side effects; another 25 percent die, for a net survival of 50 percent). Those still alive, to have a reasonable chance of long-term survival, then need bone marrow transplantation (hospitalization for two months and whole-body irradiation, with complete killing of the bone marrow, infectious complications, and the possibility for graft-versus-host disease—with a survival of approximately 50 percent, or 25 percent of the original group). Though hematologists may argue over the exact percentages, they don't argue about the outcome of no treatment—certain death in days, weeks, or at most a few months.

Believing that delay was dangerous, our oncologist broke the news to Diane and began making plans to insert a Hickman catheter and begin induction chemotherapy that afternoon. When I saw her shortly thereafter, she was enraged at his presumption that she would want treatment, and devastated by the finality of the diagnosis. All she wanted to do was go home and be with her family. She had no further questions about treatment and in fact had decided that she wanted none. Together we lamented her tragedy and the unfairness of life. Before she left, I felt the need to be sure that she and her husband understood that there was some risk in delay, that the problem was not going to go away, and that we needed to keep considering the options over the next several days. We agreed to meet in two days.

She returned in two days with her husband and son. They had talked extensively about the problem and the options. She remained very clear about her wish not to undergo chemotherapy and to live whatever time she had left outside the hospital. As we explored her thinking further, it became clear that she was convinced she would die during the period of treatment and would suffer unspeakably in the process (from hospitalization, from lack of control over her body, from the side effects of chemotherapy, and from pain and anguish). Although I could offer support and my best effort to minimize her suffering if she chose treatment, there was no way I could say any of this would not occur. In fact, the last four patients with acute leukemia at our hospital had died very painful deaths in the hospital during various stages of treatment (a fact I did not share with her). Her family wished she would choose treatment but sadly accepted her decision. She articulated very clearly that it was she who would be experiencing all the side effects of treatment and that odds of 25 percent were not good enough for her to undergo so toxic a course of therapy, given her expectations of chemotherapy and hospitalization and the absence of a closely matched bone marrow donor. I had her repeat her understanding of the treatment, the odds, and what to expect if there were no treatment. I clarified a few misunderstandings, but she had a remarkable grasp of the options and implications.

I have been a longtime advocate of active, informed patient choice of treatment or nontreatment, and of a patient's right to die with as much control and dignity as possible. Yet there was something about her giving up a 25 percent chance of long-term survival in favor of almost certain death that disturbed me. I had seen Diane fight and use her considerable inner resources to overcome alcoholism and depression, and I half expected her to change her mind over the next week. Since the window of time in which effective treatment can be initiated is rather narrow, we met several times that week. We obtained a second hematology consultation and talked at length about the meaning and implications of treatment and nontreatment. She talked to a psychologist she had seen in the past. I gradually understood the decision from her perspective and became convinced that it was the right decision for her. We arranged for home hospice care (although at that time Diane felt reasonably well, was active, and looked healthy), left the door open for her to change her mind, and tried to anticipate how to keep her comfortable in the time she had left.

Just as I was adjusting to her decision, she opened up another area that would stretch me profoundly. It was extraordinarily important to Diane to maintain control of herself and her own dignity during the time remaining to her. When this was no longer possible, she clearly wanted to die. As a former director of a hospice program, I know how to use pain medicines to keep patients comfortable and lessen suffering. I explained the philosophy of comfort care, which I strongly believe in. Although Diane understood and appreciated this, she had known of people lingering in what was called relative comfort, and she wanted no part of it. When the time came, she wanted to take her life in the least painful way possible. Knowing of her desire for independence and her decision to stay in control, I thought this request made perfect sense. I acknowledged and explored this wish but also thought that it was out of the realm of currently accepted medical practice and that it was more than I could offer or promise. In our discussion, it became clear that preoccupation with her fear of a lingering death would interfere with Diane's getting the most out of the time she had left until she found a safe way to ensure her death. I feared the effects of a violent death on her family, the consequences of an ineffective suicide that would leave her lingering in precisely the state she dreaded so much, and the possibility that a family member would be forced to assist her, with all the legal and personal repercussions that would follow. She discussed this at length with her family.

They believed that they should respect her choice. With this in mind, I told Diane that information was available from the Hemlock Society that might be helpful to her.

A week later she phoned me with a request for barbiturates for sleep. Since I knew that this was an essential ingredient in a Hemlock Society suicide, I asked her to come to the office to talk things over. She was more than willing to protect me by participating in a superficial conversation about her insomnia, but it was important to me to know how she planned to use the drugs and to be sure that she was not in despair or overwhelmed in a way that might color her judgment. In our discussion, it was apparent that she was having trouble sleeping, but it was also evident that the security of having enough barbiturates available to commit suicide when and if the time came would leave her secure enough to live fully and concentrate on the present. It was clear that she was not despondent and that in fact she was making deep, personal connections with her family and close friends. I made sure that she knew how to use the barbiturates for sleep, and also that she knew the amount needed to commit suicide. We agreed to meet regularly, and she promised to meet with me before taking her life, to ensure that all other avenues had been exhausted. I wrote the prescription with an uneasy feeling about the boundaries I was exploring—spiritual, legal, professional, and personal. Yet I also felt strongly that I was setting her free to get the most out of the time she had left, and to maintain dignity and control on her own terms until her death.

The next several months were very intense and important for Diane. Her son stayed home from college, and they were able to be with one another and say much that had not been said earlier. Her husband did his work at home so that he and Diane could spend more time together. She spent time with her closest friends. I had her come into the hospital for a conference with our residents, at which she illustrated in a most profound and personal way the importance of informed decision making, the right to refuse treatment, and the extraordinarily personal effects of illness and interaction with the medical system. There were emotional and physical hardships as well. She had periods of intense sadness and anger. Several times she became very weak, but she received transfusions as an outpatient and responded with marked improvement of symptoms. She had two serious infections that responded surprisingly well to empirical courses of oral antibiotics. After three tumultuous months, there were two weeks of relative calm and well-being, and fantasies of a miracle began to surface.

Unfortunately, we had no miracle. Bone pain, weakness, fatigue, and fevers began to dominate her life. Although the hospice workers, family members, and I tried our best to minimize the suffering and promote comfort, it was clear that the end was approaching. Diane's immediate future held what she feared the most—increasing discomfort, dependence, and hard choices between pain and sedation. She called up her closest friends and asked them to come over to say goodbye, telling them that she would be leaving soon. As we had agreed, she let me know as well. When we met, it was clear that she knew what she was doing, that she was sad and frightened to be leaving, but that she would be even more terrified to stay and suffer. In our tearful goodbye, she promised a reunion in the future at her favorite spot on the edge of Lake Geneva, with dragons swimming in the sunset.

Two days later her husband called to say that Diane had died. She had said her final goodbyes to her husband and son that morning, and asked them to leave her alone for an hour. After an hour, which must have seemed an eternity, they found her on the couch, lying very still and covered by her favorite shawl. There was no sign of struggle. She seemed to be at peace. They called me for advice about how to proceed. When I arrived at their house, Diane indeed seemed peaceful. Her husband and son were quiet. We talked about what a remarkable person she had been. They seemed to have no doubts about the course she had chosen or about their cooperation, although the unfairness of her illness and the finality of her death were overwhelming to us all.

I called the medical examiner to inform him that a hospice patient had died. When asked about the cause of death, I said, "acute leukemia." He said that was fine and that we should call a funeral director. Although acute leukemia was the truth, it was not the whole story. Yet any mention of suicide would have given rise to a police investigation and probably brought the arrival of an ambulance crew for resuscitation. Diane would have become a "coroner's case," and the decision to perform an autopsy would have been made at the discretion of the medical examiner. The family or I could have been subject to criminal prosecution, and I to professional review, for our roles in support of Diane's choices. Although I truly believe that the family and I gave her the best care possible, allowing her to define her limits and directions as much as possible, I am not sure the law, society, or the medical profession would agree. So I said "acute leukemia" to protect all of us, to protect Diane from an invasion into her past and her body, and to continue to shield society from the knowledge of the degree of suffering that people often undergo in the process of dying. Suffering can be lessened to some extent, but in no way eliminated or made benign, by the careful intervention of a competent, caring physician, given current social constraints.

Diane taught me about the range of help I can provide if I know people well and if I allow them to say what they really want. She taught me about life, death, and honesty and about taking charge and facing tragedy squarely when it strikes. She taught me that I can take small risks for people that I really know and care about. Although I did not assist in her suicide directly, I helped indirectly to make it possible, successful, and relatively painless. Although I know we have measures to help control pain and lessen suffering, to think that people do not suffer in the process of dying is an illusion. Prolonged dying can occasionally be peaceful, but more often the role of the physician and family is limited to lessening but not eliminating severe suffering.

I wonder how many families and physicians secretly help patients over the edge into death in the face of such severe suffering. I wonder how many severely ill or dying patients secretly take their lives, dying alone in despair. I wonder whether the image of Diane's final aloneness will persist in the minds of her family, or if they will remember more the intense, meaningful months they had together before she died. I wonder whether Diane struggled in that last hour, and whether the Hemlock Society's way of death by suicide is the most benign. I wonder why Diane, who gave so much to so many of us, had to be alone for the last hour of her life. I wonder whether I will see Diane again, on the shore of Lake Geneva at sunset, with dragons swimming on the horizon.

The Philosophers' Brief

Ronald Dworkin, Thomas Nagel, Robert Nozick, John Rawls, Thomas Scanlon, and Judith Jarvis Thomson

In 1997, the U.S. Supreme Court decided two cases, *State of Washington v. Glucksberg* and *Vacco v. Quill*, that dealt with the question of whether dying patients have a right to physician-assisted suicide. The original cases were brought by patients and physicians in Washington (*State of Washington v. Glucksberg*) and New York (*Vacco v. Quill*) who wanted state laws prohibiting doctors from prescribing lethal medication to be declared unconstitutional. Six moral philosophers prepared an amicus curiae brief, which is advice to the court by volunteers who are not a party to the litigation. The brief is reprinted here. In the brief the authors argue that individuals have a constitutionally protected liberty interest in making deeply personal decisions for themselves about how to approach death, and they should be free to make these decisions without imposition of any religious or philosophical orthodoxy by the court or legislature. They acknowledge that states have important interests that justify regulating physician-assisted suicide, but these interests do not justify an absolute prohibition.

Amici are six moral and political philosophers who differ on many issues of public morality and policy. They are united, however, in their conviction that respect for fundamental principles of liberty and justice, as well as for the American constitutional tradition, requires that the decisions of the Courts of Appeals be affirmed.

Introduction and Summary of Argument

These cases do not invite or require the Court to make moral, ethical, or religious judgments about how people should approach or confront their death or about when it is ethically appropriate to hasten one's own death or to ask others for help in doing so. On the contrary, they ask the Court to recognize that individuals have a constitutionally protected interest in making those grave judgments for themselves, free from the imposition of any religious or philosophical orthodoxy by court or legislature. States have a constitutionally legitimate interest in protecting individuals from irrational, ill-informed, pressured, or unstable decisions to hasten their own death. To that end, states may regulate and limit the assistance that doctors may give individuals who express a wish to die. But states may not deny people in the position of the patient plaintiffs in these cases

Amici curiae Ronald Dworkin, Thomas Nagel, Robert Nozick, John Rawls, Thomas Scanlon, and Judith Jarvis Thomson, "The Philosopher's Brief." *State of Washington et al. v. Glucksberg et al.* and *Vacco et al. v. Quill et al.*, argued January 8, 1997.

the opportunity to demonstrate, through whatever reasonable procedures the state might institute—even procedures that err on the side of caution—that their decision to die is indeed informed, stable, and fully free. Denying that opportunity to terminally ill patients who are in agonizing pain or otherwise doomed to an existence they regard as intolerable could only be justified on the basis of a religious or ethical conviction about the value or meaning of life itself. Our Constitution forbids government to impose such convictions on its citizens.

Petitioners [i.e., the state authorities of Washington and New York] and the amici who support them offer two contradictory arguments. Some deny that the patient-plaintiffs have any constitutionally protected liberty interest in hastening their own deaths. But that liberty interest flows directly from this Court's previous decisions. It flows from the right of people to make their own decisions about matters "involving the most intimate and personal choices a person may make in a lifetime, choices central to personal dignity and autonomy." *Planned Parenthood v. Casey*, 505 U.S. 833, 851(1992).

The Solicitor General, urging reversal in support of Petitioners, recognizes that the patient-plaintiffs do have a constitutional liberty interest at stake in these cases. *See* Brief for the United States as Amicus Curiae Supporting Petitioners at 12, *Washington v. Vacco* (hereinafter Brief for the United States) ("The term 'liberty' in the Due Process Clause . . . is broad enough to encompass an interest on the part of terminally ill, mentally competent adults in obtaining relief from the kind of suffering experienced by the plaintiffs in this case, which includes not only severe physical pain, but also the despair and distress that comes from physical deterioration and the inability to control basic bodily functions."); *see also id.* at 13 ("*Cruzan* . . . supports the conclusion that a liberty interest is at stake in this case.").

The Solicitor General nevertheless argues that Washington and New York properly ignored this profound interest when they required the patient-plaintiffs to live on in circumstances they found intolerable. He argues that a state may simply declare that it is unable to devise a regulatory scheme that would adequately protect patients whose desire to die might be ill informed or unstable or foolish or not fully free, and that a state may therefore fall back on a blanket prohibition. This Court has never accepted that patently dangerous rationale for denying protection altogether to a conceded fundamental constitutional interest. It would be a serious mistake to do so now. If that rationale were accepted, an interest acknowledged to be constitutionally protected would be rendered empty.

Argument

I. The Liberty Interest Asserted Here is Protected by the Due Process Clause

The Due Process Clause of the Fourteenth Amendment protects the liberty interest asserted by the patient-plaintiffs here.

Certain decisions are momentous in their impact on the character of a person's life—decisions about religious faith, political and moral allegiance, marriage, procreation, and death, for example. Such deeply personal decisions pose controversial questions about how and why human life has value. In a free society, individuals must be allowed to make those decisions for themselves, out of their own faith, conscience, and convictions. This Court has insisted, in a variety of contexts and circumstances, that this great freedom is among those protected by the Due Process Clause as essential to a community of "ordered liberty." *Palko v. Connecticut*, 302 U.S. 319, 325 (1937). In its recent decision in *Planned Parenthood v. Casey*, 505 U.S. 833, 851 (1992), the Court offered a paradigmatic statement of that principle:

> matters [ ] involving the most intimate and personal choices a person may make in a lifetime, choices central to a person's dignity and autonomy, are central to the liberty protected by the Fourteenth Amendment.

That declaration reflects an idea underlying many of our basic constitutional protections. As the Court explained in *West Virginia State Board of Education v. Barnette*, 319 U.S. 624, 642 (1943):

> If there is any fixed star in our constitutional constellation, it is that no official . . . can prescribe what shall be orthodox in politics, nationalism, religion, or other matters of opinion or force citizens to confess by word or act their faith therein.

A person's interest in following his own convictions at the end of life is so central a part of the more general right to make "intimate and personal choices" for himself that a failure to protect that particular interest would undermine the general right altogether. Death is, for each of us, among the most significant events of life. As the Chief Justice said in *Cruzan v. Missouri*, 497 U.S. 261, 281 (1990), "[t]he choice between life and death is a deeply personal decision of obvious and overwhelming finality." Most of us see death—whatever we think will follow it—as the final act of life's drama, and we want that last act to reflect our own convictions, those we have tried to live by, not the convictions of others forced on us in our most vulnerable moment.

Different people, of different religious and ethical beliefs, embrace very different convictions about which way of dying confirms and which contradicts the value of their lives. Some fight against death with every weapon their doctors can devise. Others will do nothing to hasten death even if they pray it will come soon. Still others, including the patient-plaintiffs in these cases, want to end their lives when they think that living on, in the only way they can, would disfigure rather than enhance the lives they had created. Some people make the latter choice not just to escape pain. Even if it were possible to eliminate all pain for a dying patient and frequently that is not possible—that would not end or even much alleviate the anguish some would feel at remaining alive, but intubated, helpless, and often sedated near oblivion.

None of these dramatically different attitudes about the meaning of death can be dismissed as irrational. None should be imposed, either by the pressure of doctors or relatives or by the fiat of government, on people who reject it. Just as it would be intolerable for government to dictate that doctors never be permitted to try to keep someone alive as long as possible, when that is what the patient wishes, so it is intolerable for government to dictate that doctors may never, under any circumstances, help someone to die who believes that further life means only degradation. The Constitution insists that people must be free to make these deeply personal decisions for themselves and must not be forced to end their lives in a way that appalls them, just because that is what some majority thinks proper.

II. This Court's Decisions in Casey and Cruzan Compel Recognition of a Liberty Interest Here

A. *Casey Supports the Liberty Interest Asserted Here.* In *Casey*, this Court, in holding that a state cannot constitutionally proscribe abortion in all cases, reiterated that the Constitution protects a sphere of autonomy in which individuals must be permitted to make certain decisions for themselves. The Court began its analysis by pointing out that "[a]t the heart of liberty is the right to define one's own concept of existence, of meaning, of the universe, and of the mystery of human life." 505 U.S. at 851. Choices flowing out of these conceptions, on matters "involving the most intimate and personal choices a person may make in a lifetime, choices central to personal dignity and autonomy, are central to the liberty protected by the Fourteenth Amendment." *Id.* "Beliefs about these matters," the Court continued, "could not define the attributes of personhood were they formed under compulsion of the State." *Id.*

In language pertinent to the liberty interest asserted here, the Court explained why decisions about abortion fall within this category of "personal and intimate" decisions. A decision whether or not to have an abortion, "originat[ing] within the zone of conscience and belief," involves conduct in which "the liberty of the woman is at stake in a sense unique to the human condition and so unique to the law." *Id.* at 852. As such, the decision necessarily involves the very "destiny of the woman" and is inevitably "shaped to a large extent on her own conception of her spiritual imperatives and her place in society." *Id.* Precisely because of these characteristics of the decision, "the State is [not] entitled to proscribe [abortion] in all instances." *Id.* Rather, to allow a total prohibition on abortion would be to permit a state to impose one conception of the meaning and value of human existence on all individuals. This the Constitution forbids.

The Solicitor General nevertheless argues that the right to abortion could be supported on grounds other than this autonomy principle, grounds that would not apply here. He argues, for example, that the abortion right might flow from the great burden an unwanted child imposes on its mother's life. Brief for the United States at 14–15. But whether or not abortion rights could be defended on such grounds, they were not the grounds on which this Court in fact relied. To the contrary, the Court explained at length that the right flows from the constitutional protection accorded all individuals to "define one's own concept of existence, of meaning, of the universe, and of the mystery of human life." *Casey*, 505 U.S. at 851.

The analysis in *Casey* compels the conclusion that the patient-plaintiffs have a liberty interest in this case that a state cannot burden with a blanket prohibition. Like a woman's decision whether to have an abortion, a decision to die involves one's very "destiny" and inevitably will be "shaped to a large extent on [one's] own conception of [one's] spiritual imperatives and [one's] place in society." *Id.* at 852. Just as a blanket prohibition on abortion would involve the improper imposition of one conception of the meaning and value of human existence on all individuals, so too would a blanket prohibition on assisted suicide. The liberty interest asserted here cannot be rejected without undermining the rationale of *Casey*. Indeed, the lower court opinions in the Washington case expressly recognized the parallel between the liberty interest in *Casey* and the interest asserted here. *See Compassion in Dying v. Washington*, 79 F.3d 790, 801(9th Cir. 1996) (en banc) ("In deciding right-to-die cases, we are guided by the Court's approach to the abortion cases. *Casey* in particular provides a powerful precedent, for in that case the Court had the opportunity to evaluate its past decisions and to determine whether to adhere to its original judgment."), *aff'g.* 850 F. Supp. 1454, 1459 (W. D. Wash. 1994) ("[T]he reasoning in *Casey* [is] highly instructive and almost prescriptive. . . ."). This Court should do the same.

B. *Cruzan Supports the Liberty Interest Asserted Here.* We agree with the Solicitor General that this Court's decision in "*Cruzan* . . . supports the conclusion that a liberty interest is at stake in this case." Brief for the United States at 8. Petitioners, however, insist that the present cases can be distinguished because the right at issue in *Cruzan* was limited to a right to reject an unwanted invasion of one's body. But this Court repeatedly has held that in appropriate circumstances a state may require individuals to accept unwanted invasions of the body. *See, e.g., Schmerber v. California*, 384 U.S. 757 (1966) (extraction of blood sample from individual suspected of driving while intoxicated, notwithstanding defendant's objection, does not violate privilege against self-incrimination or other constitutional rights); *Jacobson v. Massachusetts*, 197 U.S. 11 (1905) (upholding compulsory vaccination for smallpox as reasonable regulation for protection of public health).

The liberty interest at stake in *Cruzan* was a more profound one. If a competent patient has a constitutional right to refuse life-sustaining treatment, then, the Court implied, the state could not override that right. The regulations upheld in *Cruzan* were designed only to ensure that the individual's wishes were ascertained correctly. Thus, if *Cruzan* implies a right of competent patients to refuse life-sustaining treatment, that implication must be understood as resting not simply on a right to refuse bodily invasions but on the more profound right to refuse medical intervention when what is at stake is a momentous personal decision, such as the timing and manner of one's death. In her concurrence, Justice O'Connor expressly recognized that the right at issue involved a "deeply personal decision" that is "inextricably intertwined" with our notion of "self-determination." 497 U.S. at 287–89.

Cruzan also supports the proposition that a state may not burden a terminally ill patient's liberty interest in determining the time and manner of his death by prohibiting doctors from terminating life support. Seeking to distinguish *Cruzan*, Petitioners insist that a state may nevertheless burden that right in a different way by forbidding doctors to assist in the suicide of patients who are not on life-support machinery. They argue that doctors who remove life support are only allowing a natural process to end in death whereas doctors who prescribe lethal drugs are intervening to cause death. So, according to this argument, a state has an independent justification for forbidding doctors to assist in suicide that it does not have for forbidding them to remove life support. In the former case though not the latter, it is said, the state forbids an act of killing that is morally much more problematic than merely letting a patient die.

This argument is based on a misunderstanding of the pertinent moral principles. It is certainly true that when a patient does not wish to die, different acts, each of which foreseeably results in his death, nevertheless have very different moral status. When several patients need organ transplants and organs are scarce, for example, it is morally permissible for a doctor to deny an organ to one patient, even though he will die without it, in order to give it to another. But it is certainly not permissible for a doctor to kill one patient in order to use his organs to save another. The morally significant difference between those two acts is not, however, that killing is a positive act and not providing an organ is a mere omission, or that killing someone is worse than merely allowing a "natural" process to result in death. It would be equally impermissible for a doctor to let an injured patient bleed to death, or to refuse antibiotics to a patient with pneumonia—in each case the doctor would have allowed death to result from a "natural" process—in order to make his organs available for transplant to others. A doctor violates his patient's rights whether the doctor acts or refrains from acting, against the patient's wishes, in a way that is designed to cause death.

When a competent patient does want to die, the moral situation is obviously different, because then it makes no sense to appeal to the patient's right not to be killed as a reason why an act designed to cause his death is impermissible. From the patient's point of view, there is no morally pertinent difference between a doctor's terminating treatment that keeps him alive, if that is what he wishes, and a doctor's helping him to end his own life by providing lethal pills he may take himself, when ready, if that is what he wishes except that the latter may be quicker and more humane. Nor is that a pertinent difference from the doctor's point of view. If and when it is permissible for him to act with death in view, it does not matter which of those two means he and his patient choose. If it is permissible for a doctor deliberately to withdraw medical treatment in order to allow death to result from a natural process, then it is equally permissible for him to help his patient hasten his own death more actively, if that is the patient's express wish.

It is true that some doctors asked to terminate life support are reluctant and do so only in deference to a patient's right to compel them to remove unwanted invasions of his body. But other doctors, who believe that their most fundamental professional duty is to act in the patient's interests and that, in certain circumstances, it is in their patient's best interests to die, participate willingly in such decisions: they terminate life support to cause death because they know that is what their patient wants. *Cruzan* implied that a state may not absolutely prohibit a doctor from deliberately causing death, at the patient's request, in that way and for that reason. If so, then a state may not prohibit doctors from deliberately using more direct and often more humane means to the same end when that is what a patient prefers. The fact that failing to provide life-sustaining treatment may be regarded as "only letting nature take its course" is no more morally significant in this context, when the patient wishes to die, than in the other, when he wishes to live. Whether a doctor turns off a respirator in accordance with the patient's request or prescribes pills that a patient may take when he is ready to kill himself, the doctor acts with the same intention: to help the patient die.

The two situations do differ in one important respect. Since patients have a right not to have life-support machinery attached to their bodies, they have, in principle, a right to compel its removal. But that is not true in the case of assisted suicide: patients in certain circumstances have a right that the state not forbid doctors to assist in their deaths, but they have no right to compel a doctor to assist them. The right in question, that is, is only a right to the help of a willing doctor.

III. State Interests Do Not Justify a Categorical Prohibition on All Assisted Suicide

The Solicitor General concedes that "a competent, terminally ill adult has a constitutionally cognizable liberty interest in avoiding the kind of suffering experienced by the plaintiffs in this case." Brief for the United States at 8. He agrees that this interest extends not only to avoiding pain, but to avoiding an existence the patient believes to be one of intolerable indignity or incapacity as well. *Id.* at 12. The Solicitor General argues, however, that states nevertheless have the right to "override" this liberty interest altogether, because a state could reasonably conclude that allowing doctors to assist in suicide, even under the most stringent regulations and procedures that could be devised, would unreasonably endanger the lives of a number of patients who might ask for death in circumstances when it is plainly not in their interests to die or when their consent has been improperly obtained.

This argument is unpersuasive, however, for at least three reasons. *First*, in *Cruzan*, this Court noted that its various decisions supported the recognition of a general liberty interest in refusing medical treatment, even when such refusal could result in death. 497 U.S. at 278–79. The various risks described by the Solicitor General apply equally to those situations. For instance, a patient kept alive only by an elaborate and disabling life-support system might well become depressed, and doctors might be equally uncertain whether the depression is curable: such a patient might decide for death only because he has been advised that he die soon anyway or that he will never live free of the burdensome apparatus, and either diagnosis might conceivably be mistaken. Relatives or doctors might subtly or crudely influence that decision, and state provision for the decision may (to the same degree in this case as if it allowed assisted suicide) be thought to encourage it.

Yet there has been no suggestion that states are incapable of addressing such dangers through regulation. In fact, quite the opposite is true. In *McKay v. Bergstedt*, 106 Nev. 808, 801 P2d 617 (1990), for example, the Nevada Supreme Court held that "competent adult patients desiring to refuse or discontinue medical treatment" must be examined by two nonattending physicians to determine whether the patient is mentally competent, understands his prognosis and treatment options, and appears free of coercion or pressure in making his decision. *Id.* at 827–28, 801 P2d at 630. See also: *id.* (in the case of terminally ill patients with natural life expectancy of less than six months, [a] patient's right of self-determination shall be deemed to prevail over state interests, whereas [a] non-terminal patient's decision to terminate life-support systems must first be weighed against relevant state interests by trial judge); [and] *In re Farrell*, 108 N.J. 335, 354, 529 A.2d 404, 413 (1987) ([which held that a] terminally-ill patient requesting termination of life-support must be determined to be competent and properly informed about [his] prognosis, available treatment options and risks, and to have made decision voluntarily and without coercion). Those protocols served to guard against precisely the dangers that the Solicitor General raises. The case law contains no suggestion that such protocols are inevitably insufficient to prevent deaths that should have been prevented.

Indeed, the risks of mistake are overall greater in the case of terminating life support. *Cruzan* implied that a state must allow individuals to make such decisions through an advance directive stipulating either that life support be terminated (or not initiated) in described circumstances when the individual was no longer competent to make such a decision himself, or that a designated proxy be allowed to make that decision. All the risks just described are present when the decision is made through or pursuant to such an advance directive, and a grave further risk is added: that the directive, though still in force, no longer represents the wishes of the patient. The patient might have changed his mind before he became incompetent, though he did not change the directive, or his proxy may make a decision that the patient would not have made himself if still competent. In *Cruzan*, this Court held that a state may limit these risks through reasonable regulation. It did not hold—or even suggest—that a state may avoid them through a blanket prohibition that, in effect, denies the liberty interest altogether.

Second, nothing in the record supports the [Solicitor General's] conclusion that no system of rules and regulations could adequately reduce the risk of mistake. As discussed above, the experience of states in adjudicating requests to have life-sustaining treatment removed indicates the opposite. The Solicitor General has provided no persuasive reason why the same sort of procedures could not be applied effectively in the case of a competent individual's request for physician-assisted suicide.

Indeed, several very detailed schemes for regulating physician-assisted suicide have been submitted to the voters of some states and one has been enacted. In addition, concerned groups, including a group of distinguished professors of law and other professionals, have drafted and defended such schemes. *See, e.g.,* Charles H. Baron, *et al., A Model State Act to Authorize and Regulate Physician-Assisted Suicide,* 33 Harv. J. Legis. 1 (1996). Such draft statutes propose a variety of protections and review procedures designed to insure against mistakes, and neither Washington nor New York attempted to show that such schemes would be porous or ineffective. Nor does the Solicitor General's brief: it relies instead mainly on flat and conclusory statements. It cites a New York Task Force report, written before the proposals just described were drafted, whose findings have been widely disputed and were implicitly rejected in the opinion of the Second Circuit below. *See generally Quill v. Vacco,* 80 F.3d 716 (2d Cir. 1996). The weakness of the Solicitor General's argument is signaled by his strong reliance on the experience in the Netherlands which, in effect, allows assisted suicide pursuant to published guidelines. Brief for the United States at 23–24. The Dutch guidelines are more permissive than the proposed and model American statutes, however. The Solicitor General deems the Dutch practice of ending the lives of people like neonates who cannot consent particularly noteworthy, for example, but that practice could easily and effectively be made illegal by any state regulatory scheme without violating the Constitution.

The Solicitor General's argument would perhaps have more force if the question before the Court were simply whether a state has any rational basis for an absolute prohibition; if that were the question, then it might be enough to call attention to risks a state might well deem not worth running. But as the Solicitor General concedes, the question here is a very different one: whether a state has interests sufficiently compelling to allow it to take the extraordinary step of altogether refusing the exercise of a liberty interest of constitutional dimension. In those circumstances, the burden is plainly on the state to demonstrate that the risk of mistakes is very high, and that no alternative to complete prohibition would adequately and effectively reduce those risks. Neither of the Petitioners has made such a showing.

Nor could they. The burden of proof on any state attempting to show this would be very high. Consider, for example, the burden a state would have to meet to show that it was entitled altogether to ban public speeches in favor of unpopular causes because it could not guarantee, either by regulations short of an outright ban or by increased police protection, that such speeches would not provoke a riot that would result in serious injury or death to an innocent party. Or that it was entitled to deny those accused of crime the procedural rights that the Constitution guarantees, such as the right to a jury trial, because the security risk those rights would impose on the community would be too great. One can posit extreme circumstances in which some such argument would succeed. *See, e.g., Korematsu v. United States*, 323 U.S., 214 (1944) (permitting United States to detain individuals of Japanese ancestry during wartime). But these circumstances would be extreme indeed, and the *Korematsu* ruling has been widely and severely criticized.

Third, it is doubtful whether the risks the Solicitor General cites are even of the right character to serve as justification for an absolute prohibition on the exercise of an important liberty interest. The risks fall into two groups. The first is the risk of medical mistake, including a misdiagnosis of competence or terminal illness. To be sure, no scheme of regulation, no matter how rigorous, can altogether guarantee that medical mistakes will not be made. But the Constitution does not allow a state to deny patients a great variety of important choices, for which informed consent is properly deemed necessary, just because the information on which the consent is given may, in spite of the most strenuous efforts to avoid mistake, be wrong. Again, these identical risks are present in decisions to terminate life support, yet they do not justify an absolute prohibition on the exercise of the right.

The second group consists of risks that a patient will be unduly influenced by considerations that the state might deem it not in his best interests to be swayed by, for example, the feelings and views of close family members. Brief for the United States at 20. But what a patient regards as proper grounds for such a decision normally reflects exactly the judgments of personal ethics—of why his life is important and what affects its value—that patients have a crucial liberty interest in deciding for themselves. Even people who are dying have a right to hear and, if they wish, act on what others might wish to tell or suggest or even hint to them, and it would be dangerous to suppose that a state may prevent this on the ground that it knows better than its citizens when they should be moved by or yield to particular advice or suggestion in the exercise of their right to make fateful personal decisions for themselves. It is not a good reply that some people may not decide as they really wish—as they would decide, for example, if free from the "pressure" of others. That possibility could hardly justify the most serious pressure of all—the criminal law which tells them that they may not decide for death if they need the help of a doctor in dying, no matter how firmly they wish it.

There is a fundamental infirmity in the Solicitor General's argument. He asserts that a state may reasonably judge that the risk of "mistake" to some persons justifies a prohibition that not only risks but insures and even aims at what would undoubtedly be a vastly greater number of "mistakes" of the opposite kind—preventing many thousands of competent people who think that it disfigures their lives to continue living, in the only way left to them, from escaping that—to them—terrible injury. A state grievously and irreversibly harms such people when it prohibits that escape. The Solicitor General's argument may seem plausible to those who do not agree that individuals are harmed by being forced to live on in pain and what they regard as indignity. But many other people plainly do think that such individuals are harmed, and a state may not take one side in that essentially ethical or religious controversy as its justification for denying a crucial liberty.

Of course, a state has important interests that justify regulating physician-assisted suicide. It may be legitimate for a state to deny an opportunity for assisted suicide when it acts in what it reasonably judges to be the best interests of the potential suicide, and when its judgment on that issue does not rest on contested judgments about "matters involving the most intimate and personal choices a person may make in a lifetime, choices central to personal dignity and autonomy." *Casey*, 505 U.S. at 851. A state might assert, for example, that people who are not terminally ill, but who have formed a desire to die, are, as a group, very likely later to be grateful if they are prevented from taking their own lives. It might then claim that it is legitimate, out of concern for such people, to deny any of them a doctor's assistance [in taking their own lives].

This Court need not decide now the extent to which such paternalistic interests might override an individual's liberty interest. No one can plausibly claim, however—and it is noteworthy that neither Petitioners nor the Solicitor General does claim—that any such prohibition could serve the interests of any significant number of terminally ill patients. On the contrary, any paternalistic justification for an absolute prohibition of assistance to such patients would of necessity appeal to a widely contested religious or ethical conviction many of them, including the patient-plaintiffs, reject. Allowing *that* justification to prevail would vitiate the liberty interest.

Even in the case of terminally ill patients, a state has a right to take all reasonable measures to insure that a patient requesting such assistance has made an informed, competent, stable and uncoerced decision. It is plainly legitimate for a state to establish procedures through which professional and administrative judgments can be made about these matters, and to forbid doctors to assist in suicide when its reasonable procedures have not been satisfied. States may be permitted considerable leeway in designing such procedures. They may be permitted, within reason, to err on what they take to be the side of caution. But they may not use the bare possibility of error as justification for refusing to establish any procedures at all and relying instead on a flat prohibition.

Conclusion

Each individual has a right to make the "most intimate and personal choices central to personal dignity and autonomy." That right encompasses the right to exercise some control over the time and manner of one's death.

The patient-plaintiffs in these cases were all mentally competent individuals in the final phase of terminal illness and died within months of filing their claims.

Jane Doe described how her advanced cancer made even the most basic bodily functions such as swallowing, coughing, and yawning extremely painful and that it was "not possible for [her] to reduce [her] pain to an acceptable level of comfort and to retain an alert state." Faced with such circumstances, she sought to be able to "discuss freely with [her] treating physician [her] intention of hastening [her] death through the consumption of drugs prescribed for that purpose." *Quill v. Vacco*, 80 F.2d 716, 720 (2d Cir. 1996) (quoting declaration of Jane Doe).

George A. Kingsley, in advanced stages of AIDS which included, among other hardships, the attachment of a tube to an artery in his chest which made even routine functions burdensome and the development of lesions on his brain, sought advice from his doctors regarding prescriptions which could hasten his impending death. *Id.*

Jane Roe, suffering from cancer since 1988, had been almost completely bedridden since 1993 and experienced constant pain which could not be alleviated by medication. After undergoing counseling for herself and her family, she desired to hasten her death by taking prescription drugs. *Compassion in Dying v. Washington*, 850 F. Supp. 1454, 1456 (1994).

John Doe, who had experienced numerous AIDS-related ailments since 1991, was "especially cognizant of the suffering imposed by a lingering terminal illness because he was the primary caregiver for his long-term companion who died of AIDS" and sought prescription drugs from his physician to hasten his own death after entering the terminal phase of AIDS. *Id.* at 1456–57.

James Poe suffered from emphysema which caused him "a constant sensation of suffocating" as well as a cardiac condition which caused severe leg pain. Connected to an oxygen tank at all times but unable to calm the panic reaction associated with his feeling of suffocation even with regular doses of morphine, Mr. Poe sought physician-assisted suicide. *Id.* at 1457.

A state may not deny the liberty claimed by the patient-plaintiffs in these cases without providing them an opportunity to demonstrate, in whatever way the state might reasonably think wise and necessary, that the conviction they expressed for an early death is competent, rational, informed, stable, and uncoerced.

Affirming the decisions by the Courts of Appeals would establish nothing more than that there is such a constitutionally protected right in principle. It would establish only that some individuals, whose decisions for suicide plainly cannot be dismissed as irrational or foolish or premature, must be accorded a reasonable opportunity to show that their decision for death is informed and free. It is not necessary to decide precisely which patients are entitled to that opportunity. If, on the other hand, this Court reverses the decisions below, its decision could only be justified by the momentous proposition—a proposition flatly in conflict with the spirit and letter of the Court's past decisions—that an American citizen does not, after all, have the right, even in principle, to live and die in the light of his own religious and ethical beliefs, his own convictions about why his life is valuable and where its value lies.

Note

1. In that case, the parents of Nancy Cruzan, a woman who was in a persistent vegetative state following an automobile accident, asked the Missouri courts to authorize doctors to end life support and therefore her life. The Supreme Court held that Missouri was entitled to demand explicit evidence that Ms. Cruzan had made a decision that she would not wish to be kept alive in those circumstances, and to reject the evidence the family had offered as inadequate. But a majority of justices assumed, for the sake of the argument, that a competent patient has a right to reject life-preserving treatment, and it is now widely assumed that the Court would so rule in an appropriate case.

Motivations for Physician-Assisted Suicide:
Patient and Family Voices

Robert A. Pearlman, Clarissa Hsu, Helene Starks,
Anthony L. Back, Judith R. Gordon, Ashok J. Bharucha,
Barbara A. Koenig, and Margaret P. Battin

Robert A. Pearlman and collaborators represent a multidisciplinary group involving medicine (Pearlman), oncology (A. Back), anthropology (C. Hsu and B. Koenig), public health (H. Starks), psychology (J. Gordon), psychiatry (A. Bharucha), nursing (B. Koenig), and philosophy (M. Battin). Pearlman's work focuses on decision making near the end of life and quality improvement in clinical and organization ethics. In this qualitative study the authors characterize the motivations (and the interactions between motivating factors) for pursuit of physician-assisted death. The primary motivating factors reflect many of the issues identified in other studies that motivate forgoing of life-sustaining treatment and serve as a template of topics for discussion between health care providers and patients about the effects of illness, and fears about physical decline, functional losses and the dying experience.

The motivation to pursue physician-assisted suicide (PAS) has been an important issue in the debates regarding the legality and appropriate response to requests for PAS. Understanding the motivation is critically important to physicians because many of them have been asked for assistance in PAS.[1-3] Previously described explanations include inadequate treatment for pain or other symptoms (i.e., inadequate palliative care),[1-7] psychiatric problems (e.g., depression, hopelessness),[7-13] and concerns about losses (e.g., function, control, sense of community, sense of self).[5,6,14-17]

However, these explanations are principally based on three sources of evidence: physicians' impressions, patients' reports of hypothetical circumstances under which they would consider PAS, and survey data from Oregon. There is limited direct reporting from patients and family members about what drives patients to pursue PAS.

To address the current gaps in our understanding, we conducted a longitudinal qualitative study with patients who seriously pursued PAS, and their family members. This study expands the medical and ethics literature on end-of-life care by providing a detailed descriptive account of the pursuit of PAS from the perspective of patients and their family members.

Robert A. Pearlman, Clarissa Hsu, Helene Starks, Anthony L. Back, Judith R. Gordon, Ashok J. Bharucha, Barbara A. Koenig, Margaret P. Battin, "Motivations for Physician-Assisted Suicide: Patient and Family Voices," *Journal of General Internal Medicine* August 6, 2004, 234–239.

Methods

Participants

A detailed account of the recruitment and methods is described elsewhere.[18] Briefly, we recruited patients who were seriously pursuing PAS and their family members (ongoing cases), as well as family members of persons who had seriously pursued and/or died of PAS (historical cases). We recruited participants through advocacy organizations that counsel people interested in a hastened death (see Table 1 for definitions), hospices, and grief counselors. The referral sources sent information to their clients and/or verbally informed them about the study. All participants voluntarily contacted us. Patients were screened for decisional incapacity, which, if found, would have precipitated a series of actions to protect the patient. Specifically, we looked for evidence that would suggest that pursuit of a hastened death was motivated by a psychiatric disorder (e.g., severe depression, delusions).

To protect respondents' confidentiality, we destroyed all records with personal identifiers and removed identifying information from transcripts. All study procedures were reviewed and approved by the university's Institutional Review Board.

Data Collection

We conducted qualitative, semistructured interviews with patients and family members. Five investigators conducted interviews; for each family, the same investigator interviewed all members. In total, we conducted 159 interviews with 60 participants concerning 35 patients between April 1997 and March 2001. Patients and family members for ongoing cases were interviewed at enrollment and at approximately 3-month intervals until the patient's death. Family members of deceased patients (ongoing and historical cases) were interviewed on average 2.4 times.

TABLE 1   Definitions

Physician-assisted suicide (PAS). The physician prescribes medications to the patient knowing that the patient plans to self-administer them to cause death and end suffering.

Voluntary euthanasia. At the request of a patient, the physician or another person causes the patient's death by giving a medication. Herein, the term also refers to family members giving medications to patients (who have previously requested PAS) to hasten their death, and thus end suffering.

Hastened death. A lay term for physician-assisted suicide. This term was used by many of our informants who thought it represented a neutral term that differentiates aid-in-dying for people with terminal illness from suicide by those with mental illness. In the narrative of this paper, we use the term hastened death, rather than PAS and voluntary euthanasia, to reflect the views of our informants.

The interview guide included open-ended questions about the history of the illness, the reasons for and other factors influencing the pursuit of a hastened death, and the manner of death. Additional details of the interview are presented elsewhere.[18] To enhance trustworthiness of the data, all interviews were audiotaped and transcribed. In addition, all members of the multidisciplinary research team read the transcripts in their entirety, discussed them at weekly meetings, and generated follow-up questions for the interviewer, when appropriate.

Analyses

These results are based on multiple readings of the entire transcripts. Using content analysis methods,[19] the team developed primary codes, such as *reasons for a hastened death* and *catalytic factors*, to classify sections of the transcripts. The interviewer and another investigator independently coded all transcripts and met to resolve coding disagreements. Significant coding discrepancies were discussed and resolved at weekly team meetings.

For each case, two investigators (R.A.P. and C.H.) independently reread the relevant sections of the transcripts, and identified the apparent motivations for the patients' pursuit of a hastened death. Respondents frequently volunteered motivations in the context of their stories, and when they did not, the interviewers specifically probed for explanations of their interest in hastening their death. These two investigators developed detailed memos for each patient, rating the relative importance of each identified issue based on the context of the patient's overall story and the emphasis given to the issues in the narrative (see Table 3 footnote for a description of the rating process). In the cases with both patient and family member interviews ($n = 12$) and the 5 historical cases with multiple family members, we found that respondents reported similar issues. Thus, our judgments of importance were informed by these multiple reports.

The two internists (R.A.P. and A.L.B.) independently reviewed the narratives of the 25 patients who hastened their death to estimate each patient's life expectancy at the time of death. They concurred in their assessments for 23 of the 25 cases, and after discussion resolved the two disagreements. The psychiatrist (A.J.B.) reviewed the transcripts for evidence of major depression and formulated a psychiatric profile for each case.[20] The interview guide did not include a depression questionnaire due to concerns that this would decrease subject participation. A more detailed description of this review is reported elsewhere.[20]

Results

Participant Characteristics

We studied 35 cases of patients who pursued a hastened death (Table 2). These participants are described in detail elsewhere.[18,20] In brief, all patients were white, over half were married or living with a partner, and nearly half were widowed or divorced. Approximately one third of the participants were Protestant, 17% reported no religious preference, and 16% reported being atheist. Two thirds of

TABLE 2   Patient Characteristics

Characteristic	Historical Cases* (N = 23)	Ongoing Cases* (N = 12)	Total Patients (N = 35)
Mean age, y (range)	66 (33–99)	72 (60–89)	68 (33–99)
Gender (n, female)	11	6	17
Underlying illness, n			
Cancer	14	8	22
AIDS	4	1	5
Neurological diseases	4	1	5
Other[†]	1	2	3
Life expectancy at time of hastened death[‡]			
< 1 week	6	4	10
1–4 weeks	7	0	7
1–6 months	4	1	5
> 6 months	1	2	3
Past history of probable or possible major depression[§]	5	6	11
Possible major depression without decisional incapacity during the planning phase[§]	3	0	3

* The data for 23 historical cases were obtained from 52 interviews with 28 family members of patients who had seriously pursued and/or died of patient-assisted suicide (PAS). The data for ongoing cases were obtained by interviewing 12 patients actively pursuing PAS and 20 of their family members prior to the patient's death (35 and 41 interviews, respectively). After these patients died, we conducted 31 additional interviews with their family members.

† Other includes the following diagnoses: autoimmune disease, bronchiolitis obliterans, and a debilitating unexplained pain syndrome.

‡ Estimated life expectancy death based on independent assessments by two internists (R.A.P. and A.L.B.).

§ Basing on available data from multiple participants, the psychiatrist (A.J.B.) formulated a clinical impression of possible or probable major depression and decisional incapacity for each case. Probable depression was inferred when the narrative indicated that the patient met Diagnostic and Statistical Manual of Mental Disorders, Fourth Edition (DSM-IV) symptom and duration criteria, and the narrative provided additional support including reported effects on functioning. Possible depression was inferred when symptom criteria were met, but the symptom duration was uncertain. One of the 11 past histories of major depression was thought to be probable and it occurred in a retrospective case; the rest were classified as possible. All cases of major depression during the planning stage were thought to be possible.

the patients drew strength from their spiritual beliefs; the remainder did not consider spirituality in their deliberations about hastening their death. Nineteen patients had received disease-modifying treatment or attempts at curative treatment earlier in the course of their illnesses.

While all patients seriously pursued a hastened death, 17 patients self-administered medications, 8 patients were too ill so family members administered the medications, and 1 patient used a shotgun after he was unable to obtain medications. Eight patients died of their underlying illness and 1 patient was alive at the study's conclusion.

Motivating Issues

Our analysis identified 7 common influential issues within 3 categories: illness-related experiences, changes in the person's sense of self, and fears about the future (Table 3). The case in Table 4 illustrates how these issues evolve and interact over time.

Illness-related experiences

Feeling weak, tired, and uncomfortable. In 24 cases, physical changes and symptoms were judged by the investigators as influential in the individual's pursuit of hastened

TABLE 3   Motivating Issues for Pursuing a Hastened Death*

Motivating Issues	Historical Cases ($N = 23$)	Ongoing Cases ($N = 12$)	Total Patients ($N = 35$)
Illness-related experiences			
Feeling weak, tired, and uncomfortable	18	6	24
Pain and/or unacceptable side effects of pain medication	10	4	14
Loss of function	18	5	23
Sense of self			
Loss of sense of self	17	5	22
Desire for control	15	6	21
Fears about the future			
Fears about future quality of life and dying	12	9	21
Negative past experiences with dying	11	6	17

* Motivating issues were rated independently by two investigators (R.A.P. and C.H.) as to their role in influencing the pursuit of a hastened death. Four response categories were recorded: not mentioned, present but not judged to be influential, influential, and very influential. The latter two ratings are combined and reported in the table.

TABLE 4   Illustrative Case Demonstrating Dynamic and Interactive Motivations
for Pursuing a Hastened Death*

When Anna was 62 years old she was diagnosed with metastatic ovarian cancer. Throughout
her life, Anna was organized, energetic, and athletic. She was actively involved in community
activities; she was a professional, and an involved grandparent. Her philosophy about her
illness was, "I'm trying not to change my life and let cancer steal any more of it than it has
to." She also had long-standing beliefs about having control over her life and death, saying,
". . . it should be up to me to decide . . . when I've had enough suffering. . . . One of my
landmarks, if I'm at the point where all I can do is lie on a bed all day long, then to me that's
probably not living anymore."

Anna tried numerous different anticancer treatments. Over four years she underwent
multiple surgeries and rounds of chemotherapy and radiation. Many of her treatments
were quite uncomfortable. She reported, "[I was] deathly ill after every [chemo]
treatment, just not even able to read or barely even watch television. I would wake up
in the morning with dry heaves and being incontinent and rolling out of bed so I
wouldn't get the bed messed up. It was really wretched." She talked about weighing
the burdens of treatment against the burdens of disease, and decided to obtain the
medication needed for hastening death. Anna talked to her family and after receiving
their permission, obtained medications so she could hasten her death if necessary. She
reported, "I felt I had more energy to fight the cancer and just to live in the present
time. It just took a big weight off my shoulders somehow, knowing at least that that
was one thing that maybe I didn't have to worry about." It was another three years
before Anna used the medication to hasten her death.

Anna also experienced painful complications, including bowel obstruction and spinal
cord compression. Despite enduring significant amounts of pain, Anna never cited pain
as the motivation for a hastened death. Her primary concern was dying in a hospital,
"away from my home with familiar people and familiar surroundings and some privacy
and some control."

After exhausting her anticancer treatment options, Anna became very weak, sleeping
much of the day, and unable to perform many of her routine functions. She started
bleeding uncontrollably from her bowel as a late side effect of abdominal radiation and
was told that the only treatment was repeat transfusions. Her physical frailty, prospect
of a future in the hospital receiving continuous transfusions, and the loss of control all
felt like her worst fears were coming true. After returning home from a transfusion
only to immediately begin bleeding again, she told her husband, "Honey, this is it, I
can't do this anymore." Over the next 36 hours, Anna gathered her family together to
say good-bye. She ingested the medication to hasten her death with 12 loved ones in
attendance and died within 2 hours.

* Adapted with permission from New York State Task Force on Life and the Law.

death. Many different symptoms became unacceptable (e.g., shortness of breath, fatigue, diarrhea). The effects of medications and treatments also were an issue. One respondent reported that his partner, who had severe thrush due to AIDS, lost 3 days each week to Amphotericin B, which was "horrible for him." Another patient described her response to steroids:

> The side effects of the treatment are unacceptable . . . the Prednisone destroys you. For example, it destroys your muscles. My thighs are so weak I can't get up from the floor, and I don't have the energy to exercise. The whole thing is a vicious circle. . . . My face . . . looks like a melon. . . . I look like a frog in heat. (Case 23)

The participants' symptoms often shared several qualities: they caused suffering, were expected to get worse, interfered with the patient's functioning and quality of life, and contributed to undermining the person's identity and sense of self. As one woman with ovarian cancer stated, ". . . the terrible weakness and the nausea and just not feeling like you can do anything. . . . And it's kind of like goals that I actually have or things that I want to accomplish are slowly being taken away . . . it's kind of like the realm of the possible . . . is shrinking" (Case 2).

Pain and/or unacceptable side effects of pain medications. Pain, judged to be influential in 14 cases, functioned as a motivation in several ways: it could be unbearable, preoccupying, or consuming. One patient reported,

> The pain could happen immediately or it could happen an hour or two later. And then I have to see about seeing [my provider] again. It is a treadmill that I'm on; I can't get off of it, and I've had it. And I can't live like this anymore. (Case 30)

In addition, a few patients worried about the unacceptable, mind-altering effects of pain medications. A woman with cancer explained,

> Well, the pain that I had before with the rheumatoid arthritis I knew that I could handle—. . . . But this pain that I have, I'm not sure—I can't get rid of it with the pain medicine always. . . . To give me enough to keep that pain under control, they'd have to put me out, and I don't want my son to have to take care of a bed patient. (Case 6)

Loss of function. For two thirds of our participants, loss of function, ranging from losing the ability to read the newspaper or socialize with friends to the inability to eat and go to the bathroom, motivated patients' interest in a hastened death. These losses were inextricably intertwined with these patients' physical changes.

Patients and their families viewed functional losses as markers of the patient's transition from life to death. A number of patients viewed the onset of incontinence or the inability to get to the bathroom as a sentinel event in their decision-making process. A daughter described her mother's experience stating,

> She was totally bedridden. She was messing her sheets and stuff like this, and Mother just—I mean, she's just—she was a very fastidious person. And she just—she—well, basically, she thought the quality of her life was appalling. She couldn't do anything. All she could do was lie in bed. (Case 26)

Many patients accommodated over time to functional losses. Eventually, however, the losses became too great. As one family member explained,

> [My husband had] let go of so many things along the way and kind of made do. [He'd say], "Okay, well, now I can't walk anymore. Well, I sure like being on this couch." Then he lost something else. . . . But when he could no longer take in fluids, I think that really kind of pissed him off, because he had just been saying, "God, I'm so glad I have this Gatorade. This is the best. This is keeping me alive." . . . I think he couldn't find any more pleasure. (Case 21)

One woman's account exemplifies how these losses affected her mother's sense of self and attitudes about dying: "The things that were meaningful to [my mother] in her life were her art, her ability to do her art and her friends, and spending time with her friends and cooking and eating. And she was . . . very convinced that when she couldn't do any of those things anymore, her life would be meaningless, and she wouldn't want to live anymore" (Case 7).

Sense of self

Loss of sense of self. Almost two thirds of participants pursued a hastened death because they were concerned about how dying was eroding their sense of self. Patients expressed concern about losing their personality, "source of identity," or "essence." Without the ability to maintain aspects of their life that defined them as individuals, life lost its meaning and personal dignity was jeopardized. "I'm not comfortable, and I can't do anything, so as far as I'm concerned in quality of life I'm not living; I'm existing as a dependent non-person. I've lost, in effect, my essence" (Case 23). One family member explained that her mother realized that "she was going to lose significant ability to be the person she was" (Case 1). The partner of a patient with AIDS stated:

> He didn't want to kill himself; he didn't want to die. It was about finding any method to be vital and the list was narrowed down to the most—the simplest things, and when they were gone, he didn't have a reason. . . . So it wasn't just the diarrhea or the lack of driving; it was just losing, like, his definition—what his sense of vitality was. And when that was gone, then he was ready. (Case 19)

Several patients mentioned that they did not want to be remembered as ill and frail. One patient reported, ". . . not wanting to be seen by those that love me as this skin-and-bone frail, demented person. In other words, I don't want that image of me for me, and I don't want that image to be kind of a last image that my daughters and loved ones have of me. And that's just a dignity issue" (Case 4).

For some, being cared for and losing independence was an assault on their sense of self. For these individuals, sense of self was closely linked to their desire for control. One daughter described her mother's reaction to her favorite hospice nurse's care for a fecal impaction:

> The nurse was over there, basically, manually helping her along. . . . And she just said, "This is not worth it." . . . And a lot [had] to do with her as a person where

she just was so independent. The whole idea of nursing to her was just abhorrent. (Case 34)

Desire for control. In 21 cases, the patient's desire for a hastened death was linked to a long-standing sense of independence and desire to maintain control over future events. One family member described her mother as "an extraordinarily independent person, absolutely needing to be in control of her life all the time and already felt—how shall I put it—she had problems with feeling not in control" (Case 7). Another woman with lung cancer described her attitude toward hastened death as,

> I will do things my way and the hell with everything and everybody else. Nobody is going to talk me in or out of a darn thing . . . what will be, will be; but what will be, will be done my way. I will always be in control. (Case 3)

Fears about the future

Fears about future quality of life and dying. While many motivational issues were based on current experiences, another common motivation was fear about the future. We judged this as influential in 21 cases. Such concerns were often affected by past experiences. For example, one patient's fears about pain and pain management were rooted in her past experience with pain due to a lifetime of severe arthritis. She told us, "I don't want to get to the place where I'm rum-dummy with morphine, because I almost reached that spot . . . and I couldn't even make out a check" (Case 6).

Fears were usually associated with other motivating issues, such as loss of control, physical and functional decline, becoming a burden on family (noted to be influential in 3 cases), and loss of one's sense of self. However, what separated the fears from the other issues was their anticipatory nature. As one family member stated, "He said that he doesn't want to just turn into this vegetable kind of person where you're not aware of what's going on, and that everybody around you is affected; everybody's having to take care of you, feed you, clean you, give your medication" (Case 4). Often these fears pertained to lingering or prolonging death through the use of medical technologies. One family reported, "Living there and existing for three, four, five, six months. Living with tubes coming out of every orifice . . . that's what frightened her" (Case 14).

Negative past experiences with dying. In half the cases, negative personal experiences with the death of a loved one added to patients' interest in hastened death. One patient reported the following reaction to his mother's dying experience:

> [T]here's no question about wanting to make provisions for a hastened death should conditions become so unbearable. I want to spare my family as much of that grief as I can. . . . [My mother] died of cancer, and we were constantly frustrated by not being able to do anything for her. . . . And just watched her waste away. And what a terrible way to go. (Case 24)

Discussion

Our data suggest that the pursuit of a hastened death was motivated by multiple, interactive factors in the context of progressive, serious illness. These patients considered a hastened death over prolonged periods of time and repeatedly assessed the benefits and burdens of living versus dying. None of the participants cited responding to bad news, such as the diagnosis of cancer, or a depressed mood as motivations for interest in hastened death. Lack of access to health care and lack of palliative care also were not mentioned as issues of concern. These findings are comparable to those reported in Oregon.[6,16]

This report emphasizes the importance of 3 general sets of issues: the effects of illness (e.g., physical changes, symptoms, functional losses), the patient's sense of self (e.g., loss of sense of self, desire for control), and fears about the future. The cases also illustrate that pain is often not the most salient motivating factor. Thus, this report corroborates and expands known findings.[2,5,6,15,16,21–23] This research adds to the literature by providing rich descriptions from patients and family members about interactions between these issues and the meaning that patients ascribe to current and/or anticipated illness experiences.

Many participants identified the effects of the illness on two very personal attributes that often give life meaning: a person's desire for control and sense of self. When the effects of the illness and/or treatment attack these deeply personal values, a hastened death is viewed as a means to stop this process and minimize the damage. These feelings have been reported among patients with AIDS.[17,24]

The influence of some of the issues in this study differed from previous reports. For example, while the effect of pain on patients' decisions to hasten death has been widely discussed, our participants mentioned pain much less frequently than they mentioned the loss of meaningful activities and physical functioning.[4,25] Similarly, burden on family was influential in only 3 cases, although this may reflect that family members were the reporters for two thirds of our cases.[2,15,21]

Depression and hopelessness have been suggested as causal factors in the pursuit of a hastened death[7–13] because they often precede suicide attempts among patients who are not terminally ill,[26] and studies of depressed patients with HIV and cancer have documented interest in PAS.[9,11,27–30] Depression and hopelessness were not significant issues for our sample, although fears about future quality of life and dying may reflect hopelessness when it is understood to mean negative expectancies about the future and one's ability to change it. In the 3 patients with possible depression, their interest in a hastened death preceded any alteration in mood, and thus, in our judgment, their possible depression did not impinge on decisional capacity.

Importantly, other forms of psychological suffering motivated the patients in this study toward a hastened death. They experienced severe losses (e.g., bodily integrity, functioning, control) as existential suffering that undermined their personal sense of who they were.[31] This loss of sense of self (often described in terms of a loss of vitality, essence, personal definition) highlights the threats of dying to the social construction of life's meaning.[32] This may be especially salient among individuals living in a secular culture.

Two minor differences between the ongoing and historical cases are noteworthy. Patients seeking a hastened death more frequently expressed their fears and expressed their ongoing deliberations about decisions. Family members presented more of a complete story about the patients' illness. These differences are not surprising based on the different vantage points of the participants. Overall, however, similar issues were reported, lending support to the validity of the motivating issues we identified.

The results should be viewed in the context of the study's limitations. Our participants were a highly self-selected group: they were recruited from advocacy organizations that counsel patients interested in PAS and agreed to participate. Thus, these patients may not be representative of others who pursue a hastened death. In addition, depression may be underrepresented because 1) depressed patients may volunteer less for research, 2) our indirect, informal assessment may have been insufficient, and 3) depression may have served as an exclusion by the advocacy organizations for providing support.

Several important implications for clinicians emerge from these cases. First, the dynamic and interactive nature of the motivations challenges health care providers to understand the holistic illness and dying experience of patients. These data confirm the recommendation, espoused in high-quality palliative care, that providers repeatedly assess the patient's concerns about losses and dying in order to understand and tailor end-of-life care to the patient's changing personal experience.[4] Second, the motivating issues can serve as an outline of topics for talking to patients about the far-reaching effects of illness, including the quality of the dying experience. Clinicians should explore a patient's fears, and how the patient sees herself in light of current and future physical decline and functional losses.[33] A patient's request for assistance with a hastened death should generate a thorough evaluation of the patient's motives and attempts at ameliorating the patient's suffering.

We especially wish to thank the participants. The Greenwall Foundation and the Walter and Elise Haas Fund provided funding for this research. The Veterans Health Administration (VHA) and the Health Services Research and Development Service of the Department of Veterans Affairs provided additional support. Kathleen Foley, Ezekiel Emanuel, and Susan Block gave valuable guidance and/or feedback on earlier drafts. Drs. Pearlman, Bach, and Koenig were Faculty Scholars in the Project on Death in America (PDIA) of the Open Society Institute. The views expressed in this article are those of the authors and do not necessarily represent the views of the funding sources, Department of Veterans Affairs, Project on Death in America, University of Washington, University of Pittsburgh, or persons mentioned above.

References

1. Doukas DJ, Waterhouse D, Gorenflo DW, Seid J. Attitudes and behaviors on physician-assisted death: a study of Michigan oncologists. J Clin Oncol. 1995;13:1055–61.

2. Meier DE, Emmons CA, Wallenstein S, Quill T, Morrison RS, Cassel CK. A national survey of physician-assisted suicide and euthanasia in the United States. N Engl J Med. 1998;338:1193–201.

3. Emanuel EJ, Fairclough D, Clarridge BC, et al. Attitudes and practices of U.S. oncologists regarding euthanasia and physician-assisted suicide. Ann Intern Med. 2000;133:527–32.

4. Foley KM. Competent care for the dying instead of physician-assisted suicide. N Engl J Med. 1997;336:54–8.

5. Ganzini L, Nelson HD, Schmidt TA, Kraemer DF, Delorit MA, Lee MA. Physicians' experiences with the Oregon Death with Dignity Act. N Engl J Med. 2000;342:557–63.

6. Sullivan AD, Hedberg K, Fleming DW. Legalized physician-assisted suicide in Oregon—the second year. N Engl J Med. 2000;342:598–604.

7. Quill TE, Meier DE, Block SD, Billings JA. The debate over physician-assisted suicide: empirical data and convergent views. Ann Intern Med. 1998; 128:552–8.

8. Block SD, Billings JA. Patient requests to hasten death. Evaluation and management in terminal care. Arch Intern Med. 1994;154:2039–47.

9. Breitbart W, Rosenfeld B, Pessin H, et al. Depression, hopelessness, and desire for hastened death in terminally ill patients with cancer. JAMA. 2000; 284:2907–11.

10. Chochinov HM, Wilson KG, Enns M, Lander S. Depression, hopelessness, and suicidal ideation in the terminally ill. Psychosomatics. 1998;39:366–70.

11. Emanuel EJ, Fairclough DL, Daniels ER, Clarridge BR. Euthanasia and physician-assisted suicide: attitudes and experiences of oncology patients, oncologists, and the public. Lancet. 1996;347:1805–10.

12. Emanuel EJ, Fairclough DL, Emanuel LL. Attitudes and desires related to euthanasia and physician-assisted suicide among terminally ill patients and their caregivers. JAMA. 2000;284:2460–8.

13. Ganzini L, Johnston WS, McFarland BH, Tolle SW, Lee MA. Attitudes of patients with amyotrophic lateral sclerosis and their care givers toward assisted suicide. N Engl J Med. 1998;339:967–73.

14. Bachman JG, Doukas DJ, Lichtenstein RL, Alcser KH. Assisted suicide and euthanasia in Michigan. N Engl J Med. 1994;331:812–3.

15. Back AL, Wallace JI, Starks HE, Pearlman RA. Physician-assisted suicide and euthanasia in Washington State. Patient requests and physician responses. JAMA. 1996;275:919–25.

16. Chin AE, Hedberg K, Higginson GK, Fleming DW. Legalized physician-assisted suicide in Oregon—the first year's experience. N Engl J Med. 1999; 340:577–83.

17. Lavery JV, Boyle J, Dickens BM, Maclean H, Singer PA. Origins of the desire for euthanasia and assisted suicide in people with HIV-1 or AIDS: a qualitative study. Lancet. 2001;358:362–7.

18. Back AL, Starks H, Hsu C, Gordon JR, Bharucha A, Pearlman RA. Clinician-patient interactions about requests for physician-assisted suicide: a patient and family view. Arch Intern Med. 2002;162:1257–65.

19. Morse J, Field PA. Qualitative Research Methods for Health Professionals. Thousand Oaks, CA: Sage Publications; 1995.
20. Bharucha A, Pearlman RA, Back AL, Gordon JR, Starks H, Hsu C. The pursuit of physician-assisted suicide: role of psychiatric factors. J Palliat Med. 2003;6:873–83.
21. van der Maas PJ, van Delden JJ, Pijnenborg L, Looman CW. Euthanasia and other medical decisions concerning the end of life. Lancet North Am Ed. 1991;338:669–74.
22. Ganzini L, Harvath TA, Jackson A, Goy ER, Miller LL, Delorit MA. Experiences of Oregon nurses and social workers with hospice patients who requested assistance with suicide. N Engl J Med. 2002;347:582–8.
23. Ganzini L, Dobscha SK, Heintz RT, Press N. Oregon physicians' perceptions of patients who request assisted suicide and their families. J Palliat Med. 2003;6:381–90.
24. Kohlwes RJ, Koepsell TD, Rhodes LA, Pearlman RA. Physicians' responses to patients' requests for physician-assisted suicide. Arch Intern Med. 2001;161:657–63.
25. New York State Task Force on Life and the Law. When Death Is Sought: Assisted Suicide and Euthanasia in the Medical Context. Albany, NY, 1994.
26. Beck AT, Steer RA, Kovacs M, Garrison B. Hopelessness and eventual suicide: a 10-year prospective study of patients hospitalized with suicidal ideation. Am J Psychiatry. 1985;142:559–63.
27. Chochinov HM, Wilson KG. The euthanasia debate: attitudes, practices and psychiatric considerations. Can J Psychiatry. 1995;40:593–602.
28. Breitbart W, Rosenfeld BD, Passik SD. Interest in physician-assisted suicide among ambulatory HIV-infected patients. Am J Psychiatry. 1996;153:238–42.
29. Humphry D. Final Exit: The Practicalities of Self-deliverance and Assisted Suicide for the Dying, 2nd ed. New York, NY: Dell; 1997.
30. Rosenfeld B, Breitbart W. Physician-assisted suicide and euthanasia. N Engl J Med. 2000;343:151; discussion 151–3.
31. Cassell EJ. The Nature of Suffering and the Goals of Medicine, 2nd ed. New York, NY: Oxford University Press; 2004.
32. Kaufman SR. The Ageless Self: Sources of Meaning in Late Life. Madison: University of Wisconsin Press; 1986.
33. Bascom PB, Tolle SW. Responding to requests for physician-assisted suicide: "These are uncharted waters for both of us. . . ." JAMA. 2002;288:91–8.
34. Pearlman RA, Starks H. Why do people seek physician-assisted death? In: Quill T, Battin MP, eds. Physician-Assisted Dying: The Case for Palliative Care and Patient Choice. Baltimore, MD: Johns Hopkins University Press; 2004;92–3.

QUESTIONING THE PRACTICE
OF BIOETHICS

૨♠

Western Bioethics on the Navajo Reservation:
Benefit or Harm?

Joseph A. Carrese and Lorna A. Rhodes

Joseph Carrese is a general internist and Associate Professor of Medicine at the Johns Hopkins University School of Medicine and the Phoebe R. Berman Bioethics Institute. He practiced medicine for four years in the U.S. Public Health Service on the Navajo Indian reservation in Northeast Arizona. Lorna Rhodes is an anthropologist (Ph.D. from Cornell University) and Professor of Anthropology at the University of Washington. This article examines traditional Navajo perspectives regarding discussion of negative information in order to identify the limits of Western bioethical analysis in dealing with patients of non-Western cultures.

The United States is a pluralistic society, consisting of people from many different traditions and from diverse cultural backgrounds. Accordingly, not all patients share the values and moral perspectives of dominant society as currently reflected in mainstream Western biomedicine and bioethics.[1,2] Surprisingly, little research has been done on the variability of patients' values and moral perspectives by community as compared with prevailing societal, biomedical, and bioethical views. Yet, as demonstrated by two recent studies, one comparing Mexican-American and Anglo-American attitudes toward autopsies[3] and the other comparing attitudes about end-of-life care among African Americans, Hispanics, and non-Hispanic whites,[4] there are important differences to be appreciated. . . .

In the culture of Western biomedicine[5] and bioethics, the principles of autonomy and patient self-determination are centrally important.[6] Consequently, explicit and direct discussion of negative information between health care providers and patients is the current standard of care. For example, informed consent requires disclosing the risks of medical treatment, truth telling requires disclosure of bad news, and advance care planning requires patients to consider the possibility of a serious future illness. Physicians have been criticized for failing to meet these standards.[7]

In traditional Navajo culture, it is held that thought and language have the power to shape reality and to control events. Discussing the potential complications of diabetes with a newly diagnosed Navajo patient may, in the view of the traditional patient, result in the occurrence of such complications. . . .

In this study we were interested to learn how health care providers should approach the discussion of negative information with Navajo patients. Achieving a better understanding of the Navajo perspective on these issues might result in more culturally appropriate medical care for Navajo patients in Western hospitals and clinics. Finally, this inquiry provided the opportunity to consider the limitations of dominant Western bioethical perspectives.

Methods

Design

The study was a focused ethnography.[8] Ethnography is helpful in understanding the differences between various cultures and systems of meaning.[9] As a set of qualitative research methods, ethnography uses techniques such as participant observation and in-depth interviewing to generate, rather than test, hypotheses[10]; the intent of these approaches is to minimize the possibility that the nature and conduct of the inquiry itself will miss or exclude relevant information.[11]

Study Site and Population

Fieldwork was conducted between February 1993 and March 1994 on four trips to the Navajo Indian reservation, which is located primarily in northeast Arizona, but also includes portions of southern Utah and northwest New Mexico. The Navajo reservation is approximately 25,000 square miles, about the size of West Virginia; it is the largest Indian reservation in the United States.[12] . . .

Historically, the Navajo relationship with dominant society has been marked by conflict. Prominent examples include the military campaign of Kit Carson in 1863, the 300-mile Long Walk and subsequent incarceration of tribal members at the Bosque Redondo in New Mexico from 1864 to 1868, and the livestock reduction program of the 1930s.[13]

Sampling

Two sampling strategies were used in this study. First, purposeful or judgment sampling was used in the recruitment of several key informants, such as public health nurses, community health representatives, a Veterans Affairs representative, a social worker, and mental health workers. Judgment sampling was also used by the key informants, who were primarily responsible for the selection of study informants. Second, in a few cases informants themselves identified others who should be interviewed; this represents network or snowball sampling.[14]

These approaches generated a group of informants from different tribal clans and having different locations of residence and medical problems.

Informants

Thirty-four Navajo informants were interviewed, 16 men and 18 women. The age range was 26 to 87 years, with a median of 58.5 years and a mean of 60 years. We deliberately included a subgroup of eight Navajo biomedical health care providers because they were in a unique position to comment on the traditional Navajo culture as well as the Western biomedical culture.

Of the remaining 26 informants, at least six functioned in some capacity as traditional diagnosticians or healers, and all had received traditional medical services. Seventeen (65%) spoke only Navajo during the interviews, requiring the use of an interpreter. . . .

Data Sources

In-depth, open-ended interviews were conducted by the principal investigator (J.A.C.), primarily in the homes of the informants. The interviews, which averaged 1 to 2 hours each, were audiotaped and then transcribed. Second interviews were conducted if informants were willing and if the first session indicated that more could be learned from an additional meeting; this was the case for six informants.

Informants were asked to talk about their experiences with Western biomedical providers and institutions. Particular attention was paid to informants' views about the disclosure of risk and bad news as these issues emerged in the context of their stories. Probe questions were asked to clarify and further explore confusing or seemingly contradictory information.

We solicited the views of 22 of the 34 informants about the idea of advance care planning. The 12 informants who were not asked included seven patients, four (of eight) Navajo biomedical health care providers, and one traditional provider. The decision to ask patients and traditional providers about advance care planning was made after the first five interviews were conducted. These interviews included four patients. Subsequently, the decision was made to ask Navajo biomedical health care providers about advance care planning as well, because the data we were gathering from informants led us to ask not only how information about advance care planning should be discussed, but whether it should be discussed at all. The traditional health care provider was not asked about advance care planning because an interpreter was not available for the entire interview. Finally, three patients who had cancer or who were being evaluated for a diagnosis of cancer were not asked to comment on advance care planning because the investigators felt that this line of inquiry might be too upsetting.

Non-interview-related observations and reflections were recorded in a journal on a daily basis during fieldwork. Finally, Indian Health Service, federal government, and state of Arizona advance directive documents and policies were reviewed.

Data Analysis

As the transcripts were read, observations and reflections about ideas in the text were written in the margins of the transcripts, a process referred to as coding. Text with similar codes was examined and compared across interviews, leading to the identification of several major themes. Throughout the period of data analysis emerging themes were continually reviewed, alternative interpretations considered, and revisions made. Also, comments discordant with dominant themes were identified and examined.

Trustworthiness

Several steps were taken to ensure trustworthiness of the findings, a concept in qualitative research comparable to validity and reliability in quantitative research.[15] Briefly, these included (1) conducting four group feedback sessions with a total of 24 Navajo informants, three of whom were part of the original sample, to verify that the information we obtained and its interpretation made sense; (2) independent review of interpreter performance by three other Navajo speakers; (3) independent coding of a small sample of transcripts by two other physicians with fellowship training in qualitative research; (4) review of analysis with an anthropologist who is an expert in Navajo culture and with a Navajo graduate student in anthropology; (5) comparing our findings with the work of others on the Navajo; and (6) soliciting peer review in a variety of settings, such as formal work-in-progress sessions, graduate seminars, and invited presentations. . . .

Results

Transcript analysis identified several themes; only a portion of the findings will be presented here. First, we describe two themes that emerged from the open-ended interviews, followed by data regarding informants' opinions about advance care planning.

Think and Speak in a Positive Way:
Hózhoojí Nitsihakees/Hózhoojí Saad

Informants commented often that it was important to "think and speak in a positive way." This theme is encompassed by the Navajo phrases *Hózhoojí nitsihakees* and *Hózhoojí saad*. The literal translations are "think in the Beauty Way" and "talk in the Beauty Way." The prominence of these themes reflects the Navajo view that thought and language have the power to shape reality and control events.

A public health nurse referred us to a woman who had resisted attending a prenatal clinic. In the woman's experience the risks of pregnancy were discussed at the clinic, a practice she found troubling. She made the following remarks:

> I've always thought in a positive way, ever since I was young. And even when the doctors talked to me like that, I always thought way in the back of my mind: I'm not going to have a breech baby. I'm going to have a healthy baby and a real fast delivery with no complications, and that's what has happened.

This theme of thinking and speaking in a positive way often emerged when informants reflected on how doctors should communicate with patients; it reflects the Navajo view that health is maintained and restored through positive ritual language. A traditional diagnostician, commenting on how she counsels her own patients on matters of health and illness, said the following:

> In order to think positive there are plants up in the mountains that can help you. Also there are prayers that can be done. You think in these good ways, and that will make you feel better and whatever has stricken you in the deadly manner will kind of fall apart with all these good things that you put in place of it. . . . The doctor may say, "You're not going to live," but I say, "*Hózhoojí nitsihakees*"; that means "think in the Beauty Way."

Avoid Thinking or Speaking in a Negative Way: Doo'ájíniidah

Informants made a related point of requesting that providers "avoid thinking or speaking in a negative way." This theme is approximated by the Navajo phrase, "*Doo'ájíniidah.*" The literal translation is "Don't talk that way!"

Often this theme was expressed as informants recounted interactions with medical providers that upset them, the idea being that negative thoughts and words can result in harm. A middle-aged Navajo woman who is a nurse, speaking about how the risks of bypass surgery were explained to her father, said the following:

> The surgeon told him that he may not wake up, that this is the risk of every surgery. For the surgeon it was very routine, but the way that my Dad received it, it was almost like a death sentence, and he never consented to the surgery.

A second example of this theme comes from a highly regarded medicine man commenting about the special care required when healers communicate with patients:

In my practice, when I'm working with the patient, I am very careful of what I say, because any negative words could hurt the patient. So, with Western medicine, a doctor could be treating a patient, and he can mention death, and that is sharper than any needle. Therefore, with the tongue that we have, we have to be very careful of what we say at the time and point we're treating the patient.

Advance Care Planning

Given the Navajo discomfort with negative information, we were interested to learn how informants regarded the idea of advance care planning. Advance care planning requires competent patients to leave instructions (living will) or designate an agent to speak for them (durable power of attorney) to guide decisions about medical treatment in case the patient becomes unable to communicate his or her wishes owing to mental incapacity. Advance care planning requires patients to contemplate and plan for a future state characterized by profound illness.

The Patient Self-determination Act was passed into federal law effective December 1, 1991.[16] A major goal of this legislation was to increase patient participation in end-of-life decision making by encouraging adults to complete advance directive documents.[17] To facilitate this goal, two of the law's major requirements are (1) providing all adult patients admitted to hospitals with written material, at the time of admission, summarizing state law and hospital policies addressing the patient's right to formulate an advance directive, and (2) educating staff and the surrounding community about issues concerning advance directives.

In March 1992, the Indian Health Service fully adopted the requirements of the Patient Self-determination Act. Indian Health Service policy also states: "Tribal customs and traditional beliefs that relate to death and dying will be respected to the extent possible when providing information to patients on these issues."[18]

Nineteen (86%) of the 22 informants who were asked about advance care planning stated or implied that it was a dangerous violation of traditional Navajo values and ways of thinking. Nine informants stated this explicitly. For example, a 76-year-old man with chronic pain said:

> That's *doo'ájíniidah*, you don't say those things. And you don't try to bestow that upon yourself, the reason being that there are prayers for every part of life that you can put your trust on. The object is to live as long as possible here on earth. Why try to shorten it by bestowing things upon yourself?

Ten informants would not even discuss the issue because they felt that it was too dangerous. An 87-year-old male World War II veteran's response was characteristic of this group:

> In my opinion, I wouldn't recommend it, wouldn't want to talk about it, or want to comment about it.

Of the 22 informants whose opinions about advance directives were sought, only three (14%) found the idea somewhat acceptable. All of these Navajo informants were trained as Western biomedical health care providers and employed by the Indian Health Service, and their responses were consistent with Indian Health Service policy. The following comment was characteristic of this group:

> One of our responsibilities here is to work on these living wills and the power of attorney. In the Navajo philosophy you don't say these things, you don't discuss these things. But this is something that is new to the hospital, and we need to discuss it with every patient.

Comment

Our study sought to understand the perspective of Navajo informants regarding the discussion of negative information. Two closely related themes emerged from the interviews. Informants explained that patients and health care providers should think and speak in a positive way and avoid thinking or speaking in a negative way.

Ethnographers and anthropologists who have studied the Navajo people[19-21] identify *Hózhó* as the central concept in Navajo culture. Its meaning is approximated by combining the concepts of beauty, blessedness, goodness, order, harmony, and everything that is positive or ideal.[22] *Hózhó* defines the traditional Navajo way of thinking, speaking, behaving and relating to other people and the surrounding world.

It is clear that these Navajo informants have their own way of thinking about and coping with issues related to safety and danger, health and sickness, and life and death. It is a way of thinking and using language that reflects the Navajo concept *Hózhó* and the Navajo view that thought and language shape reality. Discussing negative information conflicts with the Navajo view of language and its relationship to reality and with our informants' expectation that communication between healers and patients embodies the concept of *Hózhó*. . . .

Hospital policies complying with the Patient Self-determination Act, which are intended to expose all hospitalized Navajo patients to the idea, if not the practice, of advance care planning, are ethically troublesome. For Navajo patients like our informants, an advance care planning discussion may not be viewed as beneficial, and in fact it is more likely to be regarded as potentially harmful. The study's findings question whether it is possible to comply with an Indian Health Service policy that requires providing information to all patients about advance care planning while respecting traditional views. In light of these findings, health care providers and institutions caring for Navajo patients should reevaluate their policies and procedures regarding advance care planning.

This study further demonstrates that the concepts and principles of Western bioethics are not universally held.[23-26] In our pluralistic society there are communities of patients who do not identify dominant values as their own, a fact that health care providers and institutions need to appreciate. Providers interested in understanding their patients' values and perspectives should "practice an intensive, systematic, imaginative empathy with the experiences and modes of thought of persons who may be foreign to [them] but whose foreignness [they come] to appreciate and humanly engage."[27] A deeper understanding of patients' perspectives should follow, and this in turn should be used to inform clinical interactions, research and educational activities, and institutional policies.

Finally, additional research should be done among communities of patients whose views depart from prevailing moral perspectives. We have studied one population for whom dominant Western bioethical concepts and principles are problematic. This challenges us to consider other populations for whom the routine application of these concepts and principles may pose difficulties.

References

1. Carrese J, Brown K, Jameton A. Culture, healing and professional obligations. *Hastings Cent Rep.* 1993;23:15–17.

2. Jecker NS, Carrese JA, Pearlman RA. Caring for patients in cross cultural settings. *Hastings Cent Rep.* 1995;25:6–14.

3. Perkins HS, Supik JD, Hazuda HP. Autopsy decisions: the possibility of conflicting cultural attitudes. *J Clin Ethics.* 1993;4:145–154.

4. Caralis PV, Davis B, Wright K, Marcial E. The influence of ethnicity and race on attitudes toward advance directives, life-prolonging treatments, and euthanasia. *J Clin Ethics.* 1993;4:155–165.

5. Rhodes LA. Studying biomedicine as a cultural system. In: Johnson TM, Sargent CE, eds. *Medical Anthropology: Contemporary Theory and Method.* New York, NY: Praeger; 1990:159–173.

6. Jonsen AR, Siegler M, Winslade WJ. *Clinical Ethics.* 3rd ed. New York, NY: McGraw-Hill Inc; 1992:37–38.

7. Katz J. *The Silent World of Doctor and Patient.* New York, NY: The Free Press; 1984.

8. Muecke MA. On the evaluation of ethnographies. In: Morse J, ed. *Critical Issues in Qualitative Research Methods.* Beverly Hills, Calif: Sage Publications; 1993:198–199.

9. Agar MH. *Speaking of Ethnography.* Beverly Hills, Calif: Sage Publications; 1986.

10. Spradley JP. *The Ethnographic Interview.* New York, NY: Harcourt Brace Jovanovich College Publishers; 1979.

11. Muecke MA. On the evaluation of ethnographies. In: Morse J, ed. *Critical Issues in Qualitative Research Methods.* Beverly Hills, Calif: Sage Publications; 1993:203.

12. Goodman JM. *The Navajo Atlas.* Norman: University of Oklahoma Press; 1982.

13. Locke RF. *The Book of the Navajo.* Los Angeles, Calif: Mankind Publishing; 1976.

14. Bernard HR. *Research Methods in Cultural Anthropology.* Beverly Hills, Calif: Sage Publications; 1988:97–98.
15. Lincoln YS, Guba EG. Establishing trustworthiness. In: *Naturalistic Inquiry.* Beverly Hills, Calif: Sage Publications; 1985:289–331.
16. Omnibus Budget Reconciliation Act of 1990. Pub L No. 101–508, ßß4206, 4751.
17. Wolf SM, Boyle P, Callahan D, et al. Sources of concern about the Patient Self-determination Act. *N Engl J Med.* 1991;325:1666–1671.
18. US Dept of Health and Human Services, Public Health Service, Indian Health Service. *Patient Self-determination and Advance Directives Policy.* Indian Health Service Circular 92-2, March 1992:1–5.
19. Wyman LC. *Blessingway.* Tucson: University of Arizona Press; 1970.
20. Reichard GA. *Navaho Religion: A Study of Symbolism.* New York, NY: Bollingen Foundation; 1950.
21. Kluckhohn CK. The philosophy of the Navaho Indians. In: Northrop FSC, ed. *Ideological Differences and World Order.* New Haven, Conn: Yale University Press; 1949.
22. Witherspoon G. *Language and Art in the Navajo Universe.* Ann Arbor: University of Michigan Press; 1977:24.
23. Beyene Y. Medical disclosure and refugees: telling bad news to Ethiopian refugees. *West J Med.* 1992;157:328–332.
24. Meleis AI, Jonsen AR. Ethical crises and cultural differences. *West J Med.* 1983;138:889–893.
25. Surbone A. Truth telling to the patient. *JAMA.* 1992;268:1661–1662.
26. Swinbanks D. Japanese doctors keep quiet. *Nature.* 1989;339:409.
27. Kleinman A. *The Illness Narratives: Suffering, Healing, and the Human Condition.* New York, NY: Basic Books Inc; 1988:230.

Communication Through Interpreters in Healthcare:

Ethical Dilemmas Arising from Differences in Class, Culture, Language, and Power

Joseph M. Kaufert and Robert W. Putsch

Joseph Kaufert received his Ph.D. in anthropology and is a professor in the Department of Community Health Sciences at University of Manitoba. Robert Putsch is a physician, a Clinical Professor of Medicine at the University of Washington, and founder of the Cross Cultural Health Care Program based in the Seattle community. In this article they characterize the complex relationships between interpreters, patients, and health care providers. The authors maintain that interpreters are more than mere translators; they are both witnesses to and participants in health care delivery. The authors suggest that more detailed studies are needed before codifying interpreters' ethical responsibilities.

. . . Patients and healthcare providers . . . often come from different educational, cultural, or class backgrounds. They may hold disparate views of illness and treatment, and may not agree on a common set of cultural values regarding decision making. This is the interface at which interpreters in healthcare do their work, and disparate views as well as disparate expectations are the basis for the dilemmas that they often face. The work of the interpreter is further complicated by the unequal distribution of power in the relationship between the parties involved.

Power and Dominance in Clinical Communication

The ethnomedical literature has been criticized for its failure to consider the dynamic of power between healthcare providers and their clients.[1] Medical sociologists have focused on issues of power and dominance in clinical communication,[2] and have described the clinician-patient relationship as one in which the clinician has the ultimate responsibility for developing conclusions and proposing alternative treatments. The dominance of healthcare providers

---

affects the role of the interpreter both directly and through its impact on patients.[3]

The dynamics of the physician-interpreter-patient relationship are also influenced by wider institutional, professional, and structural forces, such as a hospital's program directives, its funding arrangements, and *how* interpretation programs are represented at the management level. How interpreters interact with healthcare providers and patients is also influenced by health policies that may directly impact the definition of the role of, and power of, the interpreter; and these policies are subject to change.[4,5] Generally these external forces impose the values of a dominant group or class on the work-a-day process of providing healthcare. In addition, interpreter programs may have been created to meet governmental regulations or to mediate institutional problems rather than to solve problems for patients.

The literature about communication has emphasized language and culture as "barriers." Interpreters or bilingual healthcare workers are usually represented as ancillary members of the healthcare team who enable clinicians and patients to communicate using mutually intelligible terminology and concepts.[6] Some researchers recognize the power of an interpreter as a gatekeeper who has the power to elicit, clarify, translate, omit, or distort messages.[7] Most often, however, the neutrality, completeness, and accuracy of interpreters are discussed. These issues, however, fail to account for sources of power and control or dominance in this triadic relationship. Ethical guidelines that are based on neutrality, completeness, and accuracy often fail to take into account issues such as class, power, disparate beliefs, lack of linguistic equivalence, or the disparate use of language. We have observed situations that could not be resolved by the adoption of more culturally sensitive communication styles or by educating the healthcare provider on the cultural beliefs of patients and families.

The Challenge of Monolingualism in Multicultural Practice

There is great potential for value conflict when caregivers and patients come from different cultural, language, and class groups. Jecker et al. discuss the issue of diversity in multi-cultural practice settings.[8] In acknowledging that differences in values exist within a shared culture, they concurred with Ware and Kleinman, who state, "across class, caste, gender, age, religious and political lines—cross-cultural, conflicts may be more deeply rooted, for such differences embody not just different opinions or beliefs, but different ways of everyday living and different systems of meaning."[9]

If the healthcare providers' views on the ethical principles that govern decision making are in conflict with the values that are held by patients, their families, or their communities, disagreement over cultural values may lead to confrontation. These disagreements are often most intense when healthcare providers *evoke* the principles of patient's autonomy and truth-telling. . . .

Jecker et al. suggest that conflict that is based on cultural difference can be mediated using strategies that allow both the patient and the healthcare provider the opportunity to clarify their values.[10] Others have recommended the use of review panels and consultants who have special expertise in cross-cultural communication.[11] We have found that the idealized model-neutral communication that is facilitated by an interpreter often diverges from actual practice. We found that interpreters not only serve as helpers and educators, but also as cultural consultants, or mediators between conflicting value systems. We also found that interpreters sometimes function as witnesses when the conflicts in values were fundamental and could not be resolved simply by adopting "culturally sensitive" methods of communication.

Language Intermediaries, Institutional Responses, and the Use of Language

The literature on communication seems to imply that conflicts can be resolved by altering how one approaches communication, and seems to assume that language intermediaries play a neutral role that does not influence decision making. In fact, the literature that describes cross-cultural encounters treats language intermediaries as if they were invisible.

... We have found that many exchanges are triadic, rather than dyadic, in that they include an interpreter as an active participant.[12] An interpreter may hear discrepant views that reflect the alternative perceptions of "What's wrong?" An interpreter is exposed to, and is variably aware of, patients' and practitioners' criteria for decision making, and may be asked to mediate between a patient's and family's understanding and a professional's explanations of illness and proposals for treatment....[13]

... In clinical situations that involve making decisions, the role of the interpreter is not simply to interpret ethical issues in an objective and linguistically accurate manner. We have found, in our research on the work of Canadian Aboriginal interpreter/advocates, that interpreters may influence the interaction with patients more directly, serving as mediators between clinicians and patients, explaining patients' values, and assisting in negotiating an ethical contract.[14] In addition, in examining transcripts, some interpreters introduce a variety of biases into the messages that they give to patients and to providers.[15] This is a clear departure from the "ideal" concept of an objective, uninvolved intermediary. However, we believe that this is a reality that challenges the notion that interpretation can be consistently neutral.

The Role of the Interpreter

Policies that restrict the role of the interpreter and emphasize cultural neutrality and invisibility may ignore other dimensions of the interpreters' activities in healthcare. Legislation and professional codes of ethics have emphasized how important it is for interpreters to remain objective. This emphasis may limit how an interpreter serves as a cultural informant—one who explains the patient and community context, and acts as a broker or mediator in situations that involve culturally based conflict. Interpreters often facilitate trust between patients, families, communities, healthcare providers, and healthcare programs. They may improve the continuity of a patient's care, when the patient encounters multiple practitioners, and, with a patient's consent and participation, may provide caregivers information on a patient's family and community. Interpreters may also provide patients with information about the biomedical culture of the health system. Interpreters may facilitate patients' access to healthcare by providing culturally appropriate explanations of how a health system works and by explaining the patient's rights.

These roles have evolved in programs that have recognized the potential contribution of interpreters to improving health education, the compliance of patients, and increasing the effectiveness and efficacy of health services. Many innovative cross-cultural health programs in the United States, Europe, and Canada have developed roles for interpreters that support and legitimate their involvement in mediation, explaining cultural differences/practices ("culture brokerage"), and advocacy.[16] In other programs, interpreters work as bilingual medical assistants or as case managers who coordinate services, and act as advocates and counselors.[17]

The work of interpreters in culture brokerage and community advocacy has been incorporated into the interpreter services at the Harborview Medical Center in Seattle, which uses interpreter-mediators to work with patients, families, ethnic communities, and health institutions.[18] In different programs around the country, bilingual health workers/interpreters serve as mediators, advocates, culture brokers, medical assistants, and in some cases, case managers. [19]

Codes of Ethics for Interpreters in Healthcare

We collected codes of ethics developed for healthcare interpreters from more than 20 institutions and organizations. Many of these codes emphasize a mode of interpretation that calls for an objective, neutral role for interpreters that is similar to the interpreter's role that evolved in American Sign Language (ASL) programs and in court interpretation programs, where training and codes of conduct for interpreters have been legislated. Using the model of objective language interpretation from in the ASL and court systems, many programs have carried these approaches and assumptions into medical care settings. Because of the powerful influence of these codes, we will briefly discuss them using the *Washington Legal Code* as an example.

The Washington State Supreme Court's *Code of Conduct for Court Interpreters (GR 11.1)*, adopted in 1989, includes extensive commentary published by the Court Interpreter Task Force in 1986. . . . Under the following provision of GR 11.1 is a precise commentary about discourse in court:

> A language interpreter shall interpret or translate the material thoroughly and precisely, adding and omitting nothing. . . . [Then the commentary enlarges on the situation.] Unless the interpreter is faithful to this concept of accurate interpretation,

he or she may act as a filter or buffer in the communication process. This could damage the integrity of the trial process which is based on an adversarial system with vigorous examination and cross-examination [1986; abbreviated].[20]

Medical discourse has, as a basic goal, mutual understanding; it is not normally adversarial. . . . Attempts to encourage mutually shared understanding require the healthcare interpreter to engage in explanation, cultural brokerage, and mediation when these actions are necessary.

Cynthia E. Roat, editor of a training manual for medical interpreters, wrote a code of ethics for interpreters in healthcare that was first published by Region 10 of the U.S. Public Health Service in 1995 (an abbreviated version is presented in Figure 1 below).[21]

The code combines codes of ethics from three health interpretation programs in the United States. As the code is being used in training interpreters, there has been a remarkable opportunity to listen to experienced interpreters discuss each item in the code, drawing on their past experience. . . . In discussing problems of maintaining confidentiality, one interpreter presented the following problematic case example.

Confidentiality. A young patient was known to an interpreter from an interaction in a clinic for sexually transmitted diseases (STDs). While he interpreted for the patient during an evaluation and physical exam at a second clinic, the interpreter heard the patient twice respond negatively to questions about prior STDs. What if the prior diagnosis of an STD (as known to the interpreter) had an implication for the patient's current problem? What should the interpreter have done? Other interpreters who listened as he related this case recalled similar cases in which important prior diagnoses, especially alcoholism, were denied by patients. The confidentiality of a medical encounter is privileged and protected. It is clear, however, that some of the knowledge that interpreters may have about patients has been gained in other medical encounters or even outside of a medical setting. This is especially true in small communities.

Dilemmas raised by healthcare interpreters abound in other areas, and we will briefly highlight some of these here.

Accuracy and completeness. Interpreters asked: How do you deal with questions or comments by a healthcare provider or a patient that may be perceived as offensive? Rude? What if a healthcare provider is disrespectful of a patient's reference to commonly held beliefs or practices?

Non-judgmental. Interpreters recognize that the code prohibits them from imposing their own personal values on others. However, they emphasize that neutrality is difficult to maintain when, in the interpreters judgment, there seems to be a clear failure on the part of one of the parties to understand basic, relevant information.

Clients' self-determination. Discussing approaches to maintain patients' autonomy, interpreters asked: What if a patient asks for an interpreter's opinion (and patients frequently do), and rejects the interpreter's response that he or she is not allowed to express an opinion to the patient? Interpreters also commented on the need to provide supplementary information to patients and to create op-

Confidentiality	Interpreters must treat all information learned during the interpretation as confidential, divulging nothing without the full approval of the patient and his/her physician.
Accuracy	…Interpreters must transmit the messages in a thorough and faithful manner, omitting or adding nothing, giving consideration to linguistic variations in both languages and conveying the tone and spirit of the message. Word for word interpretation may not convey the intended idea. The interpreter must determine the relevant concept and say it in language that is readily understandable … to the person being helped.
Completeness	…Interpreters must interpret everything that is said by all peoples in the interaction but should inform the health professional if the content…might be perceived as offensive, insensitive, or harmful to the dignity and well-being of the patient.
Conveying Cultural Frameworks	When appropriate, interpreters shall explain cultural differences to health providers and patients.
Non-Judgmental Attitude	An interpreter's function is to facilitate communication. Just as interpreters should not omit anything being said, they should also not add their own personal opinions, advice, or judgment.
Client Self-Determination	The client may ask the interpreter for his or her opinion.…The interpreter should not influence the opinion of patients or families by telling them what action to take.
Attitude Toward Clients	The interpreter should strive to develop a relationship of trust and respect at all times with the patient by adopting a caring, attentive, …impartial attitude toward the patient, toward his or her questions, concerns, and needs.
Acceptance of Assignments	Interpreters should disclose any real or perceived conflict of interest that would affect their objectivity in delivery of service. Additionally, if level of experience or personal sentiments make it difficult to abide by any of the above conditions, the interpreter should decline or withdraw from the assignment.
Compensation	The fee or salary paid by the agency is the only compensation that the interpreter should accept. Interpreters should not accept additional money, considerations, or favors for services.

Figure 1   A Code of Ethics for Interpreters in Health Care.

portunities in which a patient is presented the opportunity to choose between clearly explained alternatives. Doing this may maximize a patient's capacity for self-determination in a medical system that the patient might otherwise find difficult to understand. In doing this, an interpreter may technically violate the ethical code—especially if he or she is prohibited from explaining medical culture and process to patients.

The code of ethics does not formally address instances when interpreters act as advocates for patients. A number of job descriptions written for bilingual healthcare workers include advocacy as an expected part of their function. It is interesting to contrast a widely held negative response to the idea that interpreters in healthcare may need to advocate for a patient with the common institutional standard that "the patient comes first." Healthcare visits that include an interpreter represent a unusual circumstance, in that the interpreter is a witness to the healthcare process. Over the past years both of the authors have looked at problematic cases that involved patients, families, and interpreters. Interpreters clearly are at risk when they act on a patient's behalf, . . . as [the] case [below,] selected from a large number of similar instances, demonstrates.

> [Case:] An elderly Russian woman commented to the interpreter that her health coverage had been terminated. "How will I pay for this?" she asked. The interpreter sequentially mentioned her inquiry to the physician, the registered nurse, and the clinic clerk. No one responded or provided direction or advice. Acting on his own, the interpreter took the patient to the finance office to discuss a discount payment program. The interpreter's action was reported to the clinic's administrator. The interpreter's contract to interpret was withdrawn on the basis that he had undertaken to advocate for the patient.

. . . Major controversies, even formal complaints against interpreters, have arisen when an interpreter has attempted to explain a culture-bound issue to a healthcare provider. These controversies increase in the areas of disclosure, informed consent, truth-telling, and in circumstances that relate to death and dying. We will illustrate these (referring to the items listed in Figure 1) in the detailed case studies that follow. The first case concerns interpretation in the negotiation of an informed-consent agreement between an Aboriginal Cree-speaking health interpreter, a patient, and a specialist in a Winnipeg hospital. The study illustrates: (1) the communication issues that arise in interpreting the different values and cultural models of illness and treatment of the patient, the practitioner, and the interpreter; and (2) the impact of the interpreter in facilitating interaction between the participants.

Case One: Mediation of the Informed-Consent and Diagnostic Processes

The patient was a 46-year-old female Cree speaker, who consulted with a gastroenterologist for a problem with anemia and to request an evaluation of possible sites of blood loss in her upper and lower bowel. The interaction involved

590 Ethical Dilemmas from Differences in Class, Culture, Language, and Power

obtaining informed consent for a gastroscopy and a colonoscopy. An interpreter/ advocate, who worked for the hospital's Native Services Program, attended the meeting between the patient and the clinician. During the visit, a real-time videotape record and verbatim translation into non-idiomatic English was made, in order to document the exchange in which the interpreter translated information from the clinician on the problem of anemia and the possible linkage to gastrointestinal blood loss. The physician was initially most concerned that the patient should understand the association between her feelings of "weakness" and the anemia. Later in the meeting, he introduced a more complex explanation that linked anemia with her loss of blood, and her darkened stool with her use of an anti-inflammatory medication. Throughout this interchange, the interpreter tried to ensure that the patient understood her options and was aware of her right to refuse treatment.[22]

> **Doctor:** She's anemic and pale, which means she must be losing blood.
> **Interpreter (in Cree):** This is what he says about you. You are pale, you have no blood.[23]
> **Doctor:** Has she had any bleeding from the bowel when she's had a bowel movement?
> **Interpreter (in Cree):** When you have a bowel movement, do you notice any blood.
> **Patient (in Cree):** I'm not sure.
> **Interpreter (in Cree):** Is your stool ever black or very light? What does it look like?
> **Patient (in Cree):** Sometimes dark.

The interpreter's statements provide links for the patient, links between her feelings of weakness and possible gastrointestinal symptoms; she also asks the patient to recall the changes she herself had observed in the color of her stool. By helping the patient understand that her anemia is based on changes she had herself observed in her body, the interpreter is laying the groundwork for an explanation of the procedures to be used by the clinician. The interpreter has injected her own knowledge of taking a gastrointestinal history by asking— uncued by the physician—about black or very light stools. In so doing, she has conveyed a message that helps interpret the physician's meaning, but has injected her own explanation of the questions, without informing the physician. Training programs for interpreters often encourage interpreters to achieve "transparency"—to make certain that monolingual participants are aware of any added material or commentary that they add during the exchanges. In this case, the interpreter could have done so by informing the physician of the meaning of the added questions about the color of the patient's stool.

> **Interpreter (in Cree):** We want to know, he says, why it is that you are lacking blood, that's why he asked you what your stool looks like. Sometimes you lose blood from there when your stool is black.

In the continuing exchange, the interpreter translated a direct statement from the clinician, but also explained why he should have asked the question about an apparently unconnected issue, namely the color of the patient's stool.

In her final statement, the interpreter provided an unprompted clarification of the relationship between blood loss and the appearance of the stool. This further explanation of meaning across boundaries of culture, language, training, and class illustrates the interpreter's role regarding accuracy and conveying cultural frameworks, that is, expanding on, and explaining, meaning. In so doing, she violated the principle of maintaining transparency (which is not included in the code).

In the next exchange (not reproduced here), the interpreter provided another unprompted explanation that linked the clinician's earlier questions about the patient's use of anti-inflammatory medications with the concept of the loss of blood. These elaborated and simplified explanations were made in Cree and were not translated into English for the physician. The pattern of communication that emerges in this exchange is clearly triadic, and illustrates the capacity of the interpreter to introduce new information. This illustrates how an interpreter can serve as a broker in the emerging relationship between a clinician and a patient. This also raises the ethical question of whether the interpreter followed formal guidelines for interpreters that require that she more systematically "back translate" her summary to the patient for the physician (thereby maintaining transparency). This strategy would have provided feedback to the physician, which would have enabled him to clarify, or further share his message, and to explore the patient's understanding of it.

On the following day, colonoscopy and radiological examination revealed that the patient had a polyp in her colon. The physician recommended that the polyp be cauterized, and worked with the interpreter to negotiate a second consent with the patient.

> *Interpreter (in Cree):* Do you want to have this procedure done? Will you consent to have this growth removed, burned?
> *Patient (in Cree):* I don't know.
> *Interpreter (in Cree):* You know, if it's not removed it may bleed. It may cause problems. *Interpreter:* Dr. __, isn't it true that if it's not removed, it can bleed and she can become anemic?
> *Physician:* That's correct, we feel that your anemia may result from the bleeding of the polyp.
> *Interpreter (in Cree):* If it's not removed, you may end up with cancer. You know? And you will not have the operation [that is, the proposed colonoscopy does not involve invasive surgery]. It's harder when a person has an operation. You know? This procedure [colonoscopy] that he's going to do will get it on time. Before it begins to bleed or starts to grow. You're lucky it's caught on time. And it will bother you when you have had a bowel movement. This way there's no danger that this growth will bleed.
> *Patient (in Cree):* I don't know.
> *Interpreter (in Cree):* Well if you want to come in for the procedure while you are here . . . [the patient lived at a distance, in a rural community]. It's all up to you to think about.

In this exchange, the interpreter became an active participant in the process of eliciting consent. She once again provided unprompted messages and interjected her opinions and views about risk. Her non-transparent commentary, however, shows how intermediaries play a pivotal role in patient/clinician interaction, based partly on their ability to be selective in interpreting what is said, but also through a process of embellishing meaning.

The interpreter in this exchange also used requests for clarification or further feedback as a way of becoming a more active participant in the process that led up to the solicitation of the patient's consent. The interpreter's un-cued intervention that introduced the association with the risk of cancer is an example of the power of the intermediary to introduce information. The interpreter's intervention was based on previous experience in working with the same gastroenterologist in similar diagnostic evaluations. She was aware of the potential association between polyps and an increased risk of cancer. She used the same explanations that clinicians use to explain risk and to persuade patients to undergo procedures. However, she introduced this information without informing the clinician of the message she had passed. From an ethical perspective, the interpreter's intervention might have been problematic, and this underscores the importance of both understanding and controlling the interpretative process. Yet, from the perspective of the interpreter/advocate, the reference to cancer was made on the basis of her own understanding of the medical context of the exchange, coupled with her understanding of the meaningfulness of cancer to the patient.

The ability to introduce additional information into the decision-making process is a critical aspect of the informal power of interpreters as mediators in cross-cultural decisions. The power of the interpreter in controlling communication is exercised within a linguistic "black box," inaccessible to the other participants in the interaction. The only comprehensive way to access this "box" is through the use of formal protocols that require the presence of at least two interpreters at each interchange, each validating the other's interpretation.[24] Another possible, but very protracted, alternative requires that the interpreter follows up each translation with a detailed, literal summary of how each component of the message was represented. Even when the patient is asked to summarize the message transmitted by the interpreter, there is no guarantee regarding the accuracy of the original interpretation. The process does, however, provide feedback to the interpreter.

Truth-Telling, Advanced Directives, and Issues Relating to Death and Dying

Recent research by medical anthropologists has examined the impact of culture in health communication that involves conflicting interpretations of values and frameworks for ethical decision making between patients, family members, and clinicians. At times, the conflict reflects culturally based differences in values and in approaches to end-of-life decision making.[25] The principle of autonomy is at the center of the bioethical literature that deals with truth-telling, advanced directives, and informed consent. Recent literature starts from the premise that bioethics is, itself, a culturally constructed body of knowledge, reflecting the core values of biomedicine and Western philosophy.[26] Bioethical principles reflect Western notions of the sovereignty of the individual person and of individual life. Similarly, the importance attached to respect for persons in consent law assumes the existence of autonomous decision makers who are, as Gostin stated, "capable of deliberation about personal goals and of acting under the direction of such deliberation."[27]

Currently, the principle of autonomy dominates other values and principles in discussions of informed consent. Alternative values, namely the good or primacy of the community and the family, take precedence over autonomy in many other cultures.[28] Despite his recognition of disparate value systems, Gostin, in a recent editorial in the *Journal of the American Medical Association*, asserted: "The right of autonomy or self-determination is broadly perceived to be a morally necessary method of demonstrating genuine respect for human integrity."[29]

In our work with interpreters in Seattle and Winnipeg, we observed situations in which the different decision makers held irreconcilable positions on the need for patients' autonomy and on truth-telling. Trained interpreters or family intermediaries were sometimes placed in an untenable position of having to provide ambiguous—but conciliatory—interpretation. This approach placed professional interpreters in situations in which they were at risk of violating their ethical contracts with the patient and the practitioner. Asked to mediate between conflicting ethical perspectives, interpreters or family intermediaries often were placed in positions that involved intense role conflict. Our final two case studies illustrate situations in which differences in interpretation involved recognition of fundamental differences in cultural values that define the role of the family, the need to respect individual autonomy, and the need for truth-telling.

Case Two: Communicating a Terminal Prognosis to an Elderly Vietnamese Patient

The second case study centers on the problem of communicating a terminal prognosis, to a 64-year-old, monolingual Vietnamese-speaking man, hospitalized with far-advanced hepatocellular carcinoma. In spite of extensive liver involvement, the patient was generally cognitively alert and competent to make decisions regarding his treatment. Over the course of his protracted illness, several English-speaking family members spoke with the oncologist as well as with the patient's primary care provider about the family's unwillingness to discuss the diagnosis with the patient. They emphasized that direct communication with the patient that focused on his terminal prognosis and inquiry about his personal choices about palliative care was unacceptable. In the same meetings, the family members clearly communicated that they wanted "everything to be done." Proxy decisions made by family members avoided making clear choices between palliative-care measures and more invasive, potentially life-prolonging interventions. On several occasions during the patient's hospital stay, family conferences were held to describe the implications of the end stage of the patient's disease. However, these meetings were held outside of the patient's room and did not include him.

A trained Vietnamese-speaking medical interpreter was involved in the case. The practitioners attempted to use the interpreter to increase direct communication with the patient, to thereby resolve the impasse. The primary care provider spoke with the interpreter, and emphasized the medical futility of the more-invasive treatments and life-support measures that had been demanded by the family members. The physician asked the interpreter to communicate directly with the patient about his terminal prognosis and to explore the individual's personal wishes for continued care. In discussing the treatment options that involved the potential referral of the patient to hospice care, the physician said, "If we do this we can spare him further discussions about therapy and can focus our efforts on comfort, on pain relief, and even begin to discuss his own wishes. For example, whether he would like to die here in hospital or at home."[30] The interpreter recognized that this action would ultimately involve more direct and explicit communication of the physician's assessment that the patient was in the final stages of dying. The interpreter stated her discomfort with a role that involved telling the patient bad news, and refused to interpret this message for the physician. The interpreter stated, "I can't tell him, it's against our culture." She explained her reluctance to communicate the terminal prognosis in terms of Vietnamese cultural beliefs, and emphasized the need to respect the family's request.

The interpreters refusal to override the proxy decision-making power assumed by the family and her respect for their position remained unchanged. The rapid progression of the patient's terminal illness ultimately precluded the use of another intermediary. The attending physicians decided that only comfort care would be provided, despite the family's continuing focus on possible curative interventions. The patient's obstructive jaundice progressed to hepatic failure and coma. The patient subsequently died without being directly informed of his diagnosis and its implications.

In deciding how to mediate between the conflicting values of the family and the healthcare providers regarding truth telling, the interpreter and healthcare providers were faced with a number of questions, some of which included: (1) Was the interpreter correct when she stated that the beliefs of traditional Vietnamese culture and North American Vietnamese immigrant culture prohibited disclosure of terminal prognosis? (2) Was this prohibition a reflection of the interpreter's personal values and culturally based explanatory framework? (3) Was the interpreter's refusal a reflection of the family's demand for proxy decision making and commitment to maintaining power and control over the revelation of "bad news"? And was this position negotiated? (4) Should the interpreter have withdrawn, or should she have been asked to withdraw? (See acceptance of assignments, Figure 1.) And was the healthcare providers' position too unyielding?

Too ethnocentric? (5) Or was this a reasonable role for the interpreter to play, considering the need for her to act as a cultural mediator, and the need for both sides to negotiate?[31] (6) What about hospital policies that asserted that family members should not be used as interpreters?

The interpreter's and the family's control over access to the patient via language placed the healthcare providers in a position of having to accommodate the family's limits. Discourse and the ethical principles that emphasized truth-telling and autonomy were subordinated to cultural values that prohibited the telling of bad news and that emphasized the family's role as decision makers.

Case Three: Communicating Terminal Prognosis to an Ibo-Speaking Patient

The third case involves the communication of a diagnosis of dangerous disease with a likely terminal prognosis, and the risks of chemotherapy, to a 52-year-old Nigerian patient who had aggressive T-cell lymphoma. The patient, who was visiting his family in the U.S., underwent emergency surgery for presumed acute cholecystitis (an acute inflammatory process involving the gall bladder); during surgery he was placed on a respirator. The identification of aggressive T-cell lymphoma during the surgery was unanticipated. The physicians had difficulty weaning the patient from the ventilator post-operatively, and it became apparent that the patient had extensive lung involvement with the lymphoma. The patient spoke only Ibo, but was responsive and cognitively competent.

The patient's bilingual son and daughter were informed of the attending physician's request that a professional interpreter be involved in communicating with the patient. They responded that they would not allow another bilingual Ibo-speaking person from outside the family to provide interpretation. They also refused a proposal that the AT&T Language Line Service be used to provide objective, external interpretation. The family members insisted that they would provide all interpretation, and asked that information on the patient's diagnosis and terminal prognosis be withheld from both the patient and from his monolingual daughter who had traveled with him from Nigeria. They explained that their sister was pregnant and needed to be protected from the trauma of knowing her father's terminal prognosis. The patient's son and daughter who lived in the U.S. also explained that disclosure would abate hope, and might hasten the patient's death (see Figure 2 [below]).

The bilingual family members varied in how they gave information about the patient's prognosis to his family members in the U.S. and to his family members in Nigeria. The son and daughter contacted elders from their father's kin group in Nigeria, and informed them of the situation and sought their advice. During consultations with the healthcare providers, they expressed the fear that if their monolingual, Ibo-speaking sister, who was often in the room with her father, learned of the prognosis, this would place her unborn child "at risk." As the patient's illness progressed, the son and daughter who lived in the U.S. informed their mother, who lived in Nigeria, of their father's diagnosis.

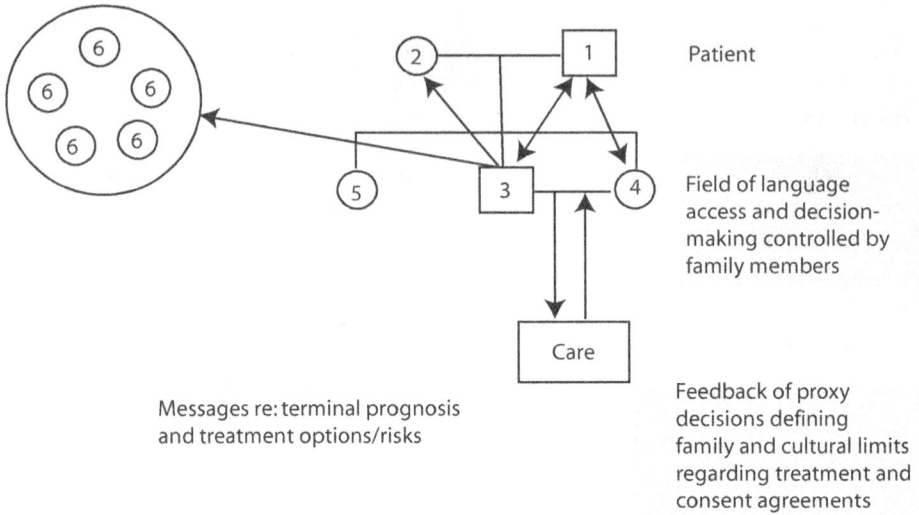

Patient

Field of language
access and decision-
making controlled by
family members

Care

Messages re: terminal prognosis
and treatment options/risks

Feedback of proxy
decisions defining
family and cultural limits
regarding treatment and
consent agreements

Key:
1. *Monolingual, Ibo-speaking Nigerian cancer patient who was visiting his family in the U.S.*
2. *Patient's wife, who resides in Nigeria, communicates with proxy decision makers in U.S. via phone.*
3. *Bilingual son who is employed in U.S. as an engineer, becomes co-decision maker with sister via control of language interpretation.*
4. *Bilingual daughter who works as a nurse in U.S. shares control of communication and decision making with her brother.*
5. *Monolingual daughter from Nigeria, accompanying her father, is frequently present in hospital room, but is excluded from communication because her family fears that bad news would harm her pregnancy and her fetus.*
6. *Members of the extended family in Nigeria who receive delayed messages, and were informed of the serious nature of the illness.*

Figure 2   Diagram showing elective interpretation and communication by family members, blocking message of dangerous illness and terminal prognosis to patient and to selected family members of an Ibo-speaking man with aggressive T-cell lymphoma.

The son and daughter's complete control over communication through their involvement as interpreters created a dilemma for the caregivers. After several days, the physicians informed the son and daughter that the patient's survival and chances of getting off the ventilator might be improved if he underwent chemotherapy. Given the son and daughter's opposition to direct communication with the patient, which would be required to obtain the patient's consent for therapy, the oncologist was initially unwilling to initiate chemotherapy. The oncologist firmly stated that the patient must know the diagnosis, as well as the risks and benefits of the proposed chemotherapy, so that treatment could be initiated in an ethical way. The son and daughter again refused to communicate either the terminal prognosis or the risks of proposed treatments to their father. Ultimately, the oncologist, after extended consultation with son and daughter and the healthcare team members, agreed that the son and daughter could act as proxy decision makers in signing the consent agreement for chemotherapy. Following the chemotherapy, the patient came off the ventilator, went into remission, and was able to return to his home in Nigeria.

The family's perspective in the third case study clearly prohibited communication of a terminal prognosis. This perspective appears to reflect cultural values that parallel those documented in recent research on the negotiation of advance directives and other consent agreements in cross-cultural situations.[32] Based on this research, questions have been raised about whether principles of autonomy are "truly respectful of all people in all cultures."[33] Carrese and Rhodes's ethnographic study that documented Navajo values about the communication of negative information during deliberations on end-of-life treatments reported a strong cultural prohibition against "the telling of bad news"; 86 percent of the subjects said that advance-care planning was a dangerous violation of traditional Navajo values. Carrese and Rhodes concluded, "Policies complying with the Patient Self-Determination Act, which are intended to expose all hospitalized Navajo patients to advance care planning are troublesome and warrant re-evaluation."[34] It is important to realize that the boundaries of these beliefs and reactions to truth-telling are not simply delineated by lines of language, education, class, and ethnicity. The literature describes similar reactions to truth-telling in a number of different ethnic communities.[35]

Conclusion

There are currently major gaps between ethical codes of conduct for interpreters in healthcare and the realities of medical interpretation. Issues such as poverty and class distinctions, the use of language, beliefs about health, and family process influence the positions and assumptions that patients and healthcare providers carry into, and through, the process of delivering healthcare. Interpreters are both witnesses and participants in this process. They often recognize the nature and consequence of medical discourse, and the risks when it breaks down. Conflict around issues such as truth-telling, obtaining informed consent, and the revelation of dangerous diagnoses may lead to dilemmas in which the institution or provider (as in cases two and three) must either accede to the interpreter's view of cultural issues, accept the family's request, or withdraw. Negotiation does not, and should not, always lead to acquiescence to Western views of informed consent, truth-telling, or patients' autonomy.

The vital role played by the interpreters in the second and third cases illustrates how power is asserted and how role conflict is inherent in cross-cultural communication that involves conflicting ethical value systems. In models of interpretation that acknowledge and legitimate the interpreter's role as a mediator, a culture broker, an advocate, or a professional team member, there has been a move toward the formal specification of the obligations and rights of intermediaries. We believe that the premature development of these specifications, in the absence of more detailed study of the broad issues involved, is risky. Medical discourse within the boundaries of a "single" culture is problematic enough.

. . . We have found in our research on obtaining informed consent and truth-telling during deliberation on end-of-life treatment that the interaction between a patient and an interpreter often involves significant trust relationships, and may be bound by cultural constraints. Since a patient's trust in an interpreter may not always be transferred to the healthcare professional, this relationship may need to be reflected in institutional process and recognition of the interpreter's roles in relating to patients, families, healthcare providers, and institutions.

Notes

1. A Young, "The Anthropology of Illness and Sickness," *Annual Review of Anthropology* 11, no. 1 (1982): 257–85.
2. E T Friedson, "Professional Dominance," *in Social Structure of Medical Care* (New York: Atherton Press, 1970); D Tuckett and A Williams, "Approaches to the Measurement of Explanation and Information Giving in Medical Consultations: A Review of Empirical Studies," *Social Science and Medicine* 18 (1985): 571–80.
3. RW Putsch and M Joyce, "Methodology in Cross-Cultural Care. In Walker HK, Hall WD, and Hurst JW, eds., *Clinical Methods, 3rd edition* (Boston: Butterworths, 1990): 1050–55.
4. JM Kaufert and WW Koolage, "Role Conflict Among Culture Brokers: The Experience of Native Canadian Medical Interpreters." *Social Science Medicine* 18, 1984: 283–86.
5. RW Putsch, "Cross-Cultural Communication: The Special Case of Interpreters in Health Care." *Journal of the American Medical Association* 254, 1985: 3344–48.
6. JM Kaufert and WW Koolage, "Role Conflict Among Culture Brokers:The Experience of Native Canadian Medical Interpreters." *Social Science and Medicine* 18, 1984: 283–86, at p. 284.
7. JM Kaufert, "Sociological and Anthropological Perspectives on the Impact of Interpreters on Clinician/Client Communication," *Sante Culture Health* 7, no. 2–3 (1990): 32–35; Putsch, "Cross-cultural Communication: The Special Case of Interpreters in Health Care." *Journal of the American Medical Association* 254, 1985: 3344–48.
8. NS Jecker, JA Carrese, and RA Pearlman, "Caring for Patients in Cross-Cultural Settings," *Hastings Center Report* 25, no. 1 (1995): 6–14.
9. NC Ware and AR Kleinman, "Culture and Somatic Experience," *Psychosomatic Medicine* 54 (1992): 546–60.
10. NS Jecker, JA Carrese, and RA Pearlman, "Caring for Patients in Cross-Cultural Settings," *Hastings Center Report* 25, no. 1 (1995): 6–14, at p. 10.
11. LO Gostin, "Informed Consent, Cultural Sensitivity and Respect for Persons" (editorial), *Journal of the American Medical Association* 274 (10), 1995: 844–45.

12. RW Putsch, "Cross-Cultural Communication: The Special Case of Interpreters in Health Care" *Journal of the American Medical Association* 254, 1985: 3344–48, at p. 3344; RW Putsch and M Joyce, "Methodology in Cross-Cultural Care. In Walker HK, Hall WD, and Hurst JW, eds., *Clinical Methods, 3rd edition* (Boston: Butterworths, 1990): 1050–55.

13. JM Kaufert, "Sociological and Anthropological Perspectives on the Impact of Interpreters on Clinician/Client Communication," *Sante Culture Health* 7 (2–3), 1990: 32–35.

14. JM Kaufert and O'Neil, "Biomedical Rituals and Informed Consent: Native Canadians and the Negotiation of Clinical Trust," in Weisz G., ed., *Social Science Perspectives on Medical Ethics* (Dordrecht, the Netherlands: Kluwer Press, 1990), 41–63.

15. JM Kaufert, "Sociological and Anthropological Perspectives on the Impact of Interpreters on Clinician/Client Communication," *Sante Culture Health* 7 (2–3), (1990): 32–35.

16. H Verrept and F Louck, Health Advocates in Belgian Health Care (unpublished report by the Medesch-Sociale Wetenschappen, Brussels, Belgium, September, 1995), 1–10; L Jackson-Carroll, E Graham, and JC Jackson, "Beyond Medical Interpretation: The Role of Interpreter-Cultural Mediators (ICMs) in Building Bridges between Ethnic Communities and Health Institutions," *Handbook for Selecting, Training, and Supporting Key Outreach Staff* (Seattle, WA: Harborview Medical Center, 1995), at pp. 3–8; JM Kaufert and WW Koolage, "Role Conflict Among Culture Brokers: The Experience of Native Canadian Medical Interpreters." *Social Science and Medicine* 18, 1984: 283–86.

17. J Westermeyer, *Psychiatric Care of Migrants: A Clinical Guide* (Washington, D.C.: American Psychiatric Press, 1989), 77–78; L Jackson-Carroll, E Graham, and JC Jackson, "Beyond Medical Interpretation."

18. L Jackson-Carroll, E Graham, and JC Jackson, "Beyond Medical Interpretation: The Role of Interpreter-Cultural Mediators (ICMs) in Building Bridges between Ethnic Communities and Health Institutions," *Handbook for Selecting, Training, and Supporting Key Outreach Staff* (Seattle, WA: Harborview Medical Center, 1995).

19. RW Putsch, JM Kaufert, and M Lavallee, "Balancing the Expectations of Institutions and Cultural Communities in Urban Health Interpretation Programs" (Presentation at the First International Conference on Community Interpreters, 5 June 1995, Toronto, Ontario).

20. Washington State Supreme Court, Rules of Court, GR 11.1 (b), *Code of Conduct for Court Interpreters,* and attached commentary of the Court Interpreter Task Force (1986).

21. CE Roat, ed., *Bridging the Gap: A Basic Training for Medical Interpreters, 40 hours, for Multilingual Groups,* vet. 2 (Seattle, Wash.: Cross-Cultural Health Care Program, January 1996, originally published by the U.S. Public Health Service Region 10, Seattle, Washington, 1995), 30–31.

22. Note on the use of language in text: in many systems, interpreters are asked to speak in the first person (as if they were the patient speaking), and health-care providers are asked to address commentary to the patient: "You are anemic . . . , etc." This construct of interpersonal exchange does not work well in languages that depend on formal identification of kinship or rela-tionships (depending on one's age or social position). In these linguistic frameworks, individuals are addressed by age-specific or family structure-specific terms, and use of these terms often implies the speaker's respect for the person addressed. For example, first-person interpretation is difficult to do in Cree; Cree interpreters frequently use third-person commentary when they interpret. A similar pattern has been reported by Vietnamese in-terpreters in Seattle.

23. The interpreter has used the term, "no blood," because there is no term for anemia in Cree. The interpreter's code addresses lack of linguistic equiva-lency under "accuracy" and "conveying cultural frameworks."

24. A comprehensive process of documenting the changes in messages was used in our research on Cree, Ojibway, and Inuit interpreters in Manitoba, Canada. It involved audiotaping or videotaping verbatim text, back-translating messages, and confirming interpretations with both clinicians and interpreters. We found little in the healthcare literature on validating comparisons that involve textual information and review of videotaped tran-scriptions by independent interpreters, clinicians, and patients' representa-tives. Additional research using more systematic linguistic analysis may be required to document the difference between the ideal and the real in inter-preter-dependent healthcare interviews.

25. JA Carrese and LA Rhodes, "Western Bioethics on a Navajo Reservation: Benefit or Harm?" *JAMA* 274, 1995: 826–29; Y Beyene, "Medical Disclo-sure and Refugees: Telling Bad News to Ethiopian patients" *Western Jour-nal of Medicine* 157, 1992: 328–32; JM Kaufert and O'Neil, "Biomedical Rituals and Informed Consent: Native Canadians and the Negotiation of Clinical Trust," in Weisz G, ed., *Social Science Perspectives on Medical Ethics* (Dordrecht, the Netherlands: Kluwer Press, 1990), 41–63.

26. P Marshall, "Anthropology and Bioethics" *Medical Anthropology Quarterly* 6 (1), 1992: 49–73.

27. LO Gostin, "Informed Consent, Cultural Sensitivity and Respect for Per-sons" (editorial), *Journal of the American Medical Association* 274, no. 10 (1995): 844–45.

28. JM Kaufert and WW Koolage, "Role Conflict Among Culture Brokers: The Experience of Native Canadian Medical Interpreters." *Soc Sci Med* 18, 1984: 283–86.

29. LO Gostin, "Informed Consent, Cultural Sensitivity and Respect for Per-sons" (editorial), *Journal of the American Medical Association* 274 (10), 1995: 844–45.

30. Cases from Washington State have been recorded using ethnographic tech-niques, conducted by participant interviews, with the goal of understanding

the participants' intents and meanings. The "trouble cases" found in this environment that we describe have involved extensive debriefing and record keeping. Although the nature of the circumstances were often viewed retrospectively, the second author's position as an attending physician allowed him to be involved in the cases reported here from the outset. The statements attributed to the physicians and interpreters are paraphrased and are not quotes from audio recordings.

31. AR Kleinman, L Eisenberg, and B Good, "Culture, Illness and Care: Clinical Lessons from Anthropologic and Cross-Cultural Research," *Annals of Internal Medicine* 88(2), 1978: 36–45; RW Putsch and M Joyce, "Methodology in Cross-Cultural Care. In Walker HK, Hall WD, and Hurst JW, eds., *Clinical Methods, 3rd edition* (Boston: Butterworths, 1990): 1050–55; E.A. Berlin and W.C. Fowkes, "A Teaching Framework for Cross-Cultural Health Care," *Western Journal of Medicine* (1983): 934–38.

32. JM Kaufert and O'Neil, "Biomedical Rituals and Informed Consent: Native Canadians and the Negotiation of Clinical Trust," in Weisz G., ed., *Social Science Perspectives on Medical Ethics* (Dordrecht, the Netherlands: Kluwer Press, 1990), 41–63; Carrese and Rhodes, "Western Bioethics on a Navajo Reservation: Benefit or Harm?" *Journal of the American Medical Association* 274, 1995: 826–29.

33. LO Gostin, "Informed Consent, Cultural Sensitivity and Respect for Persons" (editorial), *Journal of the American Medical Association* 274 (10), 1995: 844–45.

34. JA Carrese and LA Rhodes, "Western Bioethics on a Navajo Reservation: Benefit or Harm?" *Journal of the American Medical Association* 274, 1995: 826–29.

35. Y Beyene, "Medical Disclosure and Refugees: Telling Bad News to Ethiopian Patients" *Western Journal of Medicine* 157, 1992: 328–32; E Felema and M Teklemarian, "Telling Bad News: An East African Perspective," in Can You Hear ME: A Training for Health Care Providers on Working Effectively with East African Patients (Videotape presentation including representatives of Oromo, Amharic, Somali, and Tigre communities, produced by the Cross-Cultural Health Care Program, Seattle, Washington, February 22, 1995); LJ Blackhall et al., "Ethnicity and Attitudes Toward Patient Autonomy" *Journal of the American Medical Association* 274 (10), 1995: 820–25; J Muller and B Desmond, "Ethical Dilemmas in a Cross-Cultural Context: A Chinese Example," *Western Journal of Medicine* 157 (3), 1992: 323–27; P Dalla Vorgia et al., "Attitudes of a Mediterranean Population to the Truth-Telling Issue," *Journal of Medical Ethics* 18, 1992; 67–74.

Yes, There Are African-American Perspectives on Bioethics

Annette Dula

Annette Dula is a graduate of Hampton Institute and Harvard University. She is currently a Senior Research Associate in Women Studies at the University of Colorado, Boulder. Her scholarship addresses racial dimensions of bioethics. She is the editor (with Sara Goering) of *It Just Ain't Fair: The Ethics of Health Care for African Americans*. In this essay she considers how exploitation and oppression experienced by African-American women gives rise to a unique perspective on bioethics. She concludes that the context of unequal relationships stands in the way of meeting basic ethical requirements, such as informed consent.

Reproductive Rights and Sterilization

Issues centering around birth control and reproduction are central to many bioethical discussions today. Among these are family planning, sterilization, and genetic screening—questions of particular interest to African-American women because we have been exploited in each of these areas. Therefore, we may see these issues differently from white women. If we look at the history of birth control in North America, we can understand the source of one of these different perspectives.

The birth control movement in the United States is marked by three phases. The middle of the eighteenth century witnessed the beginning of the first phase of the birth control movement.[1] "Voluntary motherhood" was the rallying cry of the early feminists. Essentially, voluntary motherhood meant that women ought to be able to say no to their husbands as a means of limiting the number of children they bore. The irony of voluntary motherhood was that while white feminists were refusing their husbands' sexual demands, African-American women did not have the same right to say no to the husbands of those same early feminists. This is to say nothing of the fact that African-American women had been exploited as breeding wenches in order to produce a stock of slaves.

The second phase of the birth control movement actually gave rise to the phrase "birth control," which was coined by Margaret Sanger in 1915.[2] Initially, this stage of the movement led to the recognition that reproductive rights and political rights were intertwined. The practice of birth control would give white

---

Annette Dula. "Yes, There Are African-American Perspectives on Bioethics." Abridged from *African-American Perspectives on Biomedical Ethics*, 1992, pp. 193–203. Reprinted by permission of Georgetown University Press.

women the freedom to pursue new opportunities, which their subsequent right to vote would soon make possible.[3] White women could go to work while African-American women cared for white children and did house work in white homes.

Unfortunately, the second stage of the birth control movement coincided with the eugenics movement in the first two decades of this century. When the white birth rate began to decline, eugenicists chastised middle-class white women for contributing to the suicide of the white race. As Paul Popenoe notes: "Continued limitation of offspring in the white race simply invites the black, brown, and yellow races to finish work already begun by birth control, and reduce the whites to a subject race."[4]

Eugenicists proposed two methods for curbing race suicide. On the one hand, middle-class white women had a moral obligation to have large families; on the other hand, poor immigrant women and African-American women had a moral obligation to restrict the size of their families because they were likely to be of inferior stock. On the basis of this argument, Guy Irving Burch of the American Eugenics Society advocated birth control for African-American and immigrant women. He notes: "We must prevent the American people from being replaced by alien or Negro stock, whether it be by immigration or by overly high birth rates among others in this country."[5] In addition, poor people created a drain on the taxes and charity of the wealthy.[6]

The woman's movement then adopted the ideals of the eugenicists regarding poor and minority women. Margaret Sanger saw the chief issue of birth control as "more children from the fit and less from the unfit."[7]

The 1940s marked the beginning of the third phase of birth control which was renamed, "Planned Parenthood."

In the 1950s, several states tried to extend sterilization laws to include compulsory sterilization of mothers of illegitimate children.[8] In the 1960s, the government began subsidizing family planning clinics. The purpose of these subsidies was to reduce the number of people on welfare by checking the transmission of poverty from generation to generation. The number of family planning clinics in an area was proportional to the number of African-Americans and Hispanics in the areas.[9] In Puerto Rico, by 1965, a third of the women had been sterilized.[10] In 1972 it was reported that there was a sevenfold rise in hysterectomies over the previous year at Los Angeles Hospital. These policies were influential in arousing African-American suspicions that there were racist motives behind family planning efforts.[11]

In 1973 two sisters, twelve-year-old Mary Alice Relf and fourteen-year-old Minnie Lee were surgically sterilized without consent. In the same town where they lived, eleven other young girls about the same ages as the Relf sisters had also been sterilized. Ten of these girls were African-American. In South Carolina, of thirty-four deliveries paid for by Medicaid, eighteen included sterilizations and all eighteen were young black women.[12] In 1972, Carl Schultz, Director of HEW's Population Affairs Office, estimated that between 100,000 and 200,000 sterilizations had been funded by the government.[13]

Thus, the first phase of the woman's movement completely ignored black women's sexual subjugation to white masters. And in the second phase, the movement adopted the racist policies of eugenics philosophy. The third stage saw a number of coercive measures supported by governmental policy to contain the population of African-Americans and poor people. While birth control per se was perceived as a benefit, African-Americans have historically objected to birth control as a method of dealing with poverty. . . .

The Answer Is Yes

Yes, we do have an African-American perspective on bioethics. . . .

There is a shocking history of medical abuse against powerless people. Often the form of the abuse is violation of informed consent. Indeed, the examples that I have presented share the two common elements of powerlessness and the absence of the informed consent. Consequently, I have suggested that in unequal relationships, informed consent does not work. . . .

Though there may be an acknowledged African-American perspective on bioethics, that does not mean that our perspectives have been fully articulated. Rather we need to organize professionally to articulate our views further. . . .

References

1. Linda Gordon, *Woman's Body, Woman's Right: Birth Control in America* (New York: Penguin Books, 1976).
2. Ibid.
3. Angela Davis, *Women, Race and Class* (New York: Vintage Books, 1981).
4. Paul Popenoe, *Conservation of the Family* (Baltimore: Williams and Wilkins, 1926), 144.
5. Gordon, *Woman's Body, Woman's Right*, 283.
6. Ibid.
7. Gordon, *Woman's Body, Woman's Right*, 281.
8. Joseph L. Morrison, "Illegitimacy, Sterilization, and Racism: A North Carolina Case History," *Social Science Review* 39 (1965): 1–10.
9. Ketayun H. Gould, "Black Women in Double Jeopardy: A Perspective on Birth Control," *National Association of Social Workers* (1984): 96–105.
10. Bonnie Mass, *Population Target: The Political Economy. Population Control in Latin America* (Toronto: Women's Press, 1976).
11. Davis, *Women, Race and Class.*
12. Herbert Aptheker, "Racism and Human Experimentation," *Political Affairs* 53, 2 (1974): 27–60.
13. Les Payne, "Forced Sterilization for the Poor," *San Francisco Chronicle*, 26 Feb 1974.

Index

www.ingramcontent.com/pod-product-compliance
Lightning Source LLC
Chambersburg PA
CBHW060417220326
41598CB00021BA/2202